Communications in Computer and Information Science 1552

More information about this series at https://link.springer.com/bookseries/7899

Vladimir M. Vishnevskiy ·
Konstantin E. Samouylov ·
Dmitry V. Kozyrev (Eds.)

Distributed Computer and Communication Networks

24th International Conference, DCCN 2021
Moscow, Russia, September 20–24, 2021
Revised Selected Papers

Editors
Vladimir M. Vishnevskiy
V.A. Trapeznikov Institute of Control
Sciences of Russian Academy of Sciences
Moscow, Russia

Konstantin E. Samouylov
Peoples' Friendship University of Russia
(RUDN University)
Moscow, Russia

Dmitry V. Kozyrev
V.A. Trapeznikov Institute of Control
Sciences of Russian Academy of Sciences
Moscow, Russia

Peoples' Friendship University of Russia
(RUDN University)
Moscow, Russia

ISSN 1865-0929 ISSN 1865-0937 (electronic)
Communications in Computer and Information Science
ISBN 978-3-030-97109-0 ISBN 978-3-030-97110-6 (eBook)
https://doi.org/10.1007/978-3-030-97110-6

This Springer imprint is published by the registered company Springer Nature Switzerland AG
The registered company address is: Gewerbestrasse 11, 6330 Cham, Switzerland

Preface

This volume contains a collection of revised selected full-text papers presented at the 24th International Conference on Distributed Computer and Communication Networks (DCCN 2021), held in Moscow, Russia, during September 20–24, 2021.

The conference is a continuation of the traditional international conferences of the DCCN series, which have taken place in Sofia, Bulgaria (1995, 2005, 2006, 2008, 2009, 2014); Tel Aviv, Israel (1996, 1997, 1999, 2001); and Moscow, Russia (1998, 2000, 2003, 2007, 2010, 2011, 2013, 2015, 2016, 2017, 2018, 2019, 2020) in the last 24 years. The main idea of the conference is to provide a platform and forum for researchers and developers from academia and industry from various countries working in the area of theory and applications of distributed computer and communication networks, mathematical modeling, and methods of control and optimization of distributed systems, by offering them a unique opportunity to share their views, discuss prospective developments, and pursue collaboration in this area. The content of this volume is related to the following subjects:

1. Communication networks, algorithms, and protocols
2. Wireless and mobile networks
3. Computer and telecommunication networks control and management
4. Performance analysis, QoS/QoE evaluation, and network efficiency
5. Analytical modeling and simulation of communication systems
6. Evolution of wireless networks toward 5G
7. Internet of Things and fog computing
8. Cloud computing, distributed systems, and parallel systems
9. Machine learning, big data, and artificial intelligence
10. Probabilistic and statistical models in information systems
11. Queuing theory and reliability theory applications
12. High-altitude telecommunications platforms
13. Security in infocommunication systems

The DCCN 2021 conference gathered 151 submissions from authors from 26 different countries. From these, 105 high-quality papers in English were accepted and presented during the conference. The current volume contains 35 extended papers which were recommended by session chairs and selected by the Program Committee for the Springer post-proceedings. Thus, the acceptance rate is 33.3%.

All the papers selected for the post-proceedings volume are given in the form presented by the authors. These papers are of interest to everyone working in the field of computer and communication networks.

We thank all the authors for their interest in DCCN, the members of the Program Committee for their contributions, and the reviewers for their peer-reviewing efforts.

September 2021

Vladimir M. Vishnevskiy
Konstantin E. Samouylov
Dmitry V. Kozyrev

Organization

DCCN 2021 was jointly organized by the Russian Academy of Sciences (RAS), the V.A. Trapeznikov Institute of Control Sciences of RAS (ICS RAS), the Peoples' Friendship University of Russia (RUDN University), the National Research Tomsk State University, and the Institute of Information and Communication Technologies of the Bulgarian Academy of Sciences (IICT BAS).

Program Committee Chairs

V. M. Vishnevskiy (Chair)	ICS RAS, Russia
K. E. Samouylov (Co-chair)	RUDN University, Russia

Publication and Publicity Chair

D. V. Kozyrev	ICS RAS and RUDN University, Russia

International Program Committee

S. M. Abramov	Program Systems Institute of RAS, Russia
S. D. Andreev	Tampere University of Technology, Finland
A. M. Andronov	Transport and Telecommunication Institute, Latvia
N. Balakrishnan	McMaster University, Canada
S. E. Bankov	Kotelnikov Institute of Radio Engineering and Electronics of RAS, Russia
A. S. Bugaev	Moscow Institute of Physics and Technology, Russia
S. R. Chakravarthy	Kettering University, USA
T. Czachorski	Institute of Theoretical and Applied Informatics of the Polish Academy of Sciences, Poland
D. Deng	National Changhua University of Education, Taiwan, China
S. Dharmaraja	Indian Institute of Technology, Delhi, India
A. N. Dudin	Belarusian State University, Belarus
A. V. Dvorkovich	Moscow Institute of Physics and Technology, Russia
Yu. V. Gaidamaka	RUDN University, Russia
P. Gaj	Silesian University of Technology, Poland
D. Grace	University of York, UK

Yu. V. Gulyaev	Kotelnikov Institute of Radio-engineering and Electronics of RAS, Russia
J. Hosek	Brno University of Technology, Czech Republic
V. C. Joshua	CMS College Kottayam, India
H. Karatza	Aristotle University of Thessaloniki, Greece
I. A. Kochetkova	RUDN University, Russia
N. Kolev	University of São Paulo, Brazil
J. Kolodziej	NASK, Poland
G. Kotsis	Johannes Kepler University Linz, Austria
A. E. Koucheryavy	Bonch-Bruevich Saint-Petersburg State University of Telecommunications, Russia
Ye. A. Koucheryavy	Tampere University of Technology, Finland
T. Kozlova Madsen	Aalborg University, Denmark
U. Krieger	University of Bamberg, Germany
A. Krishnamoorthy	Cochin University of Science and Technology, India
N. A. Kuznetsov	Moscow Institute of Physics and Technology, Russia
L. Lakatos	Eötvös Loránd University, Budapest
E. Levner	Holon Institute of Technology, Israel
S. D. Margenov	Institute of Information and Communication Technologies of the Bulgarian Academy of Sciences, Bulgaria
N. Markovich	ICS RAS, Russia
A. Melikov	Institute of Cybernetics of the Azerbaijan National Academy of Sciences, Azerbaijan
E. V. Morozov	Institute of Applied Mathematical Research of the Karelian Research Centre of RAS, Russia
V. A. Naumov	Service Innovation Research Institute (PIKE), Finland
A. A. Nazarov	Tomsk State University, Russia
I. V. Nikiforov	Université de Technologie de Troyes, France
P. Nikitin	University of Washington, USA
S. A. Nikitov	Kotelnikov Institute of Radio Engineering and Electronics of RAS, Russia
D. A. Novikov	ICS RAS, Russia
M. Pagano	University of Pisa, Italy
E. Petersons	Riga Technical University, Latvia
V. V. Rykov	Gubkin Russian State University of Oil and Gas, Russia
K. E. Samouylov	RUDN University, Russia
L. A. Sevastianov	RUDN University, Russia
M. A. Sneps-Sneppe	Ventspils University College, Latvia

Organizing Committee

Organizers and Partners

Organizers

Russian Academy of Sciences (RAS), Russia
V.A. Trapeznikov Institute of Control Sciences of RAS, Russia
RUDN University, Russia
National Research Tomsk State University, Russia
Institute of Information and Communication Technologies of the Bulgarian Academy of Sciences, Bulgaria
Research and Development Company “Information and Networking Technologies”, Russia

Support

Information support was provided by the Russian Academy of Sciences. The conference was organized with the support of the RUDN University Strategic Academic Leadership Program.

Contents

Computer and Communication Networks

Multi Task Multi-UAV Computation Offloading Enabled Mobile Edge Computing Systems ... 3
Abbas Alzaghir and Andrey Koucheryavy

The Increasing of Resource Sharing Efficiency in Network Slicing Implementation ... 18
Mikhail S. Stepanov, Sergey N. Stepanov, Umer Andrabi, Dmitriy Petrov, and Juvent Ndayikunda

Analysis of Non-preemptive Scheduling for 5G Network Model Within Slicing Framework ... 36
Yves Adou, Ekaterina Markova, and A. A. Chursin

A Hybrid Clustering-Based Routing Protocol for VANET Using k-means and Cuckoo Search Algorithm ... 48
Amani A. Sabbagh and Maxim V. Shcherbakov

OpenFlow-based Software-Defined Networking Queue Model ... 62
Vyacheslav Kartashevskiy and Marina Buranova

Hybrid MCDM for Cloud Services: AHP(blocks) & Entropy, TOPSIS & MOORA (methodology Review and Advances) ... 77
Iliyan Petrov

Hybrid MCDM for Cloud Services: AHP(blocks) & Entropy, TOPSIS & MOORA (Case Study with QoS and QoE Criteria) ... 92
Iliyan Petrov

Ultra-Dense Internet of Things Model Network ... 111
Anastasia Marochkina, Alexander Paramonov, and Tatiana M. Tatarnikova

Integrity, Resilience and Security of 5G Transport Networks Based on SDN/NFV Technologies ... 123
I. Buzhin, M. Bessonov, Y. Mironov, and M. P. Farkhadov

Algorithm of Finding All Maximal Induced Bicliques of Hypergraph ... 136
Aleksandr Soldatenko and Daria Semenova

Customer Experience Model for Communication Service Provider Digital Twin 148
Vladimir Akishin, Sergey Kislyakov, and Alexander Sotnikov

Analytical Modeling of Distributed Systems

Matrix-Geometric Solutions for the Models of Perishable Inventory Systems with a Constant Retrial Rate 163
Agassi Melikov, Mamed Shahmaliyev, and János Sztrik

Analysis of Two-Way Communication Retrial Queuing Systems with Non-reliable Server, Impatient Customers to the Orbit and Blocking Using Simulation 174
Ádám Tóth, János Sztrik, Tamás Bérczes, and Attila Kuki

On a Queue with Marked Compound Poisson Input and Exponentially Distributed Batch Service 186
K. A. K. Al Maqbali, V. C. Joshua, and Achyutha Krishnamoorthy

A Two Server Queueing Inventory Model with Two Types of Customers and a Dedicated Server 201
Nisha Mathew, V. C. Joshua, and Achyutha Krishnamoorthy

On Convergence of Tabu-Enhanced Quantum Annealing Algorithm 214
A. S. Rumyantsev, D. Pastorello, E. Blanzieri, and V. Cavecchia

Semi-markov Resource Flow as a Bit-Level Model of Traffic 220
Anatoly Nazarov, Alexander Moiseev, Ivan Lapatin, Svetlana Paul, Olga Lizyura, Pavel Pristupa, Xi Peng, Li Chen, and Bo Bai

Asymptotic Diffusion Analysis of an Retrial Queueing System M/M/1 with Impatient Calls 233
Elena Danilyuk, Svetlana Moiseeva, and Anatoly Nazarov

Sufficient Stability Conditions for a Multi-orbit Retrial System with General Retrials Under Classical Retrial Policy 247
Ruslana Nekrasova

Analysis of the Probabilistic and Cost Characteristics of the Queueing Network with a Control Queue and Quarantine in Systems and Negative Requests by Means of Successive Approximations 259
Katsiaryna Kosarava and Dmitry Kopats

The Automata-Based Model for Control of Large Distributed Systems 272
Yu. S. Zatuliveter and E. A. Fishchenko

Information Spreading with Application to Non-homogeneous Evolving Networks ... 284
Natalia M. Markovich and Maksim S. Ryzhov

Machine Learning for Recognition Learning of Control Systems for Autonomous Unmanned Underwater Vehicles of Events in Hostile Environments ... 293
Vyacheslav Abrosimov and Ekaterina Panteley

Statistical Model of Graph Structure Based on "VKontakte" Social Network ... 307
A. A. Kislitsyn and Yu. N. Orlov

Response Time Estimate for a Fork-Join System with Pareto Distributed Service Time as a Model of a Cloud Computing System Using Neural Networks ... 318
A. V. Gorbunova and A. V. Lebedev

Approximation of the Two-Dimensional Output Process of a Retrial Queue with MMPP Input ... 333
Alexey Blaginin and Ivan Lapatin

Method of Analyzing the Availability Factor in a Mesh Network ... 346
Alexander Dagaev, Van Dai Pham, Ruslan Kirichek, Olga Afanaseva, and Ekaterina Yakovleva

The Importance of Conference Proceedings in Research Evaluation: A Methodology for Assessing Conference Impact ... 359
Dmitry Kochetkov, Aliaksandr Birukou, and Anna Ermolayeva

Cardiac Arrhythmia Disorders Detection with Deep Learning Models ... 371
Eugene Yu. Shchetinin, Leonid A. Sevastianov, Anastasia V. Demidova, and Anastasia G. Glushkova

Distributed Systems Applications

Autonomous Infrared Guided Landing System for Unmanned Aerial Vehicles ... 387
Mainak Mondal, S. V. Shidlovskiy, D. V. Shashev, and Mikhail Okunsky

Distributed Computing of R Applications Using RBOINC Package with Applications to Parallel Discrete Event Simulation ... 396
S. N. Astafiev and A. S. Rumyantsev

Algorithm for Calculating and Using the Characteristics of a Binary Image Intended for Implementation on RCE 408
A. S. Bondarchuk, D. V. Shashev, and S. V. Shidlovskiy

Evaluation of Trust in Computer-Computed Results 420
Alexander Grusho, Nikolai Grusho, Michael Zabezhailo, and Elena Timonina

Approaches for Creating a Digital Ecosystem of an Industrial Holding 433
A. E. Tyulin, A. A. Chursin, A. V. Yudin, and P. Yu. Grosheva

Intelligent Systems for Optimal Production Control of Unique Products 445
A. E. Tyulin, A. A. Chursin, I. N. Dubina, A. V. Yudin, and P. Yu. Grosheva

Author Index 463

Computer and Communication Networks

Multi Task Multi-UAV Computation Offloading Enabled Mobile Edge Computing Systems

Abbas Alzaghir(✉) and Andrey Koucheryavy

The Bonch-Bruevich Saint-Petersburg State University of Telecommunications, Pr. Bolshevikov, 22, St. Petersburg 193232, Russia

Abstract. Mobile edge computing (MEC) is a technology that has found effective solutions in terms of ultra-low latency, low energy consumption, high data rate and high reliability. MEC is become actual appropriate for the devices that has a limited resource such IoT, UAVs etc. In this paper, we consider an algorithm to improve and minimize the energy consumption of Unmanned Aerial Vehicle by offloading and execution some tasks of UAV into nearby UAVs or into edge cloud server. The proposed algorithm is based on dynamic programming that uses a randomization and hamming distance termination to obtain approximately optimal solution. The algorithm can minimize the energy consumption of UAV and improving the total execution time by offload some tasks to nearby UAVs or to edge cloud server when transmission data rate of the network is high. The algorithm can find a nearly optimal offloading decision within a few repetitions. The results show that the proposed algorithm achieves a minimal energy while meeting energy and time constraints.

Keywords: Unmanned Aerial Vehicle · Multi-UAV · Computation Offloading · Mobile Edge Computing

1 Introduction

The advent of 5G technology is expected to contribute to connecting the whole world by universal communication that links everybody and everything in all times and by all means regardless of the services, devices, geographical existence or networks. This will lead to the emergence of new use cases and also provide new business opportunities for telecommunication operators [1]. Furthermore, the performance criteria for low latency, enhanced reliability, high speed, peak throughput per connection, system spectral efficiency, low power consumption, connection and capacity density will be introduced by 5G technology.

This research is based on the Applied Scientific Research under the SPbSUT state assignment 2021.

V. M. Vishnevskiy et al. (Eds.): DCCN 2021, CCIS 1552, pp. 3–17, 2022.
https://doi.org/10.1007/978-3-030-97110-6_1

Unmanned Aerial Vehicles are becoming an essential part of 5G and are likely to play a major role in a further functional diversity for 5G networks. Because it characterized by their ease of deployment and observable, it's may be widely deployed both in surveillance, disaster management, data gathering from Internet of things (IoT), wireless sensor networks (WSN), and traffic management, etc. where expects the number of UAVs will reach at least 3.2 million connecting units by 2022 according to Federal Aviation Administration (FAA) [2].

However, the limited computing capability and energy storage of UAV terminals cannot provide intensive computing and complete high-energy tasks. Currently, mobile edge computing (MEC) has become an important solution to solve mentioned issues and to enhance transmission delay and transmission energy consumption. Furthermore, the edge server is deployed on the wireless LAN side in the edge server-based MEC solution, which shortens the distance between servers and the end devices [3]. This solution can be use computing offload to expand the service capabilities of UAVs, provide localized computing and storage resources for UAVs nearby. In other word, a large number of tasks that can't be processed locally on the UAV need to be offloaded to the edge server through wireless channel in order to implement the computation task at edge servers this lead to reduce data transmission costs, to achieve low latency and enhance energy efficiency and meet the needs of the fast and interactive response. Therefore, the key technology of the MEC solution-based edge server is computing offloading [3].

In this paper, an energy efficient computation offloading algorithm is proposed for a multi-UAV enabled edge computing system. To implement the multilevel offloading process, we have developed the dynamic programming with hamming distance algorithm presented in [4]. Where the offloading process is either to nearby UAVs or to edge cloud server.

The main contributions of this paper are minimizing the energy consumption and total task execution time of UAV with consideration of the deadline completion to accomplish the task and energy constraint.

The paper is organized as follows. In Sect. 2 we discussed the related work. In Sect. 3 we introduce the system model and the problem statement. The energy-efficient computation offloading algorithm presented in Sect. 4. And in Sect. 5, evaluation experiments that implemented to validate our multilevel offloading mode. Finally, the conclusion of this paper discussed in Sect. 6.

2 Related Works

In this section we introduce the literature review. The paper [5], in which the sum power minimization problem for an UAV-enabled MEC network. They proposed an algorithm to solve nonconvex sum power minimization problem by solving three subproblems iteratively. The feasible solution for this iterative algorithm, a fuzzy c-means clustering based algorithm is proposed.

The authors in paper [6] considers a decentralized optimization mechanism between MEC servers and users. The assignment mechanism is employees to assistance heterogeneous users to select different MEC operators in a distributed environment. They modeled an efficient mechanism for a computation offloading scheme to optimize price and energy consumption under latency constraints.

Furthermore, Task caching and offloading strategy for MEC investigated in paper [7] that decides which tasks should be cached and how much task should be offloaded. The proposed algorithm is to minimize the total energy consumed by mobile device while meet the users delay requirement.

The authors in [8] proposed an energy efficient computation offloading procedure for UAV-MEC systems, with an emphasis on physical-layer security. The optimal results provided to the problems formulated are found for both active and passive eavesdroppers.

The energy-and latency-aware algorithm proposed in paper [9] provides two UAV offloading methods. The first method is the air-offloading, where a UAV can offload its computing tasks to nearby UAVs that have available computing and energy resources. The second offloading method is the ground-offloading, which enables the offloading of tasks to an edge cloud server from the multi-level edge cloud units connected to ground stations. The algorithm selects the execution device and the offloading method based on the latency and energy constraints. Moreover, a multi (UAVs) enabled mobile edge computing (MEC) proposed in research [10] has been considered a several UAVs are deployed as flying MEC platform to ensure computing resource to ground user equipment (UEs). The authors formulate a mixed integer nonlinear programming to solve two issues: first how to achieve the association between multiple UEs and UAVs, second how to achieve the resource allocation from UAVs to UEs. Also, they proposed a Reinforcement Learning (RL)-based user Association and resource Allocation (RLAA) algorithm to tackle this problem efficiently and effectively.

3 System Model and Problem Statement

The system model consists of a multi-UAV network connected with a base station by a wireless channel, as well as the base station is provided with edge computing server as shown in Fig. 1. One of a UAVs has a set of N independent computation tasks denoted as $N = 1; 2;\ldots; N$ needed to be executed locally at UAV m ($UAV1;\ UAV2\ ;\ldots;\ UAVm$) or will be offloaded and executed either at one of the neighboring UAVs or remotely at edge computing server. The scenario considered that the UAV m remains unchanged during a computation offloading period, whereas it can be moved to another location while different periods [5,8].

3.1 Communication Model and Scenario

The communication model proposed in this paper has a multi-UAV network connected with a single base station through a wireless channel, as well as the base station equipped with edge computing resources. The multi-UAV network consist of a number of UAVs connected together wirelessly, one of them is considered as UAV m has N independent computation tasks that needed to be completed. The scenario that has considered is a multilevel computational offloading mode, so we considered two decision offloading variables denoted as β_i and α_i is a binary computation offloading decisions, where $\beta_i = 1$ indicates that the computation task i of UAV m is assigned to be processed (at first level, no offloading) locally by UAV m, whereas $\beta_i = 0$ indicates that the computation task i of UAV m will be offloaded and processed either at one of neighboring UAVs (second level) where $\alpha_i = 1$ or remotely at edge cloud server (third level), where $\alpha_i = 0$. So, we have $\beta_i = \beta_1, \beta_2, \ldots, \beta_N$ as the offloading decision profile and $\alpha_i = \alpha_1, \alpha_2, \ldots, \alpha_N$ as remotely processing decision profile for the computation tasks of UAV m.

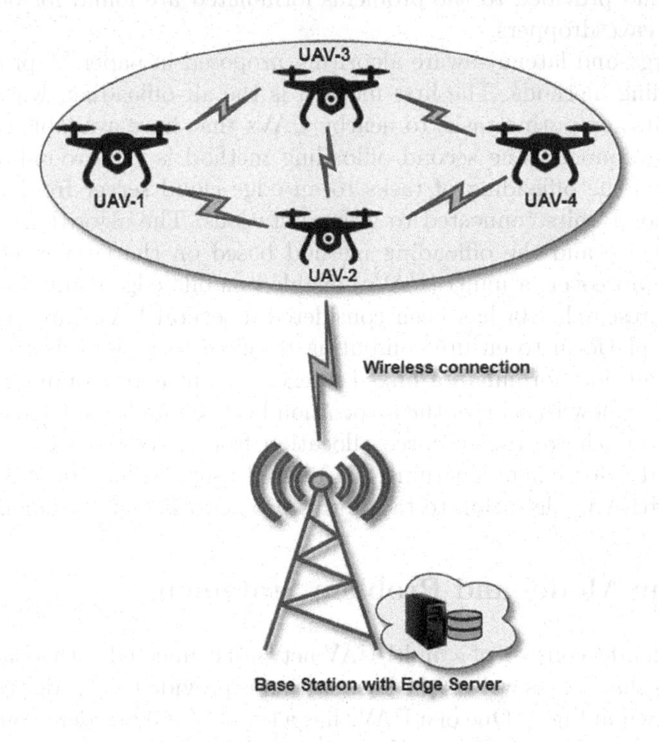

Fig. 1. System model

The maximum uplink data rate at which the computation task data can be transmitted over the wireless channel can be calculated as [3]:

$$r_m = BW * \log_2(1 + \frac{P_m G_m}{\sigma BW}) \tag{1}$$

Where BW represent the channel bandwidth, G_m is the channel gain between the UAVs and the base station, P_m denotes the transmission power of UAV m, and σ is denoting to the density of noise power.

Because the output of data size is smaller than the input data size and also the downlink data rate from the server is higher than the uplink data rate, the total overhead consumption of energy and time for transmitting output data size is neglected in this paper [11].

3.2 Computation Model

In this subsection, the computation offloading model is introduced. Firstly, as mentioned above, our model has a multi-UAV and N independent computation tasks that needed to be completed. The task requirement for each computation task i represented by a tuple $(d_i, c_i, T_i^{\max})$, where d_i represents the data size that need to be transmitted, whereas c_i represent the total number of CPU cycles, and $T_i^{\max}$ the completion deadline (time constraint) that required for task i.

3.2.1 Local Computation

For the local computation case where all the computation task i will be processed locally by UAV m, the processing time and energy consumption can be calculated locally as:

$$T_{i,m}^{local} = \frac{c_i}{f_m^{local}} \tag{2}$$

$$E_{i,m}^{local} = \epsilon(f_m^{local})^2 * c_i \tag{3}$$

Where f_m^{local} denotes the computational capability of UAV m, and ϵ denotes the energy consumption coefficient per CPU cycle [5]. We set $\epsilon = 10^{-25}$, as in [6].

3.2.2 Nearby Computation

For the nearby computation case where the computation task i of UAV m will be offloaded and processed at nearby UAVs. The UAV m looks for nearby UAVs to check if has available resources to handle the task. For this purpose, the UAV m starts a discovery process, with the objective of discovering the surrounding field, whether it contains an UAV with available resources for task handling. The offloading time for the computation task i of UAV m can be expressed as follow:

$$T_{i,m}^{offload} = \frac{d_i}{r_m} \tag{4}$$

Where r_m is the uplink data rate of UAV m in the wireless channel as mentioned in the communication model, the processing time for the computation task i at nearby UAVs, can be expressed as follow:

$$T_{i,m}^{nearby} = \frac{c_i}{f_m^{nearby}} \tag{5}$$

Where f_m^{nearby} is denote to the computational capability of nearby UAVs. From Eq. (4) and (5) we can calculate the total processing time at nearby UAVs as:

$$T_{i,m}^{nearby-UAV} = T_{i,m}^{offload} + T_{i,m}^{nearby} \tag{6}$$

$$T_{i,m}^{nearby-UAV} = \frac{d_i}{r_m} + \frac{c_i}{f_m^{nearby}} \tag{7}$$

Accordingly, the total energy consumption for offloading and processing the computation task i of UAV m at nearby UAVs can be expressed as follow:

$$E_{i,m}^{nearby-UAV} = \frac{P_m d_i}{r_m} + \frac{\rho c_i}{f_m^{nearby}} \tag{8}$$

Where P_m is transmission power of UAV m, and ρ is the power consumption for idle state [10].

3.2.3 Edge Computation

For the edge server computation case where the computation task i of UAV m will be offloaded and processed at base station server. The processing time for the computation task i of UAV m at edge server can be expressed as follow:

$$T_{i,m}^{process-BS} = \frac{c_i}{f_{edge}} \tag{9}$$

Where f_{edge} is denote to the computational capability of edge server which assigned to UAV m. Finally, the total computation for offloading and processing time at edge server can be calculated by summation Eq. (4) and (9):

$$T_{i,m}^{edge} = \frac{d_i}{r_m} + \frac{c_i}{f_{edge}} \tag{10}$$

And the energy consumption for offloading and processing the computation task i of UAV m at edge server can be expressed as follow:

$$E_{i,m}^{edge} = \frac{P_m d_i}{r_m} + \frac{\rho c_i}{f_{edge}} \tag{11}$$

The total overhead for processing the computation task i in terms of time and energy can be respectively expressed as:

$$T_i^{total} = \beta_i T_{i,m}^{local} + (1 - \beta_i)[\alpha_i T_{i,m}^{nearby-UAV} + (1 - \alpha_i) T_{i,m}^{edge}] \tag{12}$$

$$E_i^{total} = \beta_i E_{i,m}^{local} + (1 - \beta_i)[\alpha_i E_{i,m}^{nearby-UAV} + (1 - \alpha_i) E_{i,m}^{edge}] \tag{13}$$

3.3 Problem Statement

In this section, we consider the issue of achieving energy efficient computation offloading and for multi-UAV enabled edge computing systems. Regarding the above communication and computation models, the computation offloading problem is formulated as the following constrained optimization formulation problem:

$$min \sum_{i=1}^{N} E_i^{total} \tag{14}$$

$$s.t : E_i^{total} \leq E_{i,m}^{local} \tag{15}$$

$$T_i^{total} \leq T_i^{\max} \tag{16}$$

$$\beta_i \& \alpha_i \in \{0, 1\} \tag{17}$$

Our goal is to minimize the weighted sum of energy consumed by the UAV through task offloading distributions. The constraints C1 and C2 are upper bounds of energy and time consumption, respectively. Constraint C3 is guarantee that the offloading decisions variables are binary values. In the next section we introduce the proposed algorithm to solve our constraint optimization problem.

4 Energy-Efficient Computations Offloading Algorithm Based on Dynamic Programming

The proposed algorithm is expansion to "Dynamic Programming with Hamming Distance Termination (DPH) algorithm" proposed in [4]. The algorithm provides a comprehensive process for finding the optimal computation offloading decisions for a multi-UAV Enabled Mobile Edge Computing System. Initially, The UAV m decides whether it task i should be processed locally or transferred either to nearby UAVs or offloaded to the edge server. Denote this decision by the computation offloading decisions β_i and α_i where ($\beta_i = 1$), which indicates local execution, while ($\beta_i = 0$ and $\alpha_i = 1$) indicates nearby UAVs execution, whereas ($\beta_i = 0$ and $\alpha_i = 0$) indicates edge server execution. Because our system has a multilevel offloading scenario and has two decision variables, so we have developed the algorithm proposed in [4] and we use two $\mathrm{N} \times \mathrm{N}$ tables instead of one table (where N is the number of tasks that need to be complete), first table it is represent β_i bit streams and another is to represent α_i bit streams. The tables are used to store a bit streams to ensure which tasks needed to be executed locally, and which tasks needed to be offloaded and executed remotely. The generated random bit streams are filled in the first table as the ones (1s) in the next horizontal cell, and zeros (0s) in the next vertical cell where the first cell always empty. If the first bit of the stream is 1, the starting cell is (1, 2) and if the first bit of the stream is 0, the starting cell is (2, 1), whereas the second table don't depend on the value of the first bit but depends on distribution the first table. This method will avoid additional computations for joint bit strings.

Two 2D N*N tables are shown below. First table assigned to β_i bit streams, to simplify how to distribute the generated random bit stream, we assume that N = 8 and the first random β_i bit stream is 00110100 (black bits), and the second random bit stream is 10101101 (red bits). The starting cell of the first stream is (2, 1) since the first bit was 0, whereas the starting cell of the second stream was (1, 2) where the first bit was 1. The second table is assigned to α_i bit streams, where the distribution of α_i bit streams occur according to distribution of the first table and regardless the first bit is 1 Or 0. For example, Assume the first random α_i bit stream is 11000110 (black bits) and the second random bit stream is 01001011 (red bits). Although the first bit of first stream was 1 and the first bit of second stream was 0, the distribution in the second table was according to the first table and regardless the first bit is 1 Or 0. By following the aforementioned rules to fill the tables, the resulting stream as shown in the Table 1(a) and (b).

Table 1. Random bit streams

(a). β_i bit streams

	1						
0	0	1					
0	1	1/0	1	1			
		0	1	0	1		
			0				
			0				

(b). α_i bit streams

	0						
1	1	0					
1	0	0/0	1	0			
		0	1	1	1		
			1				
			0				

From the tables we calculate the energy consumption and execution time of each task by each cell has 1s in the first table (local execution) and each cell has 0s and 1s in the second table when a bit streams are randomly generated, since the cells that has 0s in the first table means moved to the second table since remotely executions. Additionally, the total energy and total execution time are also calculated. Nevertheless, if a random bit streams which has some common cells with an existing string in the table, we only calculate the total energy of new string until the first common cell and then compare this new total energy with the existing total energy at this cell [4]. If the new total energy at this specific cell is less than the previous one, we keep the new sub-string and delete the old sub-string, and replace the total-energy and cell-energy of this cell with new amounts. We then update the energy and execution time of the remaining cells for the existing bit stream, based on the new values at this common cell. Otherwise, if the total energy of the existing bit stream is less than that of the new bit stream at the common cell, we will perform the same procedure while keeping the existing stream. Every time a new stream is generated, we keep tracking the arrangement of the stream in the tables, for details see [4]. The computation offloading algorithm based on dynamic programming tables proposed in our paper as shown in the table below.

Algorithm 1. Proposed Algorithm

1: Initialize Energy and Time matrices and set the Completion deadline T_i^{max} and Transmission Rate and Max number of UAVs.
2: generate a task (randomly)
3: Loop iteration
4: generate a random bit stream
5: calculate energy and time for tasks (for all state: local, edge and nearby UAVs)
6: check the first bit to specify the starting cell in the first table
7: check the second table to specify the starting cell according to first table
8: loop i to N-1
9: **if** bit(i)== 1, in the second table **then** (nearby UAV)
10: **if** sum(Decision Matrix==1) **then**= Max number of UAVs
11: regenerate random bit (0 or 1)
12: **end if**
13: Put each bit of the bit stream in the correct position in tables
14: **if** this specific cell in tables is visited before **then** compare the new Total Energy of this cell with the previous one
15: **if** the new Total Energy of the cell is less than the previous one **then** Replace the total energy and time of this cell with the new calculated amounts.
16: Calculate the energy and time of the remaining bits of the new bit stream
17: **else**
18: Keep the previous total energy and time in the cell.
19: Calculate the Energy and time of the remaining cells of the new stream based on the existing amount of this cell
20: **end if**
21: **end if**
22: **end if**
23: **if** Number of bits in tables = N and Etotal $\leq$ Emin and Ttotal $\leq$ Tmax **then** **return** Etotal, Ttotal
24: **end if**
25: **end loop**

5 Evaluation and Results

In this section, the evaluation of the proposed Energy-Efficient Computation Offloading algorithm based on dynamic programming table is evaluated for multi-UAV network with MEC server connection.

5.1 Simulation Setup

The simulation of our work by MATLAB environment, for a multi-level offloading system scenario that consists of multi-UAV network with processing and energy capabilities connected with a base station through wireless channel. Three scenarios are used, first the UAV m can execute the task i locally. Second scenario the UAV m can offload the task i to nearby UAVs, where each UAV has connected with nearby UAVs at maximum up to five UAV simultaneously. Third scenario the UAV m can offload the task i to ground base station which equipped with MEC server through a wireless channel. The parameters are setup as shown in the Table 2.

Table 2. Simulation parameters

Parameter	Value
Number of tasks N	15
Data size d_i	[10–30] MB
CPU cycles c_i	1900 Cycle/s
Time Constraint $T_i^{\max}$	0.002 s
Transmission data rate r_m	[3–9] Mbps
Bandwidth BW	[1–5] MHz
Transmission power P_m	0.25 W
power consumption ρ	0.1 W
Computation capacity of UAV m f_m^{local}	500 MHz
Computation capacity of nearby UAVs f_m^{nearby}	500 MHz
Computation capacity of edge server f_{edge}	1000 GHz

We set the number of tasks as $N = 15$ and the number of UAVs is 10, where each UAV can transmit (offload) at maximum up to 5 nearby UAVs simultaneously, where the UAV m can select a nearby UAVs to offload the task i by depending on the distances between them. The computing capabilities of the UAVs 500 MHz, and the computing capability of MEC server is 1000 GHz. The data size randomly increased from 10 to 30 Mb. The number of CPU cycles required to accomplish the task i is 1900 cycle/s. We assume that the transmission bandwidth BW and transmitting power P_m of UAVs are [1 MHz–5 MHz] and 0.25 W, respectively, and the density of noise power of channel σ is 10^{-5}. The wireless channel gain is modeled as $G_m = 127 + 30 * log(d)$ [7], where d is the distance between UAVs and base station, and so on the transmission data rates of UAVs randomly changed between 3 Mbps and 9 Mbps according to the distances between them, in other word a shorter distance means a higher transmission rate.

The Fig. 2 describes the energy consumption and time execution versus number of UAVs, where the UAV m can offload the tasks into 5 nearby UAVs at maximum. as shown in the figure both energy consumption and time execution are decreases whenever the number of UAVs increases. Also, as shown in the Fig. 3 the energy consumption and time execution are decreasing when the transmission data rate increase.

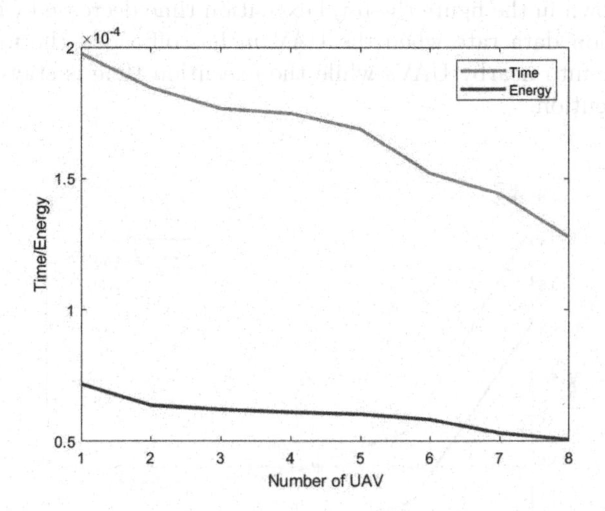

Fig. 2. Energy and time vs number of nearby UAVs

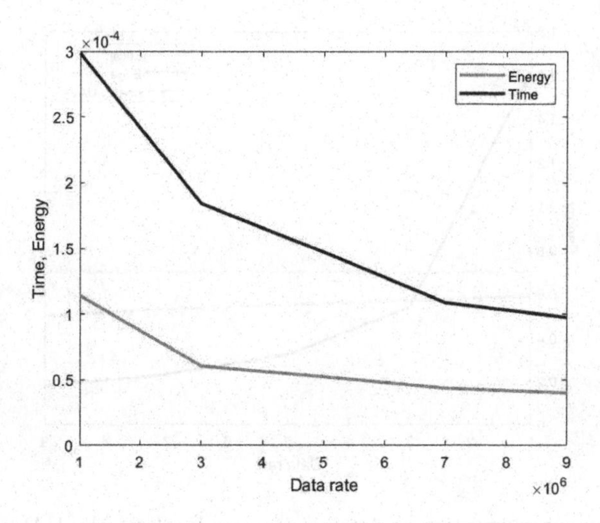

Fig. 3. Energy and time vs data rates

Figure 4 describe the total execution time versus the transmission data rate in case (i)local computing, (ii)nearby UAVs computing, and (iii)edge server computing. As shown in the figure the total execution time decreased with increased the transmission data rate when the UAV m has offloaded the tasks into the edge server or into nearby UAVs while the execution time is stays constant in case local execution.

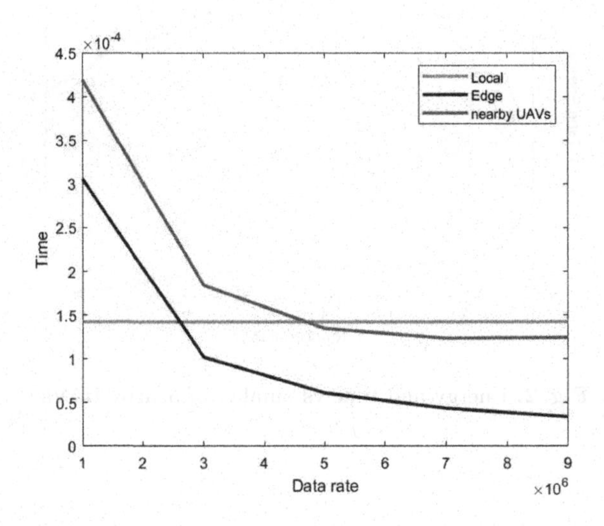

Fig. 4. The time execution vs data rate in case local, nearby UAVs, and edge server.

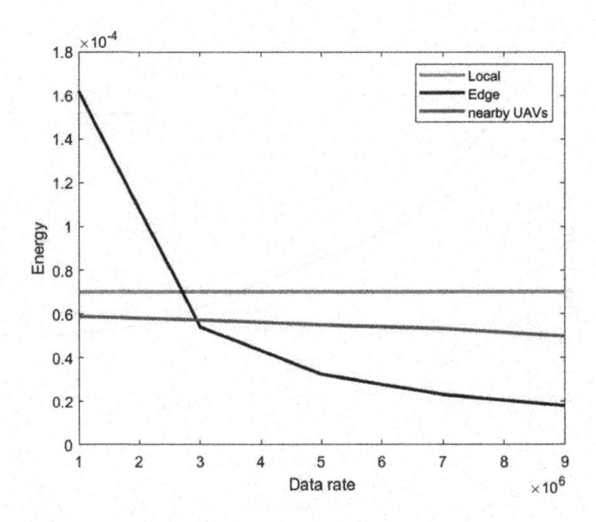

Fig. 5. The energy consumption vs data rate in case local, nearby UAVs, edge server.

Figure 5 shows transmission data rate vs the total energy consumption in case (i) local execution, (ii) edge execution, and (iii) nearby execution. The energy consumption is computed using objective function (14). Since the proposed algorithm it finds an offloading decision according to the wireless transmission rate. As shown in figure the total energy minimized when the tasks are offloaded and executed remotely at nearby UAVs or at edge server, whereas in case local execution the energy consumption is constant, since for high transmission rates, the energy consumption in case edge server execution is extremely decreased and less than the energy consumption of local and nearby executions, as well as the energy consumption of nearby UAVs is also decreased gradually with increase the transmission data rates. Figure 6 and 7 are describes the energy consumption and time execution versus the number of tasks in case all tasks are in: (1) local execution, (2) nearby UAVs execution, (3) edge server execution, where in all cases the energy consumption and time execution are approximately linearly increased with increase the number of tasks. In case edge server execution both energy consumption and time execution are much less than the energy consumption and time execution in case locally and nearby UAVs executions this is because the capability of edge server for data processing is much higher than the capability of UAVs.

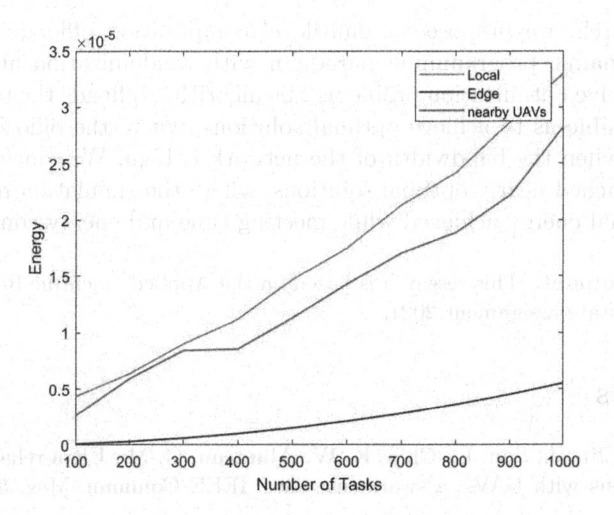

Fig. 6. Energy consumption vs number of tasks.

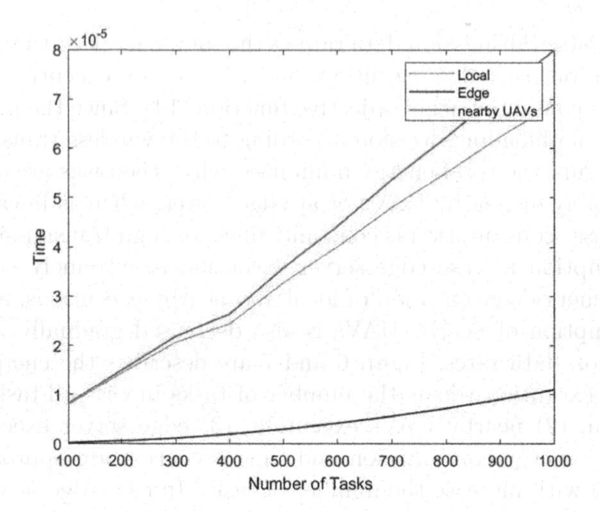

Fig. 7. Time execution vs number of tasks.

6 Conclusion

In this research, we proposed a multilevel computation offloading algorithm based on dynamic programming paradigm with randomization and hamming distance to solve optimization problem. The algorithm offloads the tasks dynamically as possible as to achieve optimal solutions, where the offloading process implements when the bandwidth of the network is high. We conclude that the algorithm reached nearly optimal solutions, where the simulation results shows that a minimal energy achieved while meeting time and energy constraints.

Acknowledgement. This research is based on the Applied Scientific Research under the SPbSUT state assignment 2021.

References

1. Yuan, Z., Jin, J., Sun, L., Chin, K.-W., Muntean, G.-M.: Ultra-reliable IoT communications with UAVs: a swarm use case. IEEE Commun. Mag. **56**(12), 90–96 (2018)
2. Khuwaja, A.A., Chen, Y., Zhao, N., Alouini, M.-S., Dobbins, P.: A survey of channel modeling for UAV communications. IEEE Commun. Surv. Tutor. **20**(4), 2804–2821 (2018)
3. Shan, N., Li, Y., Cui, X.: A multilevel optimization framework for computation offloading in mobile edge computing. Math. Prob. Eng. **2020** (2020)
4. Shahzad, H., Szymanski, T.H.: A dynamic programming offloading algorithm for mobile cloud computing, pp. 1–5 (2016)
5. Yang, Z., Pan, C., Wang, K., Shikh-Bahaei, M.: Energy efficient resource allocation in UAV-enabled mobile edge computing networks. IEEE Trans. Wirel. Commun. **18**(9), 4576–4589 (2019)

6. Rahati-Quchani, M., Abrishami, S., Feizi, M.: An efficient mechanism for computation offloading in mobile-edge computing. arXiv preprint arXiv:1909.06849 (2019)
7. Hao, Y., Chen, M., Hu, L., Hossain, M.S., Ghoneim, A.: Energy efficient task caching and offloading for mobile edge computing. IEEE Access **6**, 11365–11373 (2018)
8. Bai, T., Wang, J., Ren, Y., Hanzo, L.: Energy-efficient computation offloading for secure UAV-edge-computing systems. IEEE Trans. Veh. Technol. **68**(6), 6074–6087 (2019)
9. Ateya, A.A.A., Muthanna, A., Kirichek, R., Hammoudeh, M., Koucheryavy, A.: Energy-and latency-aware hybrid offloading algorithm for UAVs. IEEE Access **7**, 37587–37600 (2019)
10. Wang, L., et al.: Rl-based user association and resource allocation for multi-UAV enabled mec, pp. 741–746 (2019)
11. Elgendy, I.A., Zhang, W., Tian, Y.-C., Li, K.: Resource allocation and computation offloading with data security for mobile edge computing. Future Gener. Comput. Syst. **100**, 531–541 (2019)

The Increasing of Resource Sharing Efficiency in Network Slicing Implementation

Mikhail S. Stepanov[1], Sergey N. Stepanov[1](✉), Umer Andrabi[2], Dmitriy Petrov[2], and Juvent Ndayikunda[1]

[1] Department of communication networks and commutation systems, Moscow Technical University of Communications and Informatics, 8A, Aviamotornaya str., Moscow 111024, Russia

[2] Moscow Institute of Physics and Technology (State University), 9 Institutskiy per., Dolgoprudny, Moscow Region 141701, Russia

umer.andrabi@phytech.edu, petrov.ds@phystech.edu

Abstract. Network slicing has wide range of applications in resource sharing especially for creating differentiated servicing scenarios for heterogeneous traffic environments supported by 5G networks. Despite of such potential features there are numerous problems in implementation of network slicing and hence present ETSI standards only give definitions and key principles of corresponding architecture. Such situations signifies the importance of mathematical modeling of network slicing procedures to develop efficient instruments for their implementation. An analytical framework to model the resource allocation procedures for transmission of multiservice traffic has been constructed and analyzed. The model consists of arbitrary number of traffic streams originating from different types of real time applications. All random variables used in the model have exponential distribution with corresponding mean values. Two different scenario of resource sharing has been taken into consideration for incoming traffic streams: Network Slicing, when resources are strictly divided among incoming traffic streams, and Filtering, when the access to resource is restricted depending on the amount of resource occupied by all traffic streams. It has been shown how to use both scenarios for creating conditions for differentiated servicing of heterogeneous traffic. Given numerical assessment proves that scenario based on Filtering is more efficient to solve the formulated task in contrast to the analogous scenario based on slicing.

Keywords: Network Slicing · Resource allocation and sharing · Restricted access · Recursive algorithm

U. Andrabi—The publication has been supported by the Russian Foundation for Basic Research, project No. 20-37-90048. Postgraduates.

V. M. Vishnevskiy et al. (Eds.): DCCN 2021, CCIS 1552, pp. 18–35, 2022.
https://doi.org/10.1007/978-3-030-97110-6_2

1 Introduction

Earlier developed standards for serving multiservice traffic streams followed the "one size fits all" paradigm. The main idea behind this approach is to provide communication services to a shared resource without taking into account some specific features of the traffic being served. Because of high level of diversity in traffic parameters and quality of service indicators such approach is not suitable for upcoming 5G era. This follows from the fact that implementation of such scenario leads to uncontrolled allocation of transmission resources in favor of traffic flows with relatively small data rate requirements.

This conclusion is intuitively clear and is illustrated numerically in Fig. 1 and Fig. 2, where performance measures of conjoint servicing of $n = 3$ Poissonian flows of requests are shown vs ρ the intensity of offered load per resource unit (r.u.). In order to show the negative effect of uncontrolled allocation of resource, the case of serving heterogeneous traffic is considered with resource requirement b_k for kth flow defined as follows $b_1 = 1$ r.u., $b_2 = 10$ r.u., $b_3 = 20$ r.u. Total number of resource units is $v = 200$. Offered load expressed in resource units is chosen as follows $a_1b_1 = a_2b_2 = a_3b_3$, where a_k is intensity of offered load of kth flow expressed in Erlangs. It can be seen clearly that calculated performance measures i.e. the ratios of lost requests (Fig. 1) and the mean values of resource usage (Fig. 2) are explicitly dependent on the values of resource requirement. This property manifests itself for moderate values of offered load when $\rho \leq 1$

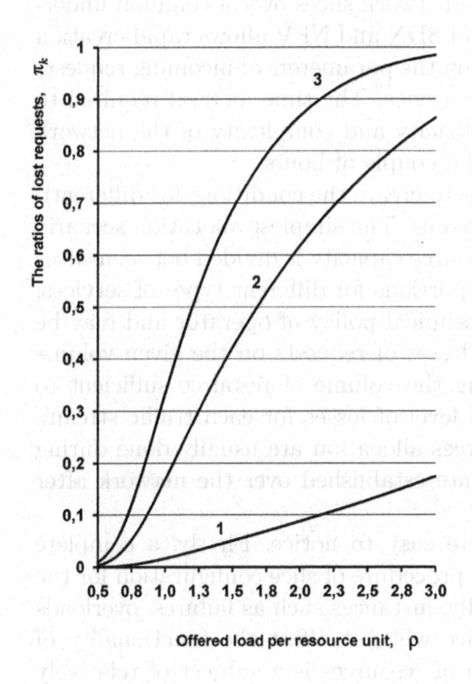

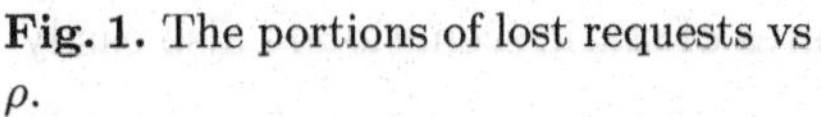

Fig. 1. The portions of lost requests vs ρ.

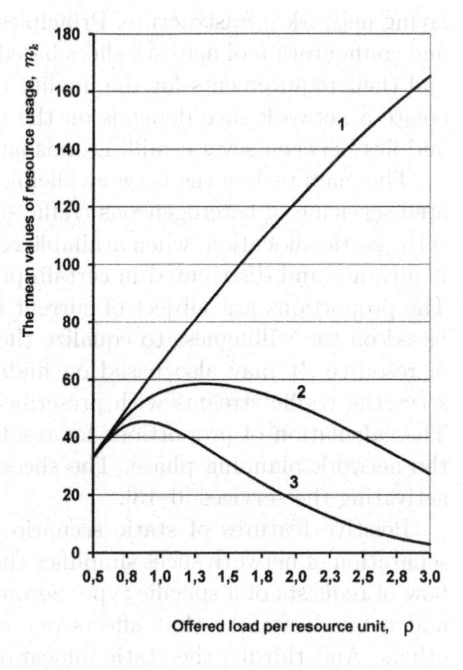

Fig. 2. The mean values of resource usage vs ρ.

and especially evident in case of overload when $\rho > 1$. In the last case, requests for "heavy" traffic transmission can be replaced while servicing by requests for "light" traffic transmission.

To overcome mentioned such challenges and create conditions for differentiated servicing of heterogeneous traffic, the concept of network slicing has been introduced. This technique gives the rules to realize distribution of transmission resources in the form of separate logical network group (slice) of traffic streams with close resource requirements [1–3,5]. Grouping of the traffic streams can also be based on the achievement of certain indicators of economic efficiency and in some cases the "slice per service" procedure may be applied, e.g. when a service requires special quality of servicing or uses unique scheme of charging by the CSP (Communications Service Provider) [4]. Another realization of the concept "slice per service" takes place when a couple of similar service instances are associated with the same service model created by the CSP during the service design phase. In this case a slice can be reserved per service but will serve many similar service instances. The choice depends on the technical policy of an operator and degree of heterogeneousness of traffic flows. Today the concept of network slicing is considered as forward-looking policy, especially for upcoming 5G era. It worth to mention that principles of network slicing are also applicable to LTE and LTE-A networks.

The implementation of network slicing is based on the concepts of software-defined networking (SDN) and network function virtualization (NFV) that gives the rules for creating flexible and scalable network slices over a common underlaying network infrastructure. Principles of SDN and NFV allows rapid creation and configuration of network slices based on the parameters of incoming requests and their requirements for the quality of service. The time interval required to create a network slice depends on the capacity and complexity of the network and lies between several milliseconds and a couple of hours.

The main task of the network slicing is to create the conditions for differentiated servicing of heterogeneous traffic streams. The simplest allocation scenario is the static allocation, when available resource capacity is divided between slices in advance and distributed in certain proportions for different types of services. The proportions are subject of current technical policy of operator and may be based on the willingness to equalize the losses of requests on the given volume of resource. It may also based on finding the volume of resource sufficient to serve the traffic streams with prescribed level of losses for each traffic stream. The calculation of proportions for resources allocation are usually done during the network planning phase. The slices are established over the network after activating the services [6–13].

Positive features of static scenario are easy to notice. Firstly, a complete separation of network slices simplifies the procedure of slice configuration for the flow of requests of a specific type. Secondly, instances such as failures, overloads and network attacks that affects one slice will not affect the functionality of others. And thirdly, the static allocation of resources is a subject of relatively simple mathematical analysis. Negative features of static scenario are also easily predicted. Main among them is inefficient use of the resources reserved for each slice, which cannot be reallocated to other slices because of the principles of static allocation scenario [1–3,5,10,11,13].

In contrast to static distribution of resources is dynamic scenario. In dynamic allocation, available transmission resources are given to incoming traffic flows without taking into account traffic requirements and performance indicators. This approach increases the efficiency of resource usage but at the expense of bad losses of requests for "heavy" traffic transmission (see, Fig. 1 and 2). To combine positive features of static and dynamic scenarios various forms of dynamic allocation procedures with restricted access have already been suggested [7,10,13]. These procedures are based on the idea to allocate some part of the available capacity for common use, and distribute the remaining part of the resources between chosen slices. This form of resource distribution requires more complex mathematical modeling, but such scenarios have more efficient form of utilization of network resources. Another positive feature of partly dynamic allocation procedures lies in the fact that they make the network more robust to handle fluctuations of incoming traffic [10,13].

It is necessary to emphasize that despite of the fact that network slicing procedures has potential to create conditions for differentiated servicing of heterogeneous traffic there are numerous challenges in network slicing implementation. Presently known ETSI standards (e.g., TS 138.913 [1], TS 123.501 [2], etc.) give only definitions and key principles of corresponding architecture. This situation makes usage of such a promising concept complicated and proprietary. All these points highlights the importance of mathematical study of network slicing procedures for developing efficient instruments for their implementation in network planning.

In this paper we have constructed and analyzed an analytical framework to model the resource allocation procedures for transmission of multiservice traffic. The model consists of arbitrary number of traffic streams created by variety of real time applications. All random variables used in the model have exponential distribution with corresponding mean values. Two scenarios of resource sharing has been considered for incoming traffic streams, these are: Network Slicing when resources are strictly divided among incoming traffic streams, and Filtering, when the access to resource is restricted depending on the amount of resource occupied by all traffic streams. The proposed model generalizes the results of [10,13] by considering more efficient procedure of creating conditions for differentiated servicing of heterogeneous traffic.

The rest of the paper is organized as follows. In Sect. 2 the mathematical description of the model used for realization of Network Slicing procedure will be presented. The model's performance measures are introduced and recursive algorithm of characteristics calculation is formulated. It will be shown how model can be used to create the conditions for differentiated servicing of heterogeneous traffic. In Sect. 3 the differentiated servicing will be obtained by Filtering incoming requests arriving for servicing depending on the amount of resource occupied by all traffic streams. Numerical assessment of the suggested resource sharing scenarios is performed in Sect. 4. Finally, conclusions are drawn in the last section.

2 Differentiated Servicing Based on the Network Slicing

Let us represent an access node of telecommunication network as transmission link with a fixed capacity C Mbps. Requests belonging to n service classes are arriving according to Poisson process with intensity λ. With probability p_k incoming call belongs to kth class and requires link's transmission capacity in size c_k Mbps. The employed Call admission control (CAC) follows complete sharing policy, that means an incoming call is accepted if $\sum_{s=1}^{n} i_s c_s + c_k \le C$, where i_s is the number of served calls belonging to sth class at the moment of incoming call, $s = 1, \ldots, n$. The functional model of the link is shown in Fig. 3.

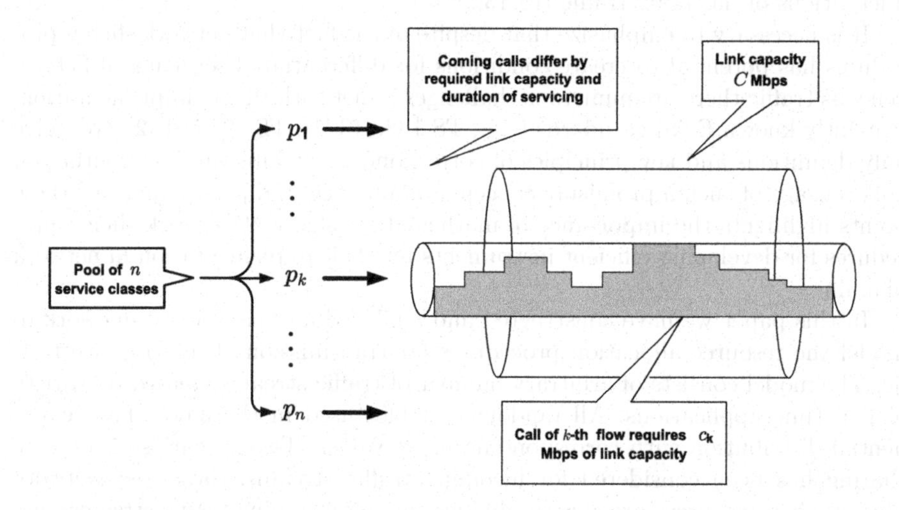

Fig. 3. The functional model of a link with a complete sharing policy.

The estimation of c_k play an important role in planning link capacity [14]. Let us consider the main approaching entities and start by taking fixed networks into consideration. Let us denote by m—the mean rate of the source expressed in bps, and by h we denote its peak rate. Let us use m and h as a capacity requirement for a chosen class of services, which provides two limiting solutions of the problem. Choosing m leads to overloading the link capacity, while h leads to underestimation of capacity. In practical applications capacity requirements are calculated in form of effective rate of source. Therefore, in our case this effective rate is denoted by d and its value expressed in bps. The correct choice of d allows to calculate maximum possible number of connection of a chosen class for fixed value of lost packets. It is clear that $m \leq d \leq h$. For some mathematically defined sources effective rate $d(s,t)$ can be found from relation [14]

$$d(s,t) = \frac{1}{st} \log_{10} M\left\{e^{sX(t)}\right\}, \quad s > 0, \quad t < \infty,$$

where s, t—are source parameters and $X(t)$ random process defining the number of packets arrived in time interval $[0,t]$. In other cases empirical relations are used. Example one of such relation is as follows [14,15]

$$d = \begin{cases} k_2\, m \left(1 + 3\, k_1 \left(1 - \frac{m}{h}\right)\right), & \text{if} \quad 3\, k_1 \leq \min(3, \frac{h}{m}); \\ k_2\, m \left(1 + 3\, k_1^2 \left(1 - \frac{m}{h}\right)\right), & \text{if} \quad 3 < 3\, k_1^2 \leq \frac{h}{m}; \\ k_2\, h & \text{in other cases,} \end{cases}$$

where $k_1 = -\frac{2\,h}{C} \log P_{\text{loss}}$; $k_2 = 1 - \frac{1}{50} \log P_{\text{loss}}$ and P_{loss}—the ratio of lost packets.

After analyzing above given considerations, we can comprehend that resource units are easily defined for fixed networks. On resource unit (or virtual channel) is a part of link's capacity (expressed in bps) chosen after analysis of information rate associated with serving request. For mobile networks based on LTE technology and its progressive development in form of 5G and beyond, the definition of resource unit in mathematical modeling of resource allocation is more complicated. In these networks, the transmission facilities in radio interface domain are represented in the form of resource blocks. Some part of these resource blocks is used to provide transmission speed required to serve the arriving call with given values of performance indicators. It is necessary to say that there is no linear relationship between number of allocated blocks and obtained value of transmission speed. Corresponding functionality has complicated character and depends on number of factors such as chosen coding scheme, acting algorithm of packet scheduling, the usage of MIMO, distance to base station, number of simultaneously served requests and their classes and so on. In this situation, the resource unit in form of transmission speed shall be determined after simulation and subsequent calibration of the obtained results with help of field measurements.

To construct a mathematical model it is necessary to convert transmission capacity into the format of resource units (r.u.). One resource unit c corresponds to the minimal capacity requirement of incoming calls. The value of c can also be calculated as greatest common divisor ($gcd(\cdot)$) of c_k, $k = 1, \dots, n$. So we have $c = gcd(c_1, \dots, c_n)$. The model's structure parameters are as follows

$$v = \left\lfloor \frac{C}{c} \right\rfloor, \quad b_k = \left\lceil \frac{c_k}{c} \right\rceil, \quad k = 1, \dots, n.$$

Results of above considerations suggests that the mathematical model of access node can be represented as a pool of v r.u. that are used for servicing n

incoming Poisson flows of requests with intensities a_k, $k = 1, \ldots, n$. Here a_k is intensity of offered load for class k calls expressed in Erlangs. A request from kth flow uses b_k r.u. for the time of connection. Without loss of generality we shall assume that all the holding times are exponentially distributed with the same mean value chosen to one. However, it should be also known that the model under consideration is insensitive to the distribution of the holding time, and each flow may furthermore have individual mean holding times [14].

Let $i_k(t)$ denote the number of calls from the kth flow served at time t. The model is described by n-dimensional markovian process of the type $r(t) = (i_1(t), \ldots, i_n(t))$ with state space S consisting of vectors $(i_1, \ldots, i_n)$, where i_k is the number of calls from the kth flow being served by the link under stationary conditions. The state space S is defined as follows: $(i_1, \ldots, i_n) \in S$, $i_k \geq 0$, $k = 1, \ldots, n$, $\sum_{k=1}^{n} i_k b_k \leq v$. Let us by $P(i_1, \ldots, i_n)$ denote the unnormalised values of stationary probabilities of $r(t)$. After normalisation the value $p(i_1, \ldots, i_n)$ denotes the mean proportion of time when exactly $i_1, \ldots, i_n$ connections are established. Later we will use upper case letters for denoting the unnormalised values of characteristics and lower case letters for denoting the normalised values of characteristics. Assume that for state $(i_1, \ldots, i_n)$ the value i denotes the total number of occupied r.u. $i = i_1 b_1 + \cdots + i_n b_n$.

The process of transmission of kth flow, $k = 1, \ldots, n$, is described by π_k the ratio of lost requests. Formal definition of π_k through values of state probabilities are as follows (here and further, summations are for all states $(i_1, \ldots, i_n) \in S$ satisfying formulated condition):

$$\pi_k = \sum_{i+b_k>v} p(i_1, \ldots, i_n). \tag{1}$$

Let us denote by m_k the mean number of r.u. occupied by kth flow requests. From Little's formula we obtain $m_k = a_k b_k (1 - \pi_k)$.

The most efficient calculation scheme for the model introduced is the recurrence algorithm first obtained in [16] and later also derived in [17,18]. The recurrence follows from the reversibility of $r(t)$. It gives the relations of detailed balance for state $(i_1, \ldots, i_n)$ in the form

$$p(i_1, \ldots, i_k, \ldots, i_n) i_k = p(i_1, \ldots, i_k - 1, \ldots, i_n) a_k. \tag{2}$$

Let

$$p(i) = \sum_{i_1 b_1 + \cdots + i_n b_n = i} p(i_1, \ldots, i_n).$$

It is clear that

$$\pi_k = \sum_{i=v-b_k+1}^{v} p(i), \quad i = 1, \ldots, v. \tag{3}$$

After summing (2) for $(i_1, \ldots, i_n) \in S$ such as $i_1\, b_1 + \ldots + i_n\, b_n = i$, subsequent multiplication on b_k and summation over $k = 1, 2, \ldots, n$ we obtain

$$p(i)\, i = \sum_{k=1}^{n} a_k\, b_k\, p(i - b_k) I(i - b_k \geq 0), \quad i = 1, \ldots, v, \tag{4}$$

where function $I(\cdot)$ equals one if the formulated condition is fulfilled else equals zero. Consistent implementation of (4) gives values of $p(i)$, $i = 0, 1, \ldots, v$ and performance measures π_k, m_k of kth traffic stream, $k = 1, \ldots, n$.

The model and algorithms developed on this basis will be used for analysis of conditions for differentiated servicing of heterogeneous traffic on a common pool of resource units. To simplify the investigation we limit our analysis to the case of conjoint servicing of $n = 2$ traffic streams. Traffic flows have fixed parameters of offered load a_k and required number of r.u. b_k, $k = 1, 2$. It is supposed that $b_1 \ll b_2$. It means that the first flow forms traffic stream with "light" requests, the second—traffic stream with "heavy" requests. As seen in Fig 1 and 2, conjoint servicing of "light" and "heavy" requests leads to the uncontrolled allocation of r.u. in favor of "light" requests. Let us implement the Network Slicing concept to create conditions for differentiated servicing. We are considering three problem settings.

1. For given pool of v (i.e. resource units r.u.), find the division of the resource into slices by equalizing the rate of losses of both flows.
2. Find the minimum value of v and the division of the common resource v into slices by equalizing the rate of losses of both flows at the prescribed level π.
3. Find the minimum value of v and the division of the common resource v into slices by providing the rate of losses for the first flow at the level π_1^* and for the second flow at the level π_2^*.

The solution of formulated problems are presented in Fig. 4 (the first task), in Fig. 5 (the second task), in Fig. 6 (the third task). The model input parameters are as follows: $v = 200$ r.u., $n = 2$, $b_1 = 1$ r.u., $b_2 = 10$ r.u., $a_1 = 100$ Erl; $a_2 = 10$ Erl. Let us denote by v_1 the number of r.u. in the first slice. The performance measures are calculated with the help of recursion (4). Numerical results show the possibility of using the Network Slicing concept for solving the formulated problems but later we will show that more efficient solution for differentiated servicing can be obtained by filtering the input flows. This results will be presented in Sect. 3.

3 Differentiated Servicing Based on the Filtering of the Input Flows

Let us consider the model of multiservice access node discussed in Sect. 2 and limit the access to servicing of requests from kth flow by integer number θ_k. The access restriction is based on the overall resource occupancy level and is implemented as follows. Let us denote by i the total number of occupied r.u. at the moment of call coming. If $i \leq \theta_k$, then arriving request from kth flow is accepted else it's is refused and is not resumed for servicing. It is clear that θ_k satisfies inequalities $0 \leq \theta_k \leq v - b_k$. Let us call this procedure as Filtering of requests from kth flow. By choosing the value of θ_k, the operator limits the access of requests from kth flow. If the node is in a state where occupied resource range from $\theta_k + 1$ to $v - b_k$, then requests from kth flow are refused and the

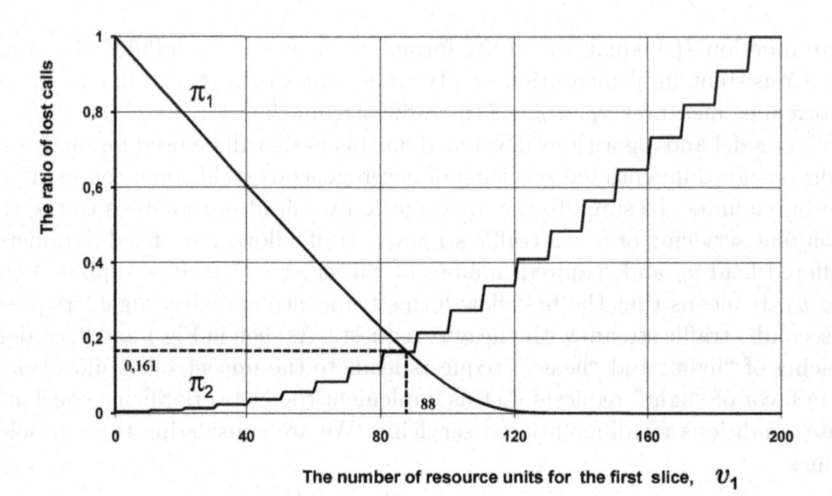

Fig. 4. The results of division of the common resource into slices equalizing the rate of losses of both flows for given value of v.

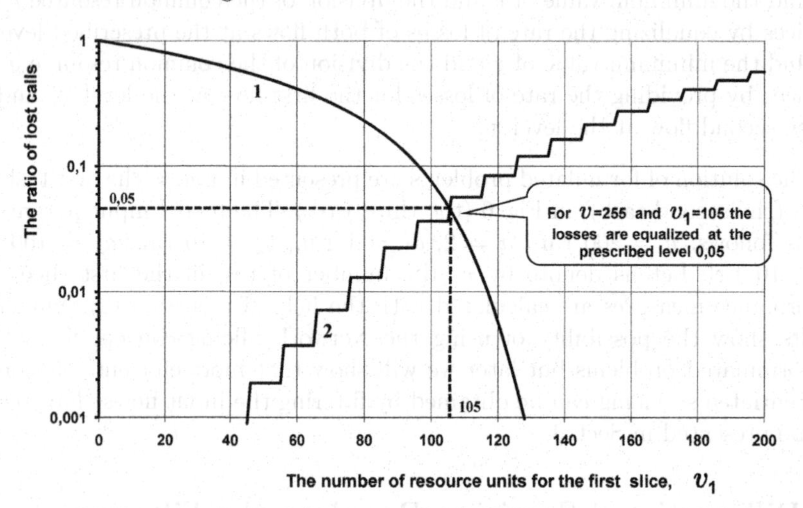

Fig. 5. The results of finding the minimum value of v and the division of the common resource into slices equalizing the rate of losses of both flows at prescribed level $\pi =$ 0,05.

available free resources are used by the other call flows with different numbers as that of k. In this manner, the considered model realizes the preferential services of these requests. The level of priority in servicing is estimated by the value of $r_k = v - b_k - \theta_k$, the number of r.u. allotted for common use to the requests whose numbers are different as that of k. If $\theta_k = v - b_k$, then Filtering will not happening. Figure 7 shows the rules of acceptance of requests of kth flow when

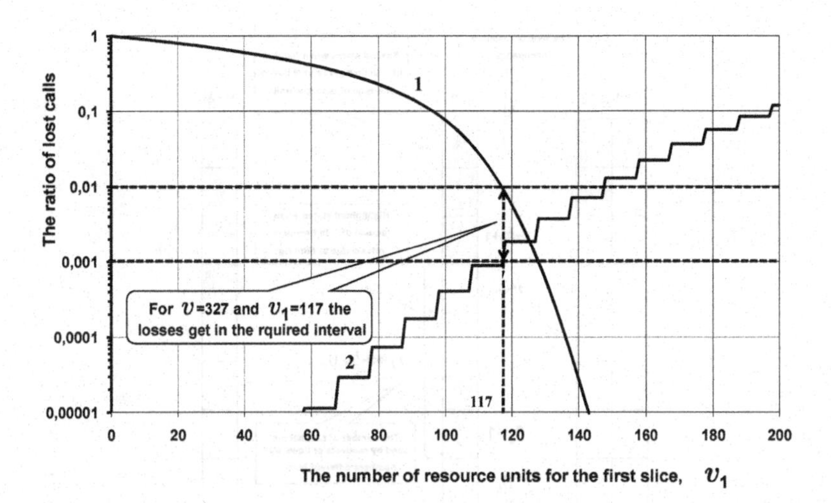

Fig. 6. The results of finding the minimum value of v and the division of the common resource into slices providing the rate of losses for the first flow at level $\pi_1^* = 0{,}01$ and the second flow at level $\pi_2^* = 0{,}001$.

the procedure of Filtering is used to create conditions for preferential service of requests with numbers different from k.

Let us describe the Filtering procedure in a more generalized way by the function $f_k(i)$ that denotes the probability of acceptance of the request from kth flow. These requests are expected to be in a state where they are occupying i resource units (r.u.). The procedure of acceptance of the request from kth flow is shown in the Fig. 7 and it can be described by choosing the following filtering function $f_k(i) = 1$, $i \le \theta_k$ and $f_k(i) = 0$, $i > \theta_k$. Let us denote for requests from kth flow by λ_k the intensity of incoming calls, by $1/\mu_k$ the mean time of call servicing and by b_k the number of r.u. required for servicing of one call. Let us suppose that all random variables used for model description have exponential distributions with corresponding mean values.

The model functioning is described by n-dimensional markovian process $r(t) = (i_1(t), \dots, i_n(t))$ where $i_k(t)$ is the number of requests from the kth flow served at time t. The state space S depends on the choice of filtering functions but for any choice the model states $(i_1, \dots, i_n)$ should satisfy the inequality $\sum_{k=1}^{n} i_k b_k \le v$. Let us denote by $P(i_1, \dots, i_n)$, the unnormalized values of stationary probabilities of $r(t)$. After normalization the values $p(i_1, \dots, i_n)$ can be used for estimation of performance measures of kth flow: the portion of lost calls π_k and the mean number of occupied resource units m_k. Assume that for state $(i_1, \dots, i_n)$ the value i denotes the total number of occupied r.u. $i = i_1 b_1 + \dots + i_n b_n$. The values π_k and m_k are defined as follows

$$\pi_k = \sum_{(i_1,\dots,i_n)\in S} p(i_1, \dots, i_n)\,(1 - f_k(i)), \quad m_k = \lambda_k b_k (1 - \pi_k)/\mu_k. \tag{5}$$

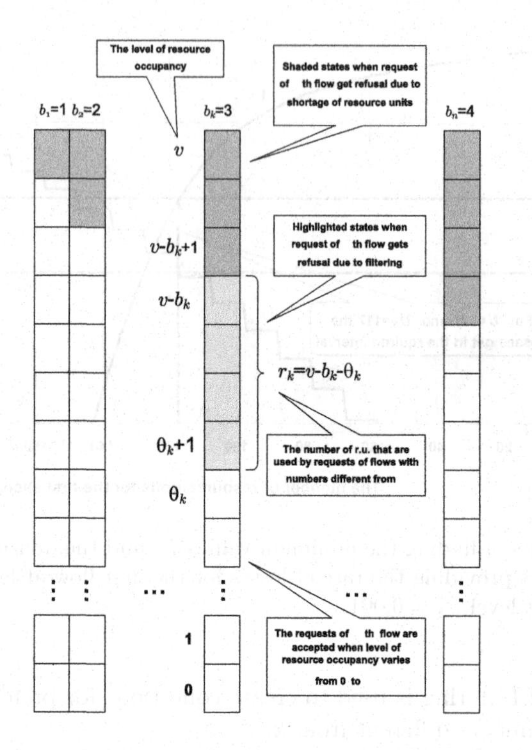

Fig. 7. The rules of acceptance of requests from kth flow when the procedure of Filtering is used for to create conditions for preferential service of requests with numbers different from k.

Because of filtering the markovian process $r(t)$ does not have a reversibility property that simplified the estimation of performance measures of the model introduced in Sect. 2. The values of $p(i_1, \ldots, i_n)$ can be found after normalizing the solution of the system of state equations by using Gauss-Zeidel iteration algorithm

$$P(i_1, \ldots, i_n) \sum_{k=1}^{n} (\lambda_k\, f_k(i) + i_k\, \mu_k) \tag{6}$$

$$= \sum_{k=1}^{n} P(i_1, \ldots, i_k - 1, \ldots, i_n)\, \lambda_k\, f_k(i - b_k)\, I(i_k > 0)$$

$$+ \sum_{k=1}^{n} P(i_1, \ldots, i_k + 1, \ldots, i_n)\, (i_k + 1)\, \mu_k\, I(i + b_k \leq v).$$

Another way to estimate π_k and m_k is to use the approximate procedure. Such algorithm can be constructed if we suppose that reversibility property is valid for $r(t)$. We get the following relation (mark ^ means the approximate character of result)

$$\hat{p}(i_1,\ldots,i_k,\ldots,i_n)\, i_k\, \mu_k = \hat{p}(i_1,\ldots,i_k-1,\ldots,i_n)\, \lambda_k\, f_k(i-b_k). \tag{7}$$

These relations are similar to the detailed balance relations (2) obtained in Sect. 2. By using the ideas that produced the recursive formula (4) we can get the same type of recursion for model with filtering.

Let us sum (7) over $(i_1,\ldots,i_n) \in S$ so that $i_1 b_1 + \ldots + i_n b_n = i$. Summation gives recursive relations for $\hat{P}(i)$, $i = 1,\ldots,v$

$$\hat{P}(i) = \frac{1}{i} \times \sum_{k=1}^{n} \lambda_k\, b_k \hat{P}(i-b_k)\, f_k(i-b_k)/\mu_k, \tag{8}$$

here

$$\hat{p}(i) = \sum_{i_1 b_1+\ldots+i_n b_n = i} \hat{p}(i_1,\ldots,i_n), \quad i = 0,1,\ldots,v.$$

After realizing the recursions (8) and subsequent normalization, the values $\hat{p}(i)$, $i = 0,1,\ldots,v$ can be used to estimate $\hat{\pi}_k$ and $\hat{m}_k$

$$\hat{\pi}_k = \sum_{i=0}^{v} \hat{p}(i)\,(1-f_k(i)), \quad \hat{m}_k = \lambda_k b_k (1-\hat{\pi}_k)/\mu_k. \tag{9}$$

The model of access node with filtering will be used to analyze the process of differentiated servicing of heterogeneous traffic on a common pool of r.u. In order to simplify the calculation and create conditions for comparison of results with the analogous results obtained in Sect. 2 (using Network Slicing), we can perform calculation in a similar fashion here and can use same input parameter values that were presented in Fig. 4, 5 and 6.

The model has two input flows of "light" and "heavy" traffic streams. The procedure of Filtering will be applied only to "light" traffic. The system of state Eq. (6) for this particular case is: (for state (i_1, i_2) the value of parameter i is calculated from expression $i = i_1 b_1 + i_2 b_2$)

$$\begin{aligned} P(i_1,i_2)\Big(\lambda_1\, f_1(i) + \lambda_2\, I(i+b_2 \le v) + i_1\,\mu_1 + i_2\,\mu_2\Big) \\ = P(i_1-1,i_2)\,\lambda_1\, f_1(i-b_1)\, I(i_1>0) + P(i_1,i_2-1)\,\lambda_2\, I(i_2>0) \\ + P(i_1+1,i_2)\,(i_1+1)\,\mu_1\, I(i+b_1 \le v) + P(i_1,i_2+1)\,(i_2+1)\,\mu_2\, I(i+b_2 \le v). \end{aligned} \tag{10}$$

To obtain all equations of (10) it is sufficient to organize cycle over i_1 and i_2 in the form $i_1 = 0,1,\ldots,\lfloor \frac{v}{b_1} \rfloor$, $i_2 = 0,1,\ldots,\lfloor \frac{v-i_1 b_1}{b_2} \rfloor$ and put values of i_1 and i_2 into (10).

Let us denote by $P^{(s)}(i_1,i_2)$ the approximation number s for unnormalised value of $P(i_1,i_2)$ obtained by Gauss-Zeidel iteration algorithm. The recursive formula that relates consecutive approximations is given as

$$P^{(s+1)}(i_1,i_2) = \frac{1}{\left(\lambda_1 f_1(i) + \lambda_2 I(i+b_2 \le v) + i_1\mu_1 + i_2\mu_2\right)} \tag{11}$$
$$\times \Bigg(P^{(s,s+1)}(i_1-1,i_2)\,\lambda_1 f_1(i-b_1)\,I(i_1>0)$$
$$+ P^{(s,s+1)}(i_1,i_2-1)\,\lambda_2 I(i_2>0)$$
$$+ P^{(s,s+1)}(i_1+1,i_2)\,(i_1+1)\,\mu_1 I(i+b_1 \le v)$$
$$+ P^{(s,s+1)}(i_1,i_2+1)\,(i_2+1)\,\mu_2 I(i+b_2 \le v)\Bigg), \quad s=1,\ldots,$$

Symbols $(s, s+1)$ means the usage of the last obtained approximation for the corresponding probability. Convergence is checked by analyzing the closeness of successful approximations.

The problem related to resource planning that were formulated in Sect. 2 to create the conditions for differentiated servicing were initially solved by using Network Slicing and the obtained results are presented in Fig. 4, 5 and 6. Now we will solve these problems by applying the Filtering procedure to the input flows. The results are presented in Fig. 8 (for the first task), in Fig. 9 (for the second task), in Fig. 10 (for the third task). The model input parameters are as follows: $v = 200$ r.u., $b_1 = 1$ r.u., $b_2 = 10$ r.u., $\lambda_1 = 100$, $\lambda_2 = 10$, $\mu_1 = 1$, $\mu_2 = 1$. As time-unit is chosen the mean time of call servicing. In the considered model by applying the Filtering procedure, we can regulate the priority of servicing for the requests of second ("heavy") stream by varying the value of r_1 the number of reserved r.u. in favor of requests of second flow. This will be done by choosing $f_1(i)$ as follows: $f_1(i) = 0$, $i \le v - b_1 - r_1$; $f_1(i) = 1$, $i > v - b_1 - r_1$. The value of r_1 varies from 0 (no reservation) to some value that will give the right answer to the task under study. The performance measures are calculated with help of recursions (11).

Numerical results show that by using Filtering we can solve the formulated problems more efficiently in contrast to the Network slicing. This can be easily concluded by comparing the data presented in Fig. 4 and Fig. 8, where we can see that the Filtering equalize the rate of losses on a given volume of r.u. at lesser level than Network Slicing (0,161 for Slicing and 0,142 for Filtering). The same conclusion follows by comparing the data presented in Fig. 5, 6 and Fig. 9, 10, where we can see that by using Filtering we provide conditions for differentiated servicing at 5–10 % lesser volume of resource than by using Network Slicing (255 and 327 r.u. for Slicing and 238 and 297 r.u. for Filtering correspondingly). In Section 4 we consider other aspects of comparison for both the procedures by providing conditions for differentiated servicing.

The results presented in Fig. 4, 5 and 6 and Fig. 8, 9 and10 are obtained for two conjoint traffic streams. The same ideas can be used in the case when $n > 2$. In this instance we can divide incoming flows into two groups with similar QoS indicators or used more complicated optimizing procedures.

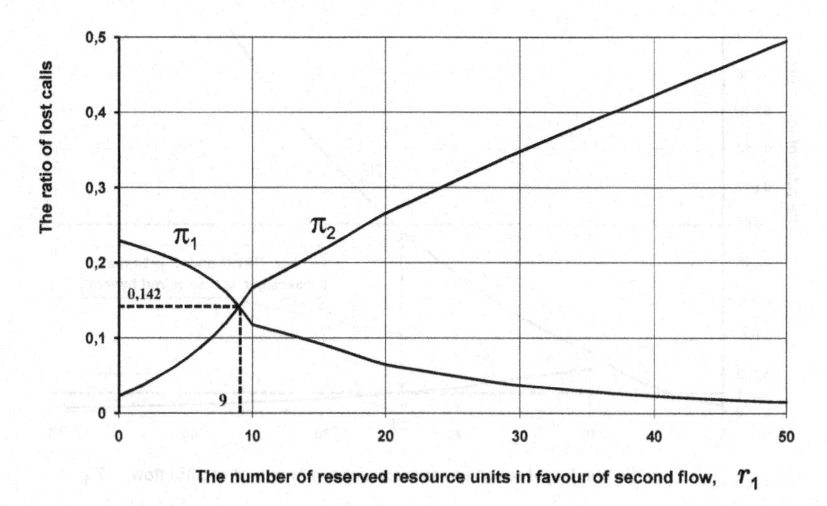

Fig. 8. The results of using the Filtering procedure to equalize the rate of losses of both the flows for a given value of v.

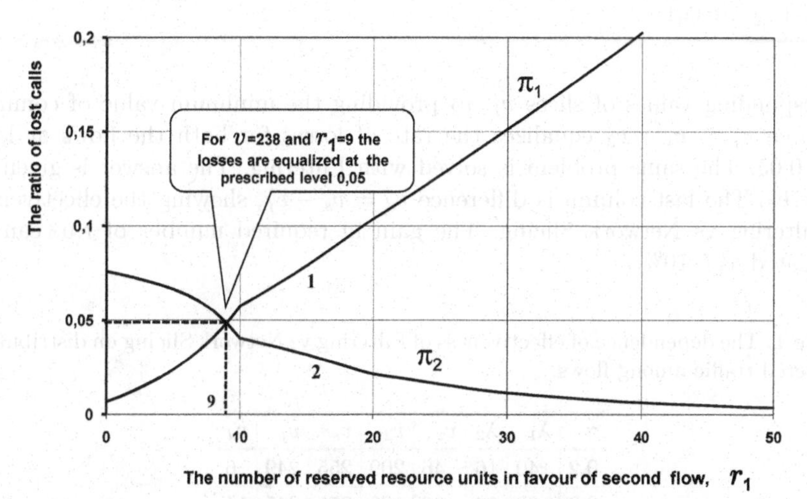

Fig. 9. The results of using the Filtering procedure to find the minimum value of v that equalize the rate of losses of both flows at prescribed level $\pi = 0{,}05$.

4 Analysis of Effectiveness of Filtering vs Network Slicing

Let us consider two examples. The dependence of effectiveness of Filtering vs Network Slicing is shown in Table 1. The model input parameters are as follows. The total offered load A is fixed, $A = 200$ r.u. The distribution of offered traffic among different flows is given by $\lambda_1 = \gamma A$, $\lambda_2 = (1 - \gamma)A$, where γ varies from 0,2 to 0,8 as shown in the Table 1. Other parameters $b_1 = 1$ r.u., $b_2 = 10$ r.u., $\mu_1 = 1$, $\mu_2 = 1$. In the Table 1 are presented the values of γ, λ_1, λ_2 and

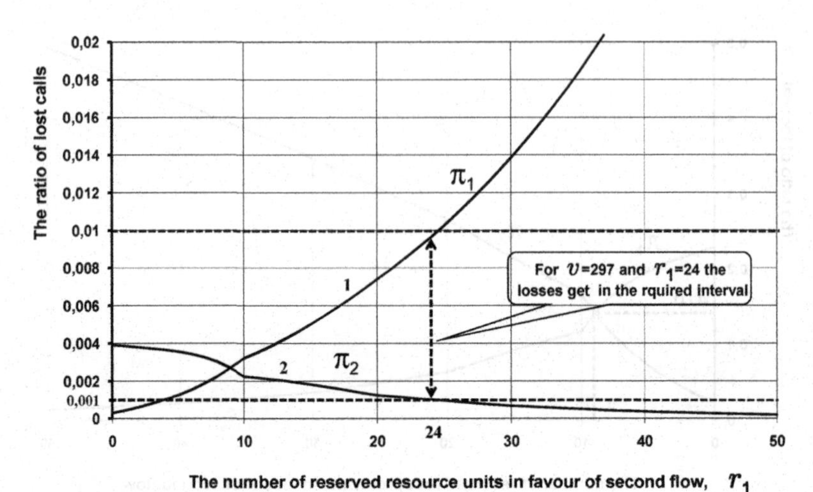

Fig. 10. The results of using the Filtering procedure to find the minimum value of v that provide the rate of losses for the first flow at level $\pi_1^* = 0{,}01$ and the second flow at level $\pi_2^* = 0{,}001$.

corresponding values of slices v_1, v_2 providing the minimum value of common resource $v_s = v_1 + v_2$ equalizes the rate of losses for both the flows at level $\pi = 0{,}05$. The same problem is solved with Filtering. The answer is given by value v_f. The last column is difference $v_d = v_s - v_f$, showing the effectiveness of Filtering vs Network Slicing. The gain in required number of r.u. can be estimated as 5-10%.

Table 1. The dependence of effectiveness of Filtering vs Network Slicing on distribution of offered traffic among flows.

γ	λ_1	λ_2	v_1	v_2	v_s	v_f	v_d
0,2	40	16	46	209	255	249	6
0,3	60	14	66	190	256	245	11
0,4	80	12	86	170	256	242	14
0,5	100	10	105	150	255	238	17
0,6	120	8	125	130	255	233	22
0,7	140	6	144	100	244	229	15
0,8	160	4	163	80	243	223	20

Another positive feature of partly dynamic allocation procedures such as Filtering lies in the fact that they make networks more robust to handle the fluctuations of incoming traffic. We can demonstrate this fact with the help of a numerical example. Let us consider the distribution of $v = 327$ r.u. in two slices $v_1 = 117$ r.u., $v_2 = 210$ r.u. by providing the value of losses for traffic streams with parameters $b_1 = 1$ r.u., $b_2 = 10$ r.u., $\lambda_1 = 100$ Erl; $\lambda_2 = 10$ Erl., $\mu_1 = \mu_2 = 1$ at level 0,01 (see Fig. 6). Let us consider the situation of overloading by increasing λ_1 from 100 to 200.

The results of estimation of increasing the rate of losses for requests forming the first flow π_1 and the mean usage of r.u. $\delta = \frac{m_1+m_2}{v}$ by both the flows are shown in the Fig. 11 and Fig. 12 respectively. The same scenario of overload is considered for the model by applying the Filtering procedure for the requests of the first flow. The input parameters are the same except the total amount v of r.u. that is equals $v = 297$ r.u. For this number of r.u. the choice $r_1 = 24$ r.u. provide the value of losses at level 0,01 (see Fig. 10). The results of estimation of π_1 and $\delta = \frac{m_1+m_2}{v}$ with increasing the values of λ_1 are shown in the Fig. 11 and Fig. 12 respectively. We can see that to create the conditions for differentiated servicing by employing Filtering procedure is more robust to handle the fluctuations of the incoming traffic in contrast to the Network Slicing procedure.

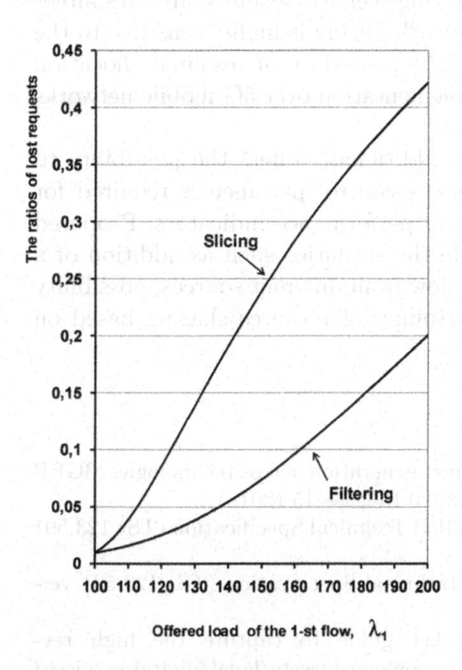

Fig. 11. The portions of lost requests vs λ_1.

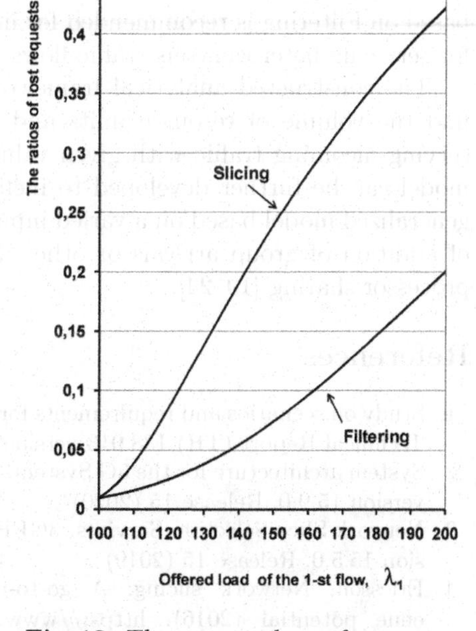

Fig. 12. The mean values of resource usage vs λ_1.

5 Conclusion

An analytical framework for to model the resource allocation procedures for transmission of multiservice traffic has been constructed and analyzed. The model consists of arbitrary number of traffic streams created by variety of real time applications. All random variables used in the model have exponential distribution with corresponding mean values. Two scenarios of resource sharing for incoming traffic streams are considered: Network Slicing when resources are strictly divided among incoming traffic streams, and Filtering, when the access to resource is restricted depending on the amount of resource occupied by all traffic streams. The mathematical description of the models used for realization of Network Slicing and Filtering procedures are presented. The model's performance measures are defined and recursive algorithm of characteristics calculation is formulated. It is shown how to use the model to create conditions for differentiated servicing of heterogeneous traffic.

The numerical assessment showed that simplest for implementation Network Slicing scenario where the available resources are strictly divided among incoming traffic streams has a number of drawbacks. The main drawback is additional requirement for number of resource units to serve incoming traffic flows with the required quality in contrast to the Filtering scenario where shared resources are not divided in separate slices. Also Network Slicing is highly sensitive to the change in the values of the offered load. The procedure of resource allocation based on Filtering is recommended for implementation over 5G mobile networks for servicing heterogeneous traffic flows.

The constructed analytical framework additionally offers the possibility to find the volume of resource units and access control parameters required for serving incoming traffic with given values of performance indicators. Proposed model can be further developed to include the scenarios such as addition of a generalized model based on a varied input flow from different sources, possibility of addition of group arrivals or other disciplines of resource sharing based on processor sharing [19–24].

References

1. Study on scenarios and requirements for next generation access technologies. 3GPP Technical Report (TR) 138.913 version 15.0.0 Release 15 (2018)
2. System architecture for the 5G System. 3GPP Technical Specification (TS) 123.501 version 15.9.0. Release 15 (2020)
3. Network Slice Selection Services. 3GPP Technical Specification (TS) 129.531 version 15.5.0. Release 15 (2019)
4. Ericsson. Network slicing: A go-to-market guide to capture the high revenue potential (2016). https://www.ericsson.com/assets/local/digital-services/network-slicing/network-slicing-value-potential.pdf
5. Nokia. Dynamic end-to-end network slicing for 5G. White Paper (2017)
6. Marquez, C., et al.: Resource sharing efficiency in network slicing. IEEE Trans. Netw. Serv. Manag. **16**(3), 909–923 (2019)
7. Dawaliby, S., Bradai, A., Pousset, Y.: Adaptive dynamic network slicing in LoRa networks. Future Gener. Comput. Syst. **98**, 697–707 (2019)
8. ElHalawany, B.M., Hashad, O., Wu, K., Tag Eldien, A.S.: Uplink resource allocation for multi-cluster internet-of-things deployment underlaying cellular networks. Mobile Netw. Appl. **25**(1), 300–313 (2019). https://doi.org/10.1007/s11036-019-01288-6

9. Malik, H., Pervaiz, H., Alam, M.M., et al.: Radio resource management scheme in NB-IoT systems. IEEE Access **6**, 15051–15064 (2018)
10. Begishev, V., Petrov, V., Samuylov, A., Moltchanov, D., Andreev, S., Koucheryavy, Y., Samouylov, K.: Resource allocation and sharing for heterogeneous data collection over conventional 3GPP LTE and emerging NB-IoT technologies. Comput. Communi. **120**(2), 93–101 (2018)
11. Stepanov, S., Stepanov, M., Tsogbadrakh, A., Ndayikunda, J., Andrabi, U.: Resource allocation and sharing for transmission of batched NB-IoT traffic over 3GPP LTE. In: The Proceedings of the 24th Conference of Open Innovations Association (FRUCT), pp. 422–429. Moscow Technical University of Communications and Informatics. Moscow, Russia (2019)
12. Stepanov, S.N., Stepanov, M.S.: Efficient algorithm for evaluating the required volume of resource in wireless communication systems under joint servicing of heterogeneous traffic for the internet of things. Autom. Remote Control **80**(11), 2017–2032 (2019). https://doi.org/10.1134/S0005117919110067
13. Stepanov, S.N., Stepanov, M.S., Andrabi, U., Ndayikunda, J.: The analysis of resource sharing for heterogenous traffic streams over 3GPP LTE with NB-IoT functionality. In: Vishnevskiy, V.M., Samouylov, K.E., Kozyrev, D.V. (eds.) DCCN 2020. LNCS, vol. 12563, pp. 422–435. Springer, Cham (2020). https://doi.org/10.1007/978-3-030-66471-8_32
14. Roberts, J., Mocci, U., Virtamo, J. (eds.): Broadband Network Traffic: Performance Evaluation and Design of Broadband Multiservice Networks. LNCS, vol. 1155. Springer, Heidelberg (1996). https://doi.org/10.1007/3-540-61815-5
15. Linderberger, K.: Dimensioning and design methods for integrated ATM networks. In: Proceedings 14th International Teletraffic Congress, Antibes Juan-les-Pins, pp. 897–906 (1994)
16. Fortet, R., Grandjean, Ch.: Congestion in a loss system when some calls want several devices simultaneously. Electric. Commun. **39**(4), 513–526 (1964)
17. Kaufman, J.S.: Blocking in a shared resource environment. IEEE Trans. Commun. COM **29**(10), 1474–1481 (1981)
18. Roberts, J.W.: A service system with heterogeneous user requirements - applications to multi-service telecommunication systems. In: Pujolle, G. (ed.) Performance of Data Communication Systems and their Applications, pp. 423–431 . North-Holland Publ. Co. (1981)
19. Iversen, V.B.: Teletraffic Engineering and Network Planning. Technical University of Denmark (2010)
20. Ross, K.W.: Multiservice Loss Models for Broadband Telecommunications Networks. Springer, Heidelberg (1995). https://doi.org/10.1007/978-1-4471-2126-8
21. Stepanov, S.N., Stepanov, M.S.: Methods for estimating the required volume of resource for multiservice access nodes. Autom. Remote Control **81**(12), 2244–2261 (2020). https://doi.org/10.1134/S0005117920120085
22. Stepanov, S.N., Stepanov, M.S.: Planning the resource of information transmission for connection lines of multiservice hierarchical access networks. Autom. Remote Control **79**(8), 1422–1433 (2018). https://doi.org/10.1134/S0005117918080052
23. Stepanov, S.N., Stepanov, M.S.: Planning transmission resource at joint servicing of the multiservice real time and elastic data traffics. Autom. Remote Control. **78**(11), 2004–2015 (2017)
24. Stepanov, S.N., Stepanov, M.S.: The model and algorithms for estimation the performance measures of access node serving the mixture of real time and elastic data. In: Vishnevskiy, V.M., Kozyrev, D.V. (eds.) DCCN 2018. CCIS, vol. 919, pp. 264–275. Springer, Cham (2018). https://doi.org/10.1007/978-3-319-99447-5_23

Analysis of Non-preemptive Scheduling for 5G Network Model Within Slicing Framework

Yves Adou(✉), Ekaterina Markova, and A. A. Chursin

Peoples' Friendship University of Russia (RUDN University),
6 Miklukho-Maklaya Street, Moscow 117198, Russian Federation
{1042205051,markova-ev,chursin-aa}@rudn.ru

Abstract. 5G networks slicing technology, recognized as one of the possible solutions to radio resources scarcity and networks inflexibility problems, is getting the most attention from researchers and standard organizations around the world. This situation was expected since slicing capabilities are very well suited for mobile networks operators allowing significant changes in their range of services and daily operations. In addition, some researches showed that priority-based scheduling mechanisms can be implemented within slicing framework. This paper analyzes the performance measures of a wireless network implementing priority-based scheduling mechanism within slicing framework and described as queueing system with buffer and retrial queue.

Keywords: 5G · Network · Slicing · Retrial queue · Non-preemptive · Priority · Scheduling · Numerical solution

1 Introduction

In recent years, new digital industries and businesses have been facing problems to be supplied with highly flexible and very controlled critical network connectivity [3,7,13]. To solution that need, leading standards organizations in the domain of modern wireless telecommunication networks developed the brand new 5G wireless networks with their main feature – slicing technology [1,9,10]. As defined by 3GPP standard, slicing allows mobile operators to provide users with customized networks called "slices" with different quality of service (QoS) capabilities [2]. Last year has seen slicing technology being intensively implemented around the globe [20,21], situation making researchers focus on investigating and proposing methods for efficient utilization of available radio resources within its framework [14,19,24]. Measuring key performance indicators (KPIs) is very essential for mobile operators, allowing them to monitor services performance,

This paper has been supported by the RUDN University Strategic Academic Leadership Program (recipients Chursin A., Adou Y.). The reported study was funded by RFBR, project numbers 19-07-00933 and 20-37-70079 (recipient Markova E.).

V. M. Vishnevskiy et al. (Eds.): DCCN 2021, CCIS 1552, pp. 36–47, 2022.
https://doi.org/10.1007/978-3-030-97110-6_3

sustain QoS and effectively support further required improvements [11,15,22]. This essentially aims at avoiding or reducing considerably service interruption, which could be materially done with introduction of the mathematical methods in retrial queueing theory [6,12]. Slicing could be implemented alongside priority-based scheduling mechanisms, i.e. non-preemptive and preemptive scheduling methods [8,23].

2 Mathematical Model

Let us consider a queueing system model with buffer and retrial queue derived from the already investigated $M/M/1/\infty$ retrial queueing system model [5,6], the main differences being the joint operation with a buffer and the implementation of a priority scheduling mechanism. We assume that two request types arrive in system according to Poisson law with rates λ_1 and λ_2 respectively. Both request types are serviced according to exponential law with parameters μ_1 and μ_2.

We suppose that services of first type requests have priority over services of second type requests. First type requests can only access server and buffer, while second type requests can access server and retrial queue.

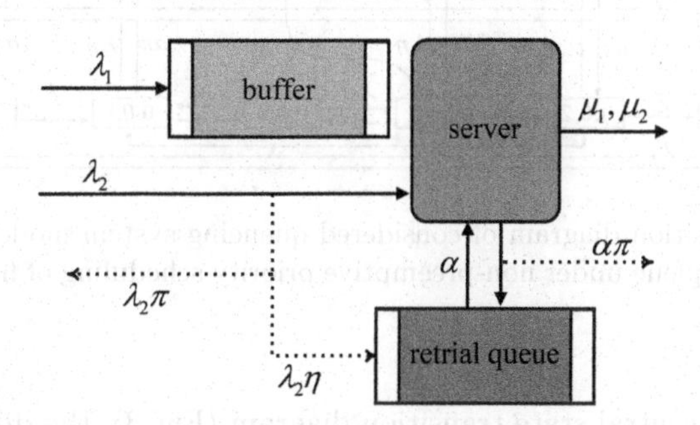

Fig. 1. Scheme model of considered queueing system with unlimited buffer and retrial queue

Radio admission control mechanism is organized as follows. For first type requests: (i) if server is idle, incoming request is immediately serviced and leaves system; (ii) otherwise, it awaits server idleness in buffer with FIFO service discipline. As for second type requests: (i) if server is idle and buffer – empty, incoming request is immediately serviced and leaves system; (ii) otherwise, it can either quit system with probability π or join retrial queue with probability $\eta = 1 - \pi$. A second type request that chooses to join retrial queue becomes a retrial second type request. Retrial second type requests, as the name suggests, retry each to get service after some random time. The number of retrials is unlimited and the time interval between two consecutive is exponentially distributed with rate α^{-1}.

The scheme model of considered queueing system with unlimited buffer and retrial queue is given in Fig. 1.

We describe system' behavior using a three-dimensional vector (n_1, n_2, s) over state space $\mathbf{X} = \{(n_1, n_2, s) : n_1 \geq 0, n_2 \geq 0, s = 0, 1, 2\}$, where n_1 represents the current quantity of first type requests in buffer, n_2 – the current quantity of second type requests in retrial queue, and s – the current state of server (i.e., $s = 0$ means server is idle, $s = 1$ – server is busy with first type request, and $s = 2$ – server is busy with second type request). The corresponding state transition diagram is shown in Fig. 2. Fig. 3 clarifies the central state transition diagram of the system.

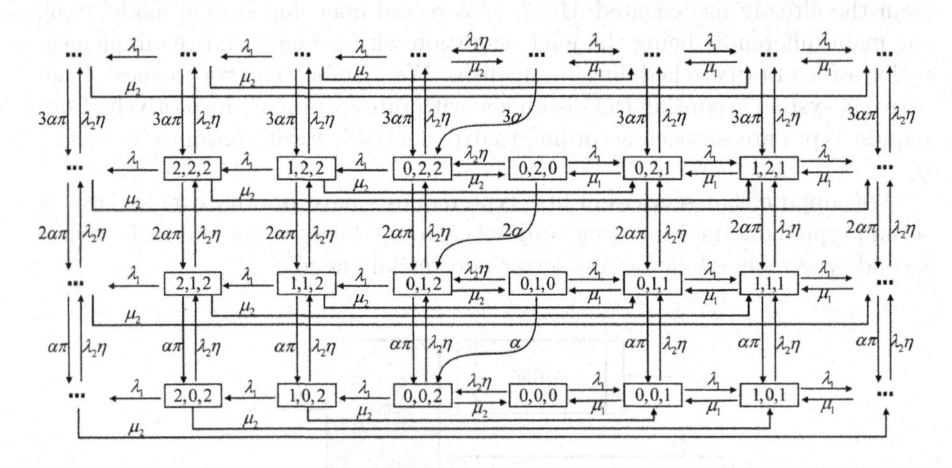

Fig. 2. State transition diagram of considered queueing system model with unlimited buffer and retrial queue under non-preemptive priority scheduling of first type requests

According to central state transition diagram (Fig. 3), the discussed Markov process of system states is described by the equilibrium equations system (1):

$$
\begin{aligned}
&[\lambda_1 + \lambda_2\eta + \mu_s I\{s \neq 0\} + n_2\alpha\,(I\{s = 0, n_1 = 0\} + \pi I\{s \neq 0, n_2 > 0\})] \\
&\times p(n_1, n_2, s) = \lambda_1 I\{s = 1, n_1 = 0\} \cdot p(n_1, n_2, s - 1) \\
&+ \lambda_1 I\{s \neq 0, n_1 > 0\} \cdot p(n_1 - 1, n_2, s) + \lambda_2\eta I\{s = 2, n_1 = 0\} \cdot p(n_1, n_2, s - 2) \\
&+ \lambda_2\eta I\{s \neq 0, n_2 > 0\} \cdot p(n_1, n_2 - 1, s) + \mu_1 I\{s = 0, n_1 = 0\} \cdot p(n_1, n_2, s + 1) \\
&+ \mu_1 I\{s = 1\} \cdot p(n_1 + 1, n_2, s) + \mu_2 I\{s = 0, n_1 = 0\} \cdot p(n_1, n_2, s + 2) \\
&+ \mu_2 I\{s = 1\} \cdot p(n_1 + 1, n_2, s + 1) + (n_2 + 1)\alpha I\{s = 2, n_1 = 0\} \\
&\times p(n_1, n_2 + 1, s - 2) + (n_2 + 1)\alpha\pi I\{s \neq 0\} \cdot p(n_1, n_2 + 1, s),
\end{aligned}
\tag{1}
$$

where $p(n_1, n_2, s), (n_1, n_2, s) \in \mathbf{X}$ represents the stationary probability distribution and I – the function indicator equaling 1 when condition is met and 0 otherwise.

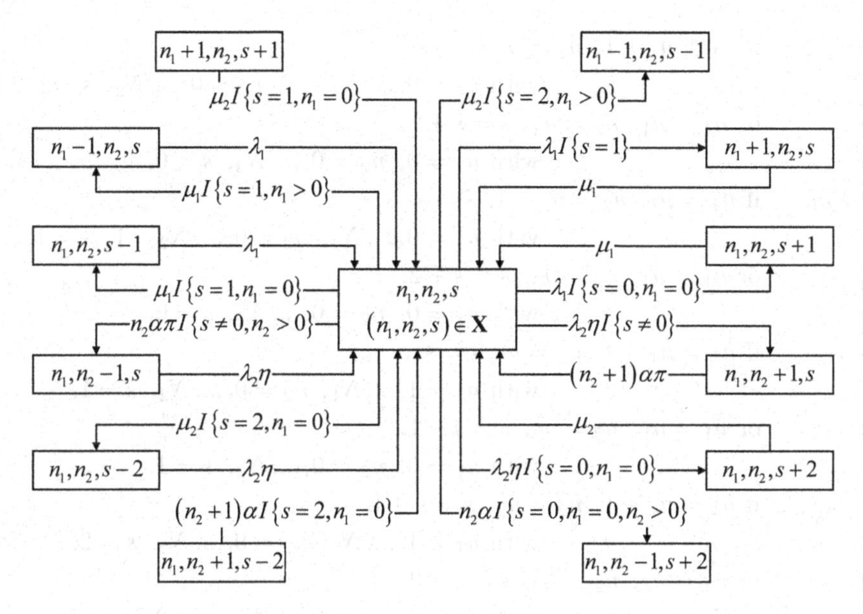

Fig. 3. Central state transition diagram of considered queueing system model with unlimited buffer and retrial queue under non-preemptive priority scheduling of first type requests

3 Stationary Probability Distribution

As already known the stationary probability distribution of the $M/M/1/\infty$ retrial queueing system model can be computed through a generating function based approach [6]. As for our model this approach is showing some difficulties related to the two request types in presence and the priority scheduling mechanism for accessing the server. The approach used in [16–18] seems to work with our particular queueing system model but is very laborious and currently under development.

Note that aside generating function based approach, the stationary probability distribution can also be computed through a numerical solution of equilibrium equations system by first fixing or setting the storage capacities to maximum values as shown in [4]. Therefore, the buffer' maximum storage capacity is set to N_1 and the retrial' queue to N_2.

The process describing considered system is not a reversible Markov process. Therefore, considering implemented non-preemptive priority scheduling discipline, one can compute system' stationary probability distribution using a numerical solution of equilibrium equations system $\mathbf{p} \cdot \mathbf{A} = \mathbf{0}, \mathbf{p} \cdot \mathbf{1}^{-1} = \mathbf{1}$, where $\mathbf{p} = p(n_1, n_2, s)_{(n_1,n_2,s)\in\mathbf{X}}$ and $\mathbf{A}$ represents the infinitesimal generator of Markov process, elements $a((n_1, n_2, s)(\hat{n}_1, \hat{n}_2, \hat{s}))$ of which are defined in (2):

$$\begin{cases}
\lambda_1, & \text{if } \hat{n}_1 = n_1 + 1,\ \hat{n}_2 = n_2,\ \hat{s} = s, \\
 & \qquad\qquad \text{with } n_1 = 0, ..., N_1 - 1,\ n_2 = 0, ..., N_2,\ s = 1, 2, \\
 & \text{or } \hat{n}_1 = n_1,\ \hat{n}_2 = n_2,\ \hat{s} = s + 1, \\
 & \qquad\qquad \text{with } n_1 = 0,\ n_2 = 0, ..., N_2,\ s = 0, \\
\lambda_2\eta, & \text{if } \hat{n}_1 = n_1,\ \hat{n}_2 = n_2 + 1,\ \hat{s} = s, \\
 & \qquad\qquad \text{with } n_1 = 0, ..., N_1,\ n_2 = 0, ..., N_2 - 1,\ s = 1, 2, \\
 & \text{or } \hat{n}_1 = n_1,\ \hat{n}_2 = n_2,\ \hat{s} = s + 2, \\
 & \qquad\qquad \text{with } n_1 = 0,\ n_2 = 0, ..., N_2,\ s = 0, \\
\mu_1, & \text{if } \hat{n}_1 = n_1 - 1,\ \hat{n}_2 = n_2, \hat{s} = s, \\
 & \qquad\qquad \text{with } n_1 = 1, ..., N_1,\ n_2 = 0, ..., N_2,\ s = 1, \\
 & \text{or } \hat{n}_1 = n_1,\ \hat{n}_2 = n_2,\ \hat{s} = s - 1, \\
 & \qquad\qquad \text{with } n_1 = 0,\ n_2 = 0, ..., N_2,\ s = 1, \\
\mu_2, & \text{if } \hat{n}_1 = n_1 - 1,\ \hat{n}_2 = n_2,\ \hat{s} = s - 1, \\
 & \qquad\qquad \text{with } n_1 = 1, ..., N_1,\ n_2 = 0, ..., N_2,\ s = 2, \\
 & \text{or } \hat{n}_1 = n_1,\ \hat{n}_2 = n_2,\ \hat{s} = s - 2, \\
 & \qquad\qquad \text{with } n_1 = 0,\ n_2 = 0, ..., N_2,\ s = 2, \\
n_2\alpha, & \text{if } \hat{n}_1 = n_1,\ \hat{n}_2 = n_2 - 1,\ \hat{s} = s + 2, \\
 & \qquad\qquad \text{with } n_1 = 0,\ n_2 = 1, ..., N_2,\ s = 0, \\
n_2\alpha\pi, & \text{if } \hat{n}_1 = n_1,\ \hat{n}_2 = n_2 - 1,\ \hat{s} = s, \\
 & \qquad\qquad \text{with } n_1 = 0, ..., N_1,\ n_2 = 1, ..., N_2,\ s = 1, 2, \\
\psi, & \text{if } \hat{n}_1 = n_1,\ \hat{n}_2 = n_2,\ \hat{s} = s, \\
 & \qquad\qquad \text{with } n_1 = 0, ..., N_1,\ n_2 = 0, ..., N_2,\ s = 0, 1, 2, \\
0, & \text{otherwise},
\end{cases} \tag{2}$$

where $\psi = -[\lambda_1 I\{n_1 < N_1, n_2 \leq N_2, s = 1, 2 \parallel n_1 = 0, n_2 \leq N_2, s = 0\} + \lambda_2\eta I\{n_1 \leq N_1, n_2 < N_2, s = 1, 2 \parallel n_1 = 0, n_2 \leq N_2, s = 0\} + \mu_1 I\{n_1 > 0, n_2 \leq N_2, s = 1 \parallel n_1 = 0, n_2 \leq N_2, s = 1\} + \mu_2 I\{n_1 > 0, n_2 \leq N_2, s = 2 \parallel n_1 = 0, n_2 \leq N_2, s = 2\} + n_2\alpha I\{n_1 = 0, n_2 > 0, s = 0\} + n_2\alpha\pi I\{n_1 \leq N_1, n_2 > 0, s = 1, 2\}]$.

4 Performance Measures

Having computed the stationary probability distribution one can compute system' main performance measures. For that, are estimated the values that can represent the limits of buffer' and retrial' queue maximum storage capacities as these approach infinity. Below are listed some of the main performance measures:

– Mean number K_1 of first type requests in buffer

$$K_1 = \lim_{N_1 \to \infty} \sum_{n_1=0}^{N_1} n_1 \left(\lim_{N_2 \to \infty} \sum_{n_2=0}^{N_2} \sum_{s=1}^{2} p(n_1, n_2, s) \right); \tag{3}$$

– Mean number K_2 of second type requests in retrial queue

$$K_2 = \lim_{N_2 \to \infty} \sum_{n_2=0}^{N_2} n_2 \left(p(0, n_2, 0) + \lim_{N_1 \to \infty} \sum_{n_1=0}^{N_1} \sum_{s=1}^{2} p(n_1, n_2, s) \right); \quad (4)$$

– Immediate service probability $Im_{1,2}$ of any new incoming type request (i.e., idleness probability of server)

$$Im_{1,2} = \lim_{N_2 \to \infty} \sum_{n_2=0}^{N_2} p(0, n_2, 0); \quad (5)$$

– Immediate service probability Im_{re2} of retrial second type request

$$Im_{re2} = Im_{1,2} - p(0, 0, 0); \quad (6)$$

– Emptiness probability E_b of buffer

$$E_b = \lim_{N_2 \to \infty} \sum_{n_2=0}^{N_2} \sum_{s=0}^{2} p(0, n_2, s); \quad (7)$$

– Emptiness probability E_{rq} of retrial queue

$$E_{rq} = p(0, 0, 0) + \lim_{N_1 \to \infty} \sum_{n_1=0}^{N_1} \sum_{s=1}^{2} p(n_1, 0, s); \quad (8)$$

– Busyness probability U_1 of server by first type request

$$U_1 = \lim_{N_1, N_2 \to \infty} \sum_{n_1=0}^{N_1} \sum_{n_2=0}^{N_2} p(n_1, n_2, 1); \quad (9)$$

– Busyness probability U_2 of server by second type request

$$U_2 = 1 - Im_{1,2} - U_1; \quad (10)$$

Let us analyze the behavior of computed main performance measures depending on system' input parameters.

5 Numerical Example

A numerical example was conducted to illustrate the behavior of system' main performance measures – mean numbers of first type requests in buffer (3) and second type requests in retrial queue (4), immediate service probabilities of any new incoming type request (5) and retrial second type request (6), emptiness probabilities of buffer (7) and retrial queue (8), busyness probabilities of server by first type request (9) and by second type request (10) – depending on average offered loads pair (ρ_1, ρ_2).

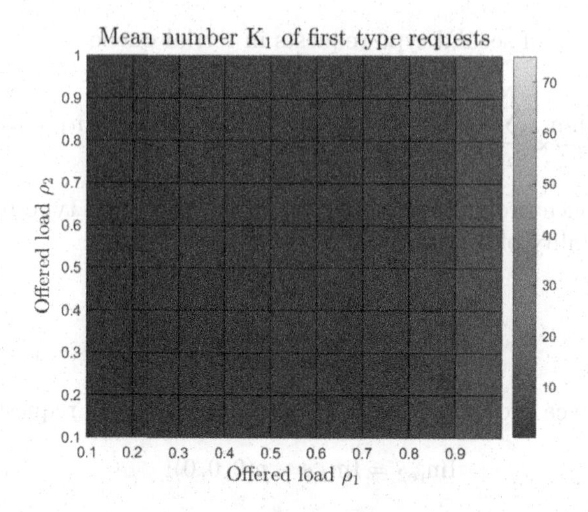

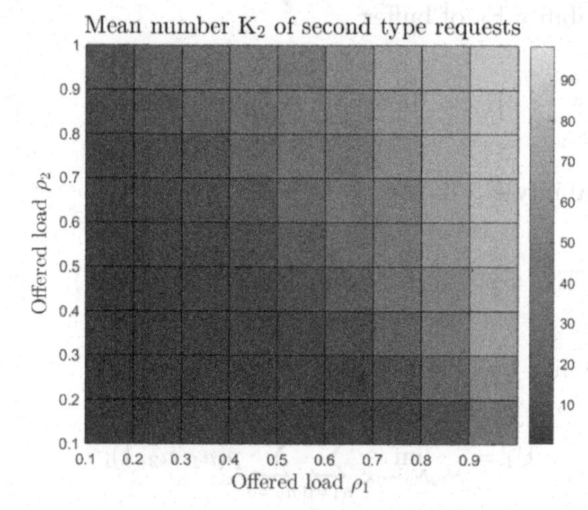

Fig. 4. Mean numbers K_1 of first type requests in buffer and K_2 of second type requests in retrial queue depending on average offered loads pair (ρ_1, ρ_2) for $\pi = 0.01$, $\mu_1 = \mu_2 = 5$, $\alpha = \rho_2\mu_2$, limits of N_1 and N_2 as they approach infinity equaling 150 with ρ_1 and ρ_2 belonging to the interval $(0, 1)$.

As illustrated in Fig. 4a the mean number K_1 of first type requests in buffer increases as their average offered load increases, and is not influenced by the value of the second' type requests average offered load. In contrary the mean number K_2 of second type requests in retrial queue increases as their average offered load increases, and is greatly influenced by the value of the first' type requests average offered load (Fig. 4b). This is indeed explained by the non-preemptive priority scheduling of first type requests over second type requests in system.

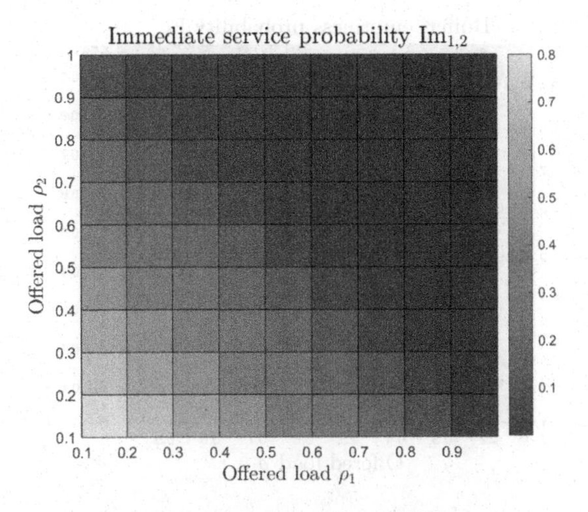

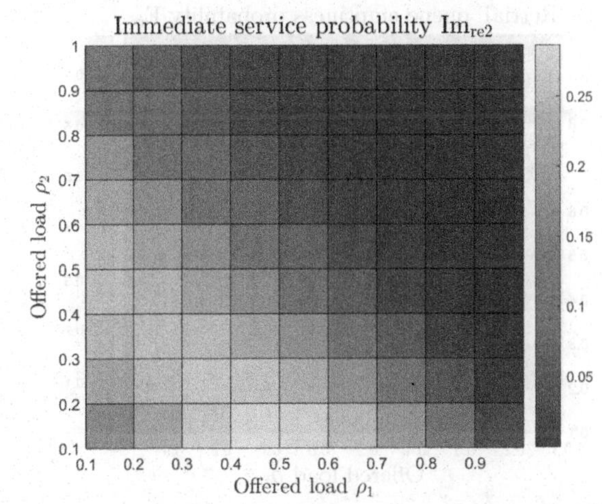

Fig. 5. Immediate service probabilities $\text{Im}_{1,2}$ of any new incoming type request and $\text{Im}_{\text{re}2}$ of retrial second type request depending on average offered loads pair (ρ_1, ρ_2) for $\pi = 0.01$, $\mu_1 = \mu_2 = 5$, $\alpha = \rho_2\mu_2$, limits of N_1 and N_2 as they approach infinity equaling 150 with ρ_1 and ρ_2 belonging to the interval $(0, 1)$.

As shown in Fig. 5a the immediate service probability $\text{Im}_{1,2}$ of any new incoming type request decreases as the average offered load of first or second type requests increases. This is explained by the augmentation of first' and second' type requests quantity in system with increase of their average' offered load values. Thus, since retrial second type requests indefinitely retry to occupy server and can quit system only with probability π after an unsuccessful attempt, it is very likely that most of them will quit system at some time after several retries. This can explain the increase of retrial' second type request immediate service probability $\text{Im}_{\text{re}2}$ up to a certain maximum value and its sudden decrease in Fig. 5b.

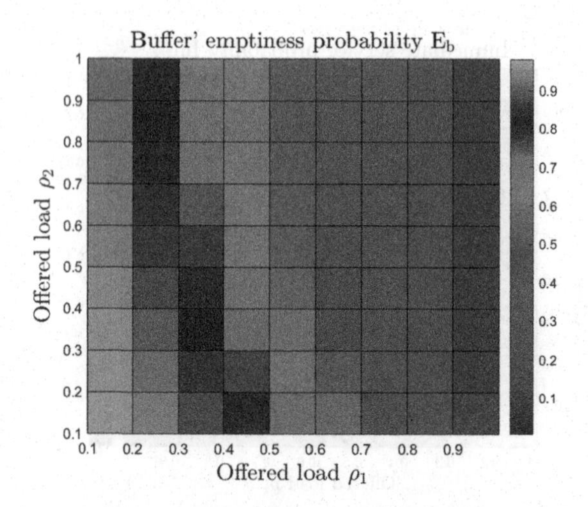

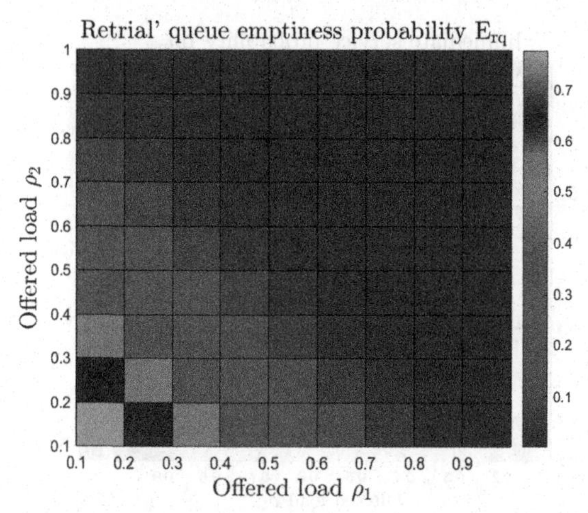

Fig. 6. Emptiness probabilities E_b of buffer and E_{rq} of retrial queue depending on average offered loads pair (ρ_1, ρ_2) for $\pi = 0.01$, $\mu_1 = \mu_2 = 5$, $\alpha = \rho_2\mu_2$, limits of N_1 and N_2 as they approach infinity equaling 150 with ρ_1 and ρ_2 belonging to the interval $(0, 1)$.

As illustrated in Fig. 6a the emptiness probability E_b of buffer decreases as the average offered load ρ_1 of first type requests increases, and is slightly affected by the value of second' type requests average offered load ρ_2. As for the emptiness probability E_{rq} of retrial queue, it decreases as the average offered loads increase (Fig. 6b). This is also explained by the non-preemptive priority scheduling mechanism implemented in system.

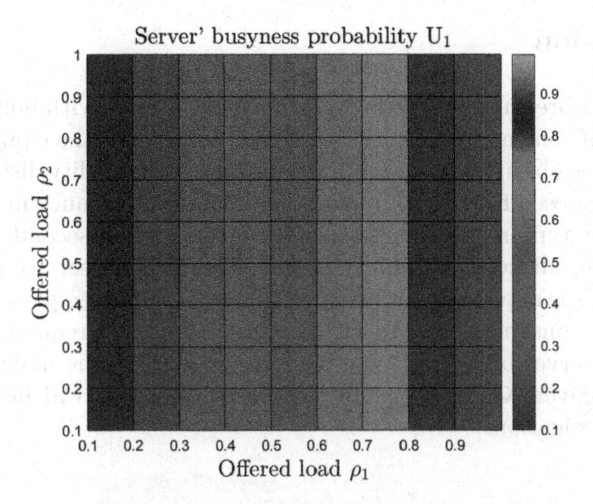

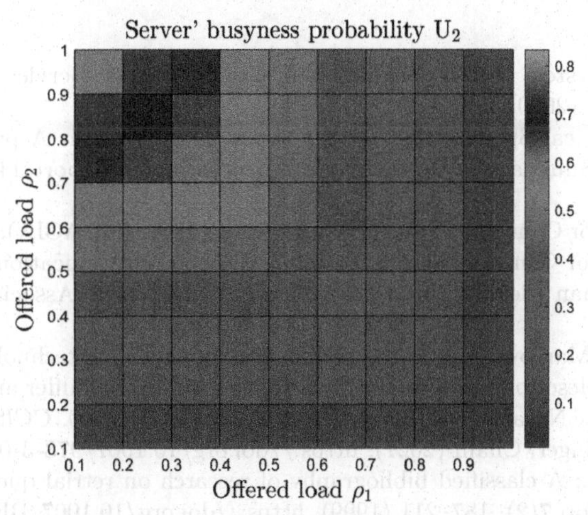

Fig. 7. Busyness probabilities U_1 of server by first type request and U_2 of server by second type request depending on average offered loads pair (ρ_1, ρ_2) for $\pi = 0.01$, $\mu_1 = \mu_2 = 5$, $\alpha = \rho_2\mu_2$, limits of N_1 and N_2 as they approach infinity equaling 150 with ρ_1 and ρ_2 belonging to the interval $(0, 1)$.

As shown in Fig. 7a the busyness probability U_1 of server by first type request increases as average offered load ρ_1 increases, and is not affected by the value of second' type requests average offered load ρ_2. As for the busyness probability U_2 of server by second type request, it increases as the average offered load ρ_2 of second type requests increases and decreases as the average offered load ρ_1 of first type requests increases (Fig. 7b). This can be explained by the non-preemptive priority scheduling technique applied in system.

6 Conclusion

We studied non-preemptive scheduling in queueing system with buffer and retrial queue within 5G slicing framework. Numerical solution of the equilibrium equations system was obtained for computing stationary probability distribution. Formulas were proposed for calculating system' main performance measures: mean number of first type requests in buffer, mean number of second type requests in retrial queue, idleness probability of server, immediate service probability of retrial second type request, emptiness probability of buffer, emptiness probability of retrial queue, busyness probability of server by first type request and busyness probability of server by second type request. Analysis of the main performance measures was given. Generating function-based approach will be implemented in further investigations.

References

1. 3GPP: 5G System Session Management Policy Control Service stage 3. White Paper, 3GPP (2020)
2. 3GPP: Technical Specification Group Services and System Aspects; Release 16 Description; Summary of Rel-16 Work Items. Technical Report TR 21.916 V1.0.0, 3GPP (2020)
3. 5G Alliance for Connected Industries and Automation (5G-ACIA): Exposure of 5G Capabilities for Connected Industries and Automation Applications. White Paper, ZVEI - German Electrical and Electronic Manufacturers' Association, 5G-ACIA (2021)
4. Adou, K.Y., Markova, E.V.: Methods for analyzing slicing technology in 5G wireless network described as queueing system with unlimited buffer and retrial group. In: Dudin, A., Nazarov, A., Moiseev, A. (eds.) ITMM 2020. CCIS, vol. 1391, pp. 264–278. Springer, Cham (2021). https://doi.org/10.1007/978-3-030-72247-0_20
5. Artalejo, J.R.: A classified bibliography of research on retrial queues: progress in 1990–1999. Top **7**(2), 187–211 (1999). https://doi.org/10.1007/BF02564721
6. Bocharov, P.P., D'Apice, C., Pechinkin, A.V.: Queueing Theory. De Gruyter (2003). https://doi.org/10.1515/9783110936025
7. Bonte, D., Saunders, J., Mavrakis, D., Martin, R.: Smart Manufacturing and How to Get Started the Implementation and ROI of Industry 4.0 Use Cases. White Paper, ABI Research, Ericsson (2020)
8. Chen, Z., Lei, H., Yang, M., Liao, Y., Qiao, L.: Blocking analysis of suspension-based protocols for parallel real-time tasks under global fixed-priority scheduling. J. Syst. Arch. **117**, 102107 (2021). https://doi.org/10.1016/j.sysarc.2021.102107
9. Mobile, C.: Categories and Services Levels of Network Slice. White Paper, Huawei and Industry Partners (2020)
10. Cisco: Ultra Cloud Core 5G Policy Control Function, Release 2021.02 - Configuration and Administration Guide. White Paper, Cisco Systems, Inc (2021)
11. Credits: Leipziger Verkehrsbetriebe: Anonymisierte Daten. https://www.telefonica.de/analytics/anonymisierte-daten.html
12. Falin, G.I., Templeton, J.G.C.: Retrial Queues, Monographs on Statistics and Applied Probability, vol. 75. Chapman & Hall/CRC, Boca Raton (1997)

13. GSMA: 5G IoT Private & Dedicated Networks for Industry 4.0: A guide to private and dedicated 5G networks for manufacturing, production and supply chains. White Paper, GSMA Alliance (2020)
14. Jmila, H., Blanc, G.: Towards security-Aware 5G slice embedding. Comput. Secur. **100**, 102075 (2021). https://doi.org/10.1016/j.cose.2020.102075
15. Manjaro, N.: Leveraging the Data Explosion. White Paper, Teradata (2021)
16. Nazarov, A., Baymeeva, G.: The $M/GI/\infty$ system subject to semi-markovian random environment. In: Dudin, A., Nazarov, A., Yakupov, R. (eds.) ITMM 2015. CCIS, vol. 564, pp. 128–140. Springer, Cham (2015). https://doi.org/10.1007/978-3-319-25861-4_11
17. Nazarov, A., Phung-Duc, T., Izmailova, Y.: Asymptotic-diffusion analysis of multi-server retrial queueing system with priority customers, pp. 236–250 (2021). https://doi.org/10.1007/978-3-030-72247-0_18
18. Nazarov, A., Sztrik, J., Kvach, A., Tóth, d.: Asymptotic analysis of finite-source M/GI/1 retrial queueing systems with collisions and server subject to breakdowns and repairs. In: Methodology and Computing in Applied Probability (2021). https://doi.org/10.1007/s11009-021-09870-w
19. Salhab, N., Langar, R., Rahim, R.: 5G network slices resource orchestration using Machine Learning techniques. Comput. Netw. **188**, 107829 (2021). https://doi.org/10.1016/j.comnet.2021.107829
20. Statista: 5G technology operational challenges to success 2020 (2020). https://www.statista.com/statistics/1178104/5g-technology-operational-challenges/
21. Statista: 5G mobile subscriptions worldwide 2019–2026, by region (2021). https://www.statista.com/statistics/521598/5g-mobile-subscriptions-worldwide/
22. Verizon: Verizon Media Analytics—Verizon Media Policies. https://www.verizonmedia.com/policies/us/en/verizonmedia/privacy/topics/analytics/index.html
23. Xu, F., Yin, Z., Gu, A., Li, Y., Yu, H., Zhang, F.: Adaptive scheduling strategy of fog computing tasks with different priority for intelligent production lines. Procedia Comput. Sci. **183**, 311–317 (2021). https://doi.org/10.1016/j.procs.2021.02.064
24. Zhang, Y., Wu, A., Chen, Z., Zheng, D., Cao, J., Jiang, X.: Flexible and anonymous network slicing selection for C-RAN enabled 5G service authentication. Comput. Commun. **166**, 165–173 (2021). https://doi.org/10.1016/j.comcom.2020.12.014

A Hybrid Clustering-Based Routing Protocol for VANET Using k-means and Cuckoo Search Algorithm

Amani A. Sabbagh(✉) and Maxim V. Shcherbakov

Volgograd State Technical University, 400005 Volgograd, Russian Federation

Abstract. The substantial growth in the automobile industry led to the exponential growth in wireless ad-hoc networks, especially, Vehicular Adhoc Networks (VANETs). VANETs are generally used in urbanized environments since they promote passenger safety in roads and prevent many accidents. Hence, we developed a hybrid routing protocol on AODV, and used clustering concept and cuckoo search algorithm to set up an efficient, and stable path between the source and destination in VANET network. This paper presented our proposed protocol, where we combined between k-means algorithm to format clusters and cuckoo search algorithm to choose a shorter and more stable path between all available paths. Further, we determined three weighted parameters in cuckoo search algorithm as a fitness function to guarantee an efficient and stable path. We performed the simulation with NS-3 simulator and Bonnmotion to evaluate the performance of the proposed protocol KMCSA and the results are compared with popular routing protocol AODV. The evaluation results show that our routing protocol KMCSA outperforms in terms of packet delivery ratio, packet loss ratio, overhead, average delay and throughput even in the case of black hole attacks.

Keywords: Routing protocols · K-means · Cuckoo search · Weighted parameters · Hybrid method · And black hole attack

1 Introduction

In recent years, In recent years, with the remarkable growth in the number of cars, many deaths were recorded due to road accidents and the driver's lack of interest in road hazards. This has led to increased interest in developing an Intelligent Transportation System (ITS) to provide vehicle drivers with accurate and timely data. Vehicular Ad hoc Network (VANET) is a new wireless network which takes an important role in building ITS. Therefore, optimizing VANETs is still having great consideration by most of the interest researchers. VANETs are generally used in urbanized environments since they promote passenger safety in roads and prevent many accidents. VANET is a subclass of Ad-Hoc Networks as Mobile Ad-Hoc Networks (MANETs). It has also a lot of

V. M. Vishnevskiy et al. (Eds.): DCCN 2021, CCIS 1552, pp. 48–61, 2022.
https://doi.org/10.1007/978-3-030-97110-6_4

characteristics which are closed to features of MANETs such as finite bandwidth, self-administration, and unstable network topology. On the other hand, VANETs are better than MANETs in overcoming the energy constrained limitations and high mobility that are predictable and road-restricted. Some of the unique characteristics of VANETs are very high mobility, heterogeneous communication range, hard delay restrictions, frequently disconnected network. VANET architecture has three types of communications: Vehicle to Vehicle (V2V), Vehicle to Roadside (V2R) and Hybrid where it combines both V2V and V2R [1], [2]. The main task VANET for establishing a connection between the vehicle for the exchange of important data such as speed, location and road condition using wireless channels to provide information about the road and avoid accidents. An efficient routing technique of data packets could ensure delivering the warning messages during collisions on-time and they also could avoid a larger number of accidents [3]. To achieve this goal, we need to develop an efficient routing protocol that take into account environmental characteristics such as the node mobility, time restrictions and scalability of vehicle communications. Hence, an efficient routing technique that is capable of swiftly delivering the data packets along with lesser packet loss is a demanding need to satisfy the objective. This ensures vehicle security and promotes satisfaction to every user [4,5]. Further, restrictions over the wireless resources, larger vehicle mobility and various losses in wireless networks pose crucial challenges towards routing source nodes to the target nodes via intermediate nodes. Various criteria decide an efficient routing including wireless links that construct a route. These networks faced various routing issues such as vehicle mobility continuously leading to topological modifications [6], in addition, expansion of networks extensively results in higher routing overheads [7,8]. There is a lot of research to improve routing in VANET, and researchers have been interested in cluster-based solutions which are able to enhance the scalability of the network. Clustering algorithm is a type of an unsupervised machine learning technique. It is used to classify the each node into a specific group (cluster) where the nodes which are in the same cluster are more similar in some specific attributes than those in other clusters. VANET Clustering can be classified into two types: dynamic depends on (V2V) and static depends on (V2R) [9]. Clustering can be used as an important method to enhance the reliability and scalability of routing algorithms in VANETs. In this study, we identified K-means clustering and Cuckoo search algorithm (CSA) to effectively solve the clustering issues. CSA is an effective method to deal with optimization issues, specifically clustering issues. Similarly, k-means is another useful technique that effectively solves clustering issues even in faster convergence. Hence, we proposed new routing protocol based on k-means algorithm to format clusters of nodes and cuckoo search algorithm to elect the cluster head of each cluster to enhance the selecting of an optimal route among known routes. We relied on, choosing a cluster head, three weight parameters (distance, angle and road id) as a fitness function to ensure a short and stable route.

2 Related Work

Routing remains as a greater challenge in VANETs as per our previous discussions. Various protocols have been discussed in the literature but they failed to fulfil the routing requirements in VANETs because of the dynamically varying nature of VANETs. In this section, we review the highly valuable attempts in constructing an efficient routing algorithm for vehicular networks. Several cluster-based studies were analysed for VANETs and MANETs [10]. Vehicular clustering is a prospective technique that is capable of enhancing network scalability in VANETs. The cluster heads (CHs) are responsible for route discovery and route path maintenance in cluster-based routing techniques. This decreases the control overheads extensively [11]. Because of the swift vehicle mobility, periodical topology modifications occur. At this stage, the maintenance cost of clusters will rapidly increase. Thus, constructing reliable clusters and maintaining their reliability and stability in case of communication are crucial challenges in the clustering protocols for VANETs. Several clustering techniques were presented on the basis of mobility parameters for the construction of clustering mechanisms. Some mobility characteristics such as vehicle direction, location and speed are considered as major features for the clustering algorithms in VANETs. A passive clustering technique has been presented by Kayis et al. in [12]. They ensured the reliability in V2V communications through cluster formation. The vehicular nodes present in the cluster are recognized and allotted some distinct tasks. The algorithm works using specific speed spells. The vehicles in the networks are organized using the same speed spells as groups. But, using the speed spell as a metric to evaluate any two vehicles moving at the same speed within the specific spell may be classified into various clusters. Distance based standard has been employed by Chen et al. [13] in cluster formation. They utilized a centralized server for managing the integration of clusters and disintegrating occurrences. Affinity propagation, a data clustering method is presented by Shea et al. [14]. This method is a distributed and mobility-based clustering technique that utilizes parameters such as position of the vehicle and mobility parameter in cluster formation by integrating the recent and upcoming positions. Few clustering techniques were presented in VANET on the basis of the sum of weighted values. A priority-based clustering technique has been presented by Wang et al. [15] where the priority is computed on the basis of predicted traveling time and varying speed. A clustering method on the basis of lane is proposed by Almalag et al. [16] that chooses the vehicle with the largest level of cluster head (CHL) will be recognized as the cluster head (CH). CHL is a parameter that contains an integrated value of the relative position, speed and traffic flow condition of the vehicle. A review of various bio-inspired clustering techniques for routing has been presented for VANETs by Bitam et al. [17]. These bio-inspired approaches are found to be highly robust and effectively adapts to network disturbances hence ensuring effective delivery of data packets with lesser complexity in VANETs. The bio-inspired routing techniques are classified into three types such as swarm intelligence algorithm, evolutionary algorithm and other bio-inspired approaches for VANET. Enhanced summary of every class in regard to scalability, complexity, Quality of Service (QoS) routing performance, robustness and mobility model was proposed. A detailed evaluation proved that these bio-inspired techniques increased the VANET routing performance concerned to the respective computational parameters. Further, Liu et al. [18] presents a survey of various position-based routing techniques. The authors stated that because of the quick changes in the network topology of VANETs, position-based routing techniques are ideal. But hybrid protocols are termed as the optimal selection for VANET routing that is suitable in urban as well as highway environments. Further, they aid in resolving the local maximum issue that occurs due to erroneous positioning.

3 Routing Protocols in VANET

Routing protocols have a major role in computer networks in order to direct data from the source to the target using the least possible network resources available and in the least possible time by choosing the best path for this data. Therefore, researchers continuously develop these protocols in order to reach the network to the best possible performance because the development of protocols do not require large costs and many work teams compared to the development of the rest of the network products that require resources and specialized centers in order to develop them. Therefore, many studies in this area seek to develop routing protocols to meet the requirements of VANET networks and provide the necessary services for vehicle safety applications and create a road safety system [19]. Routing protocols can be broadly classified according to several criteria [20]. The popular classification standard for routing protocols are divided them into three types : proactive, reactive and hybrid. The principle of proactive routing protocols is to save the information about the nodes that make up the network in the form of routing tables, and then each node sends its routing table intermittently to its neighboring nodes, which makes all the associated nodes in the network to obtain a complete map of the network and locations of distribution associated nodes. After that, each node updates its routing table by sending control messages to all associated nodes and then sending the updated schedule to the rest of the nodes periodically and without a prior routing request, which makes the routing tables present in all nodes updated continuously. Continual periodic updating routing tables nodes consumes network bandwidth, and thus, the major drawback of proactive protocols is the large loads arising from the need to flood the network management messages [21]. While reactive routing protocols have two features that differ from proactive method. Firstly, they don't store the information about the entire network but only keep the information of the next node in the used route. the second feature that they don't start route discovery by themselves, until source node requires a rout to destination node [22]. Hybrid protocols are a type of routing protocol that combines proactive and reactive routing protocols to gain their advantages and eliminate their disadvantages. they aim to reduce the overhead of maintaining information in proactive routing protocols and to reduce the initial path discovery latency in reactive routing protocols. In this paper, we studied the performance of the most popular routing protocols(AODV, DSR, OLSR, DSDV) and selected the best ones to work on developing it to suit the requirements of the VANET networks.The results we obtained in the results and discussion section show that the AODV protocol outperforms the rest of the protocols, and it is the one that will be used as a basis for building the proposed protocol.

4 Proposed Routing Protocol

4.1 Cluster Formation Algorithm

Clustering techniques are widely used in a variety of applications namely ad-hoc networks such as MANETs and VANETs and they are also used in data mining approaches. K-means clustering algorithm is one of the popular centroid based methods. It has been extensively employed since they effectively manage routing problems in ad-hoc networks. Because of the swift convergence and easy implementation, they are generally preferred in VANETs. K-means algorithm is an important type of flat clustering, uses the euclidean distance as measurement to determine the centers of clusters [23]. To start with, k-means chooses initial cluster centres, C. Let us assume c clusters are available in a data group. The main focus is to ascertain c centroids where every centroid is owned by a cluster. The centroids from a cluster have to be present to reduce the objective function of the k-means algorithm. Hence, a squared error function defines an objective function where the squared error function can be given as:

$$u_j = \sum_{a=1}^{c} \sum_{b=1}^{n} ||m_a^{(b)} - u_b||^2 \tag{1}$$

where $|m_a^{(b)} - u_b|2$ refers to the distance among the location of the node $m_a^{(b)}$ from the center of the cluster u_b. It further computed the overall distance of the nodes location n from the corresponding cluster centres. The implementation of k-means algorithm (Fig. 1) can be established as follows [23]:

1. Determine c which represents the number of clusters.
2. Randomly, initialize c nodes to use as the primary centroid group.
3. Each node is classified by computing the distance between that node and each centroid, and then classifying the node to be in the cluster which contains the nearest centroid..
4. Recompute the centroid by taking the mean of of each cluster as new centroid
5. Repeat the steps 3 and 4 for a set number of iterations or till the centroids stop their movement.

4.2 Cluster Head Selection Algorithm

Cuckoo search algorithm (CSA) is a widely used bio-inspired algorithm proposed by Yang et al. [24]. The algorithm functions on the basis of how the way of the

Algorithm 1: K-Means Cluster Formation
Step1: Obtain c number of clusters as input
Step2: Obtain $M = (m_1, m_2, ... m_n) \in L_n$ (is the location of n nodes which are the set of data points)
Step3: for $b = 1:c$ do
Step4: randomly choose($\lambda_1, \lambda_2, ... \lambda_c$)
Step5: end for
Step6: for $b = 1:c$ do
Step7: for $a = 1:n$ do
Step8: determine $\lambda_b = \{\lambda_b || max \sum_{b=1}^{c} || m_a - \lambda_b ||^2\}$
Step9: end for
Step10: end for
Step11: Assign m_a to λ_b
Step12: Once, every data point allocation is complete, re-compute the cluster centroid position.
Step13: Repeat the steps 6 to 10 till all the centroids are found to be convergent.

Fig. 1. K-Means Cluster Formation Algorithm

life of cuckoo furnished ideas for various optimization techniques. To upraise the babies of cuckoo, it lays eggs onto the other bird's nest. They take out an egg from the host birds nest and lay eggs by duplicating the eggs of the host bird. The eggs that have close correspondence with the host egg will have highly probable chances of hatching while the eggs identified by the host bird will be destroyed [25]. In this algorithm, each egg in the nest if the host bird refers to a solution to the capacity of what is calculated using few variables of adaptation function. If a new solution is found to be better than the past one, it automatically replaces the old one [24]. According to Yang and Suash [24], the CSA algorithm follows three major rules (Fig. 2):

1. Once in a time every cuckoo lays only one egg and places that egg in an arbitrarily selected host birds nest.
2. The finer nests that carry superior egg quality will be passed to the next generation.
3. The quantity of the host birds nests is predetermined. The probability of discovery that the host bird identifies cuckoos eggs is (0,1).

4.3 K-MCSA Proposed Routing Algorithm

The proposed algorithm (Fig. 3) is classified into two phases namely: (a) cluster formation and (b) cluster head (CH) selection. We utilized k-means method to format clusters and modified cuckoo search algorithm for cluster head selection. The modified CS algorithm chooses the best reliable path through smart configuration of the weights depending on the parameters namely distance factor, angle factor and road factor . The modified CS algorithm discovers the best path

Algorithm 2: Cuckoo Search Algorithm
Cuckoo Search via Lévy Flights
Step1: Generate iteration time $t = 1$
Step2: Objective function f(x), x = (x1, ..., xd)
Step3: Generate initial population of n host nests xi (i = 1, 2, ..., n)
Step4: while (t <MaxGeneration) or (stop criterion)
Step5: Get a cuckoo randomly by Lévy flights
Step6: Evaluate its quality/fitness Fi
Step7: Choose a nest among n (say, j) randomly
Step8: if (Fi < Fj),
Step9: Replace j by the new solution;
Step10: end
Step11: A fraction (pa) of worse nests is abandoned and new ones are built;
Step12: Keep the best solutions (or nests with quality solutions);
Step13: Rank the solutions and find the current best
Step14: Update the generation number $t = t + 1$
Step15: End while
Step16: End

Fig. 2. Cuckoo Search Algorithm

in a reasonable time period. The primary advantage of the CS algorithm namely its faster convergence and improved speed of execution makes it further more suitable in route discovery.

Start procedure
Source node action begins
Set N b current node and N d destination node
While N d destination Receive Packet or (stop criterion)
Current node sends Hello packet to neighbor nodes; where hello packet contains node information (position, speed, road id, heading)
Calculate the weight parameters,
Execute the k-means algorithm to find cluster centers and to form cluster group
Increment b
End while
Interrupt for receive RREQ (source address, destination address, hop, seq number, DF, AF, RF)
Destination node action begins
Initialize the size of the population to
Get available population of S paths, a = 1, 2……, s
For $a = 1 : S$ for the declared path of source to destination
While a<s or (stop criterion)
Calculate fitness value based on $W(a,b,c) = h_1 * DF(a,b,c) + h_2 * AF(b,c) + h_3 * RF(b,c)$
Execute Cuckoo Search algorithm
Calculate the fitness
Increment a
End while
Choose Optimized node as CH with best fitness value in that group
CH node sends advertisement message to group nodes
Nodes which receive the advertisement message will become the cluster member for that CH
This process continues throughout in the network
End procedure

Fig. 3. Proposed Routing Protocol

We consider the node with the smallest value of the fitness function to be CH, where:

1. DF(a,b,c) : It is the factor to choose route with shortest distance between source and destination, namely distance factor. we used the mean value of distance between three nodes to calculate it

$$distance(a,b) = sqrt[((xb - xa)^2) + ((yb - ya)^2)] \tag{2}$$

$$DF(a,b,c) = (distance(a,b) + distance(b,c) + distance(a,c))/3 \tag{3}$$

2. AF(b,c): It is the factor to choose route more stable between source and destination, where when two nodes move in the same direction, the connection will be more stable than the connection between two nodes move in different directions. we used the angle between two nodes to calculate angle factor as follow:

$$AF(b,c) = \cos(\theta) = \frac{\overrightarrow{V_b} - \overrightarrow{bc}}{||\overrightarrow{V_b}||.||\overrightarrow{bc}||} \tag{4}$$

3. RF(b,c): It is the road factor to choose route more stable between source and destination, where when two nodes move in the same road, the connection will last longer than the connection between two nodes in different roads.

$$RF(b,c) = \begin{cases} 0 \; if\ the\ vehicles\ are\ \ on\ the\ same\ road \\ +1 \; if\ the\ vehicles\ are\ \ in\ different\ roads \end{cases} \tag{5}$$

5 Modeling and Performance Evaluation

5.1 Modeling Scenario

Table 1. Simulation parameters

Parameters	Description
Simulation program	NS-3.23
Topology model	Manhattan grid road network 5 × 5 with 2000 m edge
Propagation Model	Log Distance Propagation Loss Model
Number of nodes	50, 100, 150, 200, 250
Black Hole attacks	10
Packet size	512 bytes
Packet rate	2 Kb/s
Speed range	0–50 m/s
Transmission range	250 m
Type of Traffic	CBR, UDP
topology generator	Bonn Motion
Routing Protocol	AODV, KMCSA
Mac layer	802.11 b
Simulation Time	200 s

The simulation is divided into two phases, the first to evaluate the performance of the four protocols (AODV, DSR, OLSR, DSDV), and the second to evaluate the performance of the proposed protocol. We simulated the behavior of our routing protocol and popular routing protocol AODV under various conditions of the number of nodes and 10 attacks of black hole to study the performance of the protocols; Packet-to-Delivery Ratio (PDR), throughput, overhead, packet loss rate and VANET latency. All simulations were carried out using the NS3 network simulator. We also used the Manhattan Grid mobility model using the Bonnmotion tool, which is widely used to describe the mobility of VANET nodes. The simulation parameters that we used in the scenario are shown in Table 1.

5.2 Performance Metrics

The following parameters are analysed in the simulation study [26]:

1. Packet Delivery Ratio (PDR): It is the number packets that are delivered to the destination. The higher is the packet delivery ratio, the better is the routing protocol.

$$PDR = (\sum_{a=1}^{c} rxdpackets_a / \sum_{a=1}^{c} txdpackets_a) \times 100 \tag{6}$$

   where
   txdpackets refers to the overall number of data packets transmitted.
   rxdpackets refers to the overall number of data packets received.
2. Packet Loss Ratio (PLR): It is the number packets that are not received by the destination. These packets are lost during transmission from source to destination. The packet drop may be due to signal degradation, corrupted packets or congestion, etc. The lower is the packet drop the better is the routing protocol.

$$PDR = \frac{\sum_{a=1}^{c} txdpackets_a - \sum_{a=1}^{c} rxdpackets_a}{\sum_{a=1}^{c} txdpackets_a)} \times 100 \tag{7}$$

3. Throughput: It is defined as the packets received at the destination out of total number transmitted packets. The unit used is kbps. The routing protocols with high throughput are more efficient.

$$Throughput = \frac{\sum_{a=1}^{c} numberofbytesreceived \times 8}{Starttime - Endtime} \tag{8}$$

4. Average Delay: The total time for transmitting a packet from source to the destination node is known as end to end delay. The delay performance metric includes the delays due to route discovery, packet propagation and sending time and the time of packet in queue.

$$AD = \frac{\sum_{a=1}^{c} SumDelay_a}{\sum_{a=1}^{c} rxdpackets_a} \tag{9}$$

   Where
   SumDelay refers to the sum of the delay values

5. Overhead: It represents the number of routing bytes required by the routing protocols to construct and maintain its routes. This includes all routing packet types (request, reply, error) in the network.

$$Overhead = \frac{TotalNumberofRoutingPackets}{TotalNumberofReceivedDataPackets} \quad (10)$$

5.3 Results and Discussions

The simulations were performed using Network Simulator NS-3 to compare the popular routing protocols in the first step and then to compare the proposed K-MCSA protocol and popular protocol AODV in the second step , using the simulation parameters discussed in modeling scenario section. We analyzed and evaluated the performance of the protocols in VANET for various scenarios with 50, 100, 150, 200, 250 vehicles, once with 10 attacks by the black hole and once without. We evaluated the performance by measuring the network performance metrics such as Packet delivery ratio (PDR), Packet loss ratio(PLR), Average delay, Overhead and Throughput of the entire system as a function of number of vehicles considered. Our results are represented by the following figures from 4 to 14 . the results depict that the proposed protocol is extremely dependable and scalable as the anticipated results were generated with the large number of vehicles and even with black hole attacks (Fig. 4).

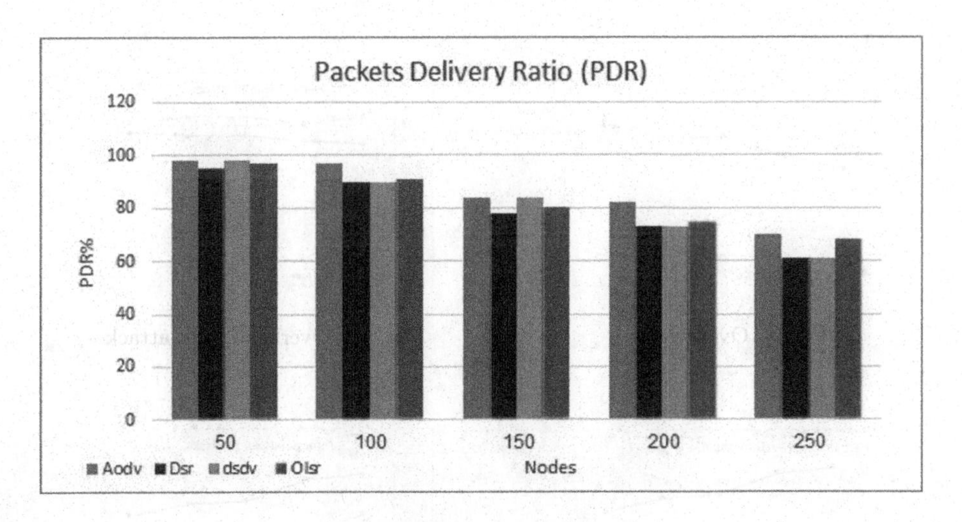

Fig. 4. Packet Delivery Ratio

We can clearly notice from Figures 5 and 7 that the K-MCSA protocol gave better results than the popular AODV protocol, with regard to the number of received packets and the number of lost packets, even with a black hole attack, as in Figures 6 and 8. This improvement is due to the use of the Fitness function,

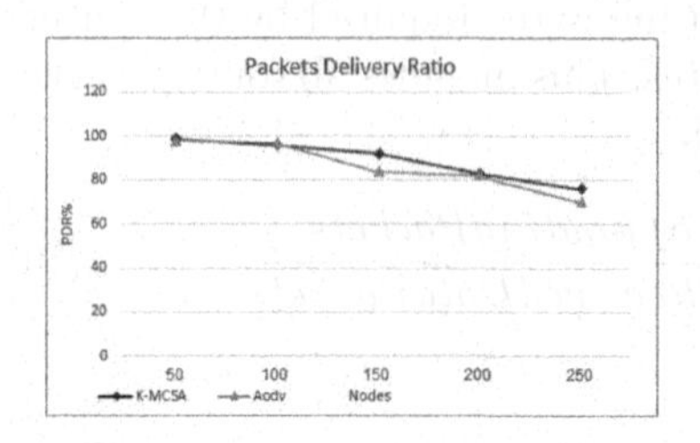

Fig. 5. Packets Delivery Ratio

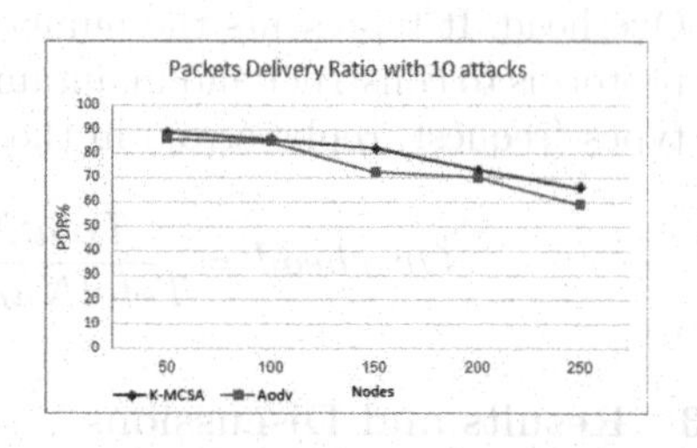

Fig. 6. Packets Delivery Ratio with attack

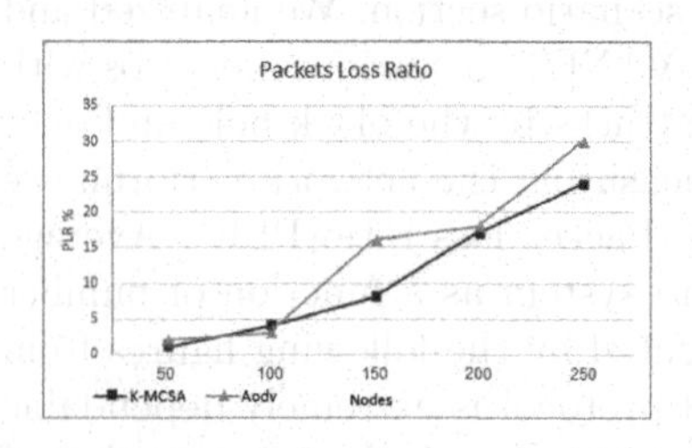

Fig. 7. Packets Loss Ratio

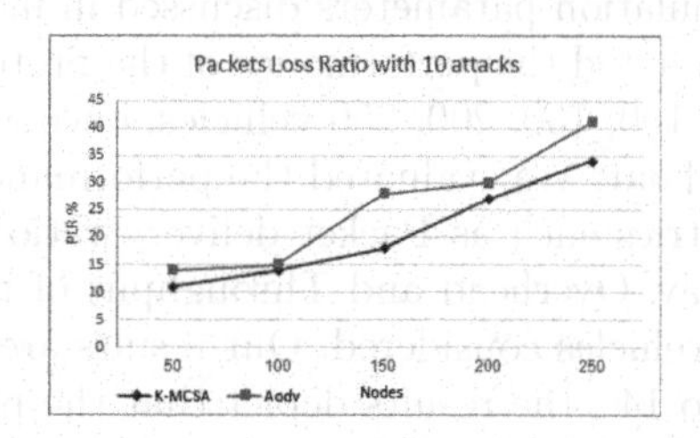

Fig. 8. Packets Loss Ratio with attack

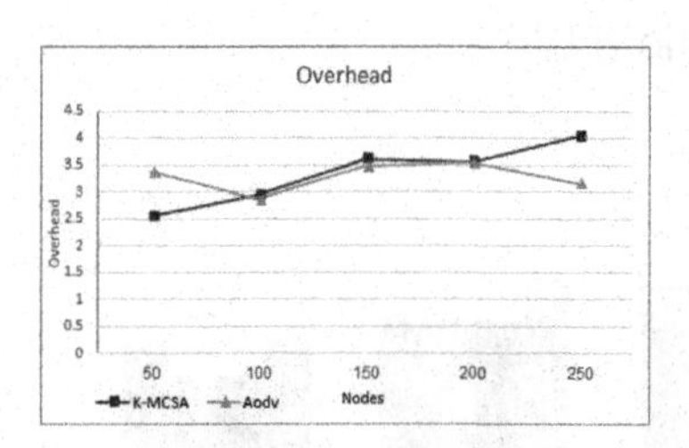

Fig. 9. Overhead

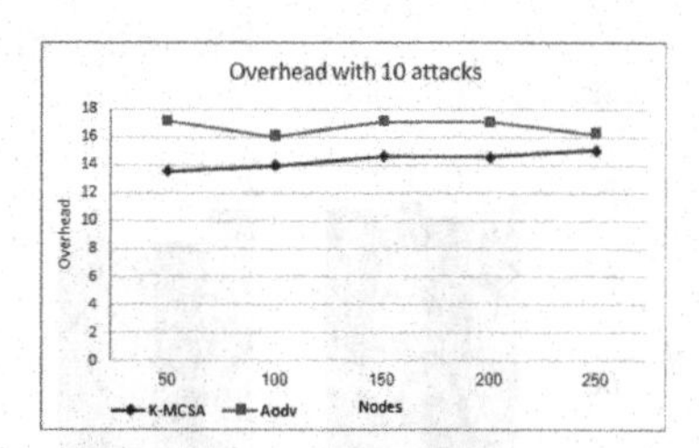

Fig. 10. Overhead with attack

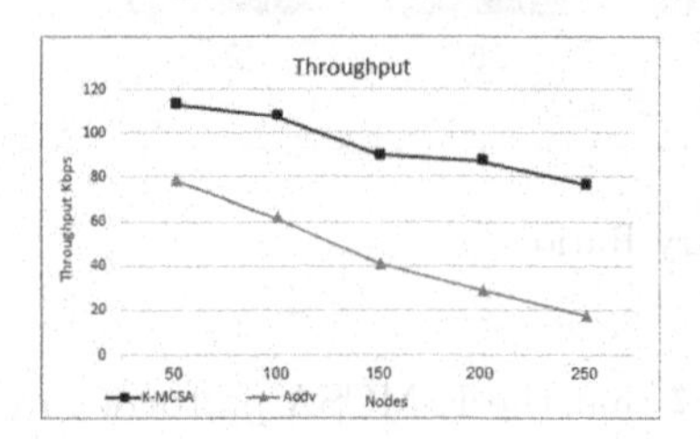

Fig. 11. Throughput

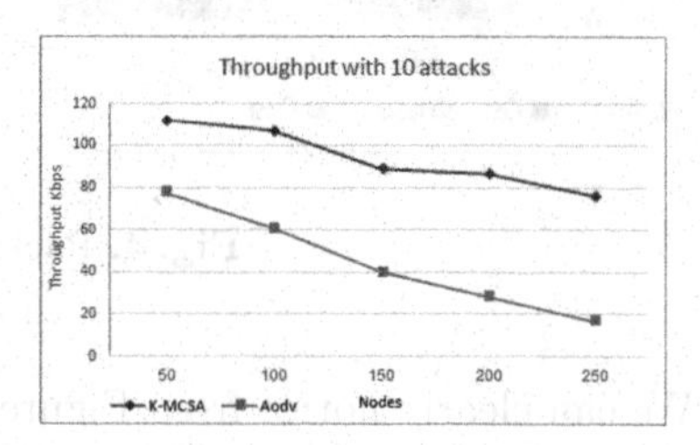

Fig. 12. Throughput with attack

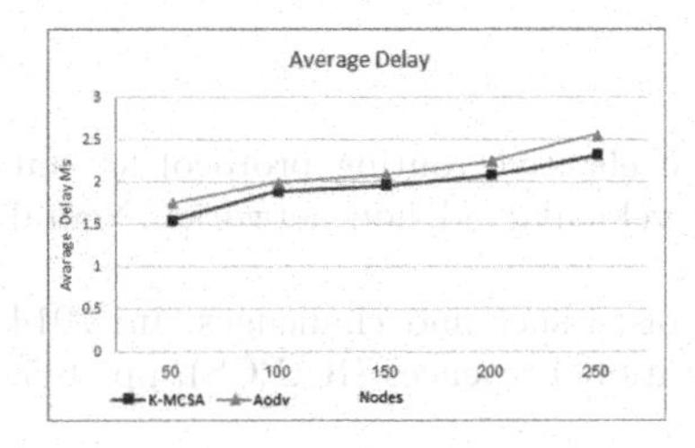

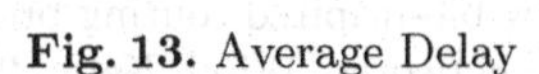

Fig. 13. Average Delay

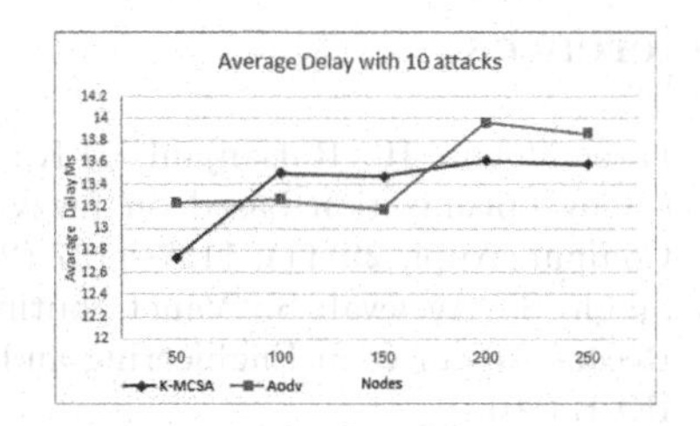

Fig. 14. Throughput with attack

which used parameters such as distance, angle and road number and took them into consideration in the path selection process, so the path consists of nodes that represent cluster heads instead of the next hop as in the AODV protocol, and this had a positive impact on the stability of the path with fewer failures, which increases percentage of lost packets in the network.

The results of the throughput and overhead are illustrated in Fig. 9, 10, 11 and 12 They show that the increased throughput of the K-MCSA protocol was due to the advantage of fast convergence rate with minimum control packets to choose the best path and CH within a short time as shown in Figure 13 and 14.

6 Conclusions

VANET networks play an important role in the development of the intelligent transportation system to transfer it to a good level that satisfies all users and reduces accidents on the roads by exchanging messages between cars, especially warning messages. VANET networks depend on routing protocols to exchange data between nodes through wireless channels, so an effective protocol that provides data exchange in a short time with fewer packet losses will improve network performance and thus improve the transmission system. This paper presented an efficient routing protocol K-MCSA that depends on cluster concept (k-means) and bio-inspired algorithm (cuckoo search). The proposed K-MCSA protocol formates cluster by k-means whereas selects cluster head(CH) by cuckoo search algorithm . The CS algorithm is one of the efficient metaheuristic algorithms particularly in higher searching space. The modified CSA identifies the optimal route from the known routes by fitness function which calculates specific weighted parameters which are distance, angle, and road. The protocols were implemented using NS3, nodes traffic generation using Bonnmotion , and results obtained using Flow Monitor. We can notice from the results that the proposed K-MCSA protocol gave better results than the AODV protocol in network performance metrics such as packet delivery ratio,packet loss ratio, average delay, overhead and throughput. The weighted parameters that used in the CSA helps to select the efficient route in short time. The proposed K-MCSA protocol was better due to its special characteristics, specifically, the swift convergence rate because of the utilization of Levy distribution function.

References

1. Fatemidokht, H., Rafsanjani, M.K.: F-ant: an effective routing protocol for ant colony optimization based on fuzzy logic in vehicular ad hoc networks. Neural Comput. Appl. **29**(11), 1127–1137 (2018)
2. Singh, S., Agrawal, S.: Vanet routing protocols: issues and challenges. In: 2014 Recent Advances in Engineering and Computational Sciences (RAECS), pp. 1–5. IEEE (2014)
3. Kout, A., Labed, S., Chikhi, S., et al.: Aodvcs, a new bio-inspired routing protocol based on cuckoo search algorithm for mobile ad hoc networks. Wirel. Netw. **24**(7), 2509–2519 (2018)
4. Al-Sultan, S., Al-Doori, M.M., Al-Bayatti, A.H., Zedan, H.: A comprehensive survey on vehicular ad hoc network. J. Netw. Comput. Appl. **37**, 380–392 (2014)
5. Jain, M., Saxena, R.: Overview of vanet: requirements and its routing protocols. In: 2017 International Conference on Communication and Signal Processing (ICCSP), pp. 1957–1961. IEEE (2017)
6. Wu, C., Ohzahata, S., Kato, T., Routing in vanets: a fuzzy constraint q-learning approach. In: 2012 IEEE Global Communications Conference (GLOBECOM), pp. 195–200. IEEE (2012)
7. Barekatain, B., Khezrimotlagh, D., Maarof, M.A., Quintana, A.A., Cabrera, A.T.: Gazelle: an enhanced random network coding based framework for efficient P2P live video streaming over hybrid WMNs. Wirel. Pers. Commun. **95**(3), 2485–2505 (2017)
8. Kumari, N.D., Shylaja, B.: AMGRP: AHP-based multimetric geographical routing protocol for urban environment of vanets. J. King Saud Univ.-Comput. Inf. Sci. **31**(1), 72–81 (2019)
9. Mukhtaruzzaman, M., Atiquzzaman, M.: Clustering in vehicular ad hoc network: algorithms and challenges. Comput. Electric. Eng. **88**, 106851 (2020)
10. Cooper, C., Franklin, D., Ros, M., Safaei, F., Abolhasan, M.: A comparative survey of vanet clustering techniques. IEEE Commun. Surv. Tutor. **19**(1), 657–681 (2016)
11. Abboud, K., Zhuang, W.: Stochastic modeling of single-hop cluster stability in vehicular ad hoc networks. IEEE Trans. Veh. Technol. **65**(1), 226–240 (2015)
12. Kayis, O., Acarman, T.: Clustering formation for inter-vehicle communication. In: 2007 IEEE Intelligent Transportation Systems Conference, pp. 636–641. IEEE (2007)
13. Chen, J., Lai, C., Meng, X., Xu, J., Hu, H.: Clustering moving objects in spatial networks. In: Kotagiri, R., Krishna, P.R., Mohania, M., Nantajeewarawat, E. (eds.) DASFAA 2007. LNCS, vol. 4443, pp. 611–623. Springer, Heidelberg (2007). https://doi.org/10.1007/978-3-540-71703-4_52
14. Shea, C., Hassanabadi, B., Valaee, S.: Mobility-based clustering in vanets using affinity propagation. In: GLOBECOM 2009–2009 IEEE Global Telecommunications Conference, pp. 1–6. IEEE (2009)
15. Wang, Z., Liu, L., Zhou, M., Ansari, N.: A position-based clustering technique for ad hoc intervehicle communication. IEEE Trans. Syst. Man Cybern. Part C (Appl. Rev.) **38**(2), 201–208 (2008)
16. Mohammad, S.A., Michele, C.W.: Using traffic flow for cluster formation in vehicular ad-hoc networks. In: IEEE Local Computer Network Conference, pp. 631–636. IEEE (2010)
17. Bitam, S., Mellouk, A., Zeadally, S.: Bio-inspired routing algorithms survey for vehicular ad hoc networks. IEEE Commun. Surv. Tutor. **17**(2), 843–867 (2014)

18. Liu, J., Wan, J., Wang, Q., Deng, P., Zhou, K., Qiao, Y.: A survey on position-based routing for vehicular ad hoc networks. Telecommun. Syst. **62**(1), 15–30 (2015). https://doi.org/10.1007/s11235-015-9979-7
19. Chandel, N., Gupta, M.V.: Comparative analysis of AODV, DSR and DSDV routing protocols for vanet city scenario. Int. J. Recent Innov. Trends Comput. Commun. **2**(6), 1380–1384 (2014)
20. Sondi, P., Wahl, M., Rivoirard, L., Cohin, O.: Performance evaluation of 802.11 p-based ad hoc vehicle-to-vehicle communications for usual applications under realistic urban mobility. Int. J. Adv. Comput. Sci. Appl. (IJACSA) **7**(5), 221–230 (2016)
21. Bhangwar, N.H., Halepoto, I.A., Khokhar, S., Laghari, A.: On routing protocols for high performance. Stud. Inf. Control **26**(4), 441–448 (2017)
22. Kalwar, S.: Introduction to reactive protocol. IEEE Pot. **29**(2), 34–35 (2010)
23. Zhang, Q., Almulla, M., Ren, Y., Boukerche, A.: An efficient certificate revocation validation scheme with k-means clustering for vehicular ad hoc network. In: 2012 IEEE Symposium on Computers and Communications (ISCC), pp. 000862–000867. IEEE (2012)
24. Yang, X.-S., Deb, S.: Cuckoo search via lévy flights. In: 2009 World Congress on Nature & Biologically Inspired Computing (NaBIC), pp. 210–214. IEEE (2009)
25. Gandomi, A.H., Yang, X.-S., Alavi, A.H.: Cuckoo search algorithm: a metaheuristic approach to solve structural optimization problems. Eng. Comput. **29**(1), 17–35 (2013)
26. Khairnar, M., Vaishali, D., Kotecha, D., et al.: Simulation-based performance evaluation of routing protocols in vehicular ad-hoc network. arXiv preprint arXiv:1311.1378 (2013)

OpenFlow-based Software-Defined Networking Queue Model

Vyacheslav Kartashevskiy and Marina Buranova(✉)

Povolzhskiy State University of Telecommunications and Informatics, 443010 23 L.Tolstoy, Samara, Russia
buranova-ma@psuti.ru

Abstract. SDN imposes requirements on manufacturers of infocommunication equipment. It concerns new scenarios support relating to network applications, for example, cloud technologies, the possibility of transmitting large volumes of traffic over long distances. It leads to the need of analyzing features of SDN networks operation, assessing their performance both at design stage and operation process. Quality of service (QoS) parameters in networks can be used as characteristics to assess the quality of operation. In this work, analytical expressions are obtained to estimate the average values of packet delay time in M/G/1 and G/G/1 systems for SDN. For M/G/1 system, solutions are given founded on two approaches. To approximate arbitrary densities (G) in G/G/1 system, an approach based on the use of hyperexponential distributions was used. When analyzing G/G/1 system, it is assumed that there are mutual independence of flows entering the system and the absence of correlations within the sequences of time intervals between packets and packet processing times. The paper presents the result of comparing the estimates of the average values of packet delay time in M/G/1 and G/G/1 systems, where the truncated normal distribution is used as probability density for service time intervals.

Keywords: Queue model · SDN · Quality of service · Hyperexponential distribution · Average package service time in the system

1 Introduction

The development trends of modern infocommunication networks such as the intensive growth of new network applications, the growth of number and types of new network devices, lead to a significant increase in the volume of transmitted traffic. In this case, there is a need to manage heterogeneous flows and provide the required level of QoS while ensuring the security of processing data transmission of flows. This leads to the fact that providers of large networks need to look for new mechanisms for network management, with the ability to quickly configure networks.

V. M. Vishnevskiy et al. (Eds.): DCCN 2021, CCIS 1552, pp. 62–76, 2022.
https://doi.org/10.1007/978-3-030-97110-6_5

Software-Defined Networking (SDN) and Network Function Virtualization (NFV) technologies allow you to control network management using customizable software, making management more intelligent and centralized. This makes the network more flexible, programmable and innovative [1,2].

The introduction of SDN networks entails the need to develop adequate models that allow you to quickly obtain accurate estimates of QoS parameters, which are necessary at the network design stage and further during operation to be able to rapidly respond to changes in network requirements and modification of the network topology [3].

It should be noted that issues of performance and scalability of SDNs are still little studied in terms of constructing analytical models. Simulations and experiments on real networks are of great importance and are widely used to evaluate performance, whereas analytical modeling has its advantages.

Most of the works devoted to the study of SDN operation efficiency, are aimed at developing simulation models and setting up experiments on real equipment. In works [4]-[6] mathematical models are presented for evaluating the performance of controllers and switches under overload conditions. The OpenFlow-based SDN analytical model in [7,8] approximates the data plane as an open Jackson network with a controller modeled as an M/M/1 queue. In [9], a model of SDN performance based on the OpenFlow protocol with several OpenFlow switches is presented, but this model was obtained under the assumption that processed flows are the simplest, while the performance of processing incoming messages of the SDN controller is calculated based on the M/G/1 model. However, it is known [10] that the flows generated by modern applications in the network are not Poisson. That requires taking into account the real parameters of traffic in the applied model.

The works [12,13] present mathematical models for systems that process non-Poissonian flows. It should be noted that only functioning parameters of individual SDN sections are considered, not the network as a whole. The development of SDN analytical model as a $G/G/1$ system remains highly relevant. The technique presented in [9] is convenient to use to expand the packet processing model in the SDN when servicing non-Poisson flows based on the $G/G/1$ mathematical model.

The SDN analytical model based on the OpenFlow protocol is considered below, since it is the most common. Such a model can be used as a basic one for other communication protocols in the SDN. As processed streams, we consider packet traffic generated in the form of bursts, which most closely corresponds to the nature of modern streams formation [14].

Similar to the approach in [9], the process of packet arrival and the procedure for forwarding packets to the switch and the OpenFlow controller are analyzed separately. Then we studied the system for forwarding packets to SDN as a whole. The M/M/1 system is considered first, followed by the G/G/1 system.

2 Model of Queuing of OpenFlow Switches

In general, the SDN network can be represented in the form of a diagram shown in Fig. 1.

All analytical models of OpenFlow networks, developed so far [9,12,13], are based on the assumption that packet traffic to the SDN switch corresponds to the Poisson distribution. At the same time, traffic studies in [10,15,16] showed that the arrival of packets in infocommunication networks has a burst character and differs from the Poisson flow [17].

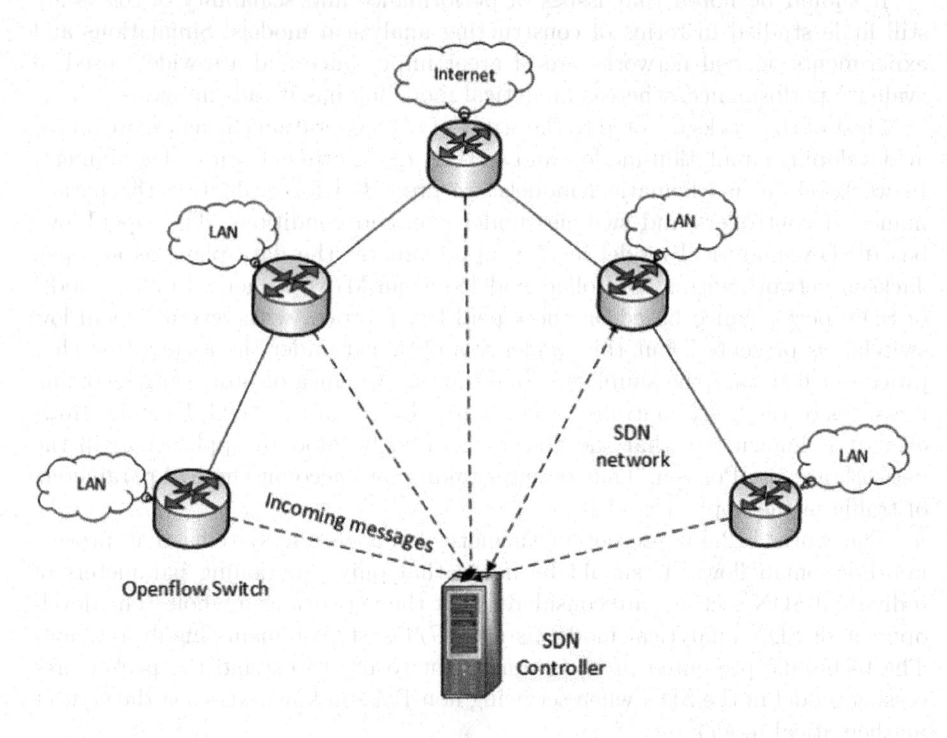

Fig. 1. Typical OpenFlow Network Scenario.

The procedure of forwarding packets to the switch is illustrated in Fig. 2. Each received packet is analyzed by the switch in order to identify a flow in regard to find the assigned rule for its processing or forwarding in its tables. If no forwarding entry is found for this stream, the switch sends a message containing the complete packet or its identifier to the SDN controller. The controller analyzes the packet, sets the flow processing rule to which the packet belongs, and sends the set rule to the switch to be added to its flow tables.

Since requests to the forwarding tables for all packets are independent of each other, the packet processing time can be represented as a random variable with an exponential distribution. Then, provided there is sufficient packet buffer capacity for all incoming queues in the switch, the packet forwarding queue model in the OpenFlow switch can be represented as the model of the queue type $M/M/1$ [9]. Further, this approach can be developed to the $G/G/1$ system.

To describe the queue in the OpenFlow switch, we introduce the following notation: $\lambda_i^{(b)}$ is the intensity of the arrival of the Poisson stream of packets in a batch to the i-th OpenFlow switch, $\lambda_i^{(p)}$ is intensity, characterizing the Poisson distribution of the number of packets in a batch,) $\mu_i^{(s)}$ is the intensity of processing the s-th packet by the switch (corresponds to the exponential distribution).

Suppose that a burst of m packets arrives at the i-th $(1 \leq i \leq k)$ OpenFlow switch, where n packets are waiting for their turn for processing. The incoming l-th packet must wait until n waiting packets in the buffer and the first $(l{-}1)$ packets from the same batch are processed. Based on these assumptions for the $M/M/1$ system, we can obtain the ratio between the average queue length and the average time spent by a packet in the queue [9].

$$\overline{L_i^{(s)}} = \sum_{n=0}^{\infty} n p_n = \lambda_i^{(b)} \lambda_i^{(p)} \overline{W_i^{(s)}}. \tag{1}$$

The average time of a packet's stay in the switch is determined in a similar way.

$$\overline{W_i^{(s)}} = \frac{\lambda_i^{(p)} + 1}{2\left(\mu_i^{(s)} - \lambda_i^{(b)} \lambda_i^{(p)}\right)}. \tag{2}$$

And based on (1) and (2), you can write an expression to estimate the average length of the switch queue

$$\overline{L_i^{(s)}} = \frac{\lambda_i^{(b)} \lambda_i^{(p)} \left(\lambda_i^{(p)} + 1\right)}{2\left(\mu_i^{(s)} - \lambda_i^{(b)} \lambda_i^{(p)}\right)}. \tag{3}$$

3 SDN Controller Queuing Model

Now let's look at the procedure of forwarding packets to the OpenFlow controller. Upon arrival of a new flow, the OpenFlow switch sends an incoming packet message as a request to configure flow processing to its SDN controller (request packet). In this regard, the incoming flow of messages from the switch to its controller corresponds to the process of the flow entering the switch. In an OpenFlow network, the controller is usually responsible for several switches and receives a stream of incoming request packets from each of them.

Based on the assumption that the arrival process on the i-th switch is Poisson with the parameter $\lambda_i^{(f)}$, and the processes on all k switches are independent of each other, we can further conclude that the total flow of message packets to the controller is also Poisson ...

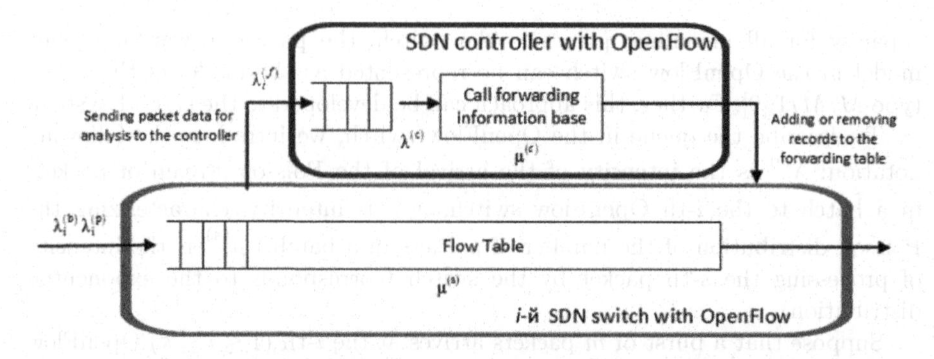

Fig. 2. SDN Controller Inbound Processing.

For an SDN controller responsible for k OpenFlow switches, the flow arriving at the i-th (i = 0, 1, 2, ..., k) switch is a Poisson flow with the parameter $\lambda_i^{(f)}$, independent of messages on other switches, then all packets of incoming messages from k switches to the controller correspond to the Poisson distribution with the parameter $\lambda_{(c)}$.

$$\lambda_{(c)} = \sum_{i=1}^{10} \lambda_i^{(f)}. \tag{4}$$

Upon arrival of a request packet from the switch, the controller determines the flow transfer rule to which the packet that has come from the switch belongs. To do this, the controller processing unit reads the incoming packet message from the queue after the last message has been processed. The incoming packet message is then processed by looking up the forwarding information base (FIB). The FIB entry contains routing information obtained using routing protocols or manual programming. Finally, the controller encapsulates the found forwarding rule in a packet message for the appropriate switch.

When processing an incoming packet-request from the switch by the SDN controller, the processing time of the incoming packet message is mainly determined by FIB lookups, which are performed by matching the longest prefix. In this regard, in [9,18], it was assumed that the search time in the FIB is considered to be distributed normally with the mean value $1/\mu^{(c)}$ and variance $\sigma^{(c)}$. The parameter $\mu^{(c)}$ is the average processing rate of incoming packet messages in the controller, and the parameter $\sigma^{(c)}$ is the standard deviation. The queue in the controller is organized according to the FIFO (First In - First Out) type, and the arrival and processing of incoming message packets are independent of each other.

In accordance with the assumptions made, it is possible to characterize the processing of message packets by the SDN controller using the $M/N/1$ queuing model, where the symbol N corresponds to a normal distribution.

To analyze a queue in a controller, it is convenient to use a semi-Markov process as a model, in which state changes occur at the moment the packet leaves. For such moments, the nested Markov chain should be defined as the number of claims present in the system at the moment the next serviced claim leaves. In this case, you can use the approach shown in [19], which is based on the application of the method of additional variables. It should be noted that the same technique was used to analyze the queue in the SDN controller in [9].

The length of the packet message queue in the controller at time t is denoted by $L(t)$. For any fixed time t, if there is a processed packet message, the distribution of the remaining processing time does not depend on time t, and the queue length $L(t), t \leq 0$ no longer has the properties of a Markov stream. Suppose that υ_n is the number of incoming message packets in the controller during the processing time x_n of the n-th message. Then $\{\upsilon_n, n \geq 1\}$ is a built-in Markov chain.

If t_n is the required processing time of the (n+1)-th message, for any t_n the probability density function, in accordance with the above assumptions, can be written as

$$f(x) = \frac{1}{\sqrt{2\pi}\sigma^{(c)}} e^{\frac{\left(x-1/\mu^{(c)}\right)^2}{2\sigma^{(c)2}}}. \tag{5}$$

Let us denote by η_n the number of new packets arriving at the controller during processing of the $(n+1)$-th packet under the assumption that the arrival process is Poisson. Then

$$P\left(\eta_n = k\right) = P\left(\eta_n = k | t_n = x\right) f(x) \mathrm{d}x = \int_0^{\infty} \frac{(\lambda^{(c)} x)^2}{k!} e^{-\lambda^{(c)} x} f(x) \mathrm{d}x, \tag{6}$$

where $P\left(\eta_n = k | t_n = x\right)$ is the probability of one-step transition, $k = 0, 1, 2, ...$

We set that $a_k = P(\eta_n = k) > 0$. Then $\{\upsilon_n, n \geq 1\}$ constitutes a Markov chain, the state change diagram of which is shown in Fig. 3, and the transition matrix will have the following form [9,19]

$$P = \begin{pmatrix} a_0\, a_1\, a_2 \cdots \\ a_0\, a_1\, a_2 \cdots \\ 0\;\, a_0\, a_1 \cdots \\ 0\;\; 0\;\, a_1 \cdots \\ \vdots\;\; \vdots\;\; \vdots\;\; \ddots \end{pmatrix}$$

Using the apparatus of generating functions for processing the sequences υ_n and η_n, according to [9,19], for the average queue length of messages from the switch to the controller in the $M/G/1$ system, we can obtain

$$\overline{L^{(c)}} = \rho^{(c)} + \frac{{\rho^{(c)}}^2 + {\lambda^{(c)}}^2 {\sigma^{(c)}}^2}{2(1-\rho^{(c)})}, \tag{7}$$

where $\rho^{(c)}$ is the load factor of the controller.

For the average residence time of packet messages in the controller, the expression will look like

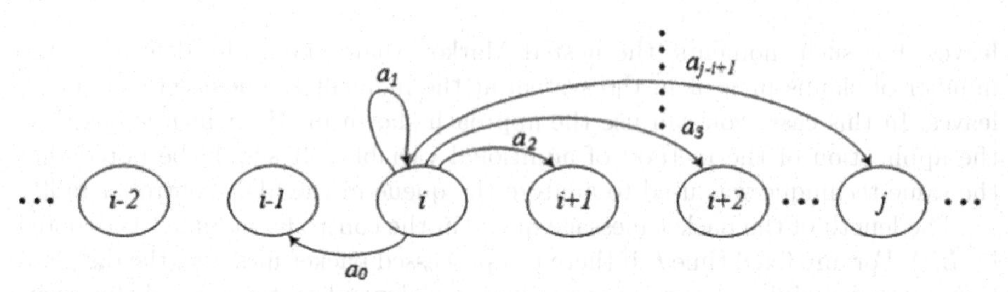

Fig. 3. Transition probability diagram for an embedded Markov chain of type M/G/1.

$$\overline{W^{(c)}} = \frac{\overline{L^{(c)}}}{\lambda^{(c)}} = \frac{1}{\mu^{(c)}} + \frac{{\rho^{(c)}}^2 + {\lambda^{(c)}}^2 {\sigma^{(c)}}^2}{2\lambda^{(c)}(1-\rho^{(c)})}. \tag{8}$$

4 SDN Network Queuing Model with OpenFlow

On OpenFlow networks, the switch maintains a queue for all packets arriving on any ingress port and forwards them according to its internal flow tables. If there is no entry in the tables corresponding to the packet, that is, the packet belongs to a new flow, the switch will send a request for flow processing rules to its SDN controller and, having received the corresponding rule from the controller, sends the packets of this flow in accordance with this rule. At the same time, the controller keeps a queue for all request packets from its slave switches. Using the above analysis of the service process for OpenFlow switches and SDN controllers, we can represent the packet processing model in OpenFlow networks as a queuing system according to Fig. 2.

In Fig. 2, the i-th OpenFlow switch with rate $\mu^{(s)}$ processes bursts of packets arriving at rate $\lambda^{(p)}\lambda^{(p)}$. Suppose that the incoming packet in the i-th switch belongs to a new flow with probability q_i, the switch sends a request packet at a rate $\lambda_i^{(f)}$ to its controller SDN. The controller receives request packets with a rate $\lambda^{(c)}$ from k OpenFlow switches with a $\mu^{(c)}$ and processes them. Replies to request packets are sent back to forward and update the flow tables in the corresponding switch at a rate

$$\lambda_i^{(f)} = q_i \lambda_i^{(b)} \lambda_i^{(p)}. \tag{9}$$

Let W_i be the time of packet forwarding through the i-th OpenFlow switch, you can calculate it taking into account two possible processing cases: direct forwarding and forwarding with the participation of the controller. The packet forwarding time in the latter case consists of two parts: the packet sojourn time in the switch $W_i^{(s)}$ and the sojourn time of the corresponding packet transmission message in the $W^{(c)}$ controller.

$$W_i = \begin{cases} W_i^{(s)} & \text{with probability } 1 - q_i \\ W_i^{(s)} + W^{(c)} & \text{with probability } q_i. \end{cases} \tag{10}$$

The average time to process a packet in SDN can be determined by the average time to forward packets through the OpenFlow switch ($\overline{W_i^{(s)}}$) according to the diagram shown in Fig. 2.

5 Results for the System $M/G/1$

It can be shown [9] that the average time for forwarding packets through the OpenFlow switch $\overline{W_i}$, where the waiting times $W_i^{(s)}$ and $W^{(c)}$ from (10) can be obtained from expressions (6) and (8) respectively, finally they will have the form

$$\begin{aligned}\overline{W_i} = E\left[W_i\right] = (1-q_i)\, E\left[W_i^{(s)}\right] + q_i \left(E\left[W_i^{(s)}\right] + E\left[W^{(c)}\right]\right) \\ = \overline{W_i^{(s)}} + q_i \overline{W_i^{(c)}} \\ = \frac{\lambda_i^{(p)}+1}{2(\mu_i^{(s)})-\lambda_i^{(p)}\lambda_i^{(c)}} + q_i\left(\frac{1}{\mu_i{}^{(c)}} + \frac{\rho^{(c)2}+\lambda^{(c)2}\sigma^{(c)2}}{2\lambda^{(c)}(1-\rho^{(c)})}\right),\end{aligned} \tag{11}$$

where $\delta_i^{(s)}$ and $\delta^{(c)}$ are the service rates in the i-th switch and controller, respectively.

The result for estimating the average packet processing time in SDN, obtained in the form of formula (11), corresponds to the mathematical model $M/N/1$.

Let us introduce the notation $\delta_i^{(s)} = 1/\overline{W_i^{(s)}}$ for the i-th switch and $\delta^{(c)} = 1/\overline{W^{(c)}}$ for the controller, where $\delta_i^{(s)}$ and $\delta^{(c)}$ are the parameters, which are defined as [19], ξ is the root of the equation $\xi = \Lambda_V(\mu - \mu\xi)$.

Here Λ_V is the Laplace transform of the density $f_V(\cdot)$, μ is the average intensity of packet processing in the system, and ${\delta_i}^{(s)}$ and $\delta^{(c)}$ are the parameters of the density of the distribution of processing time in the queue in the i-th switch and controller, respectively. The method for determining these parameters is well described in [19–21].

Based on the above, more general results for the M/G/1 system can be obtained. The expressions for the probability density of the packet processing time S in the controller $f_{(S)i}(\cdot)$ and the packet processing time C in the switch $f_{(C)}(\cdot)$ are represented in the form

$$f_{(S)i}\left(u\right) = \delta_i^{(s)} e^{-\delta_i^{(s)} u}, \tag{12}$$

$$f(x) = \frac{1}{\sqrt{2\pi}\sigma^{(c)}} e^{-\frac{\left(u-1/\delta^{(c)}\right)^2}{2\sigma^{(c)2}}}. \tag{13}$$

Assuming that C and S are independent, and taking into account (12) and (13) in expression (10), we obtain

$$w_{(S,C)i}\left(u\right) = (1-q_i)\, f_{(S)i}\left(u\right) + q_i \left[f_{(S)i}\left(u\right) \odot f_{(C)}\left(u\right)\right], \tag{14}$$

where $\odot$ is the convolution symbol The second term in expression (14) can be represented as

$$q_i \int_0^u \frac{1}{\sqrt{2\pi}\sigma^{(c)}} e^{-\frac{\left(x-1/\delta^{(c)}\right)^2}{2\sigma^{(c)2}}} \cdot \delta_i^{(s)} \cdot e^{-\delta_i^{(s)}(u-x)} \mathrm{d}x. \quad (15)$$

The upper limit of integration in (15) is u that is equal to infinity for the normal distribution and is equal to some finite value (the maximum value of the packet duration in the controller) for the truncated normal distribution.

For the $M/G/1$ system, where the normal distribution is used as an example of an arbitrary distribution, that is, for the $M/N/1$ system, the expression for the probability density of the packet processing time in the system can be written as:

$$\begin{aligned} w_{(S,C)i}(u) &= (1-q_i)\,\delta_i^{(s)} e^{-\delta_i^{(s)}u} \\ &+ q_i \frac{\delta_i{}^{(s)}}{2R} e^{-\delta_i^{(s)}u} e^{((\frac{1}{\delta^{(c)}})^2)/2\sigma^{(c)2})+\sigma^{(c)2}\delta_i^{(s)2}-2\delta_i^{(s)}\frac{1}{\delta^{(c)}}} \\ &\times[1-\Phi(\tfrac{\sigma^{(c)}}{\sqrt{2}}(\delta_i^{(s)}-\tfrac{1}{\delta^{(c)}\sigma^{(c)2}}))], \end{aligned} \quad (16)$$

where $\Phi(x) = \frac{2}{\sqrt{\pi}} \int_0^x e^{-t^2} \mathrm{d}t$ [22],

$R = \frac{1}{2}[1+\Phi(\frac{1}{\sqrt{2}\delta^{(c)}\sigma^{(c)}})]$ is normalizing constant, which is determined from the condition $\int_0^T f(u)\mathrm{d}u = 1$. For a truncated normal distribution

$R = \frac{1}{2}\Phi(\frac{\frac{1}{\delta^{(c)}}\sigma^{(c)}}{\sqrt{2}})$

Estimating the values of the integral function is associated with some difficulties that lead to the need to use various methods of approximation in the form of infinite power series [21], infinite continued fractions [23], polynomials of a special form [23] and empirical formulas [24]. In the case of expression (16), the values of the function $\Phi(x)$ can be determined in a table.

Let us compare the time densities of packet processing in the $H_2/H_2/1$ system according to (16) for different distribution parameters. The comparison result is shown in Fig. 4.

From a comparison of the graphs shown in Fig. 4, it follows that for the values of the distribution parameters $\delta_i^{(s)} = 1$ and $\delta^{(c)} = 1$, the distribution degenerates into an exponential, in other cases an insignificant peak appears.

The final expression for estimating the average packet processing time in the $M/N/1$ system for SDN will have the form

$$\bar{W}_{(S,C)i}(u) = (1-q_i)\frac{1}{\delta_i^{(s)}} + q_i M \frac{1}{\delta_i{}^{(s)2}}, \quad (17)$$

where $M = \frac{\delta_i^{(s)}}{2R} \cdot e^{((\frac{1}{\delta^{(c)}})^2/2\sigma^{(c)2})+\sigma^{(c)2}\delta_i^{(s)2}-2\delta_i^{(s)}\frac{1}{\delta^{(c)}}[1-\Phi(\frac{\sigma^{(c)}}{\sqrt{2}}(\delta_i^{(s)}-\frac{1}{\delta^{(c)}\sigma^{(c)2}}))]}$.

Taking into account formulas (1) and (17), the expression for estimating the average queue length will be written in the form

$$L = \lambda_i^{(b)}\lambda_i^{(p)} \cdot (1-q_i)\frac{1}{\sigma_i^{(s)}} + q_i M \frac{1}{\sigma_i^{(s)2}}. \quad (18)$$

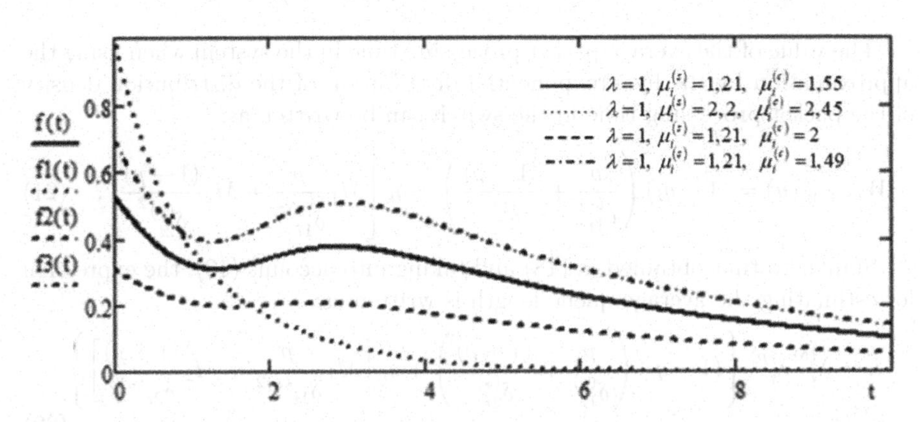

Fig. 4. Probability densities of packet processing times in SDN according to the formula (16): $f(t)$ – $\delta_i^{(s)} = 0,2, \sigma^{(c)} = 0,5$; $f1(t)$ – $\delta_i^{(s)} = 1, \sigma^{(c)} = 1$; $f2(t)$ – $\delta_i^{(s)} = 0,2, \sigma^{(c)} = 0,8$; $f3(t)$ – $\delta_i^{(s)} = 0,2, \sigma^{(c)} = 0,45$.

This approach allows us to evaluate the performance of the SDN network and the main parameters of the quality of service of traffic in the SDN network in the case of processing Poisson streams. Given that modern applications generate non-Poissonian traffic, queuing systems are better described by $G/G/1$ models.

6 SDN Network Queuing Model with OpenFlow

It is interesting to compare the values of the estimates of the SDN functioning parameters obtained for the $M/G/1$ and $G/G/1$ systems. Obviously, for this, one should use the result shown in (16), (17), and (18) for the $M/N/1$ system and find an expression for the packet waiting time density in the $G/G/1$ system. As an example of an arbitrary probability distribution for the query processing time in the controller, similarly to the approach shown for the $M/N/1$ system, we can use the normal distribution (for the random variable C) (15), and for the random variable S, we can use the probability density as an approximation by hyperexponential distribution:

$$f_{(S)i}(u) = p\delta_{1i}^{(s)} e^{-\delta_{1i}^{(s)} u} + (1-p)\,\delta_{2i}^{(s)} e^{-\delta_{2i}^{(s)} u}, \tag{19}$$

where p, $\delta_{1i}^{(s)}$, $\delta_{2i}^{(s)}$ is distribution parameters. Then the $G/G/1$ model will be approximated by the $H_2/N/1$ model, and expression (16) takes the form

$$\begin{aligned} w_{(S,C)i}(u) = {} & (1-q_i)\left(p\delta_{1i}^{(s)} e^{-\delta_{1i}^{(s)} u} + (1-p)\,\delta_{2i}^{(s)} e^{-\delta_{2i}^{(s)} u}\right) \\ & + q_i \left\{M_1 e^{-\delta_{1i}^{(s)} u} + (1-p)\,M_2 e^{-\delta_{2i}^{(s)} u}\right\}, \end{aligned} \tag{20}$$

where $M_1 = \frac{\delta_{1i}^{(s)}}{2R} \cdot e^{((\frac{1}{\delta^{(c)}})^2/2\sigma^{(c)2}) + \sigma^{(c)2}\delta_{1i}^{(s)2} - 2\delta_{1i}^{(s)}\frac{1}{\delta^{(c)}}[1-\Phi(\frac{\sigma^{(c)}}{\sqrt{2}}(\delta_{1i}^{(s)} - \frac{1}{\delta^{(c)}\sigma^{(c)2}}))]}$,
$M_2 = \frac{\delta_{2i}^{(s)}}{2R} \cdot e^{((\frac{1}{\delta^{(c)}})^2/2\sigma^{(c)2}) + \sigma^{(c)2}\delta_{2i}^{(s)2} - 2\delta_{2i}^{(s)}\frac{1}{\delta^{(c)}}[1-\Phi(\frac{\sigma^{(c)}}{\sqrt{2}}(\delta_{2i}^{(s)} - \frac{1}{\delta^{(c)}\sigma^{(c)2}}))]}$.

The value of the average packet processing time in the system when using the approximation by the hyperexponential distribution of the distribution density of the packet processing time in the switch can be written as:

$$\bar{W}_{(S,C)i}(u) = (1-q_i)\left(\frac{p}{\delta_{1i}^{(s)}} + \frac{(1-p)}{\delta_{2i}^{(s)}}\right) + q_i\left[M_1\frac{p}{\delta_{1i}^{(s)2}} + M_2\frac{(1-p)}{\delta_{2i}^{(s)2}}\right]. \quad (21)$$

Similar to that obtained in (18) and taking into account (19), the expression for estimating the average queue length is written as:

$$\bar{L} = \lambda_i^{(b)}\lambda_i^{(p)}\left\{(1-q_i)\left(\frac{p}{\delta_{1i}^{(s)}} + \frac{(1-p)}{\delta_{2i}^{(s)}}\right) + q_i\left[M_1\frac{p}{\delta_{1i}{}^{(s)2}} + M_2\frac{(1-p)}{\delta_{2i}{}^{(s)2}}\right]\right\}. \quad (22)$$

To obtain a more general result, the approximation of arbitrary densities in the $G/G/1$ system is preferable to be represented in the form of hyperexponential distributions for random variables S and C, that is, it is necessary to use expression (19) for S, and for C

$$f_{(C)}(u) = g\delta_1^{(c)}e^{-\delta_1^{(c)}u} + (1-g)\,\delta_2^{(c)}e^{-\delta_2^{(c)}}, \quad (23)$$

where g, $\delta_1^{(c)}$, $\delta_2^{(c)}$ is packet processing time density parameters in the controller. After such a replacement, the system $G/G/1$ will be approximated by the system $H_2/H_2/1$.

At the same time, for the density of the packet processing time in the $H_2/H_2/1$ system in the SDN network, one can obtain

$$f(x) = \frac{1}{\left(\sqrt{2\pi}\right)\sigma^{(c)}}e^{\frac{\left(x-1/\mu^{(c)}\right)^2}{2\sigma^{(c)2}}}, \quad (24)$$

where $A = \frac{pg\delta_{1i}^{(s)}}{\delta_{1i}^{(s)}-\delta_1^{(c)}}$, $B = \frac{(1-p)g\delta_{2i}^{(s)}}{\delta_{2i}^{(s)}-\delta_1^{(c)}}$, $L = \frac{p(1-g)\delta_{1i}^{(s)}}{\delta_{1i}^{(s)}-\delta_2^{(c)}}$, $D = \frac{(1-p)(1-g)\delta_{2i}}{\delta_{2i}^{(s)}-\delta_2^{(c)}}$.

Using (24), the analytical expression for the average packet processing time in the $H_2/H_2/1$ system for the SDN network can be written as:

$$\bar{W}_{(S,C)i}(u) = (1-q_i)\left[\frac{p}{\delta_{1i}^{(s)}} + \frac{(1-p)}{\delta_{2i}^{(s)}}\right] + q_i\left[\frac{(A+B)}{\delta_1^{(c)}} + \frac{(L+D)}{\delta_2^{(c)}}\right]. \quad (25)$$

From (25), the value of the average queue length in the SDN is written as

$$\bar{L} = \lambda_i^{(b)}\lambda_i^{(p)}\left\{(1-q_i)\left[\frac{p}{\delta_{1i}^{(s)}} + \frac{(1-p)}{\delta_{2i}^{(s)}}\right] + q_i\left[\frac{(A+B)}{\delta_1^{(c)}} + \frac{(L+D)}{\delta_2^{(c)}}\right]\right\}. \quad (26)$$

The delay variation can be determined taking into account (25) in the form of the expression

$$\sigma_{(S,C)i}(u) = \sqrt{\begin{array}{l}(1-q_i)\left[p\left(\frac{2}{\delta_{1i}^{(s)2}} - \frac{1}{\delta_{1i}^{(s)}}\right) + (1-p)\left(\frac{2}{\delta_{2i}^{(s)2}} - \frac{1}{\delta_{2i}^{(s)}}\right)\right] + \\ +q_i\left[(A+B)\left(\frac{2}{\delta_1^{(c)2}} - \frac{1}{\delta_1^{(c)}}\right) + (L+D)\left(\frac{2}{\delta_2^{(c)2}} - \frac{1}{\delta_2^{(c)}}\right)\right]\end{array}}. \quad (27)$$

As a result, formulas were obtained for the density of distributions of packet service times in the SDN network, if the queuing system is represented by the G/G/1 model for two cases:

1) the distribution density of the packet processing time in the switch is approximated by a hyperexponential distribution, the distribution density of the packet processing time in the controller is estimated by a normal distribution (the $M/H_2/1$ system);
2) the distribution density of packet processing times in the switch and controller is approximated by a hyperexponential distribution (the $H_2/H_2/1$ system).

7 Numerical Results

Formula (17) allows us to estimate the average packet processing time in the G/N/1 system for the SDN network; earlier in [9], formula (11) was obtained to approximate the average packet processing time in the M/N/1 system. To determine the accuracy of the estimates obtained, it is necessary to compare the values accepted for various models, including the results received for $M/N/1$ and $H_2/N/1$.

Let us compare the numerical estimates of the average processing times in the M/N/1 system for these two approaches. We will accept the following conditions for the operation of the network: $\lambda_i^{(b)}\lambda_i^{(p)} = 0,5$, $\lambda_i^{(p)} = 1$, $\mu^{(c)} = 1$, $\sigma^{(c)} = 1$ number of switches is 1.

The values of the functioning parameters - the average packet processing time and the average queue length for SDN, are presented in the table (Table 1).

Figure 5 shows the dependence of the delay in the SDN network depending on the network load factor for the cases considered.

Obviously, the average packet processing time in the SDN network is expected to grow with the increase in the network load factor.

Table 1. The values of the functioning parameters.

Parameter values	$\bar{W}_{(S,C)i}$, conditional units of time	$\bar{L}$, conditional units of time
System $M/N/1$ (11)	2.3	0,23
System $M/N/1$ (17)	2,9	0,29
System $H_2/N/1$ (21)	1,5	0,15
System $H_2/H_2/1$ (25)	29,1	2.9

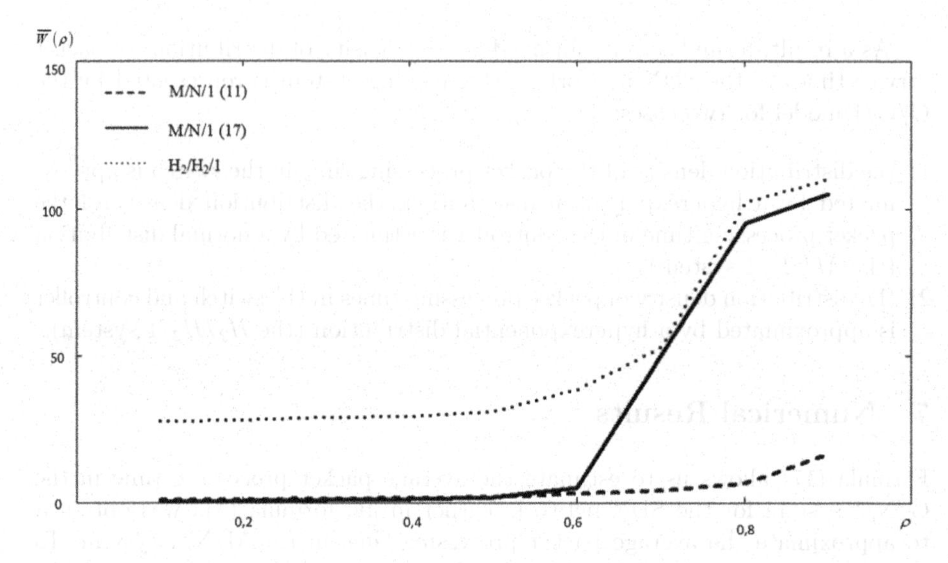

Fig. 5. The graph of the dependence of the delay on the load factor in the SDN network.

8 Conclusion

In the process of designing, deploying and operating networks, it is necessary to take into account the quality of service parameters of the flows processed in the network. In this paper, a model of the SDN network functioning is built on the basis of the mathematical apparatus of the queuing theory. As a result, analytical expressions were obtained for evaluating the main parameters of traffic QoS in SDN for the M/G/1 system and for the $G/G/1$ system, provided that the flows entering the system are mutually independent and there are no correlations within the sequences of time intervals between packets and processing times packages.

In the development of this work, it is planned to build an SDN model for the G/G/1 system when processing correlated streams. The problem of accounting for correlations within flows will improve the accuracy of the analytical model, which can further help in the development of accurate simulation models of the functioning of SDN networks. This will make it possible to quickly analyze the efficiency of the SDN network in real operating conditions.

References

1. Feamster, N., Rexford, J., Zeguraz, E.: The road to SDN: an intellectual history of programmable networks. Networks **11**(12), 20–40 (2013)
2. Astutoz, B.N., Mendonca, M.: A survey of software-defined networking: past, present, and future of programmable networks. IEEE Commun. Surv. Tutor. **16**(3), 1617–1634 (2014)

3. McKeown, N., et al.: Openow: enabling innovation in campus networks. ACM SIGCOMM Comput. Commun. Rev. (CCR) **38**(2), 69–74 (2008)
4. Bozakov, Z., Rizk, A.: Taming SDN controllers in heterogeneous hardware environments. In: Proceedings of the Second European Workshop on Software Defined Networks (EWSDN), pp. 50–55 (2013)
5. Azodolmolky, S., Wieder, P., Yahyapou, R.: Performance evaluation of a scalable software-defined networking deployment. In: Proceedings of the Second European Workshop on Software Defined Networks (EWSDN), pp. 68–74 (2013)
6. Azodolmolky, S., Nejabati, R., Pazouki, M., Wieder, P.: An analytical model for software defined networking: a network calculus-based approach. In: Proceedings of the 2013 IEEE Global Communications Conference (GLOBECOM), pp. 1397–14021 (2013)
7. Jarschel, M., Oechsner, S., Schlosser, D., Pries, R., Goll, S., Phuoc, T.G.: Modeling and performance evaluation of an openflow architecture. In: Proceedings of the Twenty-Third International Teletraffic Congress (ITC), pp. 1–7 (2011)
8. Mahmood, K., Chilwan, A., Osterbo, O., Jarschet, M.: Modelling of open-flow-based software-defined networks: the multiple node case. IET Netw. **4**(5), 278–284 (2015)
9. Xiong, B., Yang, K., Zhao, J., Li, W., Li, K.: Performance evaluation of OpenFlow-based software-defined networks based on queueing model. Comput. Netw. **102**, 174–183 (2016)
10. Sheluhin, O.I., Osin, A.V., Smolskij, S.M.: Samopodobie i fraktaly. Telekommunikacionnye prilozheniya, Fizmatlit (2008)
11. Taggu, M.S.: Self-similar processes. In: Kotz, S., Johnson, N. (eds.) Encyclopedia of Statistical Sciences, pp. 352–357. Wiley, New York (1988)
12. Samujlov, K.E., Shalimov, I.A., Buzhin, I.G., Mironov, Y.B.: Model funkcionirovaniya telekommunikacionnogo oborudovaniya programmno-konfiguriruemyh setej. Sovremennye informacionnye tekhnologii i IT-obrazovanie **14**(1), 13–26 (2018)
13. Muhizi, S., Shamshin, G., Muthanna, A., Kirichek, R., Vladyko, A., Koucheryavy, A.: Analysis and performance evaluation of SDN queue model. In: Koucheryavy, Y., Mamatas, L., Matta, I., Ometov, A., Papadimitriou, P. (eds.) WWIC 2017. LNCS, vol. 10372, pp. 26–37. Springer, Cham (2017). https://doi.org/10.1007/978-3-319-61382-6_3
14. Okamura, H., Dohi, T., Trivedi, K.S.: Markovian arrival process parameter estimation with group data. ACM Trans. Netw. **17**(4), 1326–1339 (2009)
15. Jain, R., Routhier, S.A.: Packet trains: measurements and a new model for computer network traffic. IEEE J. Sel. Areas Commun. **4**(6), 986–995 (2016)
16. Choi, B.D., Choi, D.I., Lee, Y.D., Sung, K.: Priority queueing system with fixed-length packet-train arrivals. In: Fifth International Conference on Information Technology: New Generations (ITNG 2008), vol. 145, no. 5, pp. 331–336 (1998)
17. Paxson, V., Floyd, S.: Wide area traffic: the failure of Poisson modeling. IEEE/ACM Trans. Netw. **3**(3), 226–244 (1995)
18. Liang, Z., Xu, K., Wu, J.: A novel model to analyze the performance of routing lookup algorithms. In: Proceeding of the 2003 IEEE International Conference on Communication Technology (ICCT), pp. 508–513 (2003)
19. Kleinrock, L.: Queueing Systems. Vol. I: Theory. Wiley, New York (1975)
20. Tarasov, V., Kartashevskiy, I.: Approximation of input distributions for queuing system with hyper-exponential arrival time. In: 2015 Second International Scientific-Practical Conference Problems of Infocommunications Science and Technology, pp. 15–17. IEEE, Kharkiv (2015)

21. Kartashevskiy, I., Buranova, M.: Calculation of packet jitter for correlated traffic. In: Galinina, O., Andreev, S., Balandin, S., Koucheryavy, Y. (eds.) NEW2AN/ruSMART -2019. LNCS, vol. 11660, pp. 610–620. Springer, Cham (2019). https://doi.org/10.1007/978-3-030-30859-9_53
22. Gradshtejn, I.S., Ryzhik, I.M.: Tablicy integralov, summ, ryadov i proizvedenij. 4-e izd. Fizmatgiz (1963)

Hybrid MCDM for Cloud Services: AHP(blocks) & Entropy, TOPSIS & MOORA (methodology Review and Advances)

Iliyan Petrov(✉)

Bulgarian Academy of Sciences, Institute of information and communication technologies, Akad. G. Bonchev str., bl. 2, Sofia 1113, Bulgaria
petrovindex@gmail.com, iliyan.petrov@iict.bas.bg
https://www.iict.bas.bg/ipdss/i-petrov.html

Abstract. Cloud services (CS) offer virtual, reliable, and flexible hardware and software resources as part of the distant Cloud Computing (CC) concept. The wide variety of products from different providers create the necessity for users to collect data and define criteria for comparing alternatives with reliable evaluation methods and techniques. This paper presents a combined approach based on Analytical Hierarchy Process(AHP), Information Entropy, and Multi-criteria decision making (MCDM) techniques. The set of evaluation criteria is decomposed and logically structured in a reasonable number of blocks which weights are computed more easily and effectively with a reduced number of AHP pair-comparisons. With the integration of Information Entropy, the weights of individual criteria within each block are defined objectively on basis of real primary data for quantitative and quantifiable parameters. This hybrid approach is applied in a case study based on QoS and QoE criteria with popular evaluation techniques like TOPSIS and MOORA.

Keywords: cloud computing · multi-criteria analysis · decision making · quality of services · quality of experience · TOPSIS · MOORA · AHP

1 Introduction

The rapid development of information technologies and internet communications opens large possibilities for distant Cloud Computing [1] with flexible, effective, and reliable hardware and software resources in three main models: Software as a Service (SaaS), Platform as a Service (PaaS), and Infrastructure as a Service (IaaS) [2]. The wide variety of products and plans offered by Cloud Services

This study is supported by Bulgarian FNI fund, project "Modeling and Research of Intelligent Educational Systems and Sensor Networks (ISOSeM)", contract KP-06-H74-4 from 26.11.2020.

V. M. Vishnevskiy et al. (Eds.): DCCN 2021, CCIS 1552, pp. 77–91, 2022.
https://doi.org/10.1007/978-3-030-97110-6_6

Providers (CSPs) create the need to analyze and assess a large number of parameters from the point of view of different users and applications. The selection of appropriate service has to consider diverse factors related to technical performance, pricing schemes, levels of quality, security, and support [3]. These characteristics are grouped in two main aspects Quality of Service (QoS) and Quality of Experience (QoE) [4,5]. The problem of reconciling different preferences and conflicting criteria is traditionally solved with the well-developed Multi-criteria Decision Making (MCDCM) methods as part of the "Operation analysis" in fundamental and applied scientific research [6]. One of the most challenging tasks in CS selection is the balanced aggregation of subjective expert judgements with abundant and objective technical data. The intense competition of CS increases the need to develop more accessible and effective approaches which will facilitate the integration of users in the MCDM process [7–10].

Several reviews of CS selection literature reveal that due to the complexity of cloud technologies the collected data and its analysis are mainly based on benchmarking indicators of cloud performance provided by third parties and less on real empiric data about QoS and QoE of cloud customers [11,12]. Analytical Hierarchy Process (AHP) is one of the most popular approaches for the calculation of criteria weights, relying on subjective expert opinions and large comparison matrix [13]. Unfortunately, the increasing number of criteria and DMs/experts results in decreasing consistency of comparison judgements consistency and requires additional computational and programming efforts.

To overcome such limitations this paper proposes improvements for structuring the professional opinions of experts and optimizing the objectivity of data-driven MDCM in the following directions:

- optimization of the conventional AHP approach by grouping the evaluation criteria in a logically structured and limited number of blocks for reducing the number of preference comparisons and their inconsistencies;
- application of Information Entropy (IE) concept of C. Shannon [14] for objective definition of criteria weights;
- flexible Integration of QoS and QoE assessments based on the combined "Block AHP - Entropy" approach with some of the most popular evaluation techniques (TOPSIS [15,16] and MOORA [17,18]).

The purpose of our findings is to improve the accessibility and efficiency of the combination of AHP and IE methods for providing flexible, transparent, and intuitive algorithms applicable for selections with multiple criteria in wide number of areas.

2 Review of Literature and Previous Research

To address the complexity CS several conventional and combined approaches can be applied to support the experts and Decision Makers (DM) in the selection process. Most of them focus on aspects like trust and reputation [19–21]; QoS perspective or reviewing specific aspects of servicing information systems (IS) [22,23].

Several reviews of CS selection literature reveal that due to the complexity of cloud technologies the collection data and its analysis are mainly based on benchmarking indicators of cloud performance provided by third parties and less attention to real empiric data about QoS and QoE of cloud customers [24,25]. From a methodology point of view the Analytical Hierarchy Process (AHP) [26,27] approach is one of the most popular ways for calculation of criteria weights, relying on subjective expert opinions and large comparison matrix. Unfortunately, the increasing of criteria number and DMs/experts results in decreasing consistency of comparison judgements and requires additional computational and programming efforts.

3 Decision Makers, Alternatives, Criteria

A preliminary step in any MCDM process is the constitution of the team of decision-makers (DM) and experts.

STEP 1. For simplification, we presume that the evaluation of alternatives is made based on individual or already integrated group judgements. This study does not engage in the specifics of individual or group decisions, which are well explored in several publications [28,29].

3.1 Selecting Target Alternatives.

The next step after the constitution of the decision making body is the definition of alternatives.

STEP 2. The forming of the target alternatives' set depends on the needs and plans of decision-makers and in the majority of the non-complicated situations can be made without preliminary review and selection. In the opposite case, a pre-qualification step would be very helpful for improving the selection process. In our experiment case, we use a dataset of a previous well structured very detailed study [30] in which after the pre-qualification procedure selected nine Pareto non-dominated alternatives [31] described without indicating the names of CS providers as $A_1, A_2, ..., A_9$. The tasks of the study and the profiles of target alternatives A_i define the types of criteria to be used for their assessment. Taking into account the limited volume of the current type of publications, the primary information for 8 QoS and 8 QoE criteria from the original dataset is employed shown in Table 1.

STEP 3. The target alternatives can be assessed with parameters that characterize their qualities in technical, economical and other aspects which can be classified in two main types of criteria - Benefit (Profit) and Non-Benefit (Cost), depending on how they characterize the contribution or saving of resources. The optimization for Benefit (Profit) criteria is expressed in the maximal the value, and for Non-Benefit (Cost) criteria - in the minimal values of the observed parameters.

Table 1. QoS and QoS criteria for CS selection

Quality of Service (QoS)	Type	Symbol	Quality of experience (QoE)	Type	Symbol
CPU Speed (GHz)	B	SC1	Response time	Bench (B)	EC1
Core Concurrency	B	SC2	Accessability	Bench (B)	EC2
RAM bandwidth (Gb/s)	B	SC3	Features	Bench (B)	EC3
Disk rate (Mb/s)	B	SC4	Ease of Use	Bench (B)	EC4
Disk seeks (ops/s)	B	SC5	Tech. support	Bench (B)	EC5
Latency (ms)	N-B	SC6	Customer service	Bench (B)	EC6
Availability (SLA, %)	B	SC7	Security	Bench (B)	EC7
Price (cent/h)	N-B	SC8	Trust	Bench (B)	EC8

Legend: B - Benefit; N-B - Non-Benefit; SLA - Service Level Agreement

4 Collection and Normalization of Data

STEP 4. In this step the primary data for alternatives and criteria is prepared for the selection process in the following sub-steps:

Step 4.1. Constructing of Basic Decision Matrix (BDM) comprising "m" target alternatives (A_i) which are evaluated with "n" independent criteria (C_j):

$$BDM_{m\times n} = [x_{ij}]_{m\times n}; where : i = 1, 2, ..., m; j = 1, 2, ..., n \quad (1)$$

Usually, the primary data for the observed parameters and criteria are expressed in different units and scales. To make the initial data comparable and to combine it into a consistent evaluation process it has to be normalized in a universal dimensionless format with linear or non-linear methods.

Step 4.2. Linear normalisation is achieved by calculating the proportions of relative weights "$_{ij}$" for each primary parameter value x_{ij} in all criteria $C_j(SC_j; EC_j)$ and formation of Normalized Decision Matrix (NDM) with $\Sigma_{i=1}^{n} p_{ij} = 1$:

$$p_{ij} = x_{ij}/\Sigma_{i=1}^{m} x_{ij} \quad (2)$$

$$NDM_{m\times n} = [p_{ij}]_{m\times n} \quad (3)$$

Step 4.3. In addition to the linear normalisation several types of non-linear normalization can be performed to obtain more differentiated and normalized performance ratings (ratios) r_{ij} and construct the Normalized Performance Matrix $NPM_{m\times n} = [r_{ij}]_{m\times n}$. One of them is the following non-linear variant:

$$r_{ij} = p_{ij}/\sqrt{(\Sigma_{i=1}^{m}(p_{ij})^2)} \quad (4)$$

Step 4.4. If necessary, another possible variant for normalization is related with the maximum and minimum values for "Benefit" (Eq. 5) and "Non-Benefit" (Eq. 6) criteria:

$$r_{ij} = x_{ij}/x_j^{max} \quad (5)$$

$$r_{ij} = x_j^{min}/x_{ij} \quad (6)$$

5 Objective Weighting of Criteria with Information Entropy

One of the most important aspects of any MDCM process is the definition of criteria weights and a popular approach for this task is to assign equal weights to all criteria. However, such a simplistic approach is not able to take into account from an objective point of view (supply-side) neither the role and importance of technical and economic factors nor, from a subjective point of view (demand-side), the needs and preferences of users and DMs.

One traditional approach to overcome the insufficient knowledge of users is the introduction of experts whose role is to translate the professional aspects to users and DMs. The most popular approach in this context is AHP (Sect. 5, above) which provides a transparent procedure to structure and quantify the user's preferences for different factors and criteria. Another also popular approach is the concept of Information Entropy, which is based on objective and real information about the technical, economic and other parameters available target alternatives.

As already mentioned in Sect. 2, the conventional one-sided approaches are usually not able to provide flexible and reliable methods for complex MCDM problems while the combined approaches can propose more effective solutions in specific cases [32,33].

Step 5. To ensure the objectivity of analysis this study employs the traditional Shannon Entropy (SE) approach for definition of criteria weights which can be integrated into the MCDM. The "entropy transformation of primary data is performed as follows:

Step 5.1. Calculating the values of individual Shannon Entropies "se_{ij}" for relative weights p_{ij} Eq. (2) for the evaluation criteria in NDM Eq. (3):

$$se_{ij} = -p_{ij} \times log_a(p_{ij}) \tag{7}$$

The individual entropies "se_{ij}" can be calculated in different formats depending on the logarithm basis "a" in Eq. (7). In fact, the different logarithmic formats define only different scales for the transformation of primary data (p_{ij}). Actually, after normalization to some selected level of maximum entropy the results from all formats are equivalent. In this paper is used the binomial format $log_2(p_{ij})$ which is traditionally preferred in Information Theory and in the area of Computer and Communication Science.

Step 5.2. Calculation of the total nominal Shannon Entropy for each criterion "j" as a sum of "i" individual entropies for "m" target alternatives:

$$SEnom_j = \Sigma_{i=1}^{m} se_j = -\Sigma_{i=1}^{m} w_{ij} \times log_2(w_{ij}) \tag{8}$$

Step 5.3. Calculation of maximal Shannon Entropy $SEmax$ for the set of "m" alternatives in which all elements have equal weights "$p_{ij} = 1/m$":

$$SEmax(m) = -\Sigma_{i=1}^{m} 1/m \times log_2(1/m) = log_2 m \tag{9}$$

Step 5.4. Normalization of $SEnom_j$ by comparing it to $SEmax(m)$:

$$SEnorm_j = SEnom_j / SEmax(m) = e_j = -(log_2 m)^{-1} \Sigma_{i=1}^{m} p_{ij} \times log_2(p_{ij}) \quad (10)$$

Step 5.5. Calculation of "real entropy criteria weights" $recw_j$:

$$recw_j = e_j / \Sigma_{i=1}^{m} e_j \quad (11)$$

Step 5.6. Calculation of "differed entropy criteria weights" $decw_j$. This transformation is based on an additional data treatment aiming to substitute the results from the genuine transformation of real empiric data. This treatment is an attempt to enhance the difference between criteria through a so-called "divergence value d_j" which actually represents an arithmetic difference between the normalized maximal entropy $SEmax(m)$ expressed as "1" and the normalized nominal entropy ($SEnorm_j = e_j$):

$$d_j = 1 - e_j \quad (12)$$

$$decw_j = d_j / \Sigma_{i=1}^{m} d_j = (1 - e_j) / \Sigma_{i=1}^{m} (1 - e_j) \quad (13)$$

Methodology Remark. According to Eq. (11) the criteria with higher nominal entropy in Eq. (8) or normalized entropy in Eq. (10) receive higher relative real weights "$recw_j$" than criteria with lower entropy. In other words, criteria with more even distribution of information about the resources are considered formally as more important in the further steps of evaluation and ranking of alternatives. In real terms and under conditions of free competition, the system components (criteria) tend to have similar characteristics and similar relative weights "p_i" in the system. At this stage, the MCDM process arrives at a point of uncertainty, and to overcome this situation there are two main possibilities:

Option 1. To remain in the concept of entropy and to treat the available primary data for obtaining more "differed" results. Unfortunately, this option has reduced mathematical capacity since the entropy assessments for different system configurations produce the same result. Also, this approach continues to compare the distribution of information within each criterion that may have different priorities and importance in the overall performance of alternatives for different users. It looks suspicious how in real life some residual value called "differed entropy" could create more importance without taking into account the needs and preferences of users. In our opinion, this approach is formalistic, but as it is used in several studies we present it in Step 5.6 above and the study case for comparing its behaviour and efficiency.

Option 2. To optimize the objectivity of entropy by structuring logically the professional opinions of expert(s) and taking into account the needs and preferences of users. In our opinion, a logical choice for this option is to use the most popular AHP concept for subjective criteria weighting as discussed in "Sect. 6" below.

Step 5.7. Calculation of normalised weighted values "v_{ij}" and forming the Weighted Normalized Matrix (WNM):

$$v_{ij} = r_{ij}.cw_j \tag{14}$$

$$WNM_{m\times n} = [r_{ij}.cw_j]_{m\times n} = [v_{ij}]_{m\times n} \tag{15}$$

The "real normalized weighted values" and "differed normalized weighted values" and the Real and Differed WNM are defined as:

$$rv_{ij} = r_{ij}.recw_j \tag{16}$$

$$dv_{ij} = r_{ij}.decw_j \tag{17}$$

$$RWNM_{m\times n} = [r_{ij}.recw_j]_{m\times n} = [rv_{ij}]_{m\times n} \tag{18}$$

$$DWNM_{m\times n} = [r_{ij}.dedw_j]_{m\times n} = [dv_{ij}]_{m\times n} \tag{19}$$

Here is the place to note that the transformation of the original objective "real entropy criteria weights $recw_j$" into another set of objective-like "differed entropy criteria weights $decw_j$" can produce very different outputs and distortion of results. Due to this artificial treatment the distribution of weights between is usually noticeably perturbed.

If such discussion is purely formal and theoretical, similar treatments might be regarded as experiments. Yet, in any real selection, such artificially created "differed reality" drastically transforms and reshuffles the importance of all criteria distancing them far from the situation reflected in the genuine empiric data. In practice, under the motivation for "data extraction" contained in Eqs. (14–15) substantial parts of genuine information about some criteria are "expropriated" and granted without evident reason to other criteria. Usually, this is one or two winners - the criteria which in reality have the lowest values of nominal entropy "$SEnom_j$" and its normalized entropy "$SEnorm_j = e_j$". In these types of treatments, the initial big losers are magically transformed into final big winners. Such perturbation in many cases may be regarded as "data treatment", but in our opinion, the concept of "extracting additional information" from relative weights is elusive and non-justified. In our opinion, the decision makers, experts, and scholars are not in a position to create, produce or invent additional objective information about the material facts of the real world. For that reason, in our studies we investigate the evaluations resul of alternatives in both variants (*recw* and *decw*) to explore the difference between them. Based on several previous experiments in tests in different areas, our preference is for the variant of "real entropy" as its objectivity is positioned closer to the primary information about the evaluation criteria parameters. This is confirmed by the experiments in the "case study" part of this study.

6 Subjective Weighting of Criteria: Classical AHP & AHP in Blocks

One of the most popular method for structuring and quantification of experts' preferences and defining criteria weights is the AHP which combines intuitive logic and accessible instruments for the calculation of results.

6.1 AHP Classic: Integral Set of Criteria

The classical AHP allows to structure the preferences of experts(users) under the form of "hierarchy pair-comparisons" whose total number is defined as $n(n-1)/2$ (Table 2).

Step 6A. To define subjectively the criteria weights (as an opposite approach to the objective criteria weighing in **Step 5(1-7)** a quadratic Hierarchy Preference Matrix $HPM_{n \times n}$ is constructed in which a hierarchy preference "*hp*" values reflect the preference judgement of experts or DMs expressed in the traditional AHP scale. The values of *"1-3-5-7-9"* correspond to the verbal assessments labels *"equal importance - moderate importance - strong importance - very strong importance - top importance"*. The interim values *"2-4-6-8"* are also available and can be used for more precise definition of *hp*. In addition, the problem of criteria weighting in the standard variant of AHP becomes even more complicated if in the selection participate a group of experts or DMs and each of them expresses an opinion on every "pair-comparison". Formally, the number of comparisons will be multiplied by the number of experts/DMs, but even in small groups, this will evoke a problem of increasing inconsistency and complicated reconciliation of their opinions. The rising of number of criteria and experts increases exponentially the number of comparisons and the size of $HPM_{n \times n}$, as shown in Table 4.

Table 2. Criteria, experts/DMs and comparisons in AHP

Number of criteria or blocks	2	3	4	5	6	8	10	20	25
Number of comparisons for 1 DM: $n(n-1)/2$	1	3	6	10	15	28	45	190	300
Number of comparisons for 3 DMs: $3n(n-1)/2$	3	9	18	30	45	84	135	570	900
Number of comparisons for 5 DMs: $5n(n-1)/2$	5	15	30	50	75	140	225	950	1500

Legend: DM - Decision Maker

6.2 Adjustment in AHP: Logical Systematization of Criteria in Blocks

For making the MCDM process more accessible for users is proposed a popular and pragmatic approach - to decompose the initial complex problem into smaller parts and to find for them suitable and specific solutions. In other words, to split the initial set of criteria in the context of the AHP into blocks that are logically linked with the properties of the target alternatives or the needs of the users.

Step 6B. Instead of the traditional approach described in **Step 6A**, the basic set of criteria is divided and structured in a limited number of blocks "b" (in our opinion, preferably $3 < b < 5$). On this basis, the weights of these blocks are defined with the classical AHP method. Such decomposition allows to structure the expression of expert opinions and to frame the subjectivity of their individual judgements only up to the point of defining the importance of different blocks. In other words, each block of criteria receives a defined portion of the primary information and this portion is further distributed among sub-criteria within the same block.

The "block decomposition" allows to fully exploit an important advantage of AHP - the possibility to monitor and adjust from the beginning the "hp" values in matrix $HPM_{b\times b}$ from the point of view of users preference and to check at every step the consistency of the matrix for minimizing the controversies in judgements about the importance of criteria. If needed, the procedure of "block decomposition" can be performed at one or several levels, taking into account that usually for some alternative aspects may be available more information than for other aspects. The technical criteria can be divided into several blocks - for computer systems, the blocks may be related to the main components or functions, such as CPU, memory, storage, accessibility, costs, etc. In our opinion, it would more appropriate to form blocks that contain both "benefit" and "not-benefit" criteria related to the same component of the system, rather than to isolate them in two major groups. This kind of criteria structuring can be integrated into the algorithms of all evaluation techniques - in our case TOPSIS and MOORA.

The blocks weights "bw" can be calculated on the basis of pair-comparisons as hierarchy preferences "hp" in a traditional AHP approach:

$$bw_e = \sqrt[b]{\Pi_{e=1}^{b} hp_{ee})} / \Sigma_{e=1}^{b} \sqrt[b]{\Pi_{e=1}^{b} hp_{ee})} \tag{20}$$

$$\Sigma_{e=1}^{b} bw_e = 1 \tag{21}$$

where "e" is the number of blocks of criteria (e=1, 2,..., b).

For demonstration purposes, the QoS and QoE criteria from the original dataset are decomposed into 4 blocks as shown in Table 3.

Table 3. Decomposition of QoS an QoE criteria in blocks

Computing			Storage		Accessibility		Finance
CPU speed	Cores	RAM	Disk rate	Disk seeks	Latency(Net)	Availability(SLA)	Cost
SC1	SC2	SC3	SC4	SC5	SC6	SC7	SC8
Accessibiliy		Functionality		Reliability		Security	
Resp. time	Access	Features	Ease of use	Tech.support	Cust.service	Securiy	Trust
EC1	EC2	EC3	CE4	EC5	EC6	EC7	EC8

For demonstration purposes, the criteria weights are computed within the standard AHP approach with hierarchy preferences (hp) being assigned by the author of this paper (Table 4).

Table 4. AHP (blocks) weights: Hierarchy Preference Matrix in AHP

QoS Blocks	Computing	Storage	Accessibility	Cost	$\sqrt[b]{\Pi_{e=1}^{b} hp_{ee}}$	Block weights (bw)
Computing	1	1/3	1/2	1/2	0,537	0,123
Storage	3	1	1	2	1,861	0,359
Accessibility	2	1	1	2	1,189	0,325
Cost	2	1/2	1/2	1	0,841	0,193
Consistency	0,017				4,357	1
QE Blocks	Accessibility	Functionality	Reliability	Security	$\sqrt[b]{\Pi_{e=1}^{b} h_{ee}}$	Weights
Accessibility	1	1/2	1/2	1/2	0,595	0,142
Functionality	2	1	1	2	1,414	0,337
Reliability	2	1	1	1	1,189	0,283
Security	2	1/2	1	1	1	0,238
Consistency	0,022				4,198	1

On the basis of such decomposition, the internal sub-criteria weights obtained with the objective entropy approach have to be adjusted with the AHP block weights. Logically, the blocks which contain only one criterion do not participate in further AHP, Entropy, or other transformation for calculating their weights, as their importance is already defined in Step 6B of Sect. 6.2.

7 MDCM Evaluation Techniques

7.1 Technique for Order of Preference by Similarity to Ideal Solution (TOPSIS)

In the TOPSIS evaluation technique, the evaluation and ranking in the MCDM selection are based on comparing their minimal and maximal Euclidean distances from the respective positive and negative ideal solutions.

STEP 7. Identification of TOPSIS "Value best rating" (v_j^+) and the "Value worst rating" (v_j^-) for all target alternatives in all criteria on the basis of normalized weighted values v_{ij} defined in Sect. 5, Step 5.7/Eq. (14–19).

Step 7.1. Calculation of Euclidean distances as worst and best "Separation measures" Sm_i^- or Sm_i^+ for each target alternative from respective v_j^- and v_j^+ values in each criterion:

- worst Euclidean distance for each alternative A_i:

$$Sm_i^- = \sqrt{\Sigma_{i=1}^{n}(v_{ij} - v_{ij}^-)^2} \tag{22}$$

– best Euclidean distance for each alternative A_i:

$$Sm_{-}^{+} = \sqrt{\Sigma_{i=1}^{n}(v_{ij}^{+} - v_{ij})^2} \tag{23}$$

Step 7.2. Calculation of the "Similarity value" Sv_i^* of ideal worst solution for each alternative A_i:

$$Sv_i^* = Sm_i^- / (Sm_i^+ + Sm_i^-) \tag{24}$$

STEP 8. Ranking the results for Sv_i^* from the point of view of optimization for maximal values. TOPSIS is a very logical and structured method, but being based on Euclidean distance it can produce in some situations assessments values with higher spreads of extreme values ($max - min$) and lower levels of entropy. As a result, it tends to present relatively lower competition among competitors, by enhancing the advantages of leaders and the disadvantages of losers. In some casees, these could be misleading and for that reason, at the final stage of MCDM more structured methods like TOPSIS should be used in combination with other method(s) containing different assessment logic - like MOORA in this study.

7.2 Multi-objective Optimization by Ratio Analysis (MOORA)

MOORA is one of the most accessible and universal approaches for multi-criteria optimization and decision-making. It is more transparent, intuitive, and easy for computation than other concepts, such as TOPSIS. At the same time, thanks to the compensatory concept it can handle complex tasks including multiple controversial criteria. The simplicity of evaluation allows to combine MOORA with other methods without complicated calculations. A popular combination of MOORA with the Information Entropy concept allows to address the needs for an objective definition of evaluation criteria. It is used for different MCDM problems and includes the following steps:

Step 7. Calculation of MOORA "Alternatives' Assessments" AA_j for each alternative by adding the "benefit values" bv_{ij} and subtracting the "cost values" cv_{ij} on the basis of normalized weighted values v_{ij} defines in Sect. 5, Step 5.7/Eq. (15–19) in the variants 0f of $recw_j$ and $decw_j$:

$$AA(rv_i) = \Sigma_{j=1}^{g} rv_{ij} - \Sigma_{j=g+1}^{n} rv_{ij} \tag{25}$$

$$AA(dv_i) = \Sigma_{j=1}^{g} dv_{ij} - \Sigma_{j=g+1}^{n} dv_{ij} \tag{26}$$

Step 8. Ranking the results in descending order - the best alternative has the maximal value of AA_i.

8 Integration and Sensitivity Analysis of QoS/QoE Assessments

In the majority of selection procedures, the MCDM process is usually based on one set of criteria reflecting different technical, economic, and other parameters.

This is a straightforward approach that is often used for CS selection as it looks relatively simple and easy to handle. In this respect, the combined "AHP-block & SE" approach contributes to optimizing the assessment process and producing consistent results with each of the QoS and QoE sets of criteria. However, Cloud Computing is a very specific technology sector - in an industry specialized for intensive and accurate data processing the experts and DMs have the possibility and the responsibility to ensure reliable performance and to monitor it both on the supply-side (data supplied by provider) and the demand-side (data collected from users). In reality, the simple comparison of the assessments for the two sets of QoS and QoE criteria can produce important differences and create ambiguity and uncertainty for DMs and users in the final stage of the MCDM process. To resolve this problem an additional integrations of the results can be performed in the different assessment techniques.

The generalization of these two aspects in one single number is another challenging task and in our study, we approach it by integrating the results of QoS and QoE assessments in two popular ways.

8.1 Arithmetic Average of Alternatives' Assessments

One option for accessible for integration of results is to calculate the weighted sum for $AA_{i(QoS)}$ and $AA_{i(QoE)}$ for each target alternative:

$$WS(AA)_i = a.AA_{i(QoS)} + (1 - a).AA_{i(QoE)} \tag{27}$$

where "$a \in (0, 1)$" is the preference weight for QoS; and "$1 - a$" - for QoE.

The values for "a" can be defined on the basis of DM or experts judgements taking into account the user's preferences for these aspects. The most popular and accessible approach is to assign equal weights for both QoS and QoE values as shown in Eq. (27) where it takes the form of arithmetic average:

$$A(AA_i) = (AA_{i(QoS)} + AA_{i(QoE)})/2 = 0,5.AA_{i(QoS)} + 0,5.AA_{i(QoE)} \tag{28}$$

However, such an approach is very simplistic and cannot take into account the users' preferences and the distribution of information between "AA_i" in the different techniques.

8.2 Product of Alternatives' Assessments

A second option for accessible integration of assessment results is to calculate the weighted product for $AA_{i(QoS)}$ and $AA_{i(QoE)}$ for each target alternative:

$$WP(AA_i) = a.AA_{i(QoS)} \times (1 - a)AA_{i(QoE)} \tag{29}$$

where "$a \in (0, 1)$" is the preference weight for QoS; and "$1 - a$" - for QoE.

As in Eq. (30) the values for "a" can be defined by DM's or experts' judgements taking into account the user's preferences for these aspects. The most popular approach is to assign equal coefficients for both QoS and QoE values as shown in Eq. (33):

Product of Alternatives Assessments:

$$P(AA_i) = AA_{i(QoS)} \times AA_{i(QoE)} \tag{30}$$

8.3 Sensitivity, Statistics and Entropy of Integrated Assessments

To improve the reliability of final assessments it is necessary to compare not only the ranks but also the values of "AA_i" and their main descriptive statistics (mean, minimal, maximal values) for QoS and QoE evaluations. The values of "a" and consequently of "$1 - a$" in Eq. (30) can be defined by DMs to reflect the preferences of users and, also, in our opinion, the reliability of information contained in QoS and QoE criteria. A specific task of the sensitivity analysis of $WP(AA_i)$ and $WS(AA_i)$ is to identify the different values of "a" as transition points where the preference is transferred from leaders in QoS to leaders in QoE.

Such an approach allows to explore with more details the influence of different techniques on the role of each alternative in the interim and the finally integrated assessments.

9 Conclusion and Further Research

In this paper, we discussed several methodology adjustments which are logically combined for improving the different algorithms of the MCDM approach for the selection of CS based on the combination of AHP(blocks) and Information Entropy methods. In the first stage, the set of evaluation criteria is logically decomposed into a reasonable number of blocks which weights are defined with a more compact AHP matrix with a reduced number of transparent comparisons. In the second stage, with a data-driven Information Entropy concept the weights of criteria within each block are objectively defined on basis of real values for quantitative and quantifiable parameters.

This hybrid approach is adapted for further application in a real case study based on QoS and QoE criteria with popular MCDM evaluation techniques like TOPSIS and MOORA. Finally, the assessments of QoS and QoE in each technique are integrated and ranked. A sensitivity analysis of final results allows to enhance the importance of objective empiric information and to flexibly adapt the AHP-block modelling to the specific needs and preferences of different users.

This comprehensive methodology concept is tested with a reliable and publicly available dataset in the case study part of this report in a selection of CS which is adapted for the needs of individual mass users and small educational institutions under conditions of social containment. The methodological advances proposed in this study are transparent and accessible having a universal character that can be applied for supporting the MCDM process not only for Cloud Services selection but also in a large number of other areas.

References

1. Mell, P., Grance, T.: The NIST definition of cloud computing. Commun. ACM **53**(6), (2011)
2. Whaiduzzaman, M., Gani, A., Anuar, N.B., Shiraz, M., Haque, M.N., Haque I.T.: Cloud service selection using multicriteria decision analysis. Sci. World J. Article ID 459375, 10 pages (2014). https://doi.org/10.1155/2014/459375
3. Zhang, M., Ranjan, R., Haller, A., Georgakopoulos, D., Strazdins, P.: Investigating decision support techniques for automating Cloud service selection. In: 4th IEEE International Conference on Cloud Computing Technology and Science Proceedings, Taipei, pp. 759–764 (2012). https://doi.org/10.1109/CloudCom.2012.6427501
4. Saroj, S., Dileep, V.K.: A review multi-criteria decision making methods for cloud service ranking. Int. J. Emerging Technol. Innov. Res. **3**(7), 92–94 (2016). http://www.jetir.org/papers/JETIR1607019.pdf
5. Boussoualim N., Aklouf Y.: An Approach based on user preferences for selecting SaaS product. In: International Conference on Multimedia Computing and Systems (ICMCS-2016), Marrakesh, Morocco, pp. 1182–1188 (2014). https://doi.org/10.1109/ICMCS.2014.6911278
6. Do Chung, B., Seo, K.-K.: A cloud service selection model based on analytic network process. Indian J. Sci. Technol. **8**(18) (2015). https://doi.org/10.17485/ijst/2015/v8i18/77721
7. Garg, S., Versteeg, S., Buyya, R.: A framework for ranking of cloud computing services. Futur. Gener. Comput. Syst. **29**(4), 1012–1023 (2013)
8. Rehman, Z.U., Hussain, O.K., Hussain, F.K.: Parallel cloud service selection and ranking based on QoS history. Int. J. Parallel Program **42**, 820–852 (2014)
9. Qu, L., Wang, Y., Orgun, M.A.: Cloud service selection based on the aggregation of user feedback and quantitative performance assessment. In: IEEE International Conference on Services Computing (2013), pp. 152–159, Santa Clara, CA, USA, (2013). https://doi.org/10.1109/SCC.2013.92
10. Zhao, L., Ren, Y., Li, M., Sakurai, K.: Service selection with user-specific QoS support in service-oriented architecture. J. Network Comput. Appl. **35**, 962–973 (2012)
11. Varia, J.: Best practices in architecting cloud applications in the AWS cloud. Cloud Comput. Principles Paradigms **18**, 457–490 (2011)
12. Gavvala, S.K., Jatoth, C., Gangadharan, G., Buyya, R.: Qos-aware cloud service composition using eagle strategy. Futur. Gener. Comput. Syst. **90**, 273–290 (2019). https://doi.org/10.1016/j.future.2018.07.062
13. Saaty, T.: Principia Mathematica Decernendi: Mathematical Principles of Decision Making. RWS Publ., Pittsburgh, Pensilvaniya (2010). ISBN 978-1-888603-10-1
14. Shannon, C.: A mathematical theory of communication. Bell Syst. Tech. J. **27**(3), 379–423 (1948)
15. Kwangsun, Y., Ching-Lai, C.: Multiple Attribute Decision Making: An Introduction. SAGE publ, California (1995)
16. Tiwari, R.K., Kumar, R.: G-TOPSIS: a cloud service selection framework using Gaussian TOPSIS for rank reversal problem. J. Supercomput. **77**(1), 523–562 (2020). https://doi.org/10.1007/s11227-020-03284-0
17. Chakraborty, S.: Applications of the MOORA method for decision making in manufacturing environment. Int. J. Adv. Manuf. Technol. **54**(9–12), 1155–1166 (2011)
18. Brauers, W.: Multi-objective seaport planning by MOORA decision making. Ann. Oper. Res. (1), 39–58 (2006)

19. Noor, T., Sheng, Q.: Credibility-based trust management for services in cloud environments. In: Conference Proceedings of Service-Oriented Computing - 2011, Berlin, pp. 328–343 (2011)
20. Wang, C., Wang, Y., Liu, C., Wang, X.: An audit-based trustworthiness verification scheme for monitoring the integrity of cloud servers. J. Comput. Inf. Syst. **10**(23), 9923–9937 (2014)
21. Filali, F., Yagoubi, B.: A general trust management framework for provider selection in cloud environment. In: Conference Proceedings of Advances in Databases and Information Systems - 2015, Poitiers, pp. 446–457 (2015)
22. Wu, L., Buyya, R.: Service Level Agreement (SLA) in utility computing systems. In: Performance and Dependability in Service Computing: Concepts, Techniques and Research Directions 1, pp. 1–25 (2010)
23. Bakanova, N., Atanasova, T.: Method for automated analysis of users' requests to service centre of information networks in OIS. Problems Eng. Cybern. Robot. **74**, 33–40 (2020)
24. Sidhu, J., Singh, S.: Improved TOPSIS method based on trust evaluation framework for determining trustworthiness of cloud service providers. J. Grid Comput. **15**(1), 81–105 (2017)
25. Papathanasiou, J., Kostoglou, V., Petkos, D.: A comparative analysis of cloud computing services using multicriteria decision analysis methodologies. Int. J. Inf. Decis. Sci. **7**(1), 23–35 (2014). https://doi.org/10.1504/IJIDS.2015.068117
26. Saaty, T.: Fundamentals of Decision Making and Priority Theory with Analytical Hierarchical Process. The AHP Series, vol. VI, RWS Publications, The University of Pittsburgh, Pittsburgh (1980)
27. Saaty, T.: Theory and Applications of the Analytic Network Process: Decision Making With Benefits, Opportunities, Costs, and Risks, 3rd edn. RWS Publications, Pittsburgh (2005)
28. Saaty, T., Peniwati, K.: Group Decision Making: Drawing out and Reconciling Differences. RWS Publications, Pittsburgh (2008). ISBN 978-1-888603-08-8
29. Borissova, D.: A group decision making model considering experts competency: an Application in personnel selections. Comptes-rendus de l'Academie Bulgare des Sciences **71**(11), 1520–1527 (2018)
30. Hussain, A., Chun, J., Khan, M.: A novel customer-centric Methodology for Optimal Service Selection (MOSS) in a cloud environment. Futur. Gener. Comput. Syst. **105**, 562–580 (2020)
31. Calpine, H.C., Golding, A.: Some properties of Pareto-Optimal Choices in Decision problems. Omega **4**(2), 141–147 (1976)
32. Jatoth, C., Gangadharan, G.R., Fiore, U., Buyya, R.: SELCLOUD: a hybrid multi-criteria decision-making model for selection of cloud services. Soft. Comput. **23**(13), 4701–4715 (2018). https://doi.org/10.1007/s00500-018-3120-2
33. Kirilov, L., Guliashki, V., Genova, K., Vassileva, M., Staykov, B.: Generalized scalarizing model GENS in DSS WebOptim. Int. J. Decis. Support Syst. Technol. (IJDSST) **5**(3), 1–11 (2013). https://doi.org/10.4018/jdsst.2013070101

Hybrid MCDM for Cloud Services: AHP(blocks) & Entropy, TOPSIS & MOORA (Case Study with QoS and QoE Criteria)

Iliyan Petrov(✉)

Bulgarian Academy of Sciences, Institute of information and communication technologies, Akad. G. Bonchev str., bl. 2, 1113 Sofia, Bulgaria
petrovindex@gmail.com, iliyan.petrov@iict.bas.bg
https://www.iict.bas.bg/ipdss/i-petrov.html

Abstract. In the methodology part of this study, we introduced a combined "AHP(blocks) and Entropy" approach for improving of Multi-Criteria Decision Making (MCDM) for Cloud Services (CS) selection. On this basis, we optimize the combination of professional experts' opinion and objectivity of entropy criteria weighting. In this second part, we test our findings on an existing universal dataset with Quality of Service (QoS) and Quality of Experience (QoE) criteria and adapting for the needs of individual users and small education organisations under conditions of COVID-19 pandemics.

Keywords: cloud computing · multi-criteria analysis · decision making · Quality of Services · Quality of Experience · TOPSIS · MOORA · AHP

1 Introduction

The MCDM selection of appropriate Cloud Services [1] has to consider diverse factors related with technical performance, pricing schemes, levels of quality, security, and technical support, etc. [2]. Basically, these characteristics can be grouped and assessed separately in two main aspects: Quality of Service (QoS) and Quality of Experience (QoE) [3,4].

One of the most challenging tasks in CS selection is the balanced aggregation of subjective expert judgements with the abundant and objective technical data. The competition of CSs increases the need to develop more accessible and effective approaches which will facilitate the understanding and integration of users and DMs in the MCDM process [5].

This study is supported by Bulgarian FNI fund, project "Modeling and Research of Intelligent Educational Systems and Sensor Networks (ISOSeM)", contract KP-06-H74-4 from 26.11.2020.

V. M. Vishnevskiy et al. (Eds.): DCCN 2021, CCIS 1552, pp. 92–110, 2022.
https://doi.org/10.1007/978-3-030-97110-6_7

In the first methodology part of our study, we introduced an original approach which purpose is to frame the subjectivity of experts with the Analytical Hierarchy Process (AHP) [6,7] by systematizing the evaluation criteria in blocks and on this ground to perform more structured criteria weighting within these blocksin combination with the data-driven approach of information entropy. Taking into account the complexity of CC technology and for obtaining more representative results we integrate this hybrid approach with two assessment techniques that are based on different, intuitive, and accessible mathematical logic, and at the same time are very popular for MCDM in large areas [8,9], including for selection of CS:

- Technique for Order of Preference by Similarity to Ideal Solution (TOPSIS) [10]
- Multi-Objective Optimization based on Ratio Analysis (MOORA) [11–13];

In this publication, we apply the hybrid "AHP(blocks) & Entropy" [14] approach in a case study based on two public data sets including criteria for the Quality of Services (QoS) and Quality of Experience (QoE). The case study follows the MCDM algorithm from data collection to final integration of QoS and QoE with detailed analysis of interim and final results.

The purpose of our methodological findings and practical experiments is to improve the accessibility and efficiency of the AHP and entropy approaches for providing flexible, transparent, and intuitive mechanisms, applicable for selections in wide areas by a large number of users in the Information Technologies and other sectors.

The rest of this paper has the following structure: Sect. 2 - Case study modelling for MCDM; Sect. 3 - Objective criteria weights: "real" v/s "differed" entropy; Sect. 4 - Subjective criteria weights: "classic AHP" & "AHP in blocks"; Sect. 5 - CS selection with QoS and QoE criteria; Sect. 6 - CS comparative evaluation: TOPSIS and MOORA; Sect. 7 - Integration of QoS and QoE assessments; Sect. 8 - Conclusions and future research; 9 - Acknowledgements; References.

2 Case Study Modelling for MCDM

In the Information and Communication Technologies (ICT) sector the overall design, technology performance and quality assurance are combined in the so-called "Quality of Service (QoS)" concept [15] which is composed of three components: Service Strategy; Service performance; Customer Results. In a narrower context, QoS contains mainly "Service performance" parameters reported by Cloud Service Providers (CSPs), and on this basis, they can be considered as "supply-side" criteria. On the other "demand-side" are the parameters linked to the results obtained by users, or in other words, QoE reflects the "degree of delight or annoyance of users of an application or service resulting from the fulfilment of their expectations in the light of their needs and preferences" [16,17]. The selection of CS includes the main steps of the MCDM process with adaptation to the specifics of the CC sector.

2.1 Step 1 - Decision-Makers and Experts

In this case study for simplification purposes, we presume that the evaluation is based on individual or already integrated group judgement. This paper does not engage in the organizational specifics of an individual or group decisions, which are well discussed and developed in several publications [18, 19] .

2.2 Step 2 - Target Alternatives

Our experiment includes nine alternatives of popular CSs with different technical and cost parameters. To avoid brand sympathies they are masked as $A_1, A_2, ..., A_9$ without indication for suppliers packages or names. All alternatives have universal profiles and can be used for different applications depending on the needs of users. The relatively short list of these target alternatives is a result of a preliminary selection in the a.m. study based on the concept of Pareto optimization in a non-dominated solution [20], which creates a basis for strong competition among them during the MCDM process.

2.3 Step 3 - Evaluation Criteria

The profiles of CS alternatives A_i define the types of criteria to be used for their valuation. For QoS criteria, the dataset contains Service Performance Statistics (SPS) reported by CSPs or monitored by third parties. For QoE criteria, the original dataset includes Customer Feedback Statistics (CFS) which is reported to contain feedback assessments from real cloud customers provided in a recent and comprehensive study by A. Hussain & et al. [21]. In our case study, the original data set was rearranged by dividing the criteria into logically structured blocks for addressing the specific needs of individual users and small educational institutions for distance learning and home working under conditions of social containment related to COVID-19.

Step 3.1. MCDA is usually based on one set of criteria based on technical, economic, and other parameters. The assessment criteria can be based on real parameters or transformed benchmarks values under the form of quantitative measurements or verbal qualitative judgements. In this study, we apply the 2 sets used in [21] containing 8 quantitative criteria for QoS and QoE as shown in Table 1.

Step 3.2. The classification of "Benefit (Profit)" and "Non-Benefit (Cost)" criteria depends on how they characterize the contribution or saving of resources. The optimization for Benefit (Profit) criteria is expressed in maximal values; for Non-Benefit (Cost) criteria - in minimal values.

2.4 Step 4 - Collection and Normalization of Data

The collection and preliminary normalization of primary data includes the following steps:

Table 1. QoS and QoE evaluation criteria for CS selection; B - Benefit; N-B - Non-Benefit; SLA - Service Level Agreement

QoS - Service Criteria	Type	Symbol	QoE - Experience Criteria	Type	Symbol
CPU Speed (GHz)	B	SC1	Response time	Bench (B)	EC1
Core Concurrency	B	SC2	Accessability	Bench (B)	EC2
RAM bandwidth (Gb/s)	B	SC3	Features	Bench (B)	EC3
Disk rate (Mb/s)	B	SC4	Ease of Use	Bench (B)	EC4
Disk seeks (ops/s)	B	SC5	Tech. support	Bench (B)	EC5
Latency (ms)	N-B	SC6	Customer service	Bench (B)	EC6
Availability (SLA%)	B	SC7	Security	Bench (B)	EC7
Price (cent/h)	N-B	SC8	Trust	Bench (B)	EC8

Step 4.1. Construction of Basic Decision Matrix (BDM) comprising "m" target alternatives (A_i) which are evaluated with a set of "n" criteria C_j (in our case "Service Criteria SC_j" and "Experience Criteria EC_j").

The primary data for 6 "Benefit" and 2 "Non-Benefit" QoS criteria according the to original study [21] are displayed in Table 2. The primary values of the QoE criteria are expressed in the format of benchmarks as 8 Benefit criteria according the to same original study, as displayed in Table 3.

Table 2. Basic Decision Matrix with Quality of Service (QoS) criteria

Alternatives (A_i)/QoS Criteria (SC_j)	SC1	SC2	SC3	SC4	SC5	SC6	SC7	SC8
Types of QoS Criteria (SC_j)	B	B	B	B	B	N-B	B	N-B
A1	2,16	9,3	6,8	512	135	44	99,96%	96
A2	2,13	16	6,8	512	135	47	99,95%	144
A3	1,67	7,4	5,8	538	1690	42	99,97%	23,8
A4	2,57	7,9	6,3	36	53	45	99,95%	62,4
A5	2,40	3,9	5,9	36	53	47	99,98%	31,2
A6	2,04	5,6	7,4	321	1797	47	99,99%	136
A7	2,06	3,3	7,6	321	1797	46	99,97%	68
A8	2,38	5,3	9,6	99	130	53	99,95%	56
A9	2,46	3,0	9,6	99	130	53	99,95%	28

Table 3. Basic Decision Matrix with Quality of Experience (QoE) criteria

Alternative (A_i)/QoE Criteria (EC_j)	EC1	EC2	EC3	EC4	EC5	EC6	EC7	EC8
Types of QoE Criteria (EC_j)	B	B	B	B	B	B	B	B
A1	3,05	3,20	2,56	3,62	1,60	3,73	2,25	2,61
A2	3,70	3,11	3,10	3,15	3,38	2,83	3,16	3,53
A3	3,74	3,05	3,00	2,80	3,37	2,66	3,99	3,44
A4	3,61	3,24	2,82	3,03	3,35	2,67	4,17	3,46
A5	4,67	4,02	2,51	3,70	4,70	3,06	4,27	4,57
A5	4,67	4,02	2,51	3,70	4,70	3,06	4,27	4,57
A6	4,58	4,18	2,30	4,45	4,60	2,60	4,30	4,56
A7	3,95	3,44	2,59	3,39	3,75	2,63	3,76	3,84
A8	3,64	3,13	2,51	3,23	2,73	2,54	2,83	3,32
A9	4,08	3,53	3,37	2,98	2,71	2,85	3,53	2,77

As QoS and QoE are expressed in different units and scales, they are normalized under the format of dimensionless values in the interval (0, 1), which are more suitable for further use and comparisons. Depending on the type of further evaluation the normalization can be achieved in several ways.

Step 4.2. Linear normalisation by calculating the proportions of relative weights "p_{ij}" for each parameter value (x_{ij} with $\Sigma_{i=1}^{n} x_{ij} = 1$) in all criteria "$C_j$" and formation of the Normalized Decision Matrix (NDM):

$$p_{ij} = x_{ij} / \Sigma_{i=1}^{m} x_{ij} \tag{1}$$

The outputs of this normalization are used as inputs in other non-linear normalizations and in the Information Entropy transformation in Step 5.

Step 4.3. Constructing the Normalized Performance Matrices $NPM_{m \times n} = [r_{ij}]_{m \times n}$ which will be used for obtaining a quadratic sum adjusted performance "ratings (ratios) r_{ij}" in $NPM_{m \times n} = [r_{ij}]_{m \times n}$, which is used in the TOPSIS and MOORA valuations:

$$r_{ij} = p_{ij} / \sqrt{(\Sigma_{i=1}^{m} (p_{ij})^2)} \tag{2}$$

The normalization aspect should not be underestimated, since different formats produce different values which contribute to the differentiation of results in different evaluation techniques.

3 Objective Criteria Weights: Real v/s Differed Entropy

From methodology point of view one of the main purposes of this paper is to compare the role and influence of "real entropy criteria weights $recw_j$" and "differed entropy criteria weights $decw_j$" in the selection of CS.

The main advantage of real parameters criteria is their objectivity. For that reason our MCDA starts with criteria weighting based on "real entropy" as a first approximation and reference stage for further stand-alone or combined evaluation methods. In other words, the classical "real entropy" allows to the primary data to act as objective factors for defining the weights of criteria.

3.1 Real and "Differed" Entropy in the Weights QoS Criteria

The entropy weights of QoS criteria for both "real entropy $recw_j$" and "differed entropy $decw_j$" are displayed in Table 4.

Table 4. QoS criteria weights in real and differed Shannon Entropy (SE)

QoS Criteria	CPU speed SC1	Cores SC2	RAM SC3	Disk rate SC4	Disk seeks SC5	Latency SC6	Availbility SC7	Cost SC8
$SE(recw_j)$	0,138	0,129	0,137	0,098	0,095	0,138	0,138	0,127
$SE(decw_j)$	0,004	0,083	0,009	0,382	0,417	0,002	0,00001	0,103

The distribution of information is more balanced among "real entropy criteria weights $recw_j$" indicating for active competition interactions between the CSPs and their search to configure well structured and effective cloud instances and services. Also, the "real entropy" approach is much more logical than the plain simplicity in the popular but artificial allocation of equal weights for all criteria.

The variant with "differed entropy" results in drastic perturbation of weights. Nearly 80% of genuine information is expropriated from the majority of key criteria for being granted to only two "super-champions" parameters - Disk rate (38,2%) and Disk seeks (41,7%). The peculiar logic of such treatments is that they enhance the importance of criteria with the lowest levels of nominal and normalized entropy ($SEnom_j$ and $SEnorm_j$) - in other words, for those which were initially better placed to capture the difference between the alternatives without any additional manipulation. Not surprisingly, the most deprived after such "reallocation of information" are the parameters and components with a key role in the system and the areas with the most aggressive competition among suppliers. In our case these losers are: CPU speed (0,2%), RAM (0,9%), Latency (0,2%), and Availability (0.001%). Such neglecting and discriminative attitude leads in practice to the elimination of these criteria, as their combined "differed importance" is reduced according to artificial "differed reality" to only 1,5% - a value comparable with the usually acceptable level for a statistical mistake.

3.2 Real and "Differed" Entropy in the Weights of QoE Criteria

The situation is very similar for entropy weighting of QoE criteria - the results for real ($recw_j$) and differed entropy ($decw_j$) are displayed in Table 5.

Table 5. QoE criteria weights in real and differed Shannon Entropy (SE).

QoE Criteria	Resp. time EC1	Access EC2	Features EC3	Usability EC4	T. Support EC5	C. Service EC6	Security EC7	Trust EC8
$SE(recw_j)$	0,126	0,127	0,127	0,118	0,125	0,126	0,126	0,126
$SE(decw_j)$	0,028	0,023	0,026	0,616	0,147	0,024	0,071	0,061

Being based on the primary data the real entropy weights for QoE criteria are more evenly distributed among all criteria than in the case of QoS reflecting a stronger competition among the CS alternatives from the point of view of the users (demand side). As expected, here also the "differed entropy" perturbs the importances and creates with even less reason one "super-hero" criterion - Usability (61,6%). At the same time, the majority of other key criteria are practically neglected - Response time (2,8%), Accessibility (2,3%), Features (2,6%), Customer Service (2,4%). After a similar transformation, the link with the real world is lost, and the MCDM process risks to be corrupted and to produce misleading results at a stage of final evaluations.

A logical question in this situation is: *"What to do?"*. In other words, where and how to find parameters and criteria with lower entropy - in fact, with lower competition between alternatives. Even if such abstraction is possible, it would logically lead to the practical neglecting of criteria with higher levels of entropy. If the "differed entropy" is preferred, the DMs and experts should consider what is the real content in the values of the differed entropy criteria weights "$decw_j$", and how to ensure the objectivity of the MCDM. In our opinion, the idea of "the extracting additional information" from relative weights is illusive and non-justified. The decision-makers, experts, and scholars are not in a position to create, produce or invent objective information based on simplistic arithmetic operations with entropy values. Criteria should be neglected even if they have a high level of entropy, except in the case when all alternatives have equal parameter values for some criteria. Nevertheless, in our opinion, the entropy approach is a very robust method for objective treatment of data and in our studies, the "real entropy" variant will be preferred for defining criteria weights.

4 Subjective Criteria Weights: "Classic AHP" & "AHP in Blocks"

One of the most popular approaches to defining criteria weights is AHP [22] and this is due to the fact that it provides a transparent structure with intuitive understanding and accessible tools for quantification of preferences in comparisons between criteria. The traditional AHP is a convenient option for selection cases with a limited number of criteria, alternatives, and experts. However, its application in cases with more experts and criteria results in the exponential increase of computing efforts and inconsistencies between pair-comparisons.

To solve this problem this study divides the evaluation criteria in logically structured blocks (groups). The calculation of the hierarchy importance of blocks is defined on the same principles as in the standard AHP pair-comparisons whose total number is defined as $n(n-1)/2$. As shown in Table 4, the grouping of 8 criteria in 4 blocks reduces the number of comparisons made by one expert or DM more than 4,67 times (from 28 to 6, or 28/6 = 4,67), representing an important improvement in the MCDM process (Table 6).

Table 6. Pair-comparisons by one Decision Maker: AHP (classic) v/s AHP(blocks)

	AHP (classic)	AHP (blocks)
Number of compared items (blocks or criteria): n	8	4
Number of AHP comparisons: $n(n-1)/2$	28	6

The logical structuring of criteria in blocks allows to fully exploit other important advantages of AHP: a) to model flexibly experts opinions by adjusting the hierarchy preference "*hp*" values according to experts and users preferences; b) to check the consistency of $HPM_{b \times b}$ for avoiding the serious eventual controversies in the judgements of different experts about the importance of criteria. Theoretically, the procedure of "block grouping" can be performed at one or several levels, taking into account that often more criteria or alternatives have be to considered in practice. For example, technical criteria may be divided into several sub-blocks - for hardware components, this could be CPU, RAM memory, disk storage where each sub-block can contain a reasonable number of parameters. Moreover, in such a case the experts will be facilitated to express opinions only in the areas of their specialisation.

5 CS Selection with QoS and QoE Criteria

Only few years ago the Cloud Computing (CC) business model was oriented mainly for large business and public organizations. Since 2019, under the conditions of limited social mobility related to COVID-19, the CC sector is rapidly becoming a key factor in an increasing number of public and private activities and users. In such a situation, the selection of CSs becomes more dependant on the needs of users that have different profiles - from large public and business organizations (ministries, banks, health care organization, education institutions, social services) to individual mass users (families, employees, patients, students).

CC is a capital-intensive sector based on sophisticated equipment and software which functioning is characterized by tens and hundreds of parameters in the so-called "Quality of Service" concept. The technical data announced on the supply-side from CSPs are abundant and even in normal conditions prevails in the multi-criteria analysis for selection of CS. On the demand side, the satisfaction of users about the results from CS is measured in the Quality of Experience (QoE) context.

5.1 Cloud Services Selection with QoS Criteria

To choose the appropriate configuration of CC, the users need to clarify their profiles as consumers of computing resources and services in relation to their activities and applications. It is important to understand basically how different technical components and parameters (CPU, RAM, Disk storage, Network) affect the so-called "concurrency" and "connections" in the cloud information processes. Connectivity assesses the number of requests for establishing new connections, that can be handled (per second) by a cloud server. Within HyperText Transfer Protocol (HTTP) the data transfers include the Transmission Control Protocol (TCP), which requires the so-called "three-step handshake" to be successfully established across the network between the users and the server. Such data exchanges require adequate amounts of processing capacities that depend on CPU and network properties. In brief, the network speed impacts the data flows, and CPU speed impacts the processing of these data. The constraints of these resources reduce the number of connections per second that can be established by the CS instance. Further, after a successful TCP "three-step handshake" is established this request becomes an active connection and enters in the contingent of concurrently supported client-server links. In other words, the "concurrency" is a measure for the number of connections that can be simultaneously maintained by a "loud instance", and at that stage, the storing of active connections in its state table depends on the volume and parameters of RAM available in the CS. Indeed, the connections are established with the aim to access, treat and store data, which depend on the volume and properties of the different disk storage components. To summarize, different components have different roles in CS: connectivity depends on CPU and network; concurrency - on RAM; data storage - on different disks parameters.

The social containments under conditions of COVID-19 accelerated the trend for more active utilization of CSs in education sector for "distant learning" and "home working". In the new situation, different users have various views and *rationale* for optimizing their computing resources. For example, in the education sector several players interact with different roles, needs and financial possibilities - central institutions, large universities, specialized institutes, high, middle and primary schools, teaching personnel, students from different schools and ages. The interpretation of needs and preference of the large variety of group and individual users in a challenging task [23]. In this situation, we present an attempt for generalisation of preferences of different users for the main aspects of CS performance in the education sector (Table 7).

Table 7. Users priorities in education for QoS in CS.

Users	Computing	Storage	Networking	Cost	Cost payer	Scale economies
Premium	high-mid	high-mid	high-mid	low	End-user - indirect	high
Middle	high-mid	high-mid	mid-low	low	End-user - indirect	high-mid
Small	mid-low	mid-low	mid-low	mid-low	End-user - (in)direct	mid-low
Mass end-user	low	mid-low	mid-low	mid-low	End user - direct	low

For demonstration purposes, the criteria in the QoS set are split into 4 blocks: Computing capacity, Storage space, Accessibility, and Costs. In Table 8 is presented the variant of AHP(blocks) weighting based on the need for mass users and small and mid-size educational institution who's priorities are related with reliable storage and accessibility for a predictable and limited number of clients.

Table 8. QoS criteria weighting in AHP(blocks): Hierarchy reference Matrix (HPM)

QoS Blocks	Computing	Storage	Accessibility	Cost	$\sqrt[b]{\Pi_{e=1}^{b} hp_{ee})}$	Block weights (bw)
Computing	1	1/3	1/2	1/2	0,537	0,123
Storage	3	1	1	2	1,861	0,359
Accessibility	2	1	1	2	1,189	0,325
Cost	2	1/2	1/2	1	0,841	0,193
Consistency	0,017				4,357	1

An indication for the possibilities of the AHP(blocks) method and the quality of its application in this research is the very low value of Consistency Ratio (0,017), while the maximal reference value for acceptance is 0,1. On the basis of block weights, the entropy criteria weights within the blocks in the case of QoS are displayed in Table 9 in the variants of "real entropy criteria weights $recw_j$" and "differed entropy criteria weights $decw_j$"

Table 9. QoS criteria weights: AHP(blocks) with real and differed entropy

QoS Blocks	Computing			Storage		Accessibility		Cost
Block weights (bw)	0.123			0,359		0,325		0,193
Methods/Criteria	SC1	SC2	SC3	EC4	SC5	SC6	SC7	SC8
$SEnom_j$	3,159	2,971	3,148	2,26	2,173	3,166	3,169	2,922
SEmax(9)	3,17	3,17	3,17	3,17	3,17	3,17	3,17	3,17
$SEnorm_j$	0,997	0,937	0,993	0,712	0,686	0,999	0,999	0,922
$SE(rew_j)$	0,138	0,129	0,137	0,098	0,095	0,138	0,138	0,127
$d_j = 1 - SEnorm_j$	0,003	0,063	0,007	0,288	0,314	0,001	0,0001	0,078
$SE(dew_j)$	0,004	0,083	0,009	0,382	0,417	0,002	0,00001	0,103
$AHP(block)\&SE(recw_j)$	0,042	0,039	0,042	0,200	0,15	0,162	0,162	0,193
$AHP(block)\&SE(decw_j)$	0,006	0,106	0,012	0,109	0,249	0,3249	0,0001	0,193

We can observe, that based on the AHP(blocks) weights "bw" each block is allocated a fixed "quot" of importance. Logically, the blocks which contain only one criterion do not participate in further Entropy transformation for calculating their weights, as their importance is already defined. Due to that fact, the objective entropy weighting is framed on the level of each block and the distribution of importances does not affect the criteria from other blocks. Logically, this contingency approach allows reducing an eventual drastic reallocation of weights in the variant of “differed entropy”.

5.2 Cloud Services Selection with QoE Criteria

CS can also be assessed with the so-called QoE criteria which rely on feedback from users and reflect the demand-side view on performance. Their role is important if and when their statistics are constantly available, reliable, and representative for a large number of real clients. At the same time, the QoE ismore challenging to measure than QoS, as it includes not only technical but also psychology and behaviour aspects. Very often the QoE data is not transparent about the profile of responding users.

Taking into account all the above-mentioned considerations, QoE can still be useful for small companies and mass-users, who usually do not have enough in-house knowledge to monitor and assess the combination of sophisticated technologies and conflicting criteria. In this situation, the need for more intuitive indicators that measure the satisfaction from using CS leads to framing the QoE criteria in terms of benchmarks without specific units and without the complication related with “Benefit” or “Non-Benefit” parameters. In practice benchmarks can be considered to have a “Benefit” profile - for users higher benchmark values reflect better results and higher satisfaction.

For demonstration purposes, in our study, the set of eight QoE criteria from the original study [21] is split into 4 blocks each containing two criteria and their weights are computed on the hierarchy preferences values *hp* defined by the author (Table 10). As in the case of QoS, they reflect the preferences of an individual client (family, student)and small education organizations (local schools with a smaller number of personnel and pupils). Such users are less experienced in IT, and for that reason, they are more interested in “Functionality” and “Reliability” aspects for their routine day-to-day use. In this simulation, “Security” and “Accessibility” are not considered as very top issues for education applications and programs, as it would be the case be for sensitive private or financial data in health, banking, and insurance services. As in the case of QoS, an indication for the possibilities of the AHP(blocks) method and the quality of its application in this research is the very low value of Consistency Ratio (0,022).

Table 10. QoE criteria weighting in AHP(blocks): Hierarchy reference Matrix (HPM)

QoE Blocks	Accessibility	Functionality	Reliability	Security	$\sqrt[b]{\Pi_{e=1}^{b} h_{ee})}$	Weights
Accessibility	1	1/2	1/2	1/2	0,595	0,142
Functionality	2	1	1	2	1,414	0,337
Reliability	2	1	1	1	1,189	0,283
Security	2	1/2	1	1	1	0,238
Consistency	**0,022**			Σ	**4,198**	**1**

The results of the Entropy weighting of criteria within the blocks are displayed in Table 11 in the variants of "real entropy criteria weights $recw_j$" and "differed entropy criteria weights $decw_j$". The internal criteria weights (obtained objectively with the real and differed entropy approaches) are adjusted with the AHP(blocks) weights to participate adequately in the evaluation of alternatives.

Table 11. QoE criteria weights: AHP(blocks) with real and differed entropy

QoE Blocks	Accessability		Functionality		Relibility		Security	
Block weights *(bw)*	0.142		0,337		0,283		0,238	
Methods/Criteria	EC1	EC2	EC 3	EC4	EC5	EC6	EC7	EC8
$AHP(block)\&SE(recw_j)$	0,071	0,071	0,175	0,162	0,141	0,143	0,119	0,119
$AHP(block)\&SE(decw_j)$	0,078	0,064	0,014	0,323	0,239	0,044	0,128	0,111

As in the case for QoS, the preliminary AHP(blocks) weighting frames the distribution of preferences and each block is allocated a fixed "quota" of importance. Here also the "differed entropy" inevitably provokes perturbations and creates two "differed leaders" - "Ease of use" (EC4 = 32,3%) in the "Functionality" block and Technical support (EC5 = 23,9% in the "Reliability" block. As a result the weight of the other criteria in different blocks are reduced from 17 for EC3 ("Features") to 4 times for EC6 ("Customer service"). The weights of the other two blocks ("Accessibility" and "Security") remain intact. Such a hybrid approach has several advantages compared to the classical AHP and stand-alone Entropy methods:

- the subjective experts' judgements and the data-driven entropy analysis integrate into a balanced and flexible combination the professionalism of experts and the objectivity of real data;
- the criteria and experts' opinions from different blocks do not compete with each other;
- the transparency of the MCDM process is increased, which is particularly important for the mass users of CS under conditions of social containment (COVID-19);
- the eventual perturbations due to the application of "differed entropy" are limited in scale and remain framed within the weights of respective blocks.

6 CS Comparative Evaluation: TOPSIS and MOORA

After the routine procedures of data collection and normalization the assessment of the nine target alternatives is performed with two renown techniques (TOPSIS and MOORA) in 4 variants of criteria weights: 1) stand-alone (traditional) real entropy; 2) stand-alone (traditional) differed entropy; 3) combined AHP(blocks) & real entropy; and 4) combined AHP(blocks) & differed entropy. For comparison purposes, the 8 selections for QoS and QoE are divided into 2 main groups containing 4 selections with "real entropy weights $recw_j$" and 4 selections with "differed entropy weights $decw_j$".

6.1 Evaluation and Ranking Based on QoS Criteria

The final ranking results of the nine CS alternatives based on QoS criteria are displayed in Table 12 showing that both the winners and losers are clearly defined. In the variant with real weights The top-3 best alternatives are ranked as "$A_3 \succ A_7 \succ A_6$. The 3 worst alternatives are ranked as "$A_8 \succ A_5 \succ A_4''$.

Table 12. CS ranking results based on QoS criteria.

Methods/Alternative (A_i)	A1	A2	A3	A4	A5	A6	A7	A8	A9
$SErecw_j - TOPSIS$	4	2	1	9	8	5	3	7	6
$AHPblock\&SErecw_j - TOPSIS$	4	7	1	9	6	3	2	8	5
$SErecw_j - MOORA$	4	3	1	8	9	5	2	7	6
$AHPblock\&SErecw_j - MOORA$	4	5	1	9	7	3	2	8	6
$SEdecw_j - TOPSIS$	5	4	3	9	7	2	1	8	6
$AHPblock\&SEdecw_j - TOPSIS$	6	7	1	9	5	3	2	8	4
$SEdecw_j - MOORA$	5	4	1	8	9	3	2	7	6
$AHPblock\&SEdew_j - MOORA$	4	5	1	8	6	3	2	9	7

The perturbation of results provoked by the "differed entropy criteria weights $decw_j$" are more significant in the variant of TOPSIS evaluation (due to Euclidean distance approach) and less noticeable in the MOORA variant. These results indicate also that the allocation of block weights in the variant of "AHP(blocks)" limits significantly the perturbations in the variant of "differed entropy".

6.2 Evaluation and Ranking Based on QoE Criteria

The rankings of alternatives based on QoE criteria are displayed in Table 13. In the variants of real entropy criteria weights "$recw_j$" they are the same in TOPSIS and MOORA for the "top-3" leaders ($A_5 \succ A_6 \succ A_7$). The "bottom-3" losers ($A_3 \succ A_1 \succ A_8$) are defined with small differences.

The perturbations of results in TOPSIS and MOORA provoked by the differed entropy criteria weights "$decw_j$" are very noticeable for the two best alternatives and especially in TOPSIS. Logically, the perturbations are more expressed in the stand-alone "differed entropy", while the variant AHP(blocks) substantially limits such perturbations which is especially noticeable in the case of "$decw_j$ with TOPSIS evaluation.

Table 13. CS ranking results based on QoE criteria.

Methods/Alternative (A_i)	A1	A2	A3	A4	A5	A6	A7	A8	A9
$SErecw_j - TOPSIS$	8	5	9	4	1	2	3	7	6
$AHPblock\&SErecw_j - TOPSIS$	8	4	9	5	1	2	3	7	6
$SErecw_j - MOORA$	9	5	7	4	1	2	3	8	6
$AHPblock\&SErecw_j - MOORA$	8	4	9	5	1	2	3	7	6
$SEdecw_j - TOPSIS$	9	6	1	5	2	3	4	7	8
$AHPblock\&SEdecw_j - TOPSIS$	8	4	9	5	2	1	3	7	6
$SEdecw_j - MOORA$	6	4	9	5	2	1	3	7	8
$AHPblock\&SEdecw_j - MOORA$	8	5	9	4	2	1	3	7	6

7 Integration of QoS and QoE Assessments

The comparative assessments in Sect. 6 confirmed that the combined "AHP(blocks) & SE" approach is able to produce more structured results for each of the sets of QoS and QoE criteria. Not surprisingly, however, the comparison between QoS and QoE evaluations shows differences that may create ambiguity and uncertainty for DMs and users. In the variants with "real entropy criteria weights $recw_j$" the evaluations and rankings are more perturbed for QoS in comparison with QoE. This is due to fact that the primary data in the QoS criteria reflects real parameters which are more quantitatively diversified within the alternatives. To resolve this problem we can perform an additional weighted integration for the results of the *Assessments of Alternatives* $AA_{i(QoS)}$ and $AA_{i(QoE)}$ in the different evaluation techniques:

$$CAA_i = a.AA_{i(QoS)} + (1 - a).AA_{i(QoE)} \quad (3)$$

where "$a \in (0, 1)$" is the preference weight for QoS; and "$1 - a$" - for QoE.

The value of "a" and consequently of "$1 - a$" can be assigned by experts or DMs depending on the preferences of users and, also, in our opinion, on the reliability of information contained in QoS and QoE criteria. The most common and accessible approach would be to assign equal weights for both QoS and QoE ranking values [21]. However, such an approach in some case would look as simplistic and could not take into account the distribution of information between "AA_i" in the different techniques and the preferences of users.

7.1 Comparative Statistics and Entropy of QoS and QoE Assessments

For more reliable results this study compares not only the ranks but also the real values of "AA_i" and their main descriptive statistics (mean, minimal, maximal values) for QoS and QoE evaluations. Such an approach allows exploring with more details the influence of different techniques on the behaviour of each alternative in both QoS and QoS and, subsequently, in the final integrated results. The results of QoS and QoE assessments and their descriptive statistics are presented in Table 14.

Table 14. Descriptive statistics and entropy of evaluation in TOPSIS and MOORA.

	TOPSIS			MOORA		
Alternative	QoS	QoE	*QoS/QoE*	QoS	QoE	*QoS/QoE*
A1	0,427	0,483	*0,885*	0,079	0,288	*0,274*
A2	0,371	0,591	*0,628*	0,050	0,324	*0,153*
A3	0,895	0,336	*2,665*	0,215	0,285	*0,753*
A4	0,314	0,578	*0,544*	0,006	0,324	*0,017*
A5	0,554	0,753	*0,514*	0,019	0,378	*0,1051*
A6	0,707	0,736	*0,752*	0,088	0,377	*0,233*
A7	0,347	0,631	*1,121*	0,138	0,333	*0,415*
A8	0,412	0,495	*0,700*	0,018	0,294	*0,062*
A9	0,895	0,532	*0,774*	0,036	0,320	*0,112*
Mean value	0,490	0,570	*0,860*	0,081	0,325	*0,222*
Max. value	0,895	0,753	*1,188*	0,193	0,378	*0,568*
Min. value	0,314	0,336	*0,937*	0,037	0,285	*0,019*
Max/Min	0,581	0,417	*1,39*	0,209	0,093	*2.225*
SEnom	3,08	3,14	*0,982*	2,633	3,163	*0,833*

In the case of TOPSIS the arithmetic mean for QoS is 0,86 times lower than for QoE, while the maximum value in QoS is 1,188 times higher than in QoE and the ratio of "Max/Min" in QoS is 1,39. This reflects a more intense competition of alternatives in the context of QoE, or, in other words, in the context of QoS the alternatives are more differentiated than in QoE. These observations are clearly confirmed by the higher level of nominal Shannon Entropy "$SEnom$" for the set of QoE (3,14) than QoS (3,08). Such distribution of information is bound to produce more competition at the stage of final integrated results in the field of QoE.

In the case of MOORA, the results are similar to TOPSIS, but the arithmetic mean for QoS is more than 4 times smaller than for QoE, and the maximum value of QoS is 0,568 times lower than in QoE. The ratio of "Max/Min'' is 2.225. As in the case of TOPSIS, these values reflect a more intense competition of alternatives in QoE which is also confirmed by the higher level of nominal Shannon Entropy "$SEnom''$ for the set of QoE (3,163) than in QoS (2,63). In this case MOORA produces more differentiated entropy values between QoS and QoE than TOPSIS which will be reflected in the integration of the final results.

7.2 Ranking and Sensitivity of Integrated Assessments

In this study, we explore not only the final integrated assessment of alternatives CAA_i and their ranking but also the exact transition value of "a" where the preference in the final integrated results as per eq. (14) is transferred from the leaders in QoS to leaders in QoE. The sensitivity analysis of this aspect is very important to reveal the behaviour of integrated results depending on the users' preferences between QoS and QoE and to explore them in different assessment techniques.

The integrated assessments of QoS and QoE criteria for TOPSIS are presented in Table 15 and for MOORA - in Table 16.

Table 15. TOPSIS integrated assessments of QoS and QoE criteria.

Alternative	QoS value	QoE value	QoS: QoE 25%:75%	TOPSIS Rank	QoS: QoE 56%:44%	TOPSIS Rank	QoS: QoE 75%:25%	TOPSIS Rank
A1	0,427	0,483	0,469	8	0,455	7	0,434	4
A2	0,379	0,591	0,536	4	0,481	5	0,399	7
A3	0,885	0,336	0,475	7	0,615	3	0,825	1
A4	0,314	0,578	0,512	5	0,446	8	0,347	9
A5	0,387	0,753	0,661	2	0,570	4	0,433	5
A6	0,554	0,736	0,690	1	0,645	2	0,576	3
A7	0,707	0,631	0,650	3	0,669	1	0,698	2
A8	0,347	0,495	0,458	9	0,421	9	0,365	8
A9	0,412	0,532	0,502	6	0,472	6	0,427	6

In TOPSIS the selecting of the leader is more sensitive on the proportions of QoS and QoE weights in the integrated assessment, while in MOORA the same task is less sensitive. In both TOPSIS and MOORA the sensitivity analysis of integrated results reflects the stronger competition and higher entropy in QoE compared to QoS criteria.

7.3 Discussion of Integrated Results

The final integration of assessments is another helpful instrument for providing more flexibility and considering the importance of both supply-side (QoS) and demand-side (QoE) criteria in the complex selection of SC.

Table 16. MOORA integrated assessments of QoS and QoE criteria.

Alternative	QoS value	QoE value	QoS/QoE 2%/98%	MOORA Rank	QoS/QoE 44%/56%	MOORA Rank	QoS/QoE 75%/25%	MORA Rank
A1	0,079	0,288	0,236	8	0,184	6	0,137	4
A2	0,050	0,324	0,256	5	0,187	5	0,120	5
A3	0,215	0,285	0,268	4	0,250	1	0,239	1
A4	0,006	0,324	0,244	4	0,165	8	0,085	9
A5	0,019	0,378	0,288	2	0,199	4	0,109	6
A6	0,088	0,377	0,305	1	0,232	3	0,163	3
A7	0,138	0,333	0,284	3	0,235	2	0,191	2
A8	0,018	0,294	0,225	9	0,156	9	0,088	8
A9	0,036	0,320	0,249	6	0,178	7	0,108	7

The results show a reasonable level of consensus between TOPSIS and MOORA for the top-3 alternatives despite the different logic in their algorithms. The real configuration of competing positions has a profile defined by primary data, which is "translated" similarly but still not identically by different techniques. TOPSIS is based on measuring Euclidean distances from ideal solutions, while MOORA is using a more simple additive method. in this case TOPSIS as a more structured approach produces smaller ratio of "Max/Min" and differences in the entropy values between QoS and QoE than the additive logic of MOORA.

When using more than one evaluation technique the reaching of a full consensus is practically impossible but the results can be used for cross-validating and increasing the level of confidence in the MCDM process. In this specific case, the integrated results of TOPSIS and MOORA are similar and without major critical controversies.

It is important to note that these results were obtained on the basis of the hybrid and novel "AHP (blocks) & real entropy" hybrid approach. The results based on the "differed entropy" approach reflect less logic in the evaluation and more perturbations in the ranking of alternatives. The sensitivity analysis indicates that it would be more appropriate for the experts to add efforts for clarifying the real needs of users than on creating differed realities. This sensitivity analysis also confirms the view that relying on the simplistic approach of equal distribution of weights in the MDCM (as in this case the "50/50" equalisation of weights between Qos/QoE) should not be considered as reliable and in many situations can be misleading for the users. For that reason, the DMs should define clearly their subjective preference for QoS or QoE in the selection of CSs.

8 Conclusions and Future Research

This paper presented the practical application of several methodology advances and adjustments in the MCDM for the selection of Cloud Services on the basis of QoS and QoE criteria. The logically structured "AHP(blockS)" approach allows to systematize and frame more effectively the subjective opinions of professional experts for combining them with the very robust and data-driven entropy method for objective criteria weighing. On this basis the analysis of the role of real and differed entropy demonstrated the advantages of applied studies based on real data and parameters.

The methodological findings were tested in a comprehensive case study for selecting CS based on a recent and publicly available datasets with QoS and QoE criteria. Thanks to the flexibility of the "AHP(blocks) & Entropy" approach the original datasets were arranged in blocks and adapted in a simulation for reflecting the needs and preferences of individual users and small educational organizations in the field of distance learning and home working for students and teachers under the conditions of COVID-19.

The evaluations with TOPSIS and MOORA techniques produced similar assessments and good basis for consistent integration of the results based on QoS and QoE criteria. The sensitivity analysis allowed to explore the evolution of final integrated ratings depending on the preferences of users and decision makers.

Future research may include further improvements and testing of the hybrid "AHP(bocks) & Entropy" approach with other evaluations techniques for improving the MCDM. Taking into account the sensitivity of the issue about the role real and differed entropy it is logical to continue the explore the objective weighting of criteria in the MDCM process not only with traditional indicators for diversity and disorder in terms of entropy but also with novel approaches for order and certainly in terms of concentration of risks and hierarchy of choices.

Acknowledgements. This research is supported by Bulgarian FNI fund through project "Modelling and Research of Intelligent Educational Systems and Sensor Networks (ISOSeM)", contract KP-06-H47/4 from 26.11.2020.

References

1. Mell, P., Grance, T.: The NIST Definition of Cloud Computing, Special Publication (NIST SP), National Institute of Standards and Technology, Gaithersburg (2011). https://doi.org/10.6028/NIST.SP.800-145. (Accessed 12 Feb 2022)
2. Whaiduzzaman, M., Gani, A., Anuar, N.B., Shiraz, M., Haque, M.N., Haque, I.T.: Cloud service selection using multicriteria decision analysis. Sci. World J. Article ID 459375, 10 pages (2014). https://doi.org/10.1155/2014/459375
3. Rehman Z.U., Hussain O.K., Hussain F.K. Parallel cloud service selection and ranking based on QoS history. Int. J. Parallel Program. **42**, 820–852 (2014)
4. Zhao, L., Ren, Y., Li, M., Sakurai, K.: Service selection with user-specific QoS support in service-oriented architecture. J. Networks Comput. Appl. **35**, 962–973 (2012)
5. Saroj S., Dileep V. K.: A Review Multi-criteria Decision Making Methods for Cloud Service Ranking, International Journal of Emerging Technologies and Innovative Research 3(7), 92–97 (2016). http://www.jetir.org/papers/JETIR1607019.pdf

6. Saaty, T.: Principia Mathematica Decernendi: Mathematical Principles of Decision Making. RWS Publication, Pittsburgh, Pennsylvania (2010)978-1-888603-10-1
7. Saaty, T.: Fundamentals of Decision Making and Priority Theory with Analytical Hierarchical Process. The AHP Series, vol. VI. RWS Publications, The University of Pittsburgh, Pittsburgh (1980)
8. Zavadskas, E., Turskis, Z.: Multiple criteria decision making (MCDM) methods in economics: an overview. Technol. Econ. Dev. Econ. **17**(2), 397–427 (2011)
9. Papathanasiou, J., Kostoglou, V., Petkos, D.: A comparative analysis of cloud computing services using multicriteria decision analysis methodologies. Int. J. Inf. Decis. Sci. **7**(1), 23–35 (2014). https://doi.org/10.1504/IJIDS.2015.068117
10. Kwangsun, Y., Ching-Lai, C.: Multiple Attribute Decision Making: An Introduction. SAGE Publications, California (1995)
11. Chakraborty, S.: Applications of the MOORA method for decision making in manufacturing environment. Int. J. Adv. Manuf. Technol. **54**(9–12), 1155–1166 (2011)
12. Brauers, W.: Multi-objective seaport planning by MOORA decision making. Ann. Oper. Res. **1**, 39–58 (2013)
13. Triantaphyllou, E., Mann, S.H.: An examination of the effectiveness of multidimensional decision-making methods: a decision-making paradox. Decis. Support Syst. **5**, 303–312 (1989)
14. Shannon, C.: A mathematical theory of communication. Bell Syst. Tech. J. **27**(3), 379–423 (1948)
15. Jatoth, C., Gangadharan, G.R., Fiore, U., Buyya, R.: SELCLOUD: a hybrid multicriteria decision-making model for selection of cloud services. Soft. Comput. **23**(13), 4701–4715 (2018). https://doi.org/10.1007/s00500-018-3120-2
16. ITU-T Recommendation P.10: Vocabulary for performance and quality of service, Amendment 5 (07/16). https://www.itu.int/rec/T-REC-P.10
17. Ding, S., Wang, Z., Wu, D., Olson, D.L.: Utilizing customer satisfaction in ranking prediction for personalized cloud service selection. Decis. Support Syst. **93**, 1–10 (2017)
18. Hamdani, R.W.: The complexity calculation for group decision making using TOPSIS algorithm. In: AIP Conference Proceedings (2016)
19. Borissova, D.: A group decision making model considering experts competency: an Application in personnel selections. Comptes-rendus de l'Academie Bulgare des Sciences **1**(11), 1520–1527 (2018). ISSN: 1310–1331
20. Calpine, H.C., Golding, A.: Some properties of Pareto-Optimal Choices in Decision problems. Omega **4**(2), 141–147 (1976)
21. Hussain, A., Chun, J., Khan, M.: A novel customer-centric Methodology for Optimal Service Selection (MOSS) in a cloud environment. Futur. Gener. Comput. Syst. **105**, 562–580 (2020)
22. Saaty, T.L.: Relative measurement and its generalization in decision making why pairwise comparisons are central in mathematics for the measurement of intangible factors the analytic hierarchy/network process. RACSAM **102**(2), 251–318 (2008)
23. Lee, Y.H.: A decision framework for cloud service selection for SMEs: AHP analysis. SOP Trans. Mark. Res. **1**(1), 51–57 (2014)

Ultra-Dense Internet of Things Model Network

Anastasia Marochkina[1], Alexander Paramonov[1(✉)], and Tatiana M. Tatarnikova[2]

[1] The Bonch-Bruevich Saint-Petersburg State University of Telecommunications, Prospek Bolshevikov 22, St. Petersburg, Russia
[2] Russian State Hydrometeorological University, Saint Petersburg, Russia

Abstract. The article proposes the results of the analysis of the problems in the process of modeling the Internet of Things network, the choice of the structure and composition of the set of tools for full-scale and combined modeling. Subject of study: problems of modeling the high density Internet of things network. Research method: system and statistical analysis. Main results: the main tasks of modeling the high density Internet of Things networks are formulated, a three-level structure of the model is proposed, an analysis of the available technical means is carried out and the composition of technical means at each of the levels is proposed. Practical significance: the obtained results of the analysis of the use of technical means in solving various modeling problems and the proposed set of tools for simulating the Internet of Things network can be used in solving research problems, as well as in the educational process of the university.

Keywords: Full-scale modeling · IoT network · access network · communication center · workstation · server

1 Introduction

The growth in the number of Internet of Things (IoT) devices leads to a steady increase in network density, i.e. the number of devices per unit area. In perspective, the density of nodes in the IoT networks can reach values of 1 device/m2 and higher. Such high density of nodes inevitably leads to changes in the properties of the network. First, it entails the growth of mutual influence of network nodes, intra-channel interference. Interference is created by neighboring nodes transmitting data, and limits the network bandwidth. Secondly, the structure of the network depends on the environment, reflecting its properties in some way [1,2].

However, the high density of nodes also has positive aspects, namely, the redundancy of resources, thanks to which it is possible to implement effective management of the network structure. A large number of nodes expands our ability to select various structures that are tied not to individual nodes, but

V. M. Vishnevskiy et al. (Eds.): DCCN 2021, CCIS 1552, pp. 111–122, 2022.
https://doi.org/10.1007/978-3-030-97110-6_8

to specified (target) points in space. Such structure manipulations require the development of appropriate methods and algorithms. In general, the IoT network in this case can be considered as a set of nodes (resource) that can be configured in different ways, depending on the purpose(s) of the network.

At the same time, there is a need for methods for describing and forming networks with given properties. In particular, it is necessary to be able to relate the parameters of the environment, network and traffic to the performance indicators.

The development of such methods is associated with the need to verify the created models and methods. Methods of numerical (simulation) modeling are traditionally used for this purpose. However, when modeling high-density networks, difficulties arise associated with the need to simulate thousands or more nodes, as well as to realistically simulate the propagation of signals and the environment. Common simulation systems tend to operate with a set of fewer network elements, and propagation simulations are limited to a set of standard models.

In such conditions rises a need to create a universal tool for full-scale or hybrid modeling that provides a solution to a wide range of problems in the high-density Internet of Things network modeling field. As a solution for such a job, a set of devices and programs is proposed that provides configuration of a full-scale or hybrid model for solving a specific problem. We will call this set of tools CMSIV. The aim of this work is to develop the structure and composition of the high-density CMSIV.

2 Formulation of the Problem

A high-density CMSIV should allow the implementation and study of the main features of such networks, the study of which by traditional means is complicated by the need to use excessively large amounts of equipment, or the need to solve problems of high computational complexity. Let us formulate the main tasks of the set of hardware and software that this platform represents.

1. Simulating the interaction of nodes of a wireless Internet of Things access network, taking into account their mutual influence (in-channel interference). This model should also take into account and allow investigating the impact the specific environment has on the network, traffic properties and intranet interference. It is also desirable for the model to allow us to carry out the studies for several protocols of a wireless channel organization.
2. Modeling the network structure of a sufficiently large size, i.e. modeling a fragment of the Internet of Things network with a sufficiently large number of nodes, for which the methods and approaches for managing the network structure of the Internet of Things can be applied. The number of nodes in such a network should allow the choice of the network structure, both taking into account the mutual influence of the nodes, and taking into account the influence of the network environment, i.e. on the one hand, it must be

sufficiently big. On the other hand, this number should not be too large, which will inevitably affect the complexity of deploying, rebuilding and working with the model.

3. The modeling platform should provide the ability to analyze the performance of the network in the provision of services. It should be flexible enough, both in terms of hardware and software. The equipment used to simulate the nodes of the network should provide the ability to change their functionality by replacing the software. The platform should contain both software and hardware for analyzing the interaction of network nodes both at the radio channel level and at the level of cable network interfaces.

Thus, the structure of the modeling platform will be determined by a set of hardware and software. The choice of those and other means should be based on the need to solve the problems identified above.

3 Choice of Structure and Hardware

The proposed structure of the high-density network modeling platform is shown in Fig. 1.

The composition of the platform can be conditionally divided into three levels: the access network level, the transport level and the service level.

4 Access Level Standards

At the access network level, a wireless network model is implemented, consisting of n set of nodes. The capabilities of this network are determined by the capabilities of the nodes, their total amount - n. Operation in licensed areas of the spectrum is associated with a number of restrictions, the complexity of the technical implementation of the model, taking into account these restrictions, as well as the need to obtain permits. Within the field of this work, such an approach is inappropriate. Among the most common technologies for organizing a radio channel in an IoT network in unlicensed parts of the spectrum, one can note the standards IEEE 802.11 (WiFi), IEEE 802.15.1 (Bluetooth), IEEE 802.15.4 (ZeegBee), LoRa, IEEE 802.11ah. The first three are used to build networks with a limited service radius (tens - hundreds of meters), the next two are used to build networks of a medium radius (tens of kilometers). All these standards can be used to build an IOT access network, however, the implementation of all of them within the one model significantly complicates it, and also complicates the interaction with such a model, since the combination of standards for small- and medium-scale networks (communication range) requires a change in the geometric scale of the model, which is not very convenient when using the model in a laboratory. Therefore, it is advisable to implement a model of a single scale.

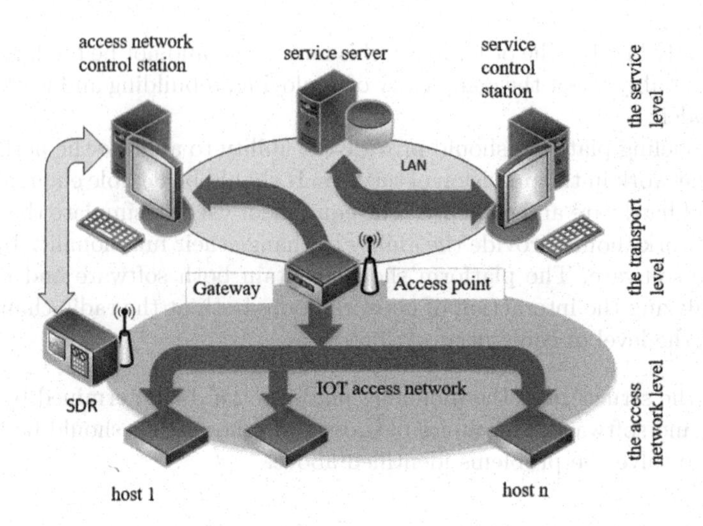

Fig. 1. Structure of CMSIV

5 Modeling the Environment

One of the tasks of this level is also modeling the fractal properties of the environment, in particular, according to [1,2], these properties can be specified by fractal curves, for example, Hilbert curves or Sierpinski carpet, Fig. 2

In this case, the line or its fragment simulates an obstacle to the propagation of a radio signal (complete or partial suppression). In a full-scale model, the implementation of such a functional can be achieved by several methods: 1. Direct physical method. The use of shielding partitions with a high degree of signal attenuation, for example, grounded metal structures in the form of lattices (nets), the selection of appropriate rooms for the placement of nodes. 2. Indirect software method. Controlling the power of the transmitters of the network nodes in such a way that its reduction is equivalent to the signal attenuation. 3. Combined method. It involves the use of both physical and software methods, which allows, on the one hand, to reduce the requirements for physical structures, and on the other hand, to simulate not only attenuation, but also reflections of signals from the surfaces of these structures. It is likely that the combined or software methods are more flexible and appropriate for use in modeling problems. To implement these methods, it is necessary that the network nodes support the functions of software power control of the transmitted radio signals. The choice of physical means for building the structure of the environment is advised to be carried out in the process of trial operation of modeling tools based on the results of field experiments.

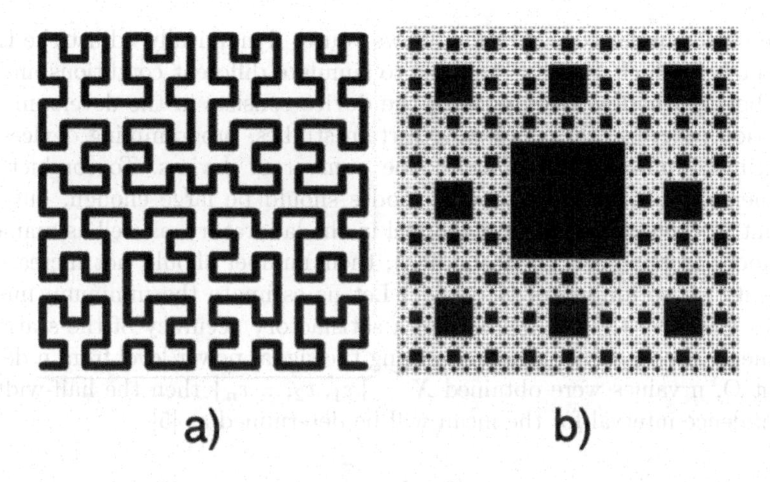

Fig. 2. Examples of environment models, a - Hilbert curve, b - Sierpinski carpet.

6 Composition of Technical Means

We assume that at the initial stage of research in the model it is sufficient to support the IEEE 802.11 (WiFi), IEEE 802.15.1 (Bluetooth) standards, as the support for the IEEE 802.15.4 standard requires additional equipment, for example [3], while WiFi and Bluetooth standards can be supported by one type of equipment [4]. The most accessible devices for which there is a sufficiently developed software are the Espressif controllers [4]. These devices support two WiFi and Bluetooth wireless communication standards, have a sufficiently powerful controller, sufficient RAM and flash memory, and also have a wired USB interface for the means of connecting to other devices and programming. The controllers also have a set of general-purpose pins that can be used in a variety of ways. The functioning of the controller is completely determined by the program written into it, which gives ample opportunities for its use both for research tasks and when using the controller in the educational process. The main technical characteristics of the ESP32 module (controller) are shown in Table 1.

Table 1. Main characteristics of the ESP32 controller

No	Characteristic	Value
1	Supported IEEE 802.11 specifications	b/g/n,150 Mbps
2	Supported IEEE 802.15.1 Specifications	V4.2 BR/EDR, BLE
3	Interfaces	SPI, I2S, I2C, UART
4	CPU	Xtensa®dual-core32-bit LX6
5	ROM	4 MB (flash)
6	RAM	512 Kbytes
7	Supply voltage	3,3 V
8	Consumption current (depending on the mode)	250 mA/20 mA/10 μA

The device under consideration allows you to dynamically adjust the transmitter power, which makes it possible to simulate different conditions and distances between nodes. Working with the device consists in the development of application programs for conducting certain studies, programming devices and conducting a full-scale experiment. The number of devices. To conduct field experiments, the number of network nodes should be large enough, but from the point of view of deploying the model in the laboratory, as well as managing these nodes (setting and programming), their number should not increase the complexity of working with the model Let us estimate the minimum number of nodes based on the achievement of a satisfactory accuracy of the statistical experiment. For example, when measuring the signal power level from n devices at point O, n values were obtained $X = \{x_1, x_2, ...x_n\}$ then the half-width of the confidence interval for the mean will be determined as [5]

$$\triangle = t_{a,n} \frac{\sigma}{\sqrt{n}} \tag{1}$$

where σ - standard deviation, n - number of measured values, - Student's coefficient [5]. The relative error can be represented as

$$\delta(n) \frac{\triangle}{\tilde{x}} 100 = \frac{t_{a,n}\sigma}{\tilde{x}\sqrt{n}} 100 = t_{a,n} C \frac{1}{\sqrt{n}100} \tag{2}$$

where C - the coefficient of variation.

Figure 3 shows the dependence of the relative error on the number of network nodes for different values of the coefficients of variation.

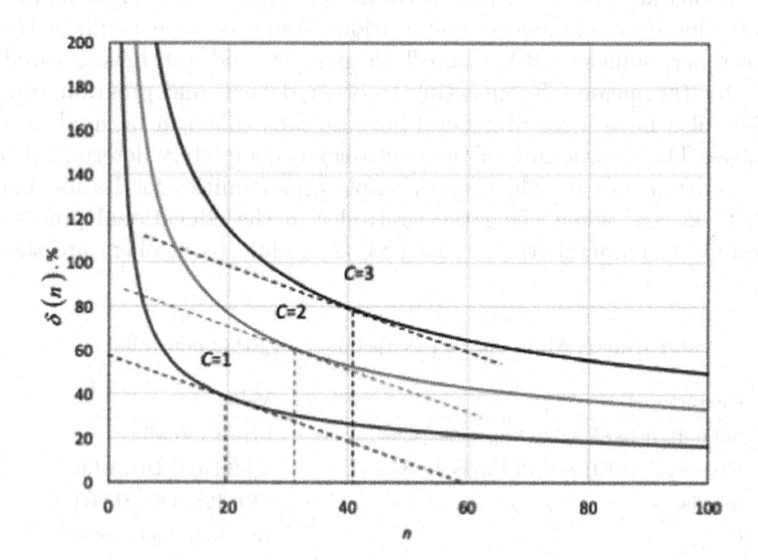

Fig. 3. Dependence of the relative error on the number of network nodes

As can be seen from the graphs above, the error decreases with increasing n and increases as the coefficient of variation rises. It should be noted that the rate of error decrease is inversely proportional to the square root of the number of nodes and the rate of its decrease slows noticeably after a certain value. Let us choose as such a value the point at which the derivative of the function is equal to one

$$\frac{d\delta(n)}{dn} = 1, \quad C = 1, \quad t_{\alpha,n} \approx 1,72, \quad n \approx 19,48 \tag{3}$$

For the coefficient of variation C = 1, the number of nodes n = 20. With high values of the coefficient of variation, this amount will be somewhat higher (at C = 2 and $C = 3$ - $n \approx 31$ and $n \approx 41$, respectively). However, the values of the coefficient of variation not exceeding one or close to it are typical for many random processes in this applied area. Therefore, at this stage, it is advisable to choose the number of nodes in the access network equal to 20.

The use of ESP32 controllers allows for sufficient flexibility of the equipment due to the possibility of implementing of various functionalities, as well as due to the possibility of using various standards for organizing the radio channel.

The lower limit in terms of the flexibility of this solution is determined by the fact that ESP32 programming is carried out at the SDK [6] (Software Development Kit) level, i.e. the set of instructions (libraries) offered by the developer of this device. For example, this does not allow changing the low-level communication protocols between devices or implementing any non-standard functions for interacting with a radio channel.

This lack of flexibility, in the structure under consideration, is compensated by the use of SDR (Software Defined Radio). SDR requirements are determined by the network nodes used at the access network model level. The frequency range is determined by the frequency range of the WiFi and Bluetooth standards, taking into account the 2.4 GHz bands supported by the ESP32 modules. The bandwidth should be capable of working with WiFi and Bluetooth devices. The largest channel bandwidth in the 2.4 GHz band is 40 MHz.

Currently, there are more than 100 different out-of-the-box solutions from various manufacturers, such as single board devices or study kits. The analysis of the main types of these devices shows, that SDR USRP B200, which is a software-controlled transceiver that allows the reception and transmission of radio signals in the frequency range from 70 MHz to 6 GHz with a bandwidth of up to 56 MHz, to the greatest extent satisfies the tasks of this platform.

SDR in this structure is used as a universal tool for monitoring the propagation medium and simulating the transmission of radio signals.

Table 2 shows the main characteristics of SDR USRP B200 [8].

Table 2. Main characteristics of USRP B200

	Characteristic	Value
1	Frequency range	70 MHz–6 GHz
2	Frequency band	up to 56 MHz
3	ADC bit width	12 Bit
4	Transmitter output power	≥10 dBm
5	Front end	USB 3.0
6	Power consumption	≤5 W
7	Transmitter output power	≥10 dBm

The cheapest and most affordable alternative to this type of SDR can be HackRF One, which differs from the USRP D200 in lower bandwidth (20 MHz), lower ADC capacity (8 bits) for about half the cost. Despite the fact that its bandwidth is half that of the 2.4 GHz WiFi standard, this device also allows you to solve most of the simulation problems.

Transport Layer. The transport layer provides interoperability between the access layer, the service server, and workstations. The basis of this level is a data transmission network, within the framework of the modeling platform, it is a local area network (LAN). The interface between the access level and the LAN is provided using an access point. In this configuration, the requirements for the access point are to support the radio channel standards used in the access network. Taking into account the selected type of ESP32 nodes - this is WiFi (IEEE 802.11b/g/n 2.4 GHz) and Bluetooth (IEEE 802.15.1 4.2), the device must provide the functions of a router and an Ethernet-to-LAN interface.

For these purposes, two options can be used:

- The first, the use of a personal computer (laptop) equipped with appropriate interfaces. WiFi and Bluetooth interfaces can be either built-in, for example, as part of a laptop, or external, connected to a personal computer;
- The second, the use of a single-board computer, for example, Rasbery Pi 4 [9] meets these requirements, as it includes WiFi and Bluetooth adapters.

These options are close in functionality and the choice of a particular one can be made at the stage of building a modeling platform, based on current needs.

The existing local area network of the organization is used as a LAN. If it is unavailable, it is enough to introduce one network device (router), which will ensure the interaction of all the necessary elements of this level and the interface with the external network.

Service Level. This level hosts servers and workstations. The server in this modeling platform is used to test scenarios for the provision of various services. The server through the transport layer and the access layer interacts with network nodes, as well as with workstations that imitate the actions to control the service and/or the actions of the client.

The hardware requirements used at this level are based on the requirements of the software used and the features of the simulated service.

Due to the fact that the modeling platform serves only to test various solutions for organizing a network and services, and does not imply the use of large amounts of data (databases of customers and services) that need to be processed at a given time, it is advisable to proceed from the requirements for a modern personal computer, used in programming tasks.

Requirements for workstations are practically the same as above. The workstation is used to simulate service clients, as well as to work with elements of the simulation platform: network nodes, access points, network devices, as well as to develop compilation of programs.

The approximate configuration of the computer for the service server and workstations is shown in Table 3.

Table 3. Approximate computer specifications

N	Characteristic	Value
1	CPU	AMD Ryzen 7 3700X
2	RAM	16 GB
3	Solid State Drive (SSD)	512 GB
4	Hard disk drive (HDD)	2 TB
5	Video card	nVidia GeForce GT 1030

Software. The software should include: operating, software development tools, traffic monitoring software, statistical analysis tools. The composition of the main software is shown in Table 4.

Table 4. Composition of softwares

N	Software purpose	Name
1	Access network nodes	Linux (Ubuntu)
2	Access point (gateway)	Linux (Ubuntu)/Windows
3	SDR	Linux
4	Server	ESP-IDF, Arduino IDE
5	Work station	C++, Python
6	Access network nodes	Wireshark, CommView
7	Access point (gateway)	GNU Radio

The general composition of the technical means of the high-density network modeling platform is shown in Table 5.

Table 5. Composition of technical means

N	Characteristic	A type	Amount
1	Access network nodes	ESP32	20
2	Access point (gateway)	Rasbery*1 Pi 4	1
3	SDR	USRP B200*2	1
4	Server	ПК*3	1
5	Work station	ПК*4	2*5

* The positions depend on the problem being solved.
1 – As noted above, an alternative can be a PC equipped with appropriate interface equipment.
2 - An alternative option is possible, depending on the tasks to be solved.
3 - The need for a server is determined by the simulation problem being solved.
4 - Depending on the tasks to be solved, one workstation can be used.
5 - Both workstations can be combined within one device.

As can be seen from Table 5, the composition of the hardware can vary, depending on the problem being solved. For example, in the tasks of studying the mutual influence of the nodes of the access network, as well as the effects of the structure of the environment on the properties of the access network, the nodes of the access network, their programming tools, monitoring and analysis tools play a primary role. In the tasks of analyzing the traffic of a service, along with the equipment of the access network, the use of equipment of the transport layer and the service layer is required.

Table 6 shows conditional indicators of the use of equipment in various tasks, obtained as expert estimates.

Table 6. Use of equipment in various tasks

N	A task	Using				
		ESP32	SDR	Gateway	R. station	Server
1	Analysis of the mutual influences of the nodes of the access network	+	+	-	+	-
2	Analysis of the impact of the environment on the access network	+	+	-	+	-
3	Designing Routing Protocols	+	+	-	+	-
4	Development of network management protocols	+	+	+	+	-
5	Analysis and development of services	+	-	+	+	+
6	Development and analysis of service management methods	+	-	+	+	+
	Average usage%	100	67	50	100	33

As can be seen from the table above, in the proposed research tasks, the equipment of the access network (network nodes) and a workstation are used to the greatest extent. Gateway and SDR are applied in half of the tasks. The server hardware is used the least.

7 Conclusions

1. The tasks of the high-density network modeling platform are the tasks of studying and analyzing the functioning of the wireless network of the high-density Internet of Things, namely: the interaction of the nodes of the wireless access network with each other, the influence of the environment on the properties of the network, the processes of providing services.
2. A high-density network modeling complex should include hardware and software at three conventional levels: access level, transport level and service level.
3. The choice of hardware and software for CMSIV is based on the principles of goal setting, compatibility and minimum sufficiency.
4. The analysis of the tasks solved by the CMSIV showed that the means of the access network level and workstations used for software development, full-scale modeling of the access network and analysis of the results are of the greatest importance and use.
5. The proposed structure of the CMSIV, the composition of hardware and software tools allow to obtain a sufficiently flexible complex for full-scale or hybrid modeling of IoT networks, including high-density IoT networks and fractal properties of the environment.
6. The proposed CMSIV can be used to solve research problems, develop protocols and services, as well as in the educational process when studying the Internet of Things networks.

Acknowledgment. This research is based on the Applied Scientific Research under the SPbSUT state assignment 2021.

References

1. Tonkikh, E.V., Paramonov, A.I., Kucheryavyy, A.E.: Self-similarity properties of the network structure and its modeling for a high-density Internet of things network. Telecommun. **8**, 18–24 (2020)
2. Paramonov, A.: Beyond 5G network architecture study: fractal properties of access network. MDPI Appl. Sci. **10**, 7191 (2020). https://doi.org/10.3390/app10207191
3. Embedded ZigBee Module. http://en.four-faith.com/f8913-embedded-zigbee-module.html?gclid=EAIaIQobChMIhrehyuar7wIVk7WyCh2vAwQLEAAYASAAEgIWu_D_BwE. Accessed 13 Mar 2021
4. Espressif. https://www.espressif.com/en/products/socs/esp32. Accessed 13 Mar 2021
5. Kendall, M., Stuart, A.: Statistical conclusions and connections. M. Science (1973). 899 C

6. Espressif SDKs & Demos. https://www.espressif.com/en/products/socs/esp32/resources. Accessed 13 Mar 2021
7. List of software-defined radios. https://en.wikipedia.org/wiki/List_of_software-defined_radios#cite_note-crosscountrywireless2-24, 13 Mar 2021
8. Ettus Research USRP B200. https://www.ettus.com/all-products/ub200-kit/. Accessed 13 Mar 2021
9. Raspberry Pi 4. https://www.raspberrypi.org/products/raspberry-pi-4-model-b/. Accessed 13 Mar 2021

Integrity, Resilience and Security of 5G Transport Networks Based on SDN/NFV Technologies

I. Buzhin[1(✉)], M. Bessonov[1], Y. Mironov[1], and M. P. Farkhadov[2]

[1] Moscow Technical University of Communications and Informatics, Moscow, Russia
{i.g.buzhin,m.a.bessonov,i.b.mironov}@mtuci.ru

[2] V.A. Trapeznikov Institute for Control Sciences of Russian Academy of Science, Moscow, Russia
mais@ipu.ru

Abstract. 5G networks are being actively implemented in many countries around the world. The goal of introducing technologies that make up the new standard is to reduce information transmission delays, increase transmission speed, and increase the number of devices served. To achieve the targets, 5G networks must be built using software-defined networking (SDN) and network function virtualization (NFV) technology. This paper analyzes the integrity, resilience and security of 5G networks based on SDN/NFV, as well as provides recommendations for the comprehensive information security of such networks.

Keywords: 5G networks · software-defined networking (SDN) · network function virtualization (NFV) · information security

1 Introduction

Information security requirements have not yet been developed for 5G networks and SDN/NFV technologies in Russia. Consider the state standards of the Russian Federation in terms of ensuring the security of telecommunication networks, we will give definitions to the concepts under consideration.

A transport network is usually understood as a set of resources that carry out transportation in telecommunication networks (transmission, control, redundancy, control systems). In mobile networks, a backhaul network is a backhaul network that links base stations to nodes in the core network, and a backbone network (Backhaul) that links core network nodes to each other. In 5G networks based on SDN/NFV (5G/SDN/NFV) technologies, the backhaul network is divided into a Fronthaul network connecting gNB-RU radio module groups and gNB-DU distributed modules, and a Midhaul network connecting gNB-CU modules and gNB-DU modules.

The Concept for the Creation and Development of 5G/IMT-2020 Networks in the Russian Federation [6] contains software and hardware tools for information

V. M. Vishnevskiy et al. (Eds.): DCCN 2021, CCIS 1552, pp. 123–135, 2022.
https://doi.org/10.1007/978-3-030-97110-6_9

security that solve the problem of ensuring confidentiality, integrity, availability of information stored, processed and transmitted in these networks: USIM manufacturing and programming equipment-cards, USIM-cards (eSIM), subscriber authentication means, data processing centers (DPC) with server equipment and orchestrators, SDN controllers, SDN switches, base stations, subscriber devices.

The Concept considers the principles of ensuring the security of 5G/SDN/NFV networks in accordance with the current specifications, provides recommendations for ensuring their information security: the use of domestic cryptographic algorithms; the use of trusted software (software); the use of a trusted electronic component base (ECB); protection against unauthorized access (AUA) to information; ensuring the stability of the system; counteraction to undocumented functionality in software; counteraction to undeclared ECB capabilities; use of "trusted" time, etc.; application of firewalling (AF); protection against computer attacks (CA) by means of IPS/IDS; protection against DDoS attacks by means of IPS/IDS; anti-malware protection by means of antivirus protection; the ability to connect to state system for detecting, preventing and eliminating the consequences of computer attacks (SSDPECCA); ensuring the integrity of software, settings and configurations; protection of channels for transmission of control messages.

It should be noted that when developing a threat model, it is advisable to focus not on the concept, but on GOSTs and documents of the Federal Service for Technical and Export Control of the Russian Federation (FSTEC) [1–5,12, 13,17].

International guidance documents in the field of detection, prevention and elimination of the consequences of computer attacks on 5G transport networks built using SDN/NFV include documents [8–11,14–16]. In addition, SDN itself and the OpenFlow protocol describe the sources [7,14]. When considering the information security of 5G/SDN/NFV networks, one should pay attention to the information security of both 5G and SDN/NFV, since in foreign sources the information security recommendations are published separately for 5G and separately for SDN/NFV. It should be noted in advance that these recommendations do not fundamentally differ from the general requirements for the protection of information and communication networks of the Russian Federation established by GOSTs and FSTEC documents, including in the field of information security threats (IST).

The document [7] describes the layers and architecture of SDN networks, [8] describes solutions for ensuring information security of virtualized NFV functions, the NFV trust model, architecture and operations, and [9,16] describes the architecture of 5G networks without reference to SDN/NFV technology. The document [10] describes mechanisms for ensuring information security of 5G systems, 5G core, 5G New Radio. Documents [11,15] describe mechanisms for ensuring information security of SDN systems.

Considering the recommendations for ensuring information security SDN/NFV, one should refer to the documents [8, 15, 49, 50]. In these sources, threats and recommendations are not described in sufficient detail; in general, they do not in any way supplement domestic guidance documents in terms of information protection.

The tasks implemented at the stage of design and deployment of SDN/NFV technologies on transport network segments are considered in [51, 52, 53].

2 5G Security Threat Model with SDN/NFV

The concept of the integrity of the transport network, the stability of the transport network, the security of the transport network are absent in the governing documents of the Russian Federation. There are concepts of information integrity, security of a telecommunication network [2], stability of the functioning of a telecommunication network [1]. Let us define the corresponding concepts.

5G/ SDN/NFV security - is defined as the ability of 5G/SDN/NFV to resist a specific set of threats, intentional or unintentional destabilizing impacts on hardware, software, firmware, communication channels and protocols included in 5G/SDN/NFV, which can lead to a deterioration in the quality of services.

In order to take into account all possible areas of manifestation of threats for 5G/SDN/NFV, it is necessary to develop a security threat model.

To take into account all possible violators, a model of the violator is developed, which is understood as an abstract (formalized or non-formalized) description of the violator of the security policy. The 5G/SDN/NFV security mechanism consists of a set of organizational, hardware, software and firmware tools, methods, methods, rules and procedures used to implement the 5G/ SDN/NFV security requirements.

The 5G/SDN/NFV security mechanism consists of a set of organizational, hardware, software and firmware tools, methods, methods, rules and procedures used to implement the 5G/SDN/NFV security requirements.

Ensuring
5G/SDN/NFV security in accordance with GOST: a) 5G/SDN/NFV protection from unauthorized access to network elements and information at all stages of its life cycle; b) countering technical intelligence; c) counteraction to network attacks and viruses; d) protection of 5G/SDN/NFV against unauthorized attacks (UA), including physical protection of hardware and firmware 5G/SDN/NFV; e) differentiation of access of users /administrators and subjects to resources in accordance with the adopted security policy; i) the use of organizational security methods. Organizational methods include: 1) developing a security policy; 2) organization of 5G/SDN/NFV security monitoring; 3) determination of the order of actions in case of emergencies and emergencies; 4) determination of the procedure for responding to information security incidents (IS); 5) training of personnel on security issues.

FSTEC of Russia plays the leading role in the regulation of information protection activities. From the guidance documents of the FSTEC of Russia, documents should be distinguished [3–5]. The methodology [3] establishes an approach to the definition of UST and the development of models of UST in information systems, as well as the development of a model of the intruder.

This approach can be applied to 5G/SDN/NFV networks, since the technique is intended for organizations that carry out work on the creation (design) of information systems in accordance with the legislation of the Russian Federation. The technique is used in conjunction with basic and typical models of information security threats in information systems (there are no such for 5G/SDN/NFV).

When determining the IST, the development of models of the UST for 5G/SDN/ NFV in accordance with the document, it is required to determine: the area to which the process for determining the IST will be applied; sources of IST; to assess the likelihood (possibility) of the implementation of IST and the degree of possible damage; develop a model of the intruder.

The basic security threat model [4] can be applied in conjunction with the methodology [3] to form a 5G/SDN/NFV threat model. When developing the model, the FSTEC of Russia Information Security Threats Databank is used [1].

3 Threats and Vulnerabilities of SDN/NFV When Use in the 5G transport Segment

Let's consider a possible classification of 5G/SDN/NFV vulnerabilities [3]. Vulnerabilities caused by the shortcomings of the organization of TPI from the UA, the presence of TCIL, are beyond the boundaries of the system under consideration. In general, SDN/NFV security issues are considered in [21,23,24,27,30,32–40].

1. Software vulnerabilities: firmware, firmware (physical hardware, virtualized physical PNF hardware); hardware drivers (physical hardware, PNF); OS (OS of automated workstations (AWP), virtualization platforms); hypervisor, VNF and other virtualized functions.
2. Vulnerabilities caused by the presence of a hardware and software bug in 5G/SDN/NFV equipment.
3. Vulnerabilities in the implementations of networking protocols and data transmission channels (IP, OpenFlow, etc.).
4. Vulnerabilities of information security systems (information security systems in the form of PNF, VNF), software and hardware.

The reasons for the emergence of vulnerabilities are [4]: errors in design and development of software, software and hardware (VNF, firmware, etc., virtualization platforms, network functions); deliberate introduction of vulnerabilities in the design and development of software, software and hardware (for example, VNF, firmware, etc., virtualization platforms, network functions); incorrect software settings (VNF, firmware, network functions, applications); unlawful change of operating modes of JSC, hypervisor, applications, etc.; unauthorized use of unreported software - applications, VNF, etc.; the introduction of malware that creates vulnerabilities in 5G/SDN/NFV software; unauthorized unintentional actions of users/administrators leading to vulnerabilities; failures in the work of JSC, PJSC, software (caused by power failures, failure of hardware elements as a result of aging and reduced reliability, external influences of electromagnetic fields of technical devices, etc.).

Sources of anthropogenic threats 5G/SDN/NFV [3] are: individuals who take actions (intentional) to access 5G/SDN/NFV or disrupt the functioning of 5G/SDN/NFV or infrastructure (intentional KILL); persons with access to 5G/SDN/NFV, whose actions (not intentional) can lead to a breach of information security (unintentional KILL).

The sources of man-made threats 5G/SDN/NFV [3] are: low reliability of hardware (RH), software and hardware (SH), software; low reliability of physical communication channels; lack/low efficiency of systems of redundancy and/or duplication; low reliability and/or lack of redundancy of power supply systems, air conditioning, security systems, etc.; low quality of technical maintenance.

The list of security threats is formed using the FSTEC of Russia Information Security Threats Databank [17], posted on the official website, taking into account the specifics of 5G/SDN/NFV. At the same time, irrelevant threats or threats that go beyond the 5G/SDN/NFV system under consideration should be excluded, namely threats to grid systems, threats to supercomputers, threats to big data storage systems.

KILL is relevant for 5G/SDN/NFV with a given structure and characteristics, if there is a possibility of implementation of KILL by an intruder with the appropriate potential, and the implementation of KILL will lead to unacceptable damage from violation of information security properties [3]. To determine the feasibility of implementing KILL, the level of 5G/SDN/NFV security should be assessed, as well as the intruder's potential required to implement KILL. At the stage of assessing 5G/SDN/NFV systems, when determining the feasibility of implementing KILL, according to [3], one should be guided by the level of design security of 5G/SDN/NFV. Analyzing the indicators [3], one should come to the conclusion that 5G/SDN/NFV at the development stage is a system with low design security.

Assessing the degree of damage from the implementation of KILL, one should conclude that as a result of a violation of one of the information security properties, significant negative consequences for 5G/SDN/NFV are possible, namely - 5G/SDN/NFV and/or the administrator/user will not be able to fulfill the assigned functions on them. Thus, the degree of damage is high. Taking into account the high level of the possibility of implementing UBI, the high possible degree of damage from the implementation of UBI, it should be concluded that the set of KILL is relevant for 5G/SDN/NFV.

Considering that it is possible to evaluate each KILL separately only at the stage of creating a specific 5G/SDN/NFV system, then many KILL should be considered relevant.

The following classes of KILL 5G/SDN/NFV are possible by the vulnerabilities used: KILL, implemented using software vulnerabilities (hypervisors, virtual functions); KILL implemented using application vulnerabilities; KILL associated with the presence of a hardware tab; KILL, the implementation of which is possible due to the vulnerabilities of network protocols and communication channels (IP, Openflow); KILL, the implementation of which is possible due to vulnerabilities associated with the shortcomings of the TPI from the UA; threats implemented using vulnerabilities that cause the presence of technical channels of information leakage; threats realized with the use of security information security vulnerabilities, man-made threats.

Analyzing international specifications, foreign literature, one should draw a conclusion about the most relevant KILL for 5G/SDN/NFV. A feature of the SDN/NFV concept is that a significant part of the data transmission infrastructure, which in a classical network is a hardware and software system, is virtualized on servers in SDN/NFV, thus generating threats typical for software: interception of information with traffic analysis; injection of malware, applications, VNF from the repository; identification of passwords; Flow rules confliction - a conflict can arise due to the fact that the 5G/SDN/NFV (OpenFlow) protocols do not distinguish between applications creating an SDN switch flow table entry from other applications that created an entry earlier, the threat leads to the creation of paths traffic bypassing the protection means in the network (in case of a compromised application); creation of a false route by unauthorized modification of the flow tables; unauthorized redirection of data flow; denial of service to the switch - denial of service to the switch can be carried out by generating traffic unknown to the switch; the switch will contact the controller to determine the rule for the flow of new traffic, which, with a large volume of requests, will lead to the exhaustion of the resources of the communication channel between the switch and the controller; denial of service to the controller - to be carried out by generating traffic unknown to the switch; the switch will contact the controller to determine the rule for the flow of new traffic, which, with a large volume of requests, will lead to the exhaustion of the controller's computing resources; overflow of event logs; overflow of data counters; overflow of the addressing table of flows of the 5G/SDN/NFV switch; substitution of a trusted object (including a switch, controller, application, etc.); obtaining a tampering device to the control channel between the switch and the controller (with the switch in the form of PNF); receipt of the UA for the 5G/SDN/NFV system and its individual elements; loss of communication with the controller by the 5G/SDN/NFV switch; scans aimed at determining the network topology, open ports and services, open connections, etc.; creation of unauthorized flow rules (Fake flow rule insertion); threat of disruption to the availability of physical equipment on which the controller is located; threat of disruption to the availability of physical equipment on which the switch is located or disruption of the switch itself when it is hardware-based; threats specific to software and virtualization tools; remote launch of applications; other. Typical man-made threats: failures and failures of control and monitoring stations; failures and failures of workstations; unintentional routing error; failures and failures of server equipment; failures and failures of disk arrays; failures and failures of network equipment; loss of communication channels; failure of the power supply system; failure of the air conditioning system; fire; flooding; natural disasters.

It should be noted that the KILL of the FSTEC base covers the already identified threats and significantly expands their list. Considering the possible consequences of the implementation of KILL, one should focus on violation of confidentiality, integrity, accessibility, accountability, and the impossibility of repudiation. In order to determine the possible consequences for each threat, one should rely on information about the functional purpose of the object of influence of the threat or the subsystem it belongs to, since the consequences of the implementation of the same threat implemented in relation to different components may differ depending on the criticality of the functions performed by each component.

4 Indicators of Integrity, Sustainability and Safety of 5G transportation Networks on the Basis of SDN/NFV

The protection of 5G/SDN/NFV transport networks should be based on an integrated approach. In this regard, it is impossible to determine any numerical indicators of integrity and safety. Resilience is assessed in terms of readiness and operational readiness indicators.

The indicators of the integrity and security of 5G transport networks based on SDN/NFV technologies are the components of a complex of legal, organizational, technical and physical measures to protect 5G transport networks based on SDN/NFV technologies.

Legal measures - consist in the application of the norms of the current legislation of the Russian Federation.

Organizational (administrative) measures - measures of the organizational nature of a legal entity (Organization) in charge of the 5G/SDN/NFV system, ensuring the regulation of the 5G/SDN/NFV functioning processes, the use of 5G/SDN/NFV resources, the activities of technical personnel, as well as the procedure for user interaction with the system in such a way as to complicate or exclude the possibility of implementing UBI or reduce the amount of losses in the event of their implementation (reputational, financial, etc.).

Technical measures - consists in the use of various software, software and hardware, hardware protection measures that are part of 5G/SDN/NFV and perform (independently or in combination with other means) protection functions (identification and authentication of users, differentiation of access to resources, registration events, cryptographic closure of information, etc.).

Physical measures - the use of specialized mechanical, electronic-mechanical devices to create physical obstacles on possible access routes for intruders to 5G/SDN/NFV components, as well as to prevent physical damage to 5G/SDN/NFV hardware due to the impact of man-made factors (fires, flooding, etc.).

Organizational measures to protect 5G/SDN/NFV transport networks should be based on a set of internal documents of the Organization, which is in charge of the 5G/SDN/NFV system, which determine the organization's IS policy, and ensure:

1. establishing the procedure for admitting employees and visitors of the Organization to the territory and premises of the Organization, providing for: - registration of the facts of entry and exit of visitors from the territory and premises of the Organization; - delimitation of access of employees to the premises of the Organization in accordance with their powers and functional responsibilities; - preventing third-party tampering in premises where 5G/SDN/NFV hardware is located;
2. establishing a procedure for accessing employees and third parties to the 5G/SDN/NFV system, providing: - assigning users access rights to 5G/SDN/NFV resources, the minimum required to perform functional duties; - revision of the access rights of employees simultaneously with a change in their status, job responsibilities (in particular, the immediate termination of access rights to 5G/SDN/NFV resources after an employee is fired or transferred to a position that does not require access to 5G/SDN/NFV resources);
3. establishing a 5G/SDN/NFV hardware and software configuration change procedure to: - coordination of the proposed changes with the responsible persons; - preliminary testing of changes that may lead to disruption of 5G/SDN/NFV functionality; - documenting the main changes in the 5G/SDN/NFV configuration in the operational documentation for 5G/SDN/NFV;
4. strict accounting of all 5G/SDN/NFV resources subject to protection by: - compilation and maintenance of lists of protected 5G/SDN/NFV resources; - documentation of all 5G/SDN/NFV hardware; - storage in libraries of software, firmware, etc. all versions of used and previously used software, firmware, etc.; - control of bringing in/taking out equipment from the premises of the Organization; - regulation of processes for performing technological operations with 5G/SDN/NFV resources; - regulation of 5G/SDN/NFV hardware maintenance processes;
5. training of employees of the Organization in the field of information security by: - familiarization of hired employees with the requirements of the current legislation and internal documents of the Organization for IS, IS 5G/SDN/NFV; - training of employees responsible for ensuring the organization's information security in specialized courses on information security;
6. conducting periodic checks of the Organization's information security: - verification of compliance by 5G/SDN/ NFV users with the requirements of the internal documents of the IS Organization; - taking punitive measures against persons guilty of violations revealed during the inspection; - conducting various types of IS audit; - conducting investigations of information security incidents: analysis of information security incidents; - elimination of the consequences of an information security incident; - identification of those responsible for the information security incident; - taking measures to prevent similar incidents in the future; - regulation of processes to ensure the continuity of 5G/SDN/NFV operation; - regulation of processes to ensure the safe operation of 5G/SDN/NFV; - regulation of 5G/SDN/NFV physical security processes; - regulation of processes to ensure the storage of documented infor-

mation; - regulation of interaction and information security processes when using third-party resources for cloud hosting 5G/SDN/NFV.

Technical protection measures for 5G/SDN/NFV transport networks should be based on the use of hardware, software and hardware and software solutions of information security that are part of 5G/SDN/NFV and perform protection functions independently and in cooperation with each other. At the same time, identification, authentication, authorization and accountability, access control - identification and authentication, provision of access rights and recording of all actions of the access subject (both users and administrators to 5G/SDN/NFV resources and components of the 5G/SDN/NFV to other components, including those located in the clouds), process to entity (in terms of operating systems).

Identification and authentication, authorization and accountability, user/ administrator access control should be ensured by: - assigning unique identifiers to users/administrators for access to 5G/SDN/NFV resources; - mandatory authentication of users/administrators when accessing 5G/SDN/NFV resources based on the use of passwords; - application of multi-factor authentication when accessing the most critical parts of 5G/SDN/NFV based on hardware identifiers (such as eToken keys); - differentiation of access to 5G/SDN/NFV resources based on the use of a mandatory essential-role DP-model (MROSL DP-model); - authorization of each user when accessing 5G/SDN/NFV information resources; - maintaining log files.

In addition, the following should be noted according to the guidance document [44]: a) given the fact that in 5G/SDN/NFV networks the network is controlled centrally, then to ensure protection, network topology rebuilding can also be used - the so-called "creating moving targets" (Moving Target Defense - MTD); b) when using cloud resources of third-party providers, it is necessary to develop separate trust models, security policies, etc. b) it is necessary to use secure servers for storing log files and copies of other information as needed, which can be used both during system recovery and during incident investigation (recording to servers should be made using unidirectional gateways).

In addition to the general security requirements outlined earlier, it is necessary to ensure [44]: 1. Verification of IP addresses: the SDN controller must be able to verify the source IP addresses. The SDN controller must restrict the list of IP addresses or range for: remote access, access from the "north" and "south" interface, neighboring controllers. 2. Security of the hypervisor: resource sharing for controllers and applications running on virtual machines should be implemented, resource sharing between virtual machines should be implemented. Once the hypervisor is attacked, the virtual machine partitioning mechanism becomes ineffective. An integrity protection mechanism based on a trusted computing architecture is required to protect the hypervisor from attacks. 3. Software Integrity: SDN controllers must support integrity checking of software (eg SDN controller software, application software) and service packs during the install/update phase. Verification methods - digital signature, enhanced MAC (Message Authentication Code) algorithm, etc. Falsified software should not be executed or installed if the integrity check is broken. 4. Protection of the integrity of the transmitted data - the integrity of the communication packets between the SDN controller and any object (for example, NE, application or OSS) must be protected by functions based on domestic algorithms. SDN controllers must support network protocols with these algorithms. 5. Protecting the Reference Data from unauthorized modification - Any change to the reference data (for example, changing the configuration and the standby flow table) in the SDN controller must be authorized. Appropriate access control can be used for related applications, switches, other SDN controllers and users. 6. Hiding the password and displaying the key - passwords, private keys of digital certificates, encryption keys, etc. in SDN controller should not be displayed on the screen in plain text. 7. Application Isolation - An SDN controller must strictly isolate data between different applications. Without authorization, one application should not be able to access or change the state of other applications, change the resource limit settings of other applications, or unsubscribe from the SDN controller for other applications. 8. Traffic separation - The SDN controller must support physical or logical separation of OSS traffic from control plane traffic. 9. Control of access to the GUI - authentication and authorization of any user who accesses the GUI should be carried out. 10. Securing Virtual Machines - Virtual machines must support features such as cgroup technology (a feature that isolates and controls resource usage for processes) to limit, collect statistics, or share the resources of process groups (CPU, memory, and disk I/O), prevent the abuse of system resources. 11. Closing unnecessary ports/services - on the SDN controller, all unnecessary ports should be closed and services disabled, and only the necessary ports/services should be opened/started. 12. Physical security of hosts. 13. Restrict Packet Forwarding from Switches - The SDN controller must restrict packet-in messages that include inconsistent, invalid packets from switches to prevent DoS attacks due to the large number of packets sent from the switches. 14. Authorization to create a thread table - The SDN controller must authorize applications when an application requests a controller to create a thread table. 15. Anti-DoS Against Compute Depletion - The SDN controller must support monitoring of compute capacity utilization and enable DoS protection mechanisms when utilization reaches a threshold (e.g. 80%). 16. Anti-DoS from Northbound/Southbound interfaces - SDN controller must support Anti-DoS northbound/southbound interfaces. The SDN controller must support monitoring of access traffic from northbound/southbound interfaces and enable DoS protection mechanisms when the amount of access traffic reaches a threshold. 17. Anti-DoS from excessive resource consumption - the SDN controller should limit the consumption of resources (eg CPU, memory) by applications. The threshold should be set according to the application-specific resolution. 18. Privileged Application Control - An SDN controller must assign privileges to each application (i.e., the controller must authorize each application) and check application privileges when they access the controller. 19. Policy Conflict Resolution - The SDN Controller must support policy conflict detection and conflict resolution. The SDN controller can set different priorities for different types of applications to ensure that policy from non-security applications cannot bypass policy from administrators or security applications. 20. Authorization to use system functionality - SDN controller must support authorization to use system functionality such as access console interface, debug interface, etc. 21. Authorization of the interface for third parties - the SDN controller must support authorization for third parties to use the interfaces to test, maintain, debug or monitor the application. 22. Hosting OS Security - The hosting OS of the SDN Controller must be secured. An intruder can control SDN controller software using hosting OS vulnerabilities. Since the SDN controller software runs on a traditional server or virtualized machine, the security requirements of the hosting OS are the same as those of the traditional OS. Therefore, traditional OS protection methods should be applied to the hosting OS of the controller. 23. The operating system must be secured on the SDN controller - attention must be paid to configuring components, deleting unused files and programs, updating operating system components, etc. 24. Vulnerabilities of the software used by the SDN controller must be monitored.

Differentiation of access of entities to other entities and processes should be based on the MROSL DP-model with log files (differentiation of access between the virtual switch and the controller, between "layers", between virtual machines).

5 Conclusion

This paper analyzes the guidance documents regulating the information security of public communication networks, foreign sources and standards in the field of providing 5G networks and SDN/NFV technologies. Recommendations are given for the comprehensive information security of 5G networks based on SDN/NFV. The complex of measures to ensure information security includes organizational, technical and physical protection measures considered in this work.

References

1. GOST R 53111–2008. The stability of the functioning of the public communication network. Requirements and verification methods
2. GOST R 52448 Information security. Ensuring the security of telecommunication networks. General Provisions
3. Methodology for determining threats to information security in information systems. FSTEC (2015)
4. The basic model of threats to the security of personal data during their processing in personal data information systems (extract). Approved by the Deputy Director of FSTEC of Russia on February 15 (2008)
5. Methodology for modeling threats to information security (draft). FSTEC of Russia (2020)
6. Concept of creation and development of 5G/IMT-2020 networks in the Russian Federation
7. RFC 7426 - Software-Defined Networking (SDN): Layers and Architecture Terminology

8. ETSI GS NFV-SEC 003 V1.1.1 (2014–12) Network Functions Virtualisation (NFV); NFV Security; Security and Trust Guidance
9. 3GPP TS 23.501 V16.4.0 (2020–03) System architecture for the 5G System (5GS)
10. 3GPP TS 33.501 V16.2.0 (2020–03). Security architecture and procedures for 5G system
11. Technical Specification SDN Security Considerations in the Data Center. ONF Solution Brief (2013)
12. GOST R 51275 Information security. Object of informatization. Factors affecting information. General Provisions
13. GOST R 50922 Information security. Basic terms and definitions
14. Open Flow Switch Specification 1.5.1, Open Networking Foundation (2015)
15. Threat Analysis for the SDN Architecture 1.0 Technical Specification, Open Networking Foundation, (2016)
16. 3GPP TS 23.502, "Procedures for the 5G System"
17. Databank of information security threats
18. Shaghaghi, M.K., Buyya, R., Jha, S.: Software-Defined Network (SDN) Data Plane Security: Issues, Solutions and Future Directions. arXiv:1804.00262v1
19. Prasad, A.R., Arumugam, S., Sheeba, B., Zugenmaier, A.: 3GPP 5G security. J. ICT Standardization **6**(1–2), 137–158 (2018)
20. Arfaoui G., et al.: A security architecture for 5G networks. IEEE Access (2018), p. 1. https://doi.org/10.1109/ACCESS.2018.2827419
21. Gao, S., Li, Z., Xiao, B., Wei, G.: Security threats in the data plane of software-defined networks. IEEE Network, pp. 1–6 (2018). https://doi.org/10.1109/MNET.2018.1700283
22. Yao, J., Han, Z., Sohail, M., Wang, L.: A robust security architecture for SDN-based 5G networks. Future Internet, pp. 1–14 (2019)
23. Casado, M.: SANE: a protection architecture for enterprise networks. In: USENIX Security Symposium (2006)
24. Scott-Hayward, S., Natarajan, S., Sezer, S.: A survey of security in software defined networks. IEEE Commun. Surv. Tutorials **18**(1), 623–654 (2016)
25. Zaidi, Z., Friderikos, V., Yousaf, Z., Fletcher, S., Dohler, M., Aghvami, H.: Will SDN be part of 5G? arXiv:1708.05096v2
26. Zhang, X., Kunz, A., Schroder, S.: Overview of 5G security in 3GPP (2017). https://doi.org/10.1109/CSCN.2017.8088619
27. Zakharov, A.A., Popov, E.F., Fuchko, M.M.: Information security aspects of SDN architecture. Bull. SibSUTI **1**, 83–92 (2016)
28. Tikhvinsky, V., Bochechka, G., Minov, A., Babin, A.: 5G networks: international standardization. CONNECT, No. 1–2, pp. 52–58 (2017)
29. Efimushkin, V.A., Ledovskikh, T.V., Korabelnikov, D.M., Yazykov, D.N.: Review of SDN/NFV solutions of foreign manufacturers. T-Comm. Telecommun. Transport. **9**(8), 5–13 (2015)
30. Kurochkin, I.I., Humenny, D.G.: Security of SDN networks. Classification of attacks. Modern Inf. Technol. IT Educ. **11**(2), 381–383 (2015)
31. Loginov, S.S.: About control levels in a software-defined network (SDN). TComm. Telecommun. Transport. **11**(3), 50–55 (2017)
32. S. Volkov, I. Kurochkin Application of machine learning methods in SDN in intrusion detection problems // International Journal of Open Information Technologies ISSN: 2307–8162 vol. 7, no. 11, 2019, pp. 49–58
33. Smelyanskiy R.L., Pilyugin P.L. Sovremennyye problemy obespecheniya bezopasnosti v SDN [Modern security problems in SDN]// REDS: Telekommunikatsionnyye ustroystva i sistemy [REDS: Telecommunication devices and systems]. 2017, No 4. p. 523–526

34. Bernardo, D.V., Chua, B.B.: Introduction and Analysis of SDN and NFV Security Architecture (SN-SECA). 796–801 (2015). https://doi.org/10.1109/AINA.2015.270
35. Feghali, K., Maroun, C.: SDN security problems and solutions analysis, pp. 1–5 (2015). https://doi.org/10.1109/NOTERE.2015.7293514
36. Liyanage, M., Ahmad, I., Ylianttila, M., Gurtov, A., Abro, A.B., de Oca, D.E.M.: Leveraging LTE security with SDN and NFV. In: Proceedings of the 10th IEEE International Conference on Industrial and InformationSystems (ICIIS), pp. 220–225 (2015)
37. Kloti, R., Kotronis, V., Smith, P.: OpenFlow: a security analysis. In: 21st IEEE International Conference on Network Protocols (ICNP), pp. 1–6, October 2013
38. Scott-Hayward, S., O'Callaghan, G., Sezer, S.: SDN security: a survey. In: IEEE SDN for Future Networks and Services (SDN4FNS), pp. 1–7 (2013)
39. Kreutz, D., Ramos, F., Verissimo, P.: Towards secure and dependable software-defined networks. In: Proceedings of the Second ACM SIGCOMM Workshop on Hot Topics in Software Defined Networking, pp. 55–60 (2013)
40. Francois, J., Festor, O.: Anomaly traceback using software defined networking. In: International Workshop on Information Forensics and Security (2014)
41. Vanbever, L., Reich, J., Benson, T., Foster, N., Rexford, J.: HotSwap: correct and efficient controller upgrades for software-defined networks. In: Proceedings of the Second ACM SIGCOMM Workshop on Hot Topics in Software Defined Networking, pp. 133–138 (2013)
42. 3GPP TS 23.101 Universal Mobile Telecommunications System (UMTS). General UMTS Architecture
43. Principles and Practices for Securing Software-Defined Networks. ONF TR-511 Open Networking Foundation (2015)
44. Security Foundation Requirements for SDN Controllers. Version 1.0. TR-529. Open Networking Foundation (2016)
45. Samuilov, K.E., Shalimov, I.A., Buzhin, I.G., Mironov, Yu.B.: Model of functioning of telecommunication equipment of software defined networks. Modern Inf. Technol. IT Educ. **14**, 13–26 (2018). No. 1
46. Tsvetkov, V.K., Oreshkin, V.I., Buzhin, I.G.: Mironov Model of restoration of the communication network using the technology of software defined networks. In: ELCONRUS 2019, pp. 1559–1563. Institute of Electrical and Electronics Engineers Inc. (2019)
47. Buzhin, I.G., Mironov, Y.B.: Evaluation of delayed telecommunication equipment of Software Defined Networks. In: SOSG 2019, p. 8706825. Institute of Electrical and Electronics Engineers Inc. (2019)

Algorithm of Finding All Maximal Induced Bicliques of Hypergraph

Aleksandr Soldatenko[1] and Daria Semenova[1,2]

[1] Siberian Federal University, Krasnoyarsk 660041, Russia
{asoldatenko,dvsemenova}@sfu-kras.ru

[2] Krasnoyarsk State Medical University, Krasnoyarsk 660022, Russia

Abstract. The problem of finding all maximal induced hypergraph bicliques is considered. The maximal bicliques have many applications and are most in demand in genetic and telecommunication networks. In a hypergraph, an edge can be a subset of any cardinality, and not just a two-element subset of a finite set of vertices. This generalization opens up additional computational capabilities for the hypergraph model applications. The paper considers the relationship between finding the most complete submatrices of the $(0, 1)$-matrix and finding all the maximum induced bicliques of the hypergraph. The main idea of the hypergraph approach is to use the properties of the hypergraph to efficiently generate solutions for the problems mentioned above. A new algorithm for finding all maximal induced hypergraph bicliques is proposed. The scheme and pseudocode of the proposed algorithm are presented together with a detailed description of its operation. A lemma is proposed that can be used to improve the performance of the algorithm on a special kind of hypergraphs. Results of computational experiments are introduced.

Keywords: Hypergraph · maximal induced bicliques · search algorithm

1 Introduction

Hypergraph is extension of classical graph theory to the case when a graph edge can contain another number of vertices instead of two [1]. Traditionally hypergraphs have found practical applying in the development of relational databases and combinatorial chemistry [2,3]. An ability to combine some vertices in one edge provides a powerful tool for researching and analyzing processes in various networks. Thus, hypergraphs are actively used in modeling road and telecommunication networks, as well as for constructing semantic networks when processing texts in natural language [4–12].

This work is supported by the Krasnoyarsk Mathematical Center and financed by the Ministry of Science and Higher Education of the Russian Federation in the framework of the establishment and development of regional Centers for Mathematics Research and Education (Agreement No. 075-02-2020-1534/1).

V. M. Vishnevskiy et al. (Eds.): DCCN 2021, CCIS 1552, pp. 136–147, 2022.
https://doi.org/10.1007/978-3-030-97110-6_10

A significant part for tasks of studying such networks is reduced to problems of determining various configurations [13]. A configuration is any system of subsets of a finite set. A subject of particular interest are problems of enumeration type, in which existence of configuations is beyond doubt, but questions are only about number of configuations and method of their representation. The problem of constructing configurations in the form of complete submatrices of $(0,1)$-matrices is one of the main tasks for practical applications. In particular, such well-known problems as finding all formal concepts in the binary context [4,8] and finding all maximal bicliques in graphs and hypergraphs [5,14] are reduced to it. In the first case, formal concept is a maximally complete submatrix of $(0,1)$-matrix of formal context [4]. For this problem proposed many algorithms, but the most famose is Close-by-One [8]. Time complexity of this algorithm does not exceed $\mathcal{O}(|\mathcal{MCS}| \cdot m \cdot n^2)$ where $\mathcal{MCS}$ is a set of all maximally complete submatrices. In the second case, if maximally complete submatrix satisfies to the special form of matrix, then it is considered as maximal biclique. There are two areas of research, the search for all maximal non-induced bicliques and for maximal induced bicliques. When each part S_0 and S_1 of biclique (S_0, S_1) is independent then biclique is induced. Set is independent when vertices in it do not adjacent. If property of set independency is ignored then biclique (S_0, S_1) will be non-induced. Algorithm for finding all maximal non-induced bicliques with the complexity $\mathcal{O}(a^3 \cdot 2^{2a} \cdot n)$ where a is arboricity of input graph [15]. A number of researchers note that in practice, only maximal non-iduced bicliques of large size are in demand [9,12,16]. In paper [9] the algorithm for search maximal non-induced bicliques with parameter p for biclique size threshold has been proposed. Complexity of this algorithm is $\mathcal{O}(n \cdot m \cdot \mathcal{N})$ where $\mathcal{N}$ is a set of all non-induced bicliques with size greater or equal to p. For finding all maximal induced bicliques of graph in paper [15] algorithm is proposed. It has complexity $\mathcal{O}\left(n \cdot k \cdot (\Delta + k) \cdot 3^{\frac{\Delta+k}{3}}\right)$ where k is degeneracy of graph. Both of these bicliques types can be efficiently applied in telecommunication network modelling [7]. In other areas of knowledge, maximum bicliques were investigated in works [3,5,17]. Let us note that problems of finding such configurations are $\sharp P$-complete [13,18–20].

The problem of finding all maximal induced bicliques for each given hypergraph, which is called Maximal Induced Bicliques Generation Problem for Hypergraphs (MIBGP for Hypergraphs) is investigated in the paper. This problem arises in various applications connected with data mining in many fields. For example, in telecommunication networks maximal bicliques used for routing organizing and defining subnets for marking them [7,12]. In metabolic and genetic networks maximal bicliques are used for represent connection between organisms and different external conditions [11,21,22]. In marketing maximal bicliques allow to form social recommendations and product bundling [11,21,22]. Maximal bicliques are used for clustering data in various fields [10].

A new algorithm for finding all maximal induced bicliques in hypergraph is proposed in the paper. The diagram and pseudocode of proposed algorithm are given as well as detailed description of its performance. Computational experiments for proposed algorithm are presented.

2 Statement of Problem of Finding All Maximal Induced Bicliques of Hypergraph

We'll introduce some definitions necessary for further presentation [1,14].

Let the (n, m)-hypergraph $H = (X, U)$ be given, where X is a finite set of vertices and U is a finite family of hyperedges of hypergraph, at the same time $|X| \geq 1$, $|U| \geq 1$ and any hyperedge of hypergraph is a subset of the set X. Let $X(u)$ is a set of all vertices that incident to the hyperedge $u \in U$ and $U(x)$ is a set of all hyperedges, which incident to the vertex $x \in X$. Maximal degree of vertex denoted as $\Delta = \max_{x \in X} |N(x)|$ where $N(x)$ is set of all vertices of hypergraph adjacent to x. One of the ways to define hypergraph is incidence $(0, 1)$-matrice I where 1 is put in case when hyperedge contains vertex and 0 otherwise.

Definition 1. *A hypergraph $H' = (X', U')$ is called the subhypergraph induced by the set of vertices X' where $U' = \{u' : X(u') = X(u) \cap X' \neq \oslash, u \in U\}$.*

Definition 2. *The subhypergraph $H' = (X', U')$ induced by the set of vertices X' is bipartite if exist such partition $S_0 \cup S_1 = X'$ that $S_0 \cap S_1 = \oslash$ and $|S_0 \cap X(u')| \leq 1$, $|S_1 \cap X(u')| \leq 1$ is true for all $u' \in U'$.*

Definition 3. *A vertex graph of hypergraph $H = (X, U)$ is called the graph $L_2(H) = (X, E)$ which set of vertices is equal to the set of vertices X of hypergraph H while two vertices of $L_2(H)$ are adjacent if and only if corresponding vertices of hypergraph H are adjacent.*

Definition 3 is shown at Fig. 1. Vertex graph $L_2(H) = (X, E)$ corresponding to hypergraph $H = (X, U)$ with sets $X = \{1, 2, 3, 4, 5\}$ and $U = \{a, b, c, d\}$.

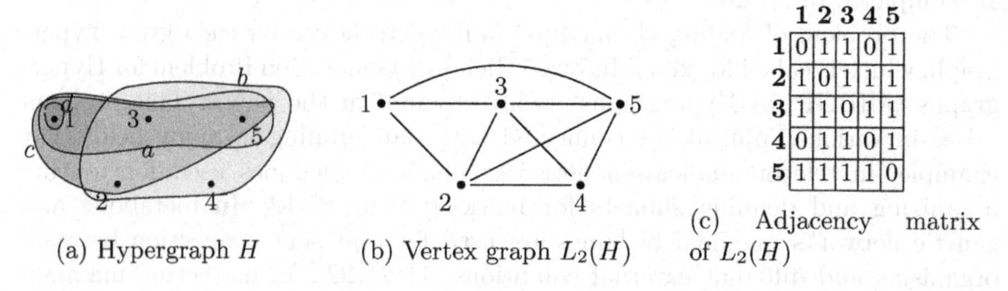

	1	2	3	4	5
1	0	1	1	0	1
2	1	0	1	1	1
3	1	1	0	1	1
4	0	1	1	0	1
5	1	1	1	1	0

(a) Hypergraph H (b) Vertex graph $L_2(H)$ (c) Adjacency matrix of $L_2(H)$

Fig. 1. Transition from hypergraph $H = (X, U)$ to vertex graph $L_2(H) = (X, E)$

We represent the following theorem without proof.

Theorem 1. *The subhypergraph $H' = (X', U')$ is bipartite if and only if vertex graph $L_2(H)$ of hypergraph H contains a bipartite subgraph induced by the set of vertices X'.*

Note that a proof of the Theorem 1 follows directly from Definitions 1–3. Let's illustrate statement of this theorem by the following Fig. 2.

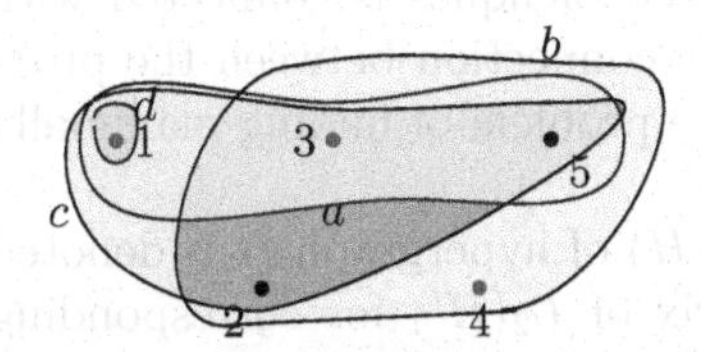

(a) Bipartite subhypergraph H' with parts $S_0 = \{1, 4\}$ and $S_1 = \{3\}$

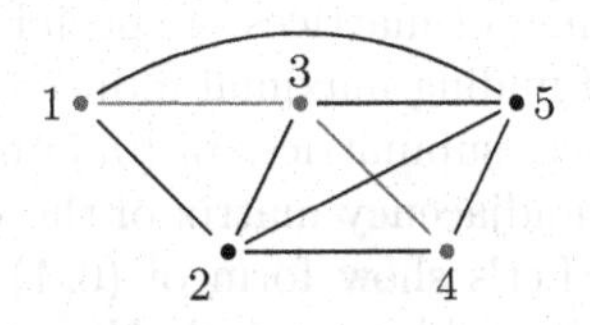

(b) Vertex bipartite subgraph $L_2(H')$ with parts $S_0 = \{1, 4\}$ and $S_1 = \{3\}$

Fig. 2. Correspondence of a bipartite subhypergraph and a bipartite subgraph

Definition 4. *A bipartite graph is called the complete bipartite graph (biclique) if each vertice of one part is connected with all vertices from the second part.*

A number of graph-theoretic problems, which belong to the class of $\sharp P$-complete or NP-complete, is reduced to the search of bicliques [19,20].

The problem of finding all maximal bicliques for graphs can be extended to hypergraphs. In particular, for bihypergraphs the problem of finding all maximal induced bicliques was studied in [17]. Hypergraph $H = (H^0, H^1)$ is called bihypergraph if each hyperedge of hypergraphs H^0 and H^1 is conteined in H. Definition of a bipartite hypergraph is introduced according to the stability of the set of vertices. Set of vertices $S \subseteq X$ is stable in H^i if S does not contain hyperedges of H^i, $i = 0, 1$. A bihypergraph $H = (H^0, H^1)$ is called bipartite if a partition $S^0 \cup S^1 = X$ and S^i is stable in H^i for $i = 0, 1$.

Definition 5. *A subhypergraph $H' = (X', U')$ induced by the set of vertices X' such that $S_0 \cup S_1 = X'$, $S_0 \cap S_1 = \oslash$ and $U' = u : s_0, s_1 \in X(u), s_0 \in S_0, s_1 \in S_1$ is called bipartite induced subhypergraph of hypergraph H.*

In what follows, by an induced bicliques of hypergraph we mean a complete bipartite induced subhypergraph of hypergraph H. A biclique that can not extended with additional adjacent vertices is called the maximal induced biclique.

The problem of finding maximal biclique is well known in graph theory. Another problem connected with maximal biclique is Maximal Bicliques Generation Problem (MBGP), which consist in finding all maximal bicliques for a graph. It is known that MBGP cannot be solved in polynomial time with respect to the size of input, since size of output set can be exponentially large [18,23].

Complexity of this problem at least hard as problem of searching of one maximal biclique which is NP-hard [22].

In this paper the problem of Maximal Induced Biclique Generation Problem for Hypergraphs (MIBGP for Hypergraphs) is researching.

MIBGP for Hypergraphs. A hypergraph $H = (X, U)$ without double hyperedges is given. It is necessary to find a set of all maximal induced bicliques.

Note that problem of finding maximal induced bicliques is connected with searching of matrices of special form [24]. Link a connection between the problem of finding maximal induced bicliques and the problem of finding maximally complete submatrices of $(0, 1)$-matrix.

An adjacency matrix of the vertex graph $L_2(H)$ of hypergraph H is denoted as A. Let's show form of $(0, 1)$ adjacency matrix of $L_2(H')$ for corresponding bipartite subhypergraph $H' = (X', U')$. Here S_0, S_1 be parts of hypergraph H with cardinality c and d respectively. Since $L_2(H')$ is the bipartite graph as well then adjacency matrix has form [24]

$$A' = \begin{pmatrix} O_c & B' \\ B'^T & O_d \end{pmatrix}, \quad (1)$$

where O_c, O_d zero matrices of sizes c and d respectively, and B' matrix of size $c \times d$, which represent the adjacency of vertices between parts S_0 and S_1. Obviously, in case when the subhypergraph H' is biclique then a matrix B is a complete submatrix of the matrix A.

From this follows the problem of finding complete submatrices of the $(0, 1)$-matrix. Note, that finding all maximally complete submatrices can be applied to searching cliques of networks, which is an important task of the analysis [25]. The problem of finding all maximally complete submatrices of $(0, 1)$-matrix is as follows.

Maximally Complete Submatrices Problem (MCSP) for Hypergraph. Hypergraph $H = (X, U)$ with incidence matrix I is given. It is required to find a set of all maximally complete submatrices of the matrix I.

The HFindMCS algorithm was previously developed for the MCSP for Hypergraph [26]. The algorithm implements a hypergraph approach. Main idea of the hypergraph approach is to use the features of the hypergraph to efficiently generate the solutions for the problems. Efficiency lies in storing not only the original hypergraph $H = (X, U)$, but its dual $H^* = (X^*, U^*)$ where $X^* = U$ and $U^* = X$. Easy to see that duality can be achieved by transposition of the incidence matrix I. In this case, the operations $X(u)$ and $U(x)$ for $x \in X$, $u \in U$ are performed in constant time. Obviously, that sizes of the configurations under study do not exceed the maximum degree of a vertex or edge Δ.

The proposed HFindMIB algorithm adapts the hypergraph approach to the MIBGP for Hypergraphs.

3 Algorithm of Finding All Maximal Induced Bicliques of Hypergraph

Let both set of vertices and set of hyperedges of hypergraph H are lexicographic ordered. In proposed algorithm a transition from hypergraph H to vertex graph $L_2(H)$ is considered. Adjacency matrix of $L_2(H)$ is represented as hypergraph $\Phi = (X_\Phi, U_\Phi)$. We'll introduce a definition of l-layer of hypergraph Φ with square matrix.

Definition 6. *Let subhypergraph Φ' induced by a set of vertices $S_0 \subseteq X_\Phi$ and a set of hyperedges $S_1 \subseteq U_\Phi$. If matrix of Φ' satisfy to form (1) then Φ' is called induced biclique and denoted (S_0, S_1). The set of all induced bicliques (S_0, S_1) where $|S_0| = l$ is called the l-layer of the hypergraph Φ.*

The main idea of the algorithm consists in generation of all induced bicliques for all l-layers and filter from them maximal induced bicliques. The HFindMIB algorithm diagram is shown in Fig. 3, as well as complexity of each phase.

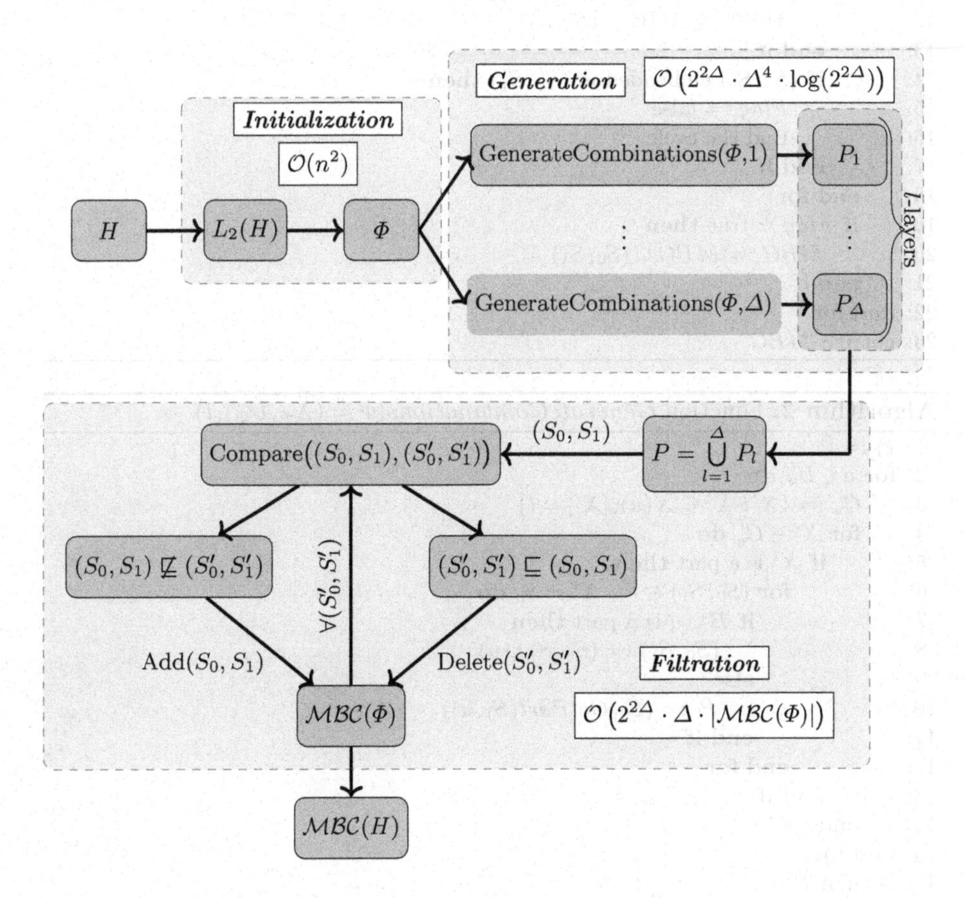

Fig. 3. Diagram of HFindMIB algorithm

In MIBGP for Hypergraphs finding all maximal induced bicliques of hypergraph H is required.

Algorithm 1. HFindMIB

```
Require: hypergraph H = (X, U).
Ensure: set MBC.
 1: Construct graph L_2(H) and represent it as a hypergraph Φ
 2: for l = 1, ..., Δ do
 3:     P_l := GenerateCombinations(Φ,l)
 4: end for
 5: P := ⋃_{l=1}^{Δ} P_l
 6: MBC := ∅
 7: for (S_0, S_1) ∈ P do
 8:     Flag := true
 9:     for (S'_0, S'_1) ∈ MBC do
10:         Compare (S_0, S_1) and (S'_0, S'_1)
11:         if (S'_0, S'_1) embedded in (S_0, S_1) then
12:             MBC := MBC \ (S'_0, S'_1)
13:         end if
14:         if (S_0, S_1) embedded in (S'_0, S'_1) then
15:             Flag := false
16:             End the cycle
17:         end if
18:     end for
19:     if Flag = true then
20:         MBC := MBC ∪ (S_0, S_1)
21:     end if
22: end for
23: return MBC
```

Algorithm 2. Function $GenerateCombinations(\Phi = (X_\Phi, U_\Phi), l)$

```
 1: P_l := ∅
 2: for u ∈ U_Φ do
 3:     C_u^l := {X' : X' ⊆ X(u), |X'| = l}
 4:     for X' ∈ C_u^l do
 5:         if X' is a part then
 6:             for (S_0, S_1) ∈ P_l : X' = S_0 do
 7:                 if B ∪ u is a part then
 8:                     (S_0, S_1) := (S_0, S_1 ∪ u)
 9:                 else
10:                     P_l := (S_0, GetPart(S_1, u))
11:                 end if
12:             end for
13:         end if
14:     end for
15: end for
16: return P_l
```

Algorithm pseudocode is presented in Algorithms 1 and 2.

Proposed HFindMIB algorithm solves this problem in three phases.

On ***initialization phase*** the transition from input hypergraph H to vertex graph $L_2(H)$ is occured.

On ***generation phase*** for adjacency matrix of vertex graph $L_2(H)$ represented as hypergraph Φ all l-layers are generated. A result of this phase is sets of all l-layers $P_l = \{(S_0, S_1) \colon S_0 \cap S_1 = \oslash\}$ where $l = 1, \ldots, \Delta$. For each value of l function $GenerateCombinations(\Phi, l)$ are executed. This function generates all possible subsets $X' \subseteq X_\Phi(u)$ for each hyperedge $u \in U_\Phi$ such that $|X'| = l$ and X' satisfy to (1). Form (1) grants that all vertices of X' does not adjacency each one with another. Set of all such subsets of l-layer of hyperedge $u \in U_\Phi$ denote as C^l_u. Each generated set X' consider as part S_0 of biclique. Since hypergraph Φ represent adjacency matrix of $L_2(H)$ that any $u \in U_\Phi$ can be treated as vertex of the hypergraph H. A part S_1 for corresponding sets $X' \in C^l_u$ is formed from hyperedges $u \in U_\Phi$ such that they do not violate (1). If addition of u to part S_1 violate (1) then current biclique split in two (S_0, S_1) and $(S_0, S_1 \sqcup u)$ where $S_1 \sqcup u$ is union elements of S_1 with u such that they not adjacent and satisfy to (1).

On ***filtration phase*** from all generated l-layers all maximal induced bicliques are selected. This grants all maximal induced bicliques for hypergraph H. A set of all induced bicliques P is formed from sets P_l which are l-layers of the hypergraph. Bicliques (S_0, S_1) and (S'_0, S'_1) where $S_0 = S'_1$ and $S_1 = S'_0$ so (S_0, S_1) and (S'_0, S'_1) that represent same biclique are generated due to the specifics of generation. To find all maximal induced bicliques it is required to determine such bicliques that are embedded in others. Comparison and detection of embedded bicliques is done in $Compare\big((S_0, S_1), (S'_0, S'_1)\big)$ procedure. This procedure is called for each element from P and compare it with all elements of $\mathcal{MBC}(\Phi)$. Let's define $(S_0, S_1) \sqsubseteq (S'_0, S'_1)$ as follows. If $S_0 \subseteq S'_0, S_1 \subseteq S'_1$ or $S_1 \subseteq S'_0, S_0 \subseteq S'_1$ then biclique (S_0, S_1) is embedded in (S'_0, S'_1). If induced biclique $(S_0, S_1) \not\sqsubseteq (S'_0, S'_1)$ where $(S'_0, S'_1) \in P$ then consider it maximal and add it to the set $\mathcal{MBC}(\Phi)$. According to Theorem 1 a set $\mathcal{MBC}(\Phi)$ is equivalent to the set of all maximal induced bicliques of the hypergraph H. After filtration phase from the set P only maximal induced bicliques has been extracted, so HFindMIB algorithm correctly solve MIBGP for Hypergraphs.

Let us illustrate the operation of the algorithm on a simple hypergraph. On Fig. 4 initialization phase (Algortihm 1 step 1) of the HFindMIB algorithm is shown. The Fig. 4a represents a input hypergraph H with incidence matrix from the Table 1a. The Fig. 4b shows vertex graph $L_2(H)$ of the input hypergraph H. The hypergraph Φ at the Fig. 4c is the result of inialization phase. Table 1b contains the incidence matrix of hypergraph Φ which aquired from adjacency matrix of vertex graph $L_2(H)$. For the hypergraph H, the maximum vertex degree is $\Delta = 4$.

It is easy to see that the maximal induced bicliques are the pairs $(\{1\}, \{3\})$, $(\{2\}, \{1, 4\})$, $(\{2\}, \{1, 5\})$, $(\{2\}, \{3, 4\})$, $(\{2\}, \{3, 5\})$, $(\{4\}, \{5\})$. For simplicity, in this example, we will write $(\{1, 4\}, \{2\})$ as $(14, 2)$, i.e. $S_0 = \{1, 4\}$ and $S_1 = \{2\}$.

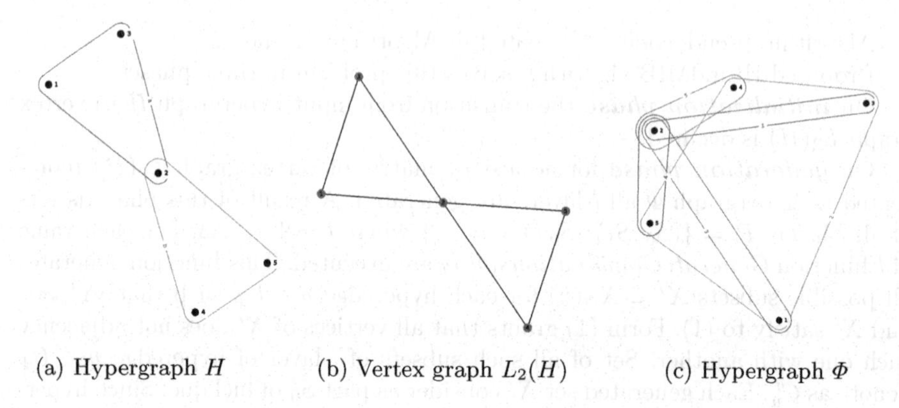

(a) Hypergraph H (b) Vertex graph $L_2(H)$ (c) Hypergraph Φ

Fig. 4. Initialization phase of HFindMIB algorithm

Table 1. Incidence matrices for initialization phase

(a) Incidence matrix of hypergraph H

	a	b
1	1	0
2	1	1
3	1	0
4	0	1
5	0	1

(b) Incidence matrix of hypergraph Φ

	1	**2**	**3**	**4**	**5**
1	0	1	1	0	0
2	1	0	1	1	1
3	1	1	0	0	0
4	0	1	0	0	1
5	0	1	0	1	0

Results of generation phase (steps 2–4) for each Δ are shown in the Table 2. On filtration phase (steps 7–22) from table the 2 rows 1, 7–9, 11–16 are removed, because this bicliques are embedded in other bicliques or equal to previously added. Note that bicliques (2, 14) and (14, 2) are equal. Thus, set $\mathcal{MBC}$ contains maximal induced bicliques (1, 3), (2, 14), (2, 15), (2, 34), (2, 35), (4, 5), which correspondence to rows 2–6, 10 of Table 2.

Earlier, we obtained the following estimate of HFindMIB algorithm

$$\mathcal{O}\left(2^{2\Delta} \cdot \Delta \cdot (|\mathcal{MBC}| + \Delta^3 \cdot \log(2^{2\Delta})) + n^2\right).$$

Time complexity of the HFindMIB algorithm depends on size of the set $\mathcal{MBC}$. This is feature of MIBGP for Hypergraphs which is an enumeration problem. Besides estimate depends on value Δ however it is overestimated since some of the subsets at each of the l-layers does not form a part of biclique. For connected hypergraphs, the this estimate can be significantly improved using the following lemma.

Lemma 1. *Let a connected hypergraph H with degree Δ be given. If an induced biclique (S_0, S_1) exists in hypergraph H, where $|S_0| = |S_1| = \Delta$, then the hypergraph H is bipartite.*

Table 2. Generated induced bicliques for all layers

Δ	№	(S_0, S_1)
1	1	(1, 2)
	2	(1, 3)
	3	(2, 14)
	4	(2, 15)
	5	(2, 34)
	6	(2, 35)
	7	(3, 1)
	8	(3, 2)
	9	(4, 2)
	10	(4, 5)
	11	(5, 2)
	12	(5, 4)
2	13	(14, 2)
	14	(15, 2)
	15	(34, 2)
	16	(35, 2)
3	—	
4	—	

A proof of this lemma follows from the definition of maximal induced biclique of hypergraph, property of connectivity of hypergraph and maximal degree of vertex Δ. Based on Lemma 1, some of the steps in the generation of induced bicliques can be skipped, which will significantly reduce the number of filtered induced bicliques.

4 Computational Experiments

To evaluate the effectiveness of solving the MIBGP for Hypergraphs problem by the proposed HFindMIB algorithm, computational experiments were performed. Experiments were done on the hypergraphs $H = (X, U)$ with different numbers of vertices $n = |X|$ and hyperedges $m = |U|$ as well as different maximum vertex degree Δ. Computational experiments were performed on a PC with an AMD Ryzen 5 3600 6-Core Processor 3.60 GHz and 16 GB of RAM. Results are given in the Table 3.

As can be seen from the Table 3 execution time of HFindMIB algorithm essentially depends on cardinality of set $\mathcal{MBC}$. This is typical for the problem of finding all maximal induced bicliques, since their number can exponentially depend on the size of the hypergraph. Other algorithms for finding all maximal induced bicliques have similar time complexity [15,16].

Table 3. Results of computational experiments

Δ	n	m	$\vert\mathcal{MBC}\vert$	t, sec
3	100	90	194	0.02
5	100	76	456	0.106
7	100	73	808	0.384
3	500	439	965	0.453
5	500	362	2218	2.378
7	500	317	4361	9.359

5 Conclusion

The solution of the MIBGP is used both for parameter estimation and for analyzing the internal structures of networks. It is well known to use maximal bicliques in the problem of determining subnets and assigning them specific labels for telecommunication networks [7]. The bicliques is also used to estimate the Shannon capacity [27], and the biclique expansion are used to determine the Hermitian rank for a Hermitian matrix represented as a graph [28]. Moreover, each individual biclique can be interpreted depending on the considered subject area.

The paper investigates the problem of finding all maximal induced bicliques of hypergraph and proposes the HFindMIB algorithm for its solution. The HFindMIB algorithm is based on the theorem of equivalence of induced bicliques of the hypergraph H and the vertex graph $L_2(H)$. Proposed HFindMIB algorithm has time complexity which has not exceeded $\mathcal{O}\Big(2^{2\Delta}\cdot\Delta\cdot(|\mathcal{MBC}|+\Delta^3\cdot\log(2^{2\Delta}))+n^2\Big)$ where Δ is the maximum vertex degree of hypergraph H.

The structure of the HFindMIB algorithm assumes the possibility of using parallel computing technologies to speed up its work. Research on improving the theoretical estimate of the time complexity of the algorithm is also promising.

References

1. Zykov, A.A.: Hypergraphs. Russian Math. Surv. **29**(6), 89–156 (1974). (in Russian)
2. Maier, D.: The Theory of Relational Databases. Mir, Moscow (1987). (in Russian)
3. Konstantinova, E.V., Skorobogatov, V.A.: Application of hypergraph theory in chemistry. Discret. Math. **235**(1–3), 365–383 (2001)
4. Ganter, B., Wille, R.: Formal Concept Analysis. Springer (1999)
5. Popov, V.B.: Extreme enumeration of the hypergraph vertex and the box clusterization problem. Din. Sist. **28**, 99–112 (2010). (in Russian)
6. Faure, N., Chretienne, P., Gourdin, E., Sourd, F.: Biclique completion problems for multicast network design. Discret. Optim. **4**(3–4), 360–377 (2007)
7. Graham, R.L., Pollak, H.O.: On the addressing problem for loop switching. Bell Syst. Tech. J. **50**(8), 2495–2519 (1971)
8. Kuznetsov, S., Obiedkov, S.: Comparing performance of algorithms for generating concept lattices. J. Exp. Theor. Artif. Intell. **14**, 189–216 (2002)

9. Li, J., Liu, G., Li, H., Wong, L.: Maximal biclique subgraphs and closed pattern pairs of the adjacency matrix: a one-to-one correspondence and mining algorithms. IEEE Trans. Knowl. Data Eng. **19**(12), 1625–1637 (2007)
10. Shaham, E., Yu, H., li, X.: On finding the maximum edge biclique in a bipartite graph: a subspace clustering approach. In: Proceedings of the 2016 SIAM International Conference on Data Mining (SDM), pp. 315–323 (2016)
11. Zhang, Y., Phillips, C.A., Rogers, G.L., Baker, E.J., Chesler, E.J., Langston, M.A.: On finding bicliques in bipartite graphs: a novel algorithm and its application to the integration of diverse biological data types. BMC Bioinform. **15**, 110 (2014)
12. Zhong-Ji, F., Ming-Xue, L., Xiao-Xin, H., Xiao-Hui, H., Xin, Z.: Efficient algorithm for extreme maximal biclique mining in cognitive frequency decision making. In: IEEE 3rd International Conference on Communication Software and Networks, pp. 25–30 (2011)
13. Tarakanov, V.E.: Kombinatornye zadachi i (0, 1)-matricy. Nauka, Moscow (1985).(In Russian)
14. Bretto, A.: Hypergraph Theory: An Introduction. Springer, Cham (2013)
15. Hermelin, D., Manoussakis, G.: Efficient enumeration of maximal induced bicliques. Discret. Appl. Math. **303**, 253–261 (2021)
16. Damaschke, P.: Enumerating maximal bicliques in bipartite graphs with favorable degree sequences. Inf. Process. Lett. **114**(6), 317–321 (2014)
17. Zverovich, I., Zverovich, I.: Bipartite bihypergraphs: a survey and new results. Discret. Math. **306**, 801–811 (2006)
18. Kuznetsov, S.O.: On computing the size of a lattice and related decision problems. Order **18**(4), 313–321 (2001)
19. Kuznetsov, S.O.: Interpretation on graphs and complexity characteristics of a search for specific patterns. Nauchno-Tekhnicheskaya Informatsiya Seriya 2 - Informatsionnye protsessy i sistemy **1**, 23–27 (1989)
20. Garey, M., Johnson, D.: Computers and Intractability. Mir, Moscow (1982).(In Russian)
21. Acuna, V., Ferreira, C.E., Freire, A.S., Moreno, E.: Solving the maximum edge biclique packing problem on unbalanced bipartite graphs. Discret. Appl. Math. **164**(1), 2–12 (2014)
22. Lyu, B., Qin, L., Lin, X., Zhang, Y., Qian, Z., Zhou, J.: Maximum biclique search at billion scale. In: Proceedings of the VLDB Endow, pp. 1359–1372 (2020)
23. Alexe, G., Alexe, S., Crama, Y., Foldes, S., Hammer, P.L., Simeone, B.: Consensus algorithms for the generation of all maximal bicliques. Discret. Appl. Math. **145**, 11–21 (2004)
24. Beineke, L., Wilson, R. J.: Topics in Algebraic Graph Theory. Mathematical Sciences Faculty Publications (2008)
25. Bykova, V.V.: The clique minimal separator decomposition of a hypergraph. J. Siberian Fed. Univ. Math. Phys. **5**(5), 36–45 (2012). (in Russian)
26. Soldatenko, A.A., Semenova, D.V.: On problem of finding all maximal induced bicliques of hypergraph. J. Siberian Fed. Univ. Math. Phys. **14**(5), 638–646 (2021). (in press)
27. Haemers, W.H.: Bicliques and eigenvalues. J. Comb. Theory **82**, 56–66 (2001)
28. Gregory, D.A., Watts, V.L., Shader, B.L.: Biclique decompositions and Hermitian rank. Linear Algebra Appl. **292**, 267–280 (1999)

Customer Experience Model for Communication Service Provider Digital Twin

Vladimir Akishin[1], Sergey Kislyakov[1,2](✉), and Alexander Sotnikov[1]

[1] The Bonch-Bruevich Saint-Petersburg State University of Telecommunications (SPbSUT), Saint-Petersburg, Russia
[2] RTC ARGUS, Saint-Petersburg, Russia
s.kislyakov@argustelecom.ru
http://www.sut.ru, http://www.argustelecom.ru

Abstract. The ultimate goal of the study is the built digital twin of a telecom operator. One of the important components of a communication service provider (CSP) digital twin is the customer's digital twin. Building a customer's digital twin is an extremely difficult task, because a number of properties of a real customer are difficult to formalize. We offer a customer experience model, which will become part of the general customer digital twin model, complementing the behavioral aspect of a real customer. General model of fuzzy cognitive maps (FCMs) was chosen as the mathematical basis for this model. For this study, we used data from two big CSPs. Based on these data, two models of FCMs were developed.

Keywords: Digital Twin · Customer Experience · Cognitive maps

1 Introduction

The global goal of this study is to develop the foundations of the theory of constructing a CSP digital twin (DT) through the construction of digital twins of CSP's different systems. The use of DT is aimed at solving a wide range of information communication management tasks. Possible DT applications are described in [1]. The DT of individual systems can be DT of info-communication infrastructure, a customer digital twin (CDT), DT of any resource. Building a CDT is an extremely difficult task, because a number of properties of a real client (for example, behavior) are difficult to formalize. Application of the customer DT will help to solve problems such as forecasting clients churn, forecasting the purchase of services, etc., better than can be solved by other methods. In this work, we suggest a model of Customer Experience (CX) as one of the parameters of the CDT. Accordingly, by simulating changes in CX, we create one of the components of the customer's digital twin. The CX model for CDT can be used to predict the movement of a customer through the stages of their life cycle. For the task we solve, we will study stages at which the client uses network resources

V. M. Vishnevskiy et al. (Eds.): DCCN 2021, CCIS 1552, pp. 148–160, 2022.
https://doi.org/10.1007/978-3-030-97110-6_11

and how they bring profit. At the moment, in the field of telecommunications, there are 9 customer life cycle stages (CLCS) [2,3].

At every single CLCS, a customer gets his new experiences - CX. The CX can be both positive and negative, but in any case, the CX of each CLCS has an impact on behavior, propensity to churn, and user loyalty at other CLCSs. In this work, we study CLCSs "Buy". As part of the client's life cycle, it is important for us to keep every single client at the "Consumption" CLCS, as well as "convince" him to buy additional services, which means even more load on the network infrastructure. In order to be able to predict the movement of the client through the CLCSs of the life cycle, we need to calculate the client's CX in real time. FCMs were used to calculate CX, and then (in the development of this study) we will calculate the probability of a client's transition to a particular CLCS of the life cycle based on the client's CX received.

2 Customer's Digital Twin as a Part of CSP Digital Twin

The development of the DT is based on the "Cross-domain model of info-communications management" (CDM) [1], based in its turn on the high-level "Domain model of info-communications" (DM) [4–6]. DM uses such terms like "information", "information interaction", "information process", "information object". Using these terms it is possible to describe the main parts (blocks) of the Info-communication system (ICS) and to describe the processes taking place in ICS. For further discussing, it is needed to put some of the new terms here. Any information system operates with information objects - images $\{\langle A\rangle, \langle B\rangle, ...\}$ of the entities $\{A, B, ...\}$.

"Information is sent" when the image transmitting from source-system A to receiver-system B signal has changed:

$$\langle A\rangle^{\xi_A} \Rightarrow \langle\langle A\rangle^{\xi_A}\rangle^{\xi_B} \tag{1}$$

"Information is perceived" when a new image of the source has arisen in the diversity of the recipient's thesaurus:

$$\langle A\rangle^{\xi_A} \longrightarrow \langle C\rangle^{\xi_C} \longrightarrow \langle\langle\langle A\rangle^{\xi_A}\rangle^{\xi_C}\rangle^{\xi_B} \tag{2}$$

"Informational impact" is the influence of a "source" A on the state of a "receiver" B, leading to a change of the image $\langle B\rangle$.

"Informational exchange" -the process of transmitting and receiving of signals, which lead to the mutual change of the images $\langle A\rangle^{\xi_B}$ and $\langle B\rangle^{\xi_A}$ to possible change of the participants thesauruses.

"Informational interaction" - is cross-change of the images of inherent systems $\langle A\rangle^{\xi_A}$ and $\langle B\rangle^{\xi_B}$, leading to change of images $\langle A\rangle^{\xi_B}$ and $\langle B\rangle^{\xi_A}$ of the other participants. "Perception of the transmitted information" - is the emergence of a new image of system A in the recipient R's variety of thesaurus of the recipient $\langle\langle A\rangle^{\xi_A}\rangle^{\xi_R}$. To understand further reasoning, we need the definitions "User",

"Potential information", "Actual information", "Information system", "Communication system", "Info-communication system". The definitions of these terms are given in references [4,5]. We suggest that you familiarize yourself with them. Within DM, in general, the elementary interaction of two information systems in the information domain is represented by the equation:

$$\langle\langle A_n\rangle^{\xi_{A_n}}\rangle^{\xi_{Cm}} \xrightarrow{Q_{22}{}^{\xi_{Cm}}{}_{Ck}} \langle\langle A_n\rangle^{\xi_{A_n}}\rangle^{\xi_{Ck}} \tag{3}$$

The DM (Fig. 1) is a universal abstract model and describes the components of an info-communication system and the information exchange within it. The model consists of three domains interconnected by relationships: a physical domain, an information domain, and a cognitive domain. The domain model is a universal abstract model and describes the components of an information-communication system and information exchange within it. The model consists of three domains interconnected by relationships: a physical domain, an information domain, and a cognitive domain. The physical domain is the domain of physical objects and energy processes. This is where physical objects interact. An information domain is a domain where information entities interact; data that is used by entities of the cognitive domain "live" here. The cognitive domain is the domain of the "intellectual function" where decisions are made. To apply the domain model to build a digital twin of a telecom operator, it is necessary to refine it taking into account the specifics of the communication industry - to introduce architecture and entities that correspond to the activities of the telecom operator.

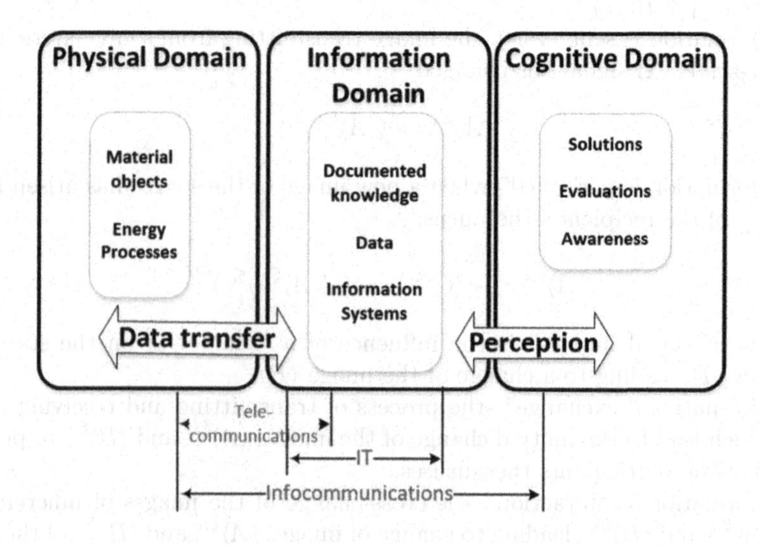

Fig. 1. The Domain Model of Info-communication System

The framework developed by the TM Forum was chosen as the main concept for finalizing the model. Today it is the most harmonious and complex system, which includes a set of interrelated standards and recommendations intended for the telecommunications, IT and digital entertainment industries and designed to facilitate the development of service-oriented software for enterprises in the industry, to increase the compatibility of its components and to simplify the interaction between the participants of the distributed value chain in the process of providing info-communication services. Frameworx (By TM Forum) seeks to describe in the form of an integrated model the activities of an info-communication company as a whole, including business processes, information exchange, IT infrastructure and interaction with partners. The TM Forum has developed a Shared Information and Data model, which has been adopted by the International Telecommunication Union (ITU) as a standard. Shared Information and Data Model (SID) - as a part of the TM Forum Frameworx - contains the definition and description of the elements and data structures involved in the business processes of an info-communication company and shared by various components of its information systems.

The new Cross-Domain Model (CDM) includes horizontal domains reflecting the specifics of info-communications: Customer Domain, Service Domain, Resource Domain, Partner Domain (Fig. 2). The CDM horizontal domains correspond to TM Forum Framewox tools: SID, enhanced Telecom Operations Map (eTOM) and Telecom Applications Map (TAM).

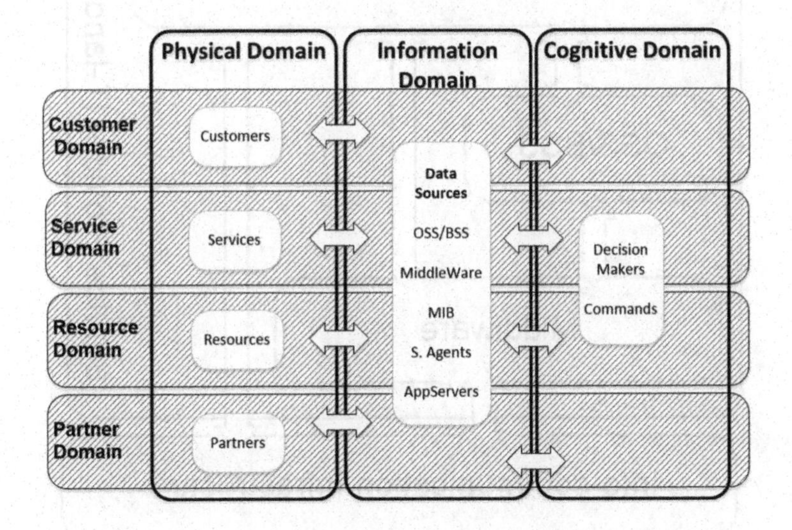

Fig. 2. The conceptual view of the Cross-Domain Model

The vertical aspects of the domain model demonstrate the unity and interaction between entities within each separate domain. These interactions, like the nature of entities, differ from domain to domain. The horizontal domains from the SID model include entities of interacting digital service provider objects. They are collected in different domains according to the principle of functional departments of the organization. For example, in the organizational structure of a service provider there is a sales department, a customer service department, a technical department, a supplier relationship department, etc. Because the new model has lots of intersecting horizontal and vertical domains, we have named it "Cross-Domain Info-communications Management model" (CDM) (Fig. 2).

In [1] it is shown that within the framework of the CSP DT a network of DTs of various systems will be formed and these DTs will interact with each other, reflecting the real processes of the CSP. The place of the CDT in the CDM is at the cross-section of the vertical ID and the horizontal Customer domain. CDM consists of vertical domains and horizontal areas. Physical Domain (PD) domain includes: operator's physical infrastructure; services with which the infrastructure is loaded; customers with whom the operator interacts with in the course of their activities; other entities of the physical world objects necessary for the realization of the business goals of the digital service provider. The Information Domain (ID) includes all sorts of data and information about PD objects obtained from a variety of available information sources (Fig. 3).

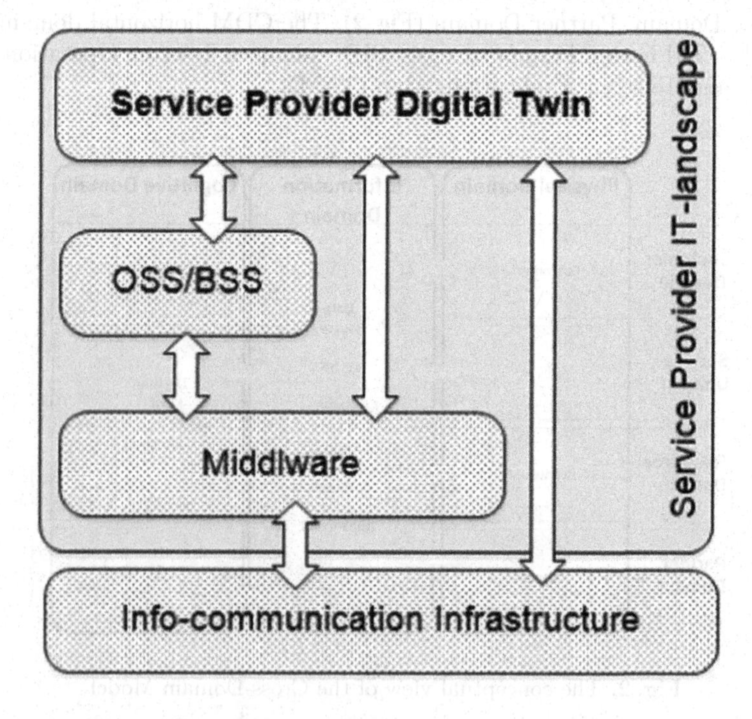

Fig. 3. The place of Digital Twin in CSP IT-landscape

In general, this is information and data presented in various formats. The Cognitive Domain (CD) in this work is not discussed, since much of it goes beyond the scope of info-communication topics. Cross-domain interaction is described in DM description. The cross-domain information exchange with feedback, which is necessary for the implementation of management and control functions. Feedback forms a closed control loop, which is introduced here into the CDM of the DT, reflecting activity of cognitive function (decision-making, analysis, etc.) in the CD.

Currently, all the data on the state of the network infrastructure, its load and events [7,8] are accumulated in the databases of Operations Support Systems (OSS) and Business Support Systems (BSS). OSS/BSS is a large class software that can be used to collect data and use it by the DT within the proposed cross-domain model. The most important feature that a CSD provides to a Service Provider is the ability to perform real-time simulations without affecting the actual physical infrastructure. That is, it becomes possible to use the DT to solve many problems, including those described at the beginning of this article. At the same time, the existing DT will not be a "black box" with unknown content. It will be a very close in properties to a real object, a complex model with input and output, which can be included in the feedback of the large "control mechanism" of post-NGN networks.

3 CX Functional Model

One of the most important components of the Customer Experience Management (CEM) is the concept of the customer life cycle [9–14]. The life cycle is the set of stages of the client's interaction with the company, starting from the moment the client becomes aware of the company and ending with the point of termination of the relationship:

- Discover (I Research, I Choose, I Validate)
- Buy (I Order, I Order, I Receive)
- Start-Up (I get set up, I am welcomed, I make my first service payment)
- Use (I Use my products and services, I manage my account, I am Valued)
- Get Support (I have a question, I have a problem, I want to escalate)
- Renew/ Leave (I renew my contract, I leave)

Based on TM Forum research [11,13], the CX functional model can be described as a structure of three levels, where each of the levels determines calculation of CX values of different levels of abstraction - from atomic metrics of CX to total values of CX (using the customer life cicle):

- Level 1 (L1) - this level describes the scenarios for collecting input data for the model, as well as how the KQIs and KPIs are calculated for operational processes that influence the formation of the customer experience model.
- Level 2 (L2) - this level describes how the quality indicators (KQI) are calculated for a specific communication channel or point of contact with the client based on the KPI/KQI results obtained at level 3. Furthermore, the obtained values of the quality of the interaction channel or point of contact are "calibrated" by subjective customer assessments of the measurement data (which are also formalized in the CX metrics model).

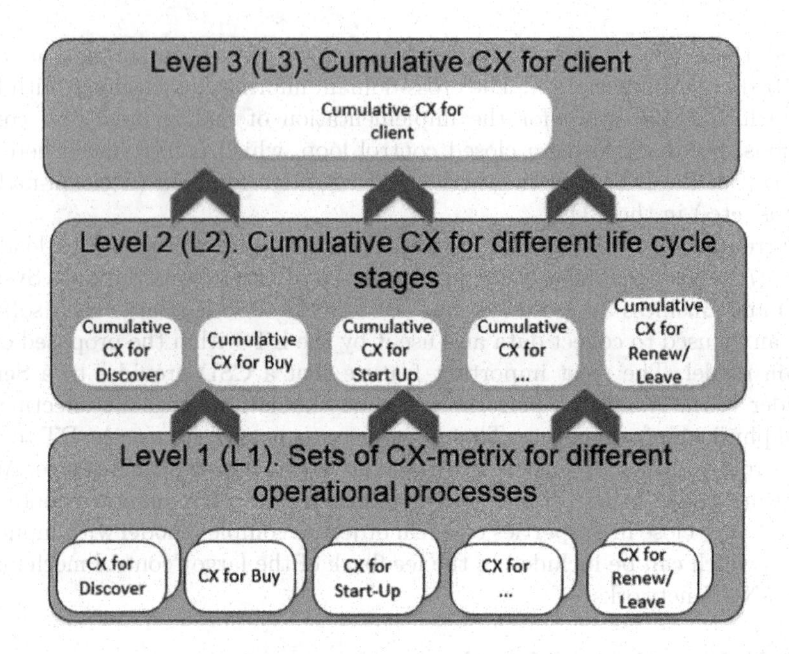

Fig. 4. 3-leveled model of CX

- Level 3 (L3) - This level describes how the Customer Experience Index is calculated in the context of a specific stage of the customer life cycle, as well as the cumulative value in the context of the entire customer life cycle.

The functional model above can be represented as an three-levelled hierarchical structure, where each level of the hierarchy will correspond to one of the levels of the functional model (Fig. 4). The hierarchical model can be decomposed into two independently computed Fuzzy Cognitive Map (FCM). The first FCM M_1, will describe an total CX scoring model for lifecycle stages, based on CX metrics. The control factors (concepts) of this FCM will be the metrics of CX. The target factors will be the values of the total CX calculated for every single CLCS. The second FCM M_2 will describe the model for assessing the total CX based on the values of CX calculated in M_2. The controlling factors of the model will be the values of the total CX per every single CLCS, and the target factor will be the final total value of CX [11,13] (Fig. 5).

4 Mathematical Model

In our model, we use FCM both for calculating the value of integral customer experience for one stage of the customer life cycle and for calculating the total CX for the entire customer life cycle. TM Forum CX metrics are used in introduced mathematical model. In general, FCM consists of nodes $(N_1, N_2, ..., N_n)$, which

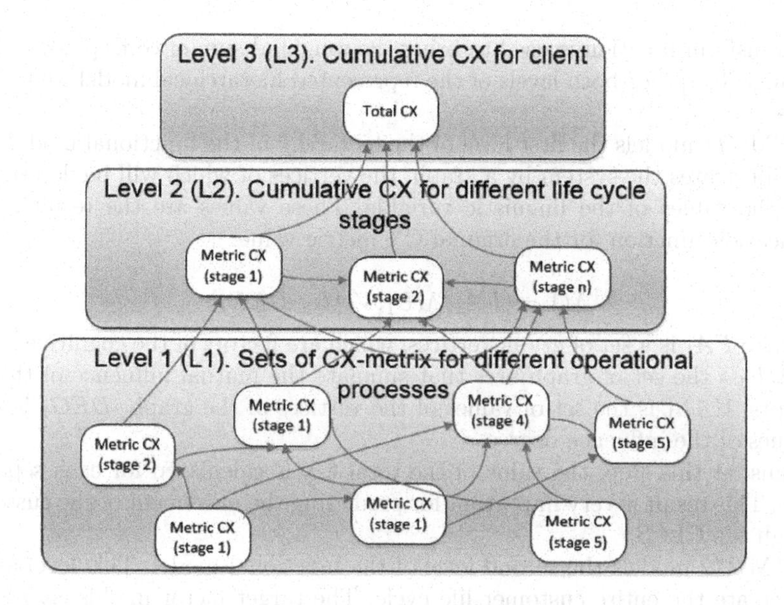

Fig. 5. Fuzzy Cognitive Map for CX calculation

are metrics of CX [15–20], and of directed arcs ($e_{i,j}$), which indicate the connection between the FCM nodes (N_i, N_j). Directional arcs are assigned fuzzy values in the interval [−1, +1]. These values show the strength of the mutual influence of factors. A positive value indicates a positive causal relationship between factors. A negative value indicates a negative relationship between two factors. Zero value corresponds to the absence of mutual influence of factors. In general, FCM is determined by the parameters N, E, C, f :

$$N = (N_1, N_2, ..., N_n) \tag{4}$$

where

– $(N_1, N_2, ..., N_n)$ are set of parameters (concepts) - the nodes of the graph. – $E : (N_i, N_j) \Rightarrow e_{i,j}$ - a function that corresponding the value of $(e_{i,j})$ to a pair of concepts (N_i, N_j), where $(e_{i,j})$ is the weight of a directed edge from N_i to N_j if $i \neq j$, and $(e_{i,j}) = 0$ if $i = j$. It means that $E(N \cdot N) = (e_{i,j})$ is connection matrix. The values of the weights on the main diagonal of the matrix are equal to zero, since changes in knowledge about the concept are not can affect the concept itself. – $C : N_i \Rightarrow C_i$ is a function that assigns to each concept N_i a sequence of its activation degrees so that for each $t \in N, C_i(t) \in L$ - is the degree of activation of the concept N_i at time t. $C(0) \in L^n$ is an initial vector containing the initial values of all concepts. $C(t) \in L^n$ is the final vector of states of concepts at a certain iteration L. – $f : R \Rightarrow L$ is the a transformation function that links $C(t+1)$ and $C(t)$ for all $t \geq 0$ so that:

$$\forall i \in \{1, 2, ..., n\}, C_i(t+1) = f(\sum_{i=1}^{n} e_{ij} \cdot C_j(t)) \tag{5}$$

The transform function is used to bring the weighted sum of concept states into the range [0; 1]. For both levels of the represented hierarchical model we use two FCMs.

FCM $M1$ models the first level of the hierarchy in the functional model. The FCM illustrates the system by a graph, the vertices of which will be determined using the values of the linguistic variable. These values are the result of the fuzzification function for the original CX metric value.

$$M1 = (FA_1, MU_1, VAL_1, DEG_1) \tag{6}$$

where FA_1 is a set of graph vertices, which are factors of the cognitive model M_1; MU_1 - the set of graph arcs that simulate the mutual influence of the CX concepts; VAL_1 is the set of values of the vertices of the graph; DEG_1 is a set of values of the influence degree.

Thus, at this step, the value of the total CX is calculated for each separate CLCS. This result is very important for predicting the movement of the customer through the CLCS.

FCM $M2$ models the second level of the functional model. This level allows to calculate the entire customer life cycle. The target factor in this case is the total CX for the entire customer's life cycle, and the model's governing factors are the values of the total CX at each CLCS, calculated as target factors of M_1. The FCM M_2:

$$M2 = (FA_2, MU_2, VAL_2, DEG_2) \tag{7}$$

Total CX for the entire customer life cycle, in essence, will be the target output parameter of the entire model, aggregating the values of CX at all CLCS.

Thus, at this step of the calculation, the solution to the problem of assessing the integral CX for the entire CLCS is simulated based on the values of CX at each CLCS.

To calculate the force of mutual influence of model factors we need to consider some "path" from the factor c_i to the factor c_j: $c_i \rightarrow c_{(i+1)} \rightarrow c_{(i+2)} \rightarrow\rightarrow c_{(j-1)} \rightarrow c_j$. This path can be defined by ordered factor indices: $(i, i+1, i+2, \ldots, j-1, j)$. Then, the indirect effect of the influence of the factor c_i on the factor c_j will be determined through the path $(i, i+1, i+2, \ldots, j-1, j)$. The overall effect of the influence of the factor c_i on the factor c_j will be determined by the set of paths N existing between these factors. The indirect effect of the influence of the factor c_i on the factor c_j:

$$C_n(c_i, c_j) = min\{e(c_p, c_{p+1}) \in (i, {k_1}^l, ..., {k_n}^l, j)\} \tag{8}$$

where C_n is the influence of the factor c_i on the factor c_j through some path n from the set of paths N; p and $p+1$ are indices of factors adjacent from left to right, through which a path is built from factor c_i to factor c_j. Operation min in this case will be equivalent to the operation of multiplication. Then, the general causal effect of the influence of the factor v_i on the factor v_j can be written as follows:

$$R_(c_i, c_j) = max(C_n(c_i, c_j)), \tag{9}$$

where $R(c_i, c_j)$ is the total influence of the factor c_i on the factor c_j through the set of paths N; $R(c_i, c_j)$ is the influence of the factor c_i on the factor c_j through the path n_i from the set of paths N.

Thus, at this step of the calculation, the problem of assessing the degree of influence of the metrics of CX and the values of the total CX is solved.

5 Results and Further Research

Models of CX have been developed to form a refined picture of a telecom operator's client in the general model of a CSP DT. The software "MentalModeller" and "Mathemematica" were used as tools for analysis and modeling of FCMs. As part of the study, 42 metrics (out of four hundred offered by TM Forum) were selected as control factors of the FCM, characterizing CX of a customer. We provide here as a result the FCM: for calculating the integral customer experience of the Buy stage for the customers of a CSP;

The initial data for building models are metrics of customer experience calculated on the basis of data from CSPs. We obtained the initial data for assessing the mutual influences of factors based on a survey of experts. Customer Experience Metrics - These are the numerical metrics of a CSP performance that experts believe impact the customer experience in the B2C segment of customers. According to the proposed model, the metrics are classified into 2 groups:

- indicators related to a specific customer and characterizing the experience of a specific customer (analogous to Per Customer Metrics in the TM Forum model);
- indicators characterizing the operating activities of the company; they are not associated with a specific customer, but they have an impact on the customer experience (analogous to Functional Metrics in the TM Forum model). This group of metrics is relevant for both models. Also, for each of the metrics, the stage of the customer's life cycle is determined to which it belongs, i.e. has the greatest impact.

The main data sources for metrics are OSS/BSS systems of a CSP: CRM, Billing, Service Desk, Work Force Management, as well as the system for collecting and analyzing customer opinions (customer opinions are collected by conducting surveys, as well as by receiving and analyzing customer comments from social networks). The range of values for each of the metrics under consideration is reduced to the range of values of the term set of the linguistic variable using the fuzzification procedure determined by experts for each metric.

To assess the mutual influence of factors in the models, two interviews of experts were conducted. In particular, the experts determined the mutual influence between the metrics of customer experience, between the metrics of customer experience and integral CX in the context of the life cycle stage, between the integral CX in the context of the life cycle stage and integral CX for the entire customer life cycle.

The model for assessing the integral customer experience for the Buy stage is based on the example of the process of connecting a new client (subscriber) from the moment when the client (subscriber) has already chosen the service that he wants to purchase and turned to one of the sales channels of the telecom operator until the moment when the service will be connected to the client. This process, in general, corresponds to the Buy stage of the TM Forum client life cycle model. Accordingly, it is possible to define a set of metrics that form an integral customer experience in the context of a given stage of the customer's life cycle. The set of metrics of the domestic telecom operator is taken as the initial data and is structured under the "reference" model of metrics from TMF. Figure 6 shows a fragment of the FCM for assessing the integrated customer experience of the Buy stage in the context of the B2C segment of the operator's customers.

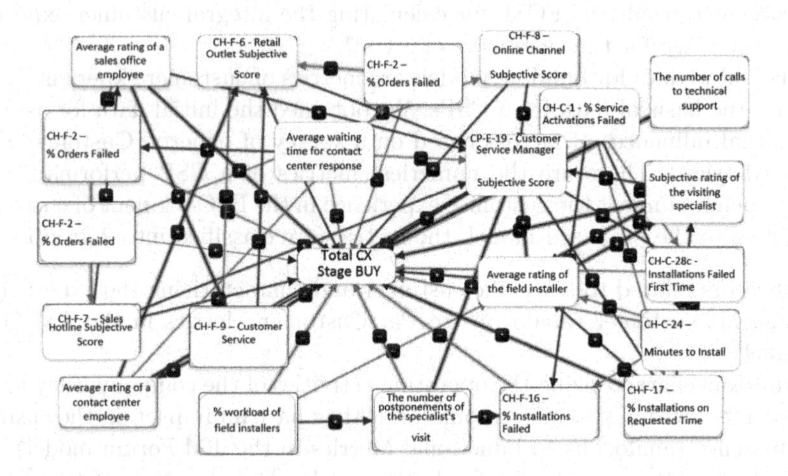

Fig. 6. Cognitive map for assessing the integrated CX of the Buy stage

6 Conclusion

The ultimate goal of the study is the built digital twin of a telecom operator. Realizing that this is a serious, large and complex task, we decomposed it into more understandable parts. We believe that the digital twin of a large system is a set of interacting digital models (small twins). By increasing the adequacy and accuracy of the models, we are gradually approaching a complex model of a telecom operator, which could be called a digital twin. The model of CX proposed in this paper is considered by us as one of the elements that make up the most accurate model of the customer - the digital twin of the customer. In the next works, we will be able to show a new, even more accurate model of CX, in which, possibly, new behavioral parameters of the client will appear.

References

1. Kislyakov, S.V.: Conceptual model of communication service provider digital twin based on infocommunication system cross-domain model. In: Proceedings of the XXth Conference Of Open Innovations Association FRUCT, 01 January 2021, vol. 28, no. 2, pp. 571–577 (2021)
2. GB962 CX Management: Introduction and Fundamentals R16.0.1. https://www.tmforum.org/resources/best-practice/gb962-customer-experience-management-introduction-and-fundamentals-r16-0-1/
3. GB962A Customer Experience Management Lifecycle Metrics R15.0.1. https://www.tmforum.org/resources/best-practice/gb962a-customer-experience-management-lifecycle-metrics-r15-0-1/
4. Sotnikov, A.D.: Infocommunication systems and their models for healthcare. Inf. Control Syst. **3**(34), 46–53 (2008)
5. Sotnikov, A.D.: Structural and functional organization of telemedicine services in applied infocommunication systems. Dissertation for the degree of Doctor of Technical Sciences, Saint-Petersburg State University of Telecommunications named after V.I. M.A. Bonch-Bruevich. Saint Petersburg (2007)
6. Sotnikov, A.D., Rogozinsky, G.G.: The multi domain infocommunication model as the basis of an auditory interfaces development for multimedia informational systems. T-Comm. **11**(5), 77–82 (2017)
7. Akishin, V.A., Kislyakov, S.V., Fenomenov, M.A.: Functional architecture of the CEM-complex for implementation in the it-landscape of a large telecom operator. T-Comm Telecommun. Transp. **10**, 12–16 (2016)
8. Akishin, V.A., Kislyakov, S.V., Phenomenov, M.A.: User experience in the operator's cognitive network management model. T-Comm. **10**(10) (2016)
9. GB962 Customer Experience Management Solution Suite. https://www.tmforum.org, https://www.tmforum.org/resources/suite/gb962-customer-experience-management-solution-suite-r17-5-0/. Accessed 2 Apr 2020
10. GB962 Customer Experience Management: Introduction and Fundamentals R16.0.1. https://www.tmforum.org, https://www.tmforum.org/resources/best-practice/gb962-customer-experience-management-introduction-and-fundamentals-r16-0-1/. Accessed 30 Apr 2020
11. GB962A Customer Experience Management Lifecycle Metrics R15.0.1. https://www.tmforum.org, https://www.tmforum.org/resources/best-practice/gb962a-customer-experience-management-lifecycle-metrics-r15-0-1/. Accessed 1 May 2020
12. RN341 Customer Experience Management Index (CEMI) Release Notes R2.0. https://www.tmforum.org, https://www.tmforum.org/resources/reference/rn341-customer-experience-management-index-cemi-release-notes-r2-0/. Accessed 1 May 2020
13. TMF066 Customer Experience Management Index Technical Specification V1.1. https://www.tmforum.org, https://www.tmforum.org/resources/technical-report-best-practice/tmf066-customer-experience-management-index-technical-specification-v1-1/. Accessed 1 Feb 2021
14. TR193 Customer Experience Management Index v1.3. https://www.tmforum.org, https://www.tmforum.org/resources/technical-report-best-practice/tr193-customer-experience-management-index-v1-3/. Accessed 10 Feb 2021
15. Goldstein, A.B., Pozharsky, N.A., Likhachev, D.A.: On cognitive maps in the management of a telecommunications operator. Informatization Commun. **1** (2016)

16. Akishin, V., Goldstein, A., Goldstein, B.: Cognitive models for access network management. In: Galinina, O., Andreev, S., Balandin, S., Koucheryavy, Y. (eds.) NEW2AN/ruSMART/NsCC -2017. LNCS, vol. 10531, pp. 375–381. Springer, Cham (2017). https://doi.org/10.1007/978-3-319-67380-6_34
17. Axelrod, R.: Structure of Decision. The Cognitive Maps of Political Elites. Princeton University Press, Princeton (1976). 405 p
18. Groumpos, P.: Fuzzy cognitive maps: basic theories and their application to complex systems. In: Glykas, M. (ed.) Fuzzy Cognitive Maps: Advances in Theory. STUDFUZZ, vol. 247, pp. 1–22. Springer, Heidelberg (2010). https://doi.org/10.1007/978-3-642-03220-2_1
19. Kosko, B.: Fuzzy cognitive maps. Int. J. Man-Mach. Stud. **24**, 65–75 (1986)
20. Kosko, B.: Fuzzy systems as universal approximators. IEEE Trans. Comput. **43**(11), 1329–1333 (1994)

Analytical Modeling of Distributed Systems

Matrix-Geometric Solutions for the Models of Perishable Inventory Systems with a Constant Retrial Rate

Agassi Melikov[1](✉), Mamed Shahmaliyev[2], and János Sztrik[3]

[1] Institute of Control Systems, National Academy of Science of Azerbaijan, Baku, Azerbaijan
[2] National Aviation Academy of Azerbaijan, Mardakan, Azerbaijan
[3] University of Debrecen, Debrecen, Hungary
sztrik.janos@inf.unideb.hu

Abstract. The model of perishable inventory system with orbit is examined under (s, S) and (s, Q) policies. The stability condition of the system is derived and the joint distribution of the number of customers in orbit and the inventory level is obtained by using matrix-geometric method. Formulas for calculation of the performance measures are developed. The behavior of performance measures under given policies are analyzed and comparative numerical results are presented.

Keywords: perishable inventory system · repeated customers · orbit · (s, S) policy · (s, Q) policy · matrix-geometric method · calculation methods · performance measures

1 Introduction

One of the important class of inventory systems (IS) is a perishable inventory systems (PIS) in which an inventory life time is a finite random quantity, for example, blood banks, systems of processing an outdated information, food provision systems, etc. In PIS the inventory level decreasing not only after its release to a customer but also due to the end of inventory life time. Note that a survey works [1–3] and a monograph [4] contain references of numerous literature sources in this direction.

Here consideration is given to models of PIS without service facility. In other words, it is assumed that inventory immediately released to primary customers (p-customers) directly from a system store house, i.e., a service time of p-customers is equal to zero. It means that formation of queue of p-customers is impossible. However, formation of an orbit of repeated p-customers (retrial customers, r-customers) is possible.

An analysis of available literature showed that PIS models without service facility had not been sufficiently studied. The paper [5] has studied a model of PIS without service station of p-customers which applies (s, Q) replenishment policy. It means that when an inventory level decreases or equals a certain level (reorder point) s a delivery order for inventory of volume $Q = S - s$, with S being the maximum volume of the system store house, is sent to a higher store house. The mentioned paper assumes that

V. M. Vishnevskiy et al. (Eds.): DCCN 2021, CCIS 1552, pp. 163–173, 2022.
https://doi.org/10.1007/978-3-030-97110-6_12

the lead time is equal to zero, and the inventory life time has an exponential distribution function (d.f.). To study the inventory level, the one-dimensional birth and death process is used. Analogous models with a positive lead time have been studied in [6,7].

In [8] the Markovian model of non-perishable IS with instant service and $(S-1, S)$ policy (i.e. one-to-one ordering policy) being investigated. It is assumed the customers that occur during the stock-out periods enter into the orbit of infinite size. The joint probability distribution of the inventory level and the number of customers in the orbit are obtained in the steady state case by applying matrix-analytical method [9]. Various system performance measures in the steady state are derived.

Note that even in an IS with instant service the queue of p-customers can be formed when inventory level is zero. Such kind models of PIS have been studied in [10–13].

This paper is close in spirit to [8]. The main contributions of this paper are as follows: (i) We extend the model investigated in [8] by considering perishable inventory items; (ii) we take into account that arrived p-customers in accordance Bernoulli scheme either join the orbit or leave the system when the inventory level is zero; (iii) we take into account that r-customers might be impatient, i.e. if upon arrivals of the r-customer the inventory level is zero, then they in accordance Bernoulli scheme either leave the system or re-join the orbit; (iv) we consider different replenishment policies, i.e. here we assume that in the system might be applied (s, Q) or (s, S) policies.

Previously, a similar model with the (s, Q) policy was studied in [14] using an approximate method based on the principles of state space merging of two-dimensional Markov chains [15]. This approach allows to find simple formulas for calculating the performance measures of the system. However, despite the simplicity and effectiveness of the specified method, it can be accepted when certain conditions are met. So, in [14], this method is applied when the following condition is met: the total intensity of the arrival of primary customers and deterioration of inventory is much higher than the intensity of the arrival of retrial customers. If this condition is not met, then the accuracy of this method is significantly reduced. Based on this, in this paper, another numerical method is developed that does not impose any conditions on the initial parameters of the system.

The rest of the paper is organized as follows. The models under study are described in Sect. 2. In Sect. 3, we perform the steady-state analysis of the system under various policies. Firstly, the stability condition of the system is derived by using Level Independent Quasi-Birth-Death Process (LIQBD) theory. Then, the joint distribution of the number of customers in orbit and the inventory level is obtained by using MGM. Main performance measures are computed in Sect. 4. Results of numerical experiments are demonstrated in Sect. 5. Conclusions are given in Sect. 6.

2 Description of the PIS Models

The inventory system has a store house of limited volume S. It is assumed that each item of the inventory, independently of the others, becomes unusable after a random time that has an exponential d.f. with parameter γ, $\gamma > 0$. Input flow of p-customers' forms Poisson stream with rate λ. If at the moment of p-customer arrival the inventory level is positive, then it is instantly serviced and leaves the system; otherwise (i.e. when inventory level is zero) the customer with probability H_p either leaves for infinity orbit to repeat its inquiry or with complementary probability $1 - H_p$ eventually leaves the system. From orbit only r-customer on the head of orbit repeat its inquiry at random time which has exponential d.f. with parameter η, i.e. retrial rate is constant value and it is independent on the number of customers in the orbit. If at the moment of a r-customer arrival inventory level is positive, then such customer is instantly serviced and leaves an orbit; otherwise the r-customer either leaves an orbit with probability H_r or with complementary probability $1 - H_r$ stays there to repeat its inquiry. Here we consider two replenishment policies: (s, S) and (s, Q). In both policies lead time is positive random variables that has exponential d.f. with the meanν^{-1}.

The problem consists in determining a joint distribution of system inventory level and the number of r-customers. This problem solution will allow us to determine performance measures as well.

3 Computation of the Steady-State Probabilities

First consider model with (s, S) policy. Mathematical model of the investigated system is two dimensional Markov chain (2-D MC). States of the indicated 2D MC are defined by 2D vectors (n, m), where n is total number of customers in orbit, $n = 0, 1, ...,$ and m is denote the inventory level, $m = 0, 1, ..., S$. State space of the indicated 2D MC is given by

$$E = \bigcup_{n=0}^{\infty} L(n), \tag{1}$$

where $L(n) = \{(n, 0), (n, 1), ..., (n, S)\}$ called the nth level, $n = 0, 1, 2, ...$

The transition rate from the state $(n_1, m_1) \in E$ to the state $(n_2, m_2) \in E$ is denoted by $q((n_1, m_1), (n_2, m_2))$. The set of all these rates forms the generator of the 2D MC. According to the accepted service scheme and replenishment policy, we obtain the following relations for the determining of the indicated transitions:

$$q((n_1, m_1), (n_2, m_2)) = \begin{cases} \lambda + m_1\gamma & \text{if } m_1 > 0,\ (n_2, m_2) = (n_1, m_1 - 1), \\ \eta & \text{if } n_1 m_1 > 0,\ (n_2, m_2) = (n_1 - 1, m_1 - 1), \\ \eta H_r & \text{if } n_1 > 0,\ m_1 = 0,\ (n_2, m_2) = (n_2 - 1, m_1), \\ \lambda H_p & \text{if } m_1 = 0,\ (n_2, m_2) = (n_1 + 1, m_1), \\ \nu & \text{if } m_1 \le s,\ (n_2, m_2) = (n_1, S), \\ 0 & \text{in other cases.} \end{cases} \tag{2}$$

Hereinafter, the equality of vectors means that their corresponding components are equal to each other.

States from the space E is renumbered in lexicographical order as follows $(0, 0), (0, 1), ..., (0, S), (1, 0), (1, 1), ..., (1, S), ...$ Then indicated 2D MC represents LIQBD for which generator has the following three diagonal form:

$$G = \begin{pmatrix} B & A_0 & \cdot & \cdot & \cdot \\ A_2 & A_1 & A_0 & \cdot & \cdot \\ \cdot & A_2 & A_1 & A_0 & \cdot \\ \cdot & \cdot & \cdot & \cdot & \cdot \\ \cdot & \cdot & \cdot & \cdot & \cdot \end{pmatrix} \tag{3}$$

All block matrices in (3) are square matrices of dimension $S + 1$. From relations (2) we conclude that entities of the block matrices $B = \left\| b_{ij} \right\|$ and $A_k = \left\| a_{ij}^{(k)} \right\|$, $i, j = 0, 1, ..., S$, are determined as follows:

$$b_{ij} = \begin{cases} \nu & \text{if } i \leq s,\ j = S, \\ \lambda + i\gamma & \text{if } i > 0,\ j = i - 1, \\ -\left(\nu + \lambda H_p\right) & \text{if } i = j = 0, \\ -(\nu + i\gamma + \lambda) & \text{if } 0 < i \leq s,\ i = j, \\ -(i\gamma + \lambda) & \text{if } s < i \leq S,\ i = j, \\ 0 & \text{in other cases;} \end{cases} \tag{4}$$

$$a_{ij}^{(0)} = \begin{cases} \lambda H_p & \text{if } i = j = 0, \\ 0 & \text{in other cases;} \end{cases} \tag{5}$$

$$a_{ij}^{(1)} = \begin{cases} \nu & \text{if } 0 \leq i \leq s,\ j = S, \\ \lambda + i\gamma & \text{if } i > 0,\ j = i - 1, \\ -\left(\nu + \lambda H_p + \eta H_r\right) & \text{if } i = j = 0, \\ -(\nu + i\gamma + \lambda + \eta) & \text{if } 0 < i \leq s,\ i = j, \\ -(i\gamma + \lambda + \eta) & \text{if } i > s,\ i = j, \\ 0 & \text{in other cases;} \end{cases} \tag{6}$$

$$a_{ij}^{(2)} = \begin{cases} \eta H_r & \text{if } i = j = 0, \\ \eta & \text{if } i > 1,\ j = i - 1, \\ 0 & \text{in other cases.} \end{cases} \tag{7}$$

Let $A = A_0 + A_1 + A_2$. Stationary distribution that correspond to the generator A is denoted by $\pi = (\pi(0), \pi(1), ..., \pi(S))$, i.e. we have

$$\pi A = \mathbf{0},\ \pi e = 1, \tag{8}$$

where $\mathbf{0}$ is null row vector of dimension $S+1$ and $\boldsymbol{e}$ is column vector of dimension $S+1$ that contains only 1's.

From relations (5)–(7) we obtain that entities of generator $A = \left\| a_{ij} \right\|$, $i, j = 0, 1, ..., S$,are determined as

$$a_{ij} = \begin{cases} -\nu & \text{if } i = j = 0, \\ \nu & \text{if } 0 \leq i \leq s,\ j = S, \\ \lambda + i\gamma + \eta & \text{if } i > 0,\ j = i - 1, \\ -(\lambda + i\gamma + \nu + \eta) & \text{if } 0 < i \leq s,\ j = i, \\ -(\lambda + i\gamma + \eta) & \text{if } i > s,\ j = i, \\ 0 & \text{in other cases.} \end{cases} \tag{9}$$

Proposition 1. Under (s, S) policy the investigated system is ergodic if and only if the following relation is fulfilled:

$$\lambda H_p \pi(0) < \eta(1 - (1 - H_r)\pi(0)). \tag{10}$$

Proof. From relations (9) we obtain that system of equations (8) has following explicit form:

$$(\nu + (m\gamma + \lambda + \eta)(1 - \delta_{m,0}))\,\pi(m) = ((m+1)\gamma + \lambda + \eta)\,\pi(m+1)\ ,\ 0 \le m \le s; \quad (11)$$

$$(m\gamma + \lambda + \eta)\,\pi(m) = ((m+1)\gamma + \lambda + \eta)\,\pi(m+1)\chi(s+1 \le m \le S-1) + \nu \sum_{m=0}^{s} \pi(m)\,\delta_{m,S},\ \ s+1 \le m \le S. \quad (12)$$

Hereinafter $\delta_{x,y}$ are denote Kronecker delta and $\chi(A)$ is indicator function of event A.

From (11) and (12) all values $\pi(m)$, $m = 1, ..., S$, are expressed by $\pi(0)$ as follows:

$$\pi(m) = \begin{cases} a_m \pi(0), \text{ if } 1 \le m \le s+1, \\ b_m \pi(0), \text{ if } s+1 < m \le S, \end{cases} \quad (13)$$

where $a_m = \prod_{i=1}^{m} \frac{\Lambda_{i-1}+\nu}{\Lambda_i}$; $b_m = \frac{\Lambda_{s+1}}{\Lambda_m} \prod_{i=1}^{s+1} \frac{\Lambda_{i-1}+\nu}{\Lambda_i}$; $\Lambda_i = \lambda + \eta + i\gamma$, $i = 1, 2, ..., S$.
The probability $\pi(0)$ is determined from normalizing condition, i.e.

$$\pi(0) = \left(1 + \sum_{m=1}^{s+1} a_m + \sum_{m=s+2}^{S} b_m\right)^{-1}.$$

In accordance to [9] (chapter 3, pages 81–83) investigated LIQBD is ergodic if and only if the following condition is fulfilled:

$$\pi A_0 e < \pi A_2 e. \quad (14)$$

By taking into account (5), (7) and (13) after some algebras from (14) we obtain that relation (10) is true.

Note 1. The ergodicity condition (10) has probabilistic meaning. Indeed, left side of (10) is equal to the rate of the primary customers in the orbit subject to inventory level is zero while right side of (10) represent weighted average total rate of the retrial customers leaving with the purchase of inventory (when the inventory level is positive) and without purchase of inventory (when inventory level is zero). Therefore, relation (10) means the following: the conditional rate of primary customers to orbit should be less than the weighted average total rate of retrial customers leaving the system.

Special Cases: 1) Fully impatient retrial customers, i.e. when $H_r = 1$ we have following ergodicity condition: $\lambda H_p \pi(0) < \eta$. 2) Patient retrial customers, when $H_r = 0$ we have following ergodicity condition: $\lambda H_p \pi(0) < \eta\ (1 - \pi(0))$. In both cases, if in addition we set $H_p = 1$, then ergodicity condition requires that the rate of primary customers to orbit should be less than the rate of retrial customers leaving the system. Note that in all cases, ergodicity condition depends on size of warehouse, as well as on perish rate of inventory and lead time.

Steady-state probabilities that corresponds to the generator matrix G we denote by $p = (p_0, p_1, p_2, ...)$, where $p_n = (p(n,0), p(n,1), ..., p(n,S))$, $n = 0, 1, ...$ Under the ergodicity condition (10) the steady-state probabilities are calculated from the following equations:

$$p_n = p_0 R^n,\ n \geq 1, \tag{15}$$

where R is nonnegative minimal solution of the following quadratic matrix equation:

$$R^2 A_2 + RA_1 + A_0 = 0.$$

Bound probabilities p_0 are determined from following system of equations with normalizing condition:

$$p_0 \left(B + RA_2\right) = \mathbf{0}.$$

$$p_0 \left(I - R\right)^{-1} e = 1, \tag{16}$$

where I is indicated identity matrix of dimension $S + 1$.

Now consider model with (s, Q)policy. State space for the model under (s, Q) policy is same with previous one, i.e. it is defined by set E and generator of appropriate 2D MC is determined by following relations:

$$q\left((n_1, m_1), (n_2, m_2)\right) = \begin{cases} \lambda + m_1\gamma & \text{if } m_1 > 0,\ (n_2, m_2) = (n_1, m_1 - 1), \\ \eta & \text{if } n_1 m_1 > 0,\ (n_2, m_2) = (n_1 - 1, m_1 - 1), \\ \eta H_r & \text{if } n_1 > 0,\ m_1 = 0,\ (n_2, m_2) = (n_2 - 1, m_1), \\ \lambda H_p & \text{if } m_1 = 0,\ (n_2, m_2) = (n_1 + 1, m_1), \\ \nu & \text{if } m_1 \leq s,\ (n_2, m_2) = (n_1, m_1 + S - s), \\ 0 & \text{in other cases.} \end{cases} \tag{17}$$

By using the indicated above lexicographical order of renumbering of states we conclude that for this model generator matrix has the following form:

$$\tilde{G} = \begin{pmatrix} \tilde{B} & A_0 & \cdot & \cdot & \cdot \\ A_2 & \tilde{A}_1 & A_0 & \cdot & \cdot \\ \cdot & A_2 & \tilde{A}_1 & A_0 & \cdot \\ \cdot & \cdot & \cdot & \cdot & \cdot \\ \cdot & \cdot & \cdot & \cdot & \cdot \end{pmatrix}$$

Entities of matrices $\tilde{B}$ and $\tilde{A}_1$ in $\tilde{G}$ are calculated as follows:

$$\tilde{b}_{ij} = \begin{cases} \nu & \text{if } i \leq s,\ j = i + S - s, \\ \lambda + i\gamma & \text{if } i > 0,\ j = i - 1, \\ -\left(\nu + \lambda H_p\right) & \text{if } i = j = 0, \\ -(\nu + i\gamma + \lambda) & \text{if } 0 < i \leq s,\ i = j, \\ -(i\gamma + \lambda) & \text{if } s < i \leq S,\ i = j, \\ 0 & \text{in other cases;} \end{cases} \tag{18}$$

$$\tilde{a}_{ij}^{(1)} = \begin{cases} \nu & \text{if } 0 \leq i \leq s,\ j = i + S - s, \\ \lambda + i\gamma & \text{if } i > 0,\ j = i - 1, \\ -\left(\nu + \lambda H_p + \eta H_r\right) & \text{if } i = j = 0, \\ -(\nu + i\gamma + \lambda + \eta) & \text{if } 0 < i \leq s,\ i = j, \\ -(i\gamma + \lambda + \eta) & \text{if } i > s,\ i = j, \\ 0 & \text{in other cases;} \end{cases} \tag{19}$$

Therefore, from relations (5), (7) and (19) we obtain that in this model entities of generator $\tilde{A} = A_0 + \tilde{A}_1 + A_2$ are determined as

$$\tilde{a}_{ij} = \begin{cases} -\nu & \text{if } i = j = 0, \\ \nu & \text{if } 0 \le i \le s,\ j = i + S - s, \\ \lambda + \eta + i\gamma & \text{if } i > 0,\ j = i - 1, \\ -(\lambda + \eta + i\gamma + \nu) & \text{if } 0 < i \le s,\ j = i, \\ -(\lambda + i\gamma + \eta) & \text{if } i > s,\ j = i, \\ 0 & \text{in other cases.} \end{cases} \quad (20)$$

Proposition 2. Under (s, Q) policy the investigated system is ergodic if and only if the condition (10) is fulfilled where $\pi(0)$ is defined as $\pi(0) = c_0\pi(s+1)$, where

$$\pi(s+1) = \left(\sum_{m=0}^{s} c_m + \sum_{m=s+1}^{S-s} d_m + \sum_{m=S-s+1}^{S} f_m \right)^{-1};$$

$$c_m = \prod_{i=m+1}^{s+1} \frac{\Lambda_i}{\nu + \Lambda_{i-1}};\ d_m = \frac{\Lambda_{s+1}}{\Lambda_m};\ f_m = \frac{\nu}{\Lambda_m} \sum_{i=m-S+s}^{s} c_i .$$

Proof. From relations (20) we obtain that balance equations for state probabilities $\pi(m)$, $0 \le m \le s$, coincide with equations (11) and balance equations for state probabilities $\pi(m)$, $s + 1 \le m \le S$, has following explicit form:

$$(m\gamma + \lambda + \eta)\pi(m) = ((m+1)\gamma + \lambda + \eta)\pi(m+1)\chi(s+1 \le m \le S - s)$$
$$+\nu\pi(m - S + s)\chi(S - s + 1 \le m \le S),\ s + 1 \le m \le S. \quad (21)$$

From (11) and (21) we obtain

$$\pi(m) = \begin{cases} c_m\pi(s+1), \text{ if } 0 \le m \le s, \\ d_m\pi(s+1), \text{ if } s+1 \le m \le S - s, \\ f_m\pi(s+1), \text{ if } S - s + 1 \le m \le S. \end{cases} \quad (22)$$

By taking into account (22) after some algebras from (14) we conclude that the fact stated above is true. Further by using system of equations (15) and (16) the steady-state probabilities for this model are calculated.

4 Performance Measures

In both models performance measures are calculated via steady-state probabilities. Main performance measures are following ones.

Average inventory level (S_{av}):

$$S_{av} = \sum_{m=1}^{S} m \sum_{n=0}^{\infty} p(n, m);$$

Average order size under (s, S) policy (V_{av}):

$$V_{av} = \sum_{m=S-s}^{S} m \sum_{n=0}^{\infty} p(n, S - m);$$

Note 2. Average order size under (s, Q) policy is constant and equal to $S - s$.

Average number of customers in orbit (L_o):

$$L_o = \sum_{n=1}^{\infty} n \sum_{m=0}^{S} p(n, m);$$

Average reorder rate (RR):

$$RR = (\lambda + (s+1)\gamma) \sum_{n=0}^{\infty} p(n, s+1) + \eta \sum_{n=1}^{\infty} p(n, s+1);$$

Loss probability of p-customers $\left(P_p\right)$:

$$P_p = \left(1 - H_p\right) \sum_{n=0}^{\infty} p(n, 0);$$

Loss probability of r-customers (P_r):

$$P_r = H_r \sum_{n=1}^{\infty} p(n, 0).$$

5 Numerical Results

In this section results of numerical experiments will be discussed and presented. The behavior of performance measures vs s under (s, S) and (s, Q) policies are depicted in Fig. 1 and Fig. 2.

We used the following parameters for numerical experiments:

$$\lambda = 40, \eta = 25, H_p = 0.7, H_r = 0.3, \nu = 10, \gamma = 25, S = 20$$

S_{av} under (s, S) policy is increasing with the increase of s as opposed to (s, Q). This behavior is expected as with higher s the inventory is replenished more frequently up to S which results in higher average inventory level. But under (s, Q) the replenishment amount is fixed $(S - s)$ and becomes lower with higher s which in turn results in lower average inventory level. Average order size V_{av} is also proportional to s which is reflected in graph. We excluded (s, Q) series from V_{av} as it is fixed for given s. RR is also lower under (s, S) policy due to higher average inventory level.

The average number of customers L_o is lower for (s, S) policy because of higher inventory level S_{av}. The higher inventory level results in more number of served customers that in turn keeps the average orbit size lower. Customer loss probabilities decrease for higher values of s due to higher S_{av} under (s, S) policy. In contrary, under (s, Q) policy for the higher values of s the slight increase in P_p and P_r is observed due to lower S_{av} which results in lesser number of served customers.

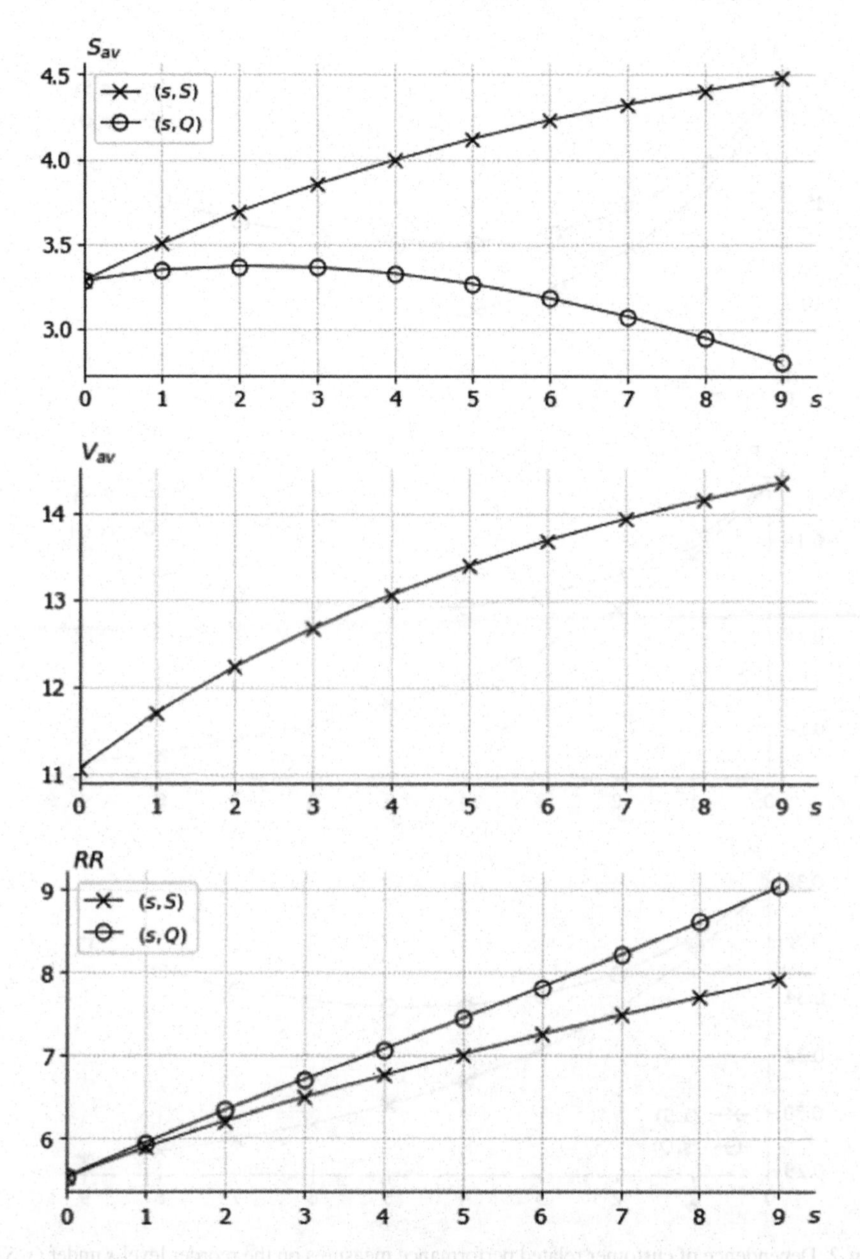

Fig. 1. Dependence of inventory related performance measures on the reorder level s under (s, S), (s, Q) policies

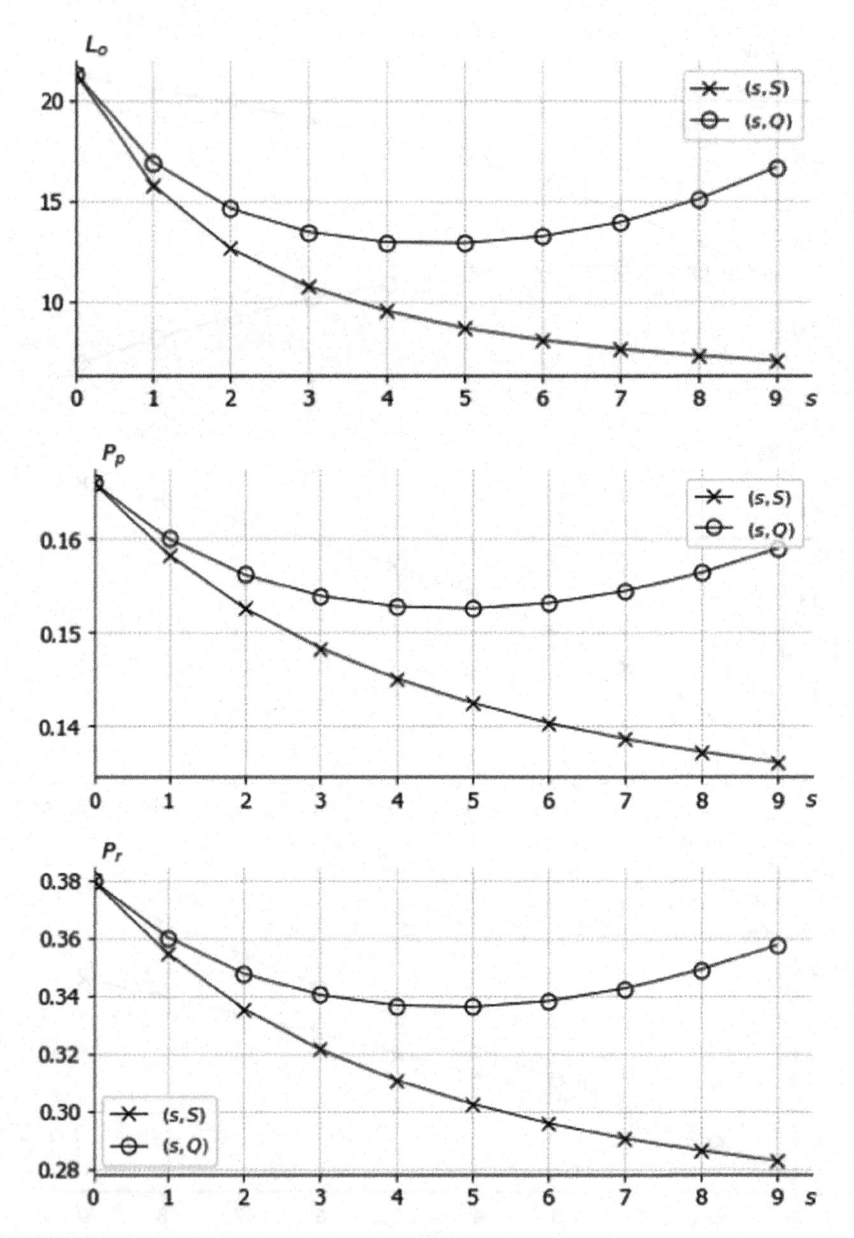

Fig. 2. Dependence of customer related performance measures on the reorder level s under (s, S), (s, Q) policies

6 Conclusion

The queuing-inventory model with perishable inventory and infinite orbit size was presented under (s, S) and (s, Q) replenishment policies. Joint distribution of the inventory level and the number of customers in the orbit was found using matrix-geometric method. Formulas for performance measures were developed. Numerical experiments were performed using developed formulas. The behavior of performance measures were analyzed under both replenishment policies and results were analyzed and presented in graphical forms.

References

1. Goyal, S., Giri, B.: Recent trends in modeling of deteriorating inventory. Eur. J. Oper. Res. **134**(1), 1–16 (2001)
2. Karaesmen, I.Z., Scheller–Wolf, A., Deniz, B.: Managing perishable and aging inventories: review and future research directions. In: Kempf, K.G., Keskinocak, P., Uzsoy, R. (eds.) Planning Production and Inventories in the Extended Enterprise. ISORMS, vol. 151, pp. 393–436. Springer, New York (2011). https://doi.org/10.1007/978-1-4419-6485-4_15
3. Bakker, M., Riezebos, J., Teunter, R.H.: Review of inventory systems with deterioration since 2001. Eur. J. Oper. Res. **221**, 275–284 (2012)
4. Nahmias, S.: Perishable Inventory Theory. ISOR, vol. 160. Springer, Heidelberg (2011). https://doi.org/10.1007/978-1-4419-7999-5
5. Liu, L.: An (s, S) continuous review models for inventory with random lifetimes. Oper. Res. Lett. **9**(3), 161–167 (1990)
6. Liu, L., Yang, T.: An (s, S) random lifetimes inventory model with positive lead time. Eur. J. Oper. Res. **113**(1), 52–63 (1999)
7. Kalpakam, S., Sapna, K.P.: Continuous review (s, S) inventory system with random lifetimes and positive lead times. Oper. Res. Lett. **16**(2), 115–119 (1994)
8. Anbazhagan, N., Wang, J., Gomathi, D.: Base stock policy with retrial demands. Appl. Math. Model. **37**, 4464–4473 (2013)
9. Neuts, M.F.: Matrix-Geometric Solutions in Stochastic Models: An Algorithmic Approach, p. 332. John Hopkins University Press, Baltimore (1981)
10. Perry, D., Stadje, W.: Perishable inventory systems with impatient demands. Math. Methods Oper. Res. **50**, 77–90 (1999)
11. Charkravarthy, S., Daniel, J.: A Markovian inventory systems with random shelf time and back orders. Comput. Ind. Eng. **47**, 315–337 (2004)
12. Ioannidis, S., et al.: Control policies for single-stage production systems with perishable inventory and customer impatience. Ann. Oper. Res. **209**, 115–138 (2012)
13. Ko, S.S., Kang, J., Kwon, E.Y.: An (s, S)-inventory model with level-dependent G/M/1-type structure. J. Ind. Manag. Optim. **12**(2), 609–624 (2016)
14. Melikov, A.Z., Ponomarenko, L.A., Shahmaliyev, M.O.: Models of perishable queuing-inventory systems with repeated customers. J. Autom. Inf. Sci. **48**(6), 22–38 (2016)
15. Ponomarenko, L., Kim, C.S., Melikov, A.: Performance Analysis and Optimization of Multi-traffic on Communication Networks, p. 208. Springer, Heidelberg (2010). https://doi.org/10.1007/978-3-642-15458-4

Analysis of Two-Way Communication Retrial Queuing Systems with Non-reliable Server, Impatient Customers to the Orbit and Blocking Using Simulation

Ádám Tóth(✉), János Sztrik, Tamás Bérczes, and Attila Kuki

University of Debrecen, Debrecen 4032, Hungary
{toth.adam,sztrik.janos,tamas.berczes,attila.kuki}@inf.unideb.hu

Abstract. The goal of this paper is to carry out a sensitivity analysis to examine the effect of different distributions of service time when blocking is applied with the help of retrial queueing systems having the property of two-way communication. This eventuates in outgoing calls (secondary customers) which are performed by the service unit after a random time in its idle state. Primary customers arrive from the finite-source according to an exponential distribution. This model does not contain queues so the service of an incoming request starts immediately if the server is functional and in an idle state. Impatience of the customers and server failures are characterized by this system which also follow an exponential distribution. The novelty of the investigation is to illustrate the effect of blocking with several figures obtained by simulation using various distributions of service time on the desired performance measures.

Keywords: Simulation · Blocking · Sensitivity analysis · Finite-source queueing system · Unreliable server · Retrial queue · Impatient customers

1 Introduction

The explosive growth of network traffic in recent years evokes the necessity of investigating communication networks to understand the behaviour of different systems. More and more communication sessions evolve partly almost every device becomes "smart" leading to higher bandwidth requirements not just in

---

The work of Ádám Tóth is supported by the ÚNKP-20-4 new national excellence program of the ministry for innovation and technology from the source of the national research, development and innovation fund. The research work of János Sztrik, Attila Kuki and Tamás Bérczes was supported by the construction EFOP-3.6.3-VEKOP-16-2017-00002. The project was supported by the European Union, co-financed by the European Social Fund.

V. M. Vishnevskiy et al. (Eds.): DCCN 2021, CCIS 1552, pp. 174–185, 2022.
https://doi.org/10.1007/978-3-030-97110-6_13

multinational companies but in our homes as well. So many unknown quantities may modify the performance of networking systems making them very complex and difficult to realize every aspect of their operation. Consequently, researchers dedicate their time to develop mathematical models describing telecommunication systems. With the help of retrial queueing systems arising real-life problems can be modelled in main telecommunication systems like telephone switching systems, call centers, or computer systems. These systems possess a virtual waiting room the so-called orbit where customers get into when the service unit is unavailable. Some examples are listed where queueing models are utilized: [1,5].

In this paper, the customer owns the impatience feature meaning that customers are able to decide to leave the system earlier without obtaining its service requisition. This is a natural occurrence of human behaviour and can be experienced in many fields of life like in healthcare applications, call centers, telecommunication networks so various works examine the effect of this phenomenon like in [11,13,15]. In these articles impatient request is portrayed: if the queue is sufficiently long balking customers choose to avoid entering the system, jockeying customers can alter queues if they encounter them may get served faster, and reneging customers leave the queue if they have waited a definite time for service.

Examining the available literature the considered models include service units that are assumed to be accessible all the time. This hypothesis does not reflect the reality as unexpected errors can take place like power outages, human negligence, or other sudden actions. Although devices are developing and become more reliable, unfortunate failures have a massive effect on the operation of the system modifying the performance measures significantly hence retrial queuing systems have been investigated in several papers recently for example in [4,8–10].

Two-way communication scheme gains ground ultimately due to its usefulness in many application fields modelling arising actual problems. One prime example is call-center where service units in an idle state may perform other activities besides satisfying the needs of incoming calls including selling, advertising, and promoting products. In other words, whenever the server is idle it may call for customers outside of the system after a random time. Utilization of such systems is always a key issue in that way many scientists are trying to optimize the service of different requests see for example [3,12,16,18].

The main focus of this paper is to carry out a sensitivity analysis inspecting the various distributions of service time of primary customers when blocking is applied on the main performance measures for instance the mean waiting time and the variance of an arbitrary, a successfully served and an impatient customer, the total utilization of the service unit, the probability of abandonment. Because giving exact formulas are difficult especially when one of the variables does not follow an exponential distribution, the obtained results are gathered by stochastic simulation based on SimPack [6] which contains the basic building blocks of the code. One of the main motivations is to develop simulation models in this way because it gives us the freedom to calculate any performance measure which we desire using various values of input parameters. The achieved results indicate the relevance of the used distributions using various parameter settings and the effect of blocking illustrated by numerous figures concentrated on the interesting phenomena of these systems.

2 System Model

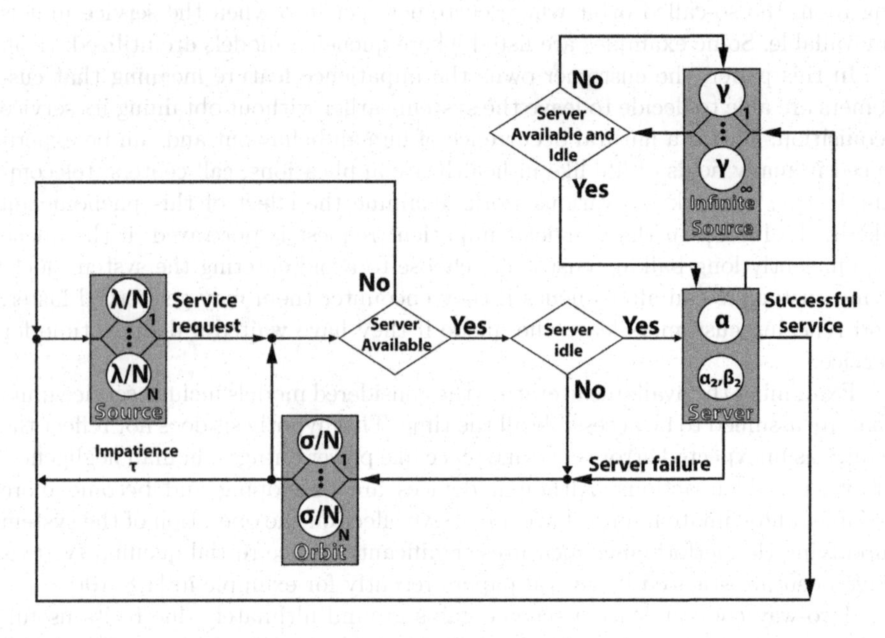

Fig. 1. System model

The regarded system is a retrial queueing system of type $M/G/1//N$ with impatient customers and an unreliable server that is capable of producing outgoing calls. N denotes the number of sources where each individual generates requests according to an exponential distribution with rate λ/N so the distribution of inter-request time is exponential with parameter λ/N (Fig. 1). There are no queues in our model in this way whenever an incoming customer finds the server in a busy state, it will be forwarded to the orbit. Otherwise, the service of an incoming customer starts instantly that follows gamma, hypo-exponential, hyper-exponential, Pareto, and lognormal distribution with different parameters but with the same mean value. During its residence in the orbit, a customer may launch an attempt to reach the service unit after an exponentially distributed time with parameter σ/N. Call generation can not occur until the end of the successful service of the individual in the source. We suppose that the service unit breaks down after an exponentially distributed time interval with parameter γ_0 when it is busy and with parameter γ_1 when idle. The repair time is also an exponentially distributed random variable with parameter γ_2 which starts instantly after a failure takes place. During a faulty period, requests can not enter the system because of blocking. Customers have impatient characteristics therefore they may decide to leave the system after waiting an exponential time in the orbit with rate τ. As mentioned earlier an idle server may perform an outgoing call towards the customers (secondary) from an infinite source after an exponentially distributed time with parameter γ. The service of secondary customers is a gamma-distributed random variable with parameters α_2 and β_2. At the time the secondary request is arriving, if the server is busy or non-operational then it will be cancelled and returns without entering the system. In the case of breakdown:

- The service of a primary request is interrupted and it is forwarded immediately towards the orbit.
- The service of a secondary request is also interrupted but it departs the system.

3 Simulation

As mentioned earlier results are obtained by a self-developed simulation program and a statistical package [7] was integrated into our code to determine the performance measures. The method of batch means is used where the useful run is divided into N batches thus $n = M - K/N$ observations are carried out in every batch. K represents the warm-up period observations at the beginning of the simulation which is rejected. M represents the length of the simulation. We just simply calculate the sample average of the whole run after the warm-up period. To have a valid estimation, batches should be long enough and the sample averages of the batches should be approximately independent. In the following articles you can find more information about this process [2,14]. The simulations are performed with a confidence level of 99.9%. The relative half-width of the confidence interval required to stop the simulation run is 0.00001. The size of a batch used to detect the initial transient duration is 1000.

Table 1 display the used values of input parameters in our scenarios.

3.1 Scenario 1

We distinguished different scenarios where the values of service times of incoming customers are different to check how the various distribution modify the operation of the system. First, the squared coefficient of variation is greater than one, and to have a valid comparison we chose the parameters that the mean and variance would be the same in every case. For this, a fitting process was performed and [17] contains detailed info about these mechanisms (Table 2).

Figure 2 displays the probability ($P(i)$) that exactly i customer is located in the system. The figure shows that there is a significant disparity among the used distributions in the average number of requests in the system. Looking carefully at the obtained curves we could state that each of them corresponds to Gaussian distribution.

Table 1. Numerical values of model parameters

N	γ_0	γ_1	σ/N	γ	α_2	β_2	τ
100	0.05	0.5	0.01	0.8	1	1	0.001

Table 2. Parameters of service time of primary customers

Distribution	Gamma	Hyper-exponential	Pareto	Lognormal
Parameters	$\alpha = 0.037$ $\beta = 0.015$	$p = 0.482$ $\lambda_1 = 0.385$ $\lambda_2 = 0.416$	$\alpha = 2.018$ $k = 1.261$	$m = -0.751$ $\sigma = 1.826$
Mean	2.5			
Variance	169			
Squared coefficient of variation	27.04			

Figure 2 demonstrates the mean waiting time of an arbitrary customer in the function of arrival intensity when the service time of the customer follows a gamma distribution. The results prove what we expected aforehand when blocking is applied lower mean waiting time is obtained especially besides higher arrival intensity. The seen ratio is true for the other used distributions as well.

After noticing the effect of blocking, the next Figure (Fig. 4) shows the comparison of mean waiting time of an arbitrary customer besides the used distributions. With increasing arrival intensity, the mean waiting time increases and then, after reaching a certain value, starts to decrease. This tendency is valid for every curve regardless of the distribution. Although having the same first two moments, maximum property characteristic of a finite-source retrial queueing system arises even with the appearance of blocking (at Fig. 3). The other noteworthy thing about the figure is that the difference between the values obtained using the different distributions is significant especially in the case of Pareto distribution.

The variance of waiting time of a successfully served customer is depicted in Fig. 5 versus arrival intensity. Interestingly the differences are significant among the used distributions in spite of the selected parameters having the same first two moments. This is especially remarkable if we compare the values at gamma distribution with the values at Pareto distribution. This performance measure starts to escalate rapidly and after λ/N reaches 0.1 variance stagnates around a certain value.

3.2 Scenario 2

In this part after observing the results of the previous scenario, we were intrigued to see the effect of another parameter setting on the performance measures. In scenario 1 the squared coefficient of variation was greater than one so this time the parameters are chosen in order the squared coefficient of variation would be less than one. Because of this, the hyper-exponential distribution can

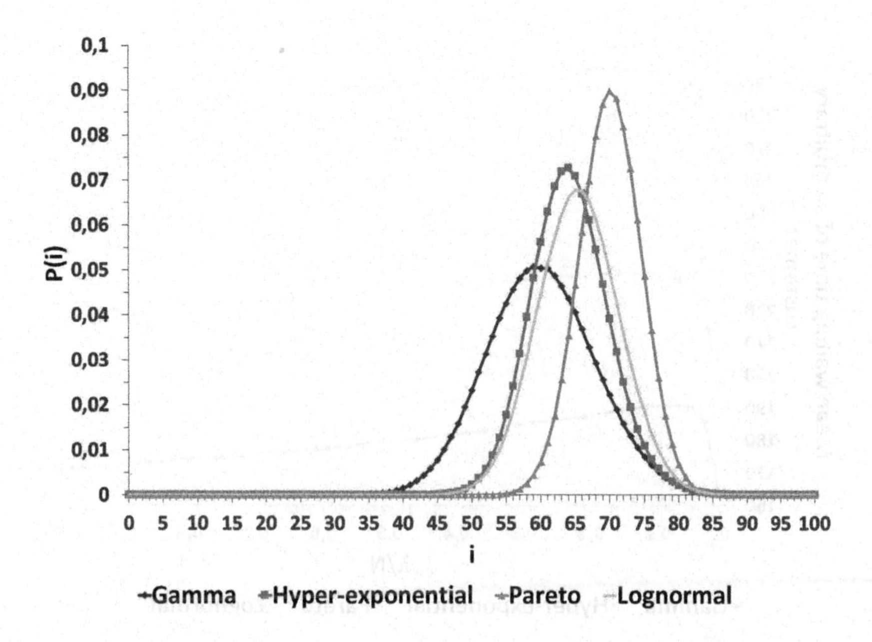

Fig. 2. Distribution of the number of customers in the system, $\lambda/N = 0.01$

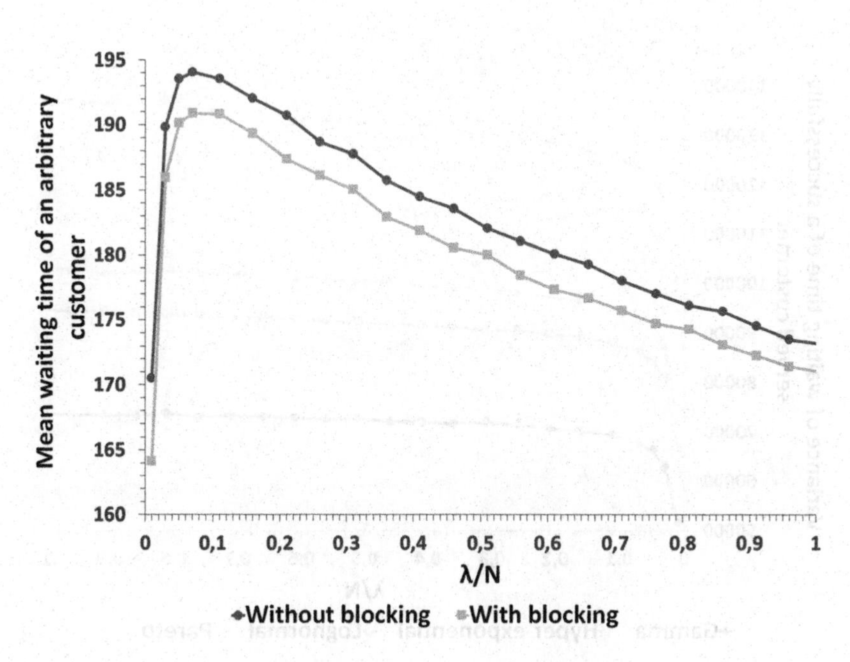

Fig. 3. The effect of blocking on the mean waiting of an arbitrary customer besides service time of gamma distribution

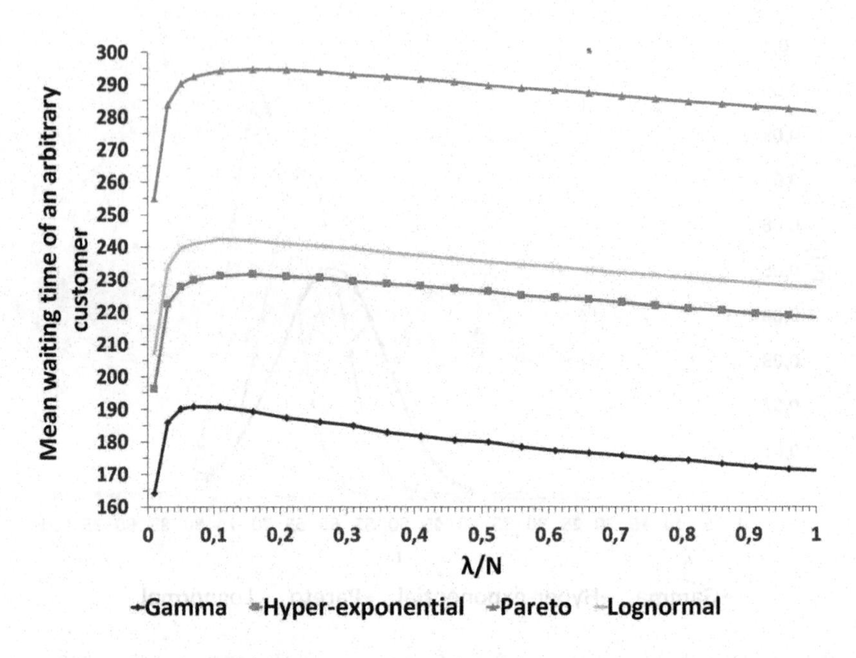

Fig. 4. The mean waiting time of an arbitrary customer

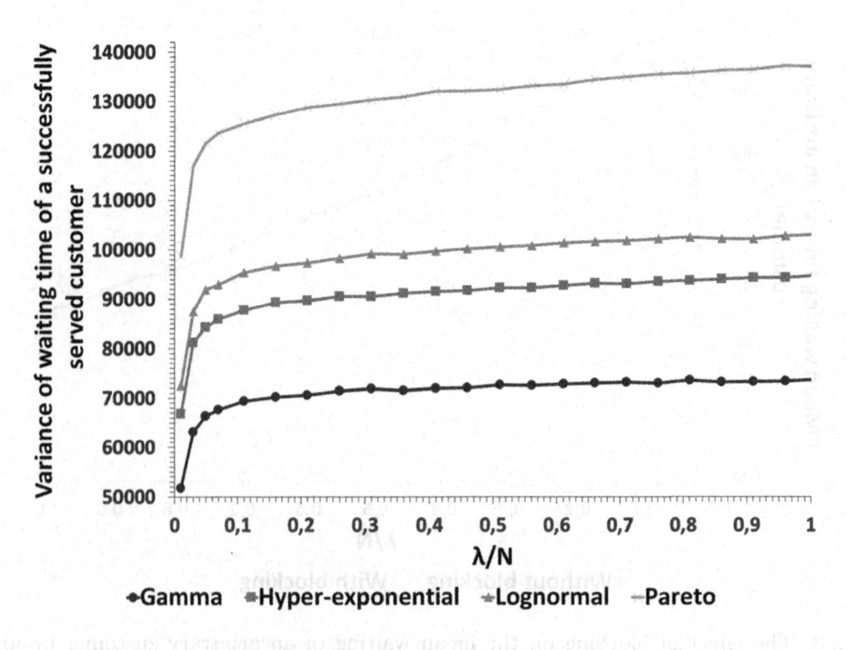

Fig. 5. The variance of waiting time of a successfully served customer

not be used that's why we replaced it with the hypo-exponential distribution. Table 3 contains the exact values of the parameters of the service time of primary customers in the case of this scenario, the other parameters remain unchanged which is shown in Table 1.

Table 3. Parameters of service time of primary customers

Distribution	Gamma	Hypo-exponential	Pareto	Lognormal
Parameters	$\alpha = 1.8$	$\mu_1 = 0.6$	$\alpha = 0.69$	$m = 2.67$
	$\beta = 0.72$	$\mu_2 = 1.2$	$k = 0.66$	$\sigma = 1.57$
Mean	2.5			
Variance	1.04			
Squared coefficient of variation	0.72222222			

The first figure (Fig. 6) shows the effect of blocking on the average waiting time of an arbitrary request as a function of the arrival intensity. With the other parameter setting, we saw that the average waiting time is lower in the blocking case, which of course applies here as well. Perhaps the only difference is that the curves are a little closer together in this scenario. However, a system with finite-source the maximum property characteristics appears even in the blocking case. It is worth mentioning that for the other distributions the difference is similar between the two cases.

How the increasing arrival intensity of the customers has an influence on the mean waiting time is illustrated in Fig. 7. Here, the mean and variance are the same again but compared to Fig. 4 the results indicate a completely different tendency. The obtained curves almost overlap each other, a minor difference can be observed at Pareto distribution but it is not significant. Similarly, after a while, the mean waiting time starts to decrease as in the previous scenario which is a characteristic of finite-source retrial queuing systems. Although the article presents results for one parameter setting, the interesting results described here were obtained are true for other settings.

After taking a closer look at the mean waiting time of an arbitrary customer, Fig. 8 demonstrates the variance of waiting time of a successfully served customer. In Scenario 1 the results in the previous scenario were significantly different from each other but here, with this parameter setting the curves are almost totally identical even for Pareto distribution. Another interesting thing about the figure is that, except for the Pareto distribution, the values obtained in this scenario are significantly higher than in the previous section.

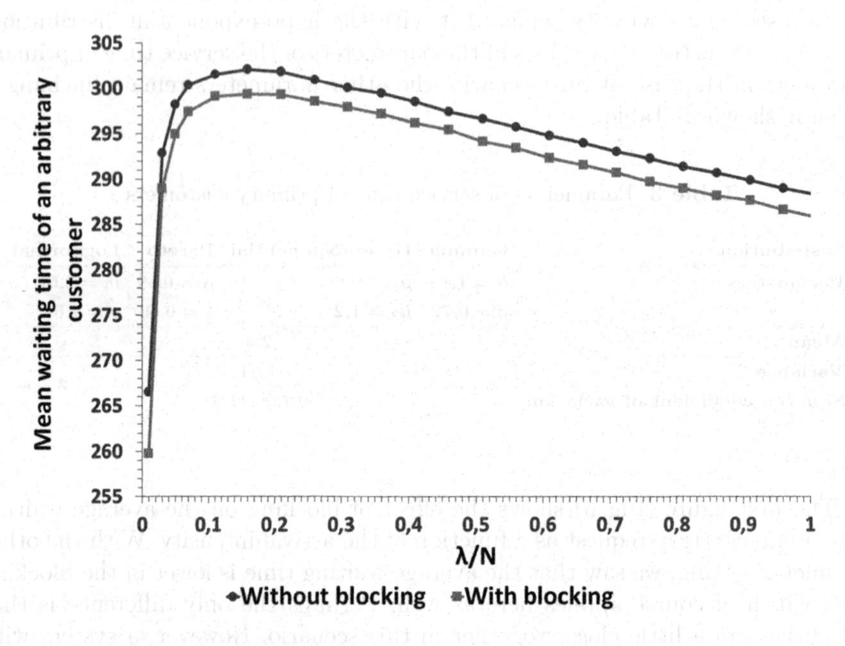

Fig. 6. The effect of blocking on the mean waiting of an arbitrary customer besides service time of gamma distribution

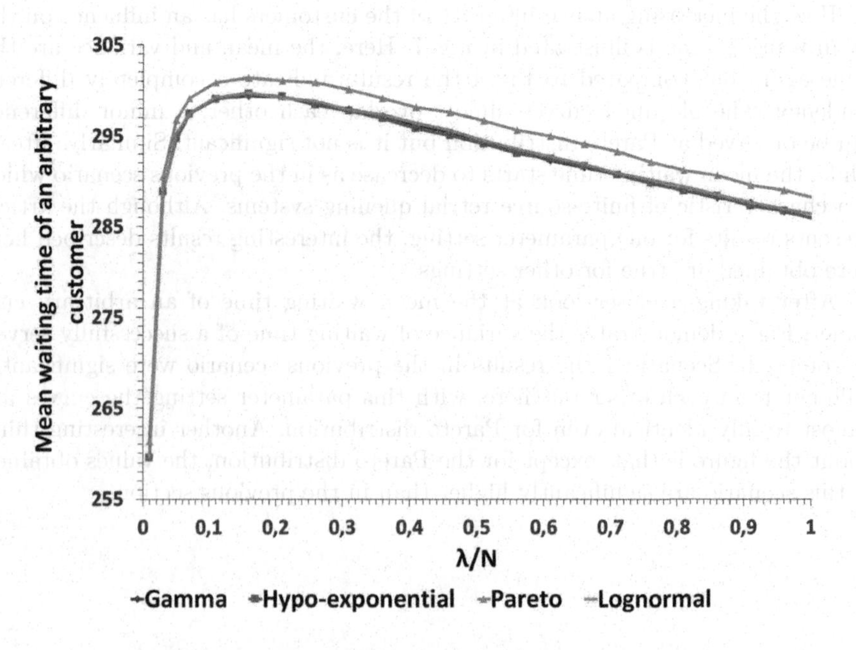

Fig. 7. The mean waiting time of an arbitrary customer

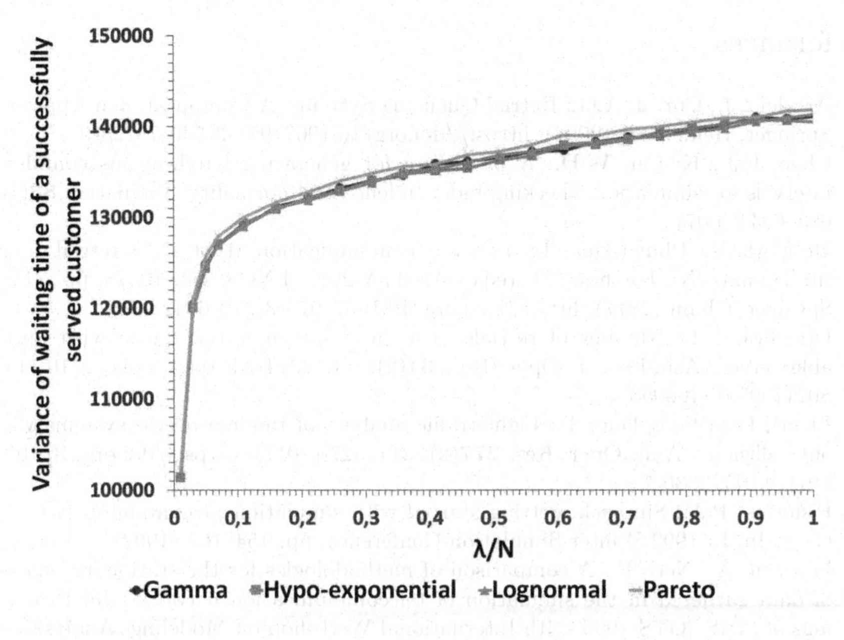

Fig. 8. The variance of waiting time of a successfully served customer

4 Conclusion

We introduced a retrial queueing system of type $M/G/1//N$ with impatient customers in the orbit and with an unreliable server having a two-way communication feature from an infinite source when blocking is implemented. Results are obtained by stochastic simulation and it is shown that the stationary probability distribution of the number of customers in the orbit tends to correspond to the Gaussian distribution despite the used distribution of service time of the primary customers. We investigated different scenarios for example when the squared coefficient of variation is greater than one the obtained values of mean waiting time of an arbitrary, successfully served customer significantly differ from each other even though the parameters are chosen that the mean and variance would be equal in case of every distribution. Results also revealed the effect of blocking which lowers the value of mean waiting time and the number of customers in the system. In our second scenario when the squared coefficient of variation is less than one interestingly the curves almost overlap each other minor disparity turns up examining all the desired performance measures. In the future, the authors intend to continue their research work, analyzing other features of the system like collisions, outgoing calls toward the customers from the orbit, or carrying out sensitivity analysis on other random variables.

References

1. Artalejo, J., Corral, A.G.: Retrial Queueing Systems: A Computational Approach. Springer, Heidelberg (2008). https://doi.org/10.1007/978-3-540-78725-9
2. Chen, E.J., Kelton, W.D.: A procedure for generating batch-means confidence intervals for simulation: checking independence and normality. Simulation **83**(10), 683–694 (2007)
3. Dragieva, V., Phung-Duc, T.: Two-way communication M/M/1//N retrial queue. In: Thomas, N., Forshaw, M. (eds.) ASMTA 2017. LNCS, vol. 10378, pp. 81–94. Springer, Cham (2017). https://doi.org/10.1007/978-3-319-61428-1_6
4. Dragieva, V.I.: Number of retrials in a finite source retrial queue with unreliable server. Asia-Pac. J. Oper. Res. **31**(2), 23 (2014). https://doi.org/10.1142/S0217595914400053
5. Fiems, D., Phung-Duc, T.: Light-traffic analysis of random access systems without collisions. Ann. Oper. Res. **277**(2), 311–327 (2017). https://doi.org/10.1007/s10479-017-2636-7
6. Fishwick, P.A.: Simpack: getting started with simulation programming in C and C++. In: In 1992 Winter Simulation Conference, pp. 154–162 (1992)
7. Francini, A., Neri, F.: A comparison of methodologies for the stationary analysis of data gathered in the simulation of telecommunication networks. In: Proceedings of MASCOTS 1996 - 4th International Workshop on Modeling, Analysis and Simulation of Computer and Telecommunication Systems, pp. 116–122, February 1996
8. Gharbi, N., Dutheillet, C.: An algorithmic approach for analysis of finite-source retrial systems with unreliable servers. Comput. Math. Appl. **62**(6), 2535–2546 (2011)
9. Gharbi, N., Ioualalen, M.: GSPN analysis of retrial systems with servers breakdowns and repairs. Appl. Math. Comput. **174**(2), 1151–1168 (2006). https://doi.org/10.1016/j.amc.2005.06.005
10. Gharbi, N., Nemmouchi, B., Mokdad, L., Ben-Othman, J.: The impact of breakdowns disciplines and repeated attempts on performances of small cell networks. J. Comput. Sci. **5**(4), 633–644 (2014)
11. Gupta, N.: Article: a view of queue analysis with customer behaviour and priorities. In: IJCA Proceedings on National Workshop-Cum-Conference on Recent Trends in Mathematics and Computing 2011 RTMC(4), May 2012
12. Kuki, A., Sztrik, J., Tóth, Á., Bérczes, T.: A contribution to modeling two-way communication with retrial queueing systems. In: Dudin, A., Nazarov, A., Moiseev, A. (eds.) ITMM/WRQ -2018. CCIS, vol. 912, pp. 236–247. Springer, Cham (2018). https://doi.org/10.1007/978-3-319-97595-5_19
13. Kumar, R., Jain, N., Som, B.: Optimization of an $M/M/1/N$ feedback queue with retention of reneged customers. Oper. Res. Decis. **24**, 45–58 (2014). https://doi.org/10.5277/ord140303
14. Law, A.M., Kelton, W.D.: Simulation Modeling and Analysis. McGraw-Hill Education, New York (1991)
15. Panda, G., Goswami, V., Datta Banik, A., Guha, D.: Equilibrium balking strategies in renewal input queue with Bernoulli-schedule controlled vacation and vacation interruption. J. Ind. Manag. Optim. **12**, 851–878 (2015). https://doi.org/10.3934/jimo.2016.12.851
16. Pustova, S.: Investigation of call centers as retrial queuing systems. Cybern. Syst. Anal. **46**(3), 494–499 (2010)

17. Sztrik, J., Tóth, Á., Pintér, Á., Bács, Z.: Simulation of finite-source retrial queues with two-way communications to the orbit. In: Dudin, A., Nazarov, A., Moiseev, A. (eds.) ITMM 2019. CCIS, vol. 1109, pp. 270–284. Springer, Cham (2019). https://doi.org/10.1007/978-3-030-33388-1_22
18. Wolf, T.: System and method for improving call center communications. US Patent App. 15/604,068, 30 November 2017

On a Queue with Marked Compound Poisson Input and Exponentially Distributed Batch Service

K. A. K. Al Maqbali(✉), V. C. Joshua, and Achyutha Krishnamoorthy

Department of Mathematics, CMS College Kottayam, Kottayam, Kerala, India
{khamis,vcjoshua,krishnamoorthy}@cmscollege.ac.in

Abstract. In this paper, we consider the batch arrival with batch service process. We assume that our queueing model has multi-server. Arrivals of customers in batches of various sizes $(1, 2, \ldots, k)$ form a marked compound Poisson process; we designate the batches as $T_1, T_2, \ldots, T_k$. The service time of T_i follows exponential distribution with parameter $\mu_i, i = 1, 2, \ldots, k$; they are served in batches of the specific size. In arrival process, waiting room of type i for T_i has finite capacity except waiting room of type 1 for T_1. In service process, server room of type i has finite capacity. T_i can go to service if server room of type i has available space. If server room of type i does not have available place, then there are two following cases. The first case, if waiting room of type i has available places, T_i must wait in this waiting room. The second case, if waiting room of type i does not have available place, then T_i must leave the system without service except T_1, who must wait in waiting room of type 1. Various performance measures are estimated with numerical solution.

Keywords: Marked Compound Poisson Process · Batch Arrival · Batch Service · Matrix Analytic Method

1 Introduction

In real-life applications, customers may arrive in different batches to the service station and each type of batch may be served with a different service rate. Niranjan and Indhira [6] reviewed some papers in the area of bulk arrival and batch service. The queueing system with batch Poisson arrival and service was analysed by Li, Fertwell and Kouvatsos [5]. Another paper was studied by Jayaramn and Matis [3] about single server markovian batch arrival model $M^{[X]}/M/1$ and single server markovian batch service model $M/M^{[X]}/1$. Baruah, Madan and Eldabi [1] studied the batch arrival queueing system with a single server queue with different service rate. Chen, Wu and Zhang [2] studied a modified Makovian bulk arrival and bulk service queues with general state dependent control.

Supported by the Indian Council for Cultural Relations (ICCR) and Ministry of Higher Education, Research and Innovation in Sultanate of Oman.

V. M. Vishnevskiy et al. (Eds.): DCCN 2021, CCIS 1552, pp. 186–200, 2022.
https://doi.org/10.1007/978-3-030-97110-6_14

Also, Krishnamoorthy, Joshua, and Kozyrev [4] studied a single server queueing inventory system with a batch Markovian arrival process and a batch Markovian service process.

In this paper, we consider the batch arrival process with batch service process. We assume that our queueing model has multi-server. In the batch arrival process, arrivals of customers in batches of various sizes $(1, 2, \ldots, k)$ form a marked compound Poisson process; we designate the batches as $T_1, T_2, \ldots, T_k$. The arrival rate of T_i is λP_i where $i = 1, 2, \ldots, k$.

The waiting room of type i is a finite capacity where $i = 2, 3. \ldots, k$. The number of batches of type T_i who are waiting in the waiting room of type i, is denoted by n_i. In other words, the size (batches) of the waiting room of type i for T_i is $0 \leq n_i \leq w_i$, where w_i is the maximum number of tables size i in the waiting room of type i for T_i where $i = 2, 3, \ldots, k$. However, the waiting room of type 1 for T_1 is an infinite capacity. The number of batches of type T_1 who are waiting in the waiting room of type 1, is denoted by n_1. Thus, $0 \leq n_1$.

In batch service process, the service time of T_i follows exponential distribution with parameter $\mu_i, i = 1, 2, \ldots, k$; they are served in batches of the specific size.

The server room of type i for T_i is a finite capacity where $i = 1, 2, \ldots, k$. The number of batches of type T_i who are in the server room of type i, is denoted by m_i. In other words, the size (batches) of the server room of type i for T_i is $0 \leq m_i \leq c_i$ where c_i is the maximum number of tables size i in the server room of type i for T_i where $i = 1, 2, 3, \ldots, k$.

On the first arrival, T_i can directly go to the server room of type i where $i = 1, 2, \ldots, k$. The second arrival T_i can go to the server room of type i if $0 \leq m_i < c_i$, which means that there is available space in the server room of type i for T_i. However, if there is no available space in the server room of type i for T_i, then T_i must wait in the waiting room of type i where $i = 1, 2, \ldots, k$.

On the next arrival, T_i can take the service in the server room of type i if T_i finds $0 \leq m_i < c_i$, which means that there is available space in the server room of type i for this T_i.

However, if T_i finds $m_i = c_i$, which means that there is unavailability space in server room of type i for this batch of type T_i where $i = 1, 2, \ldots, k$, then there are two cases for T_1 and T_i where $i = 2, 3, \ldots, k$.

The first case, T_1 must wait in the waiting room of type 1 (infinite capacity) until previous batch of type 1 completes its service in the server room of type 1. Then T_1 can go to the server room of type 1.

The second case, for T_i where $i = 2, 3, \ldots, k$ when $m_i = c_i$, T_i must wait in the waiting room (finite capacity) of type i when $0 \leq n_i < w_i$, which means that there is available space in waiting room of type i for T_i, until previous batch of type i completes its service in the server room of type i. However, if $n_i = w_i$, which means that there is no available space in the waiting room (finite capacity) of type i for T_i, then this batch must leave the system without service. So we lose current arrival T_i where $i = 2, 3, \ldots, k$. This keeps going.

2 Mathematical Description of the Model

For the analysis of the model, we introduce the following notations.

Let

$N_i(t)$ be the number of batches of type T_i in the waiting room of type i at time t where $i = 1, 2, \ldots, k$.

$M_i(t)$ be the number of batches of type T_i in the server room of type i at time t where $i = 1, 2, \ldots, k$.

$X(t) = \{(N_1(t), N_2(t), \ldots, N_{k-1}(t), N_k(t), M_1(t), M_2(t), \ldots, M_{k-1}(t), M_k(t),); t \geq 0\}$ is a continuous time Markov Chain on the state space. Therefore, this model can be studied as a level Independent Qusi-Birth-Death (LIQBD) process with state space is given by

$\Omega = \{(n_1, n_2, \ldots, n_{k-1}, n_k, m_1, m_2, \ldots, m_{k-1}, m_k)\}$ with the following conditions if $0 \leq m_i < c_i$, then $n_i = 0$ for $i = 1, 2, \ldots, k$; if $m_i = c_i$, then $0 \leq n_i \leq w_i$ for $i = 2, 3, \ldots, k$ and if $m_1 = c_1$, then $0 \leq n_1$ where c_i is the maximum number of tables size i in the server room of type i for batches of type i where $i = 1, 2, \ldots, k$ and w_i is the maximum number of tables size i in the waiting room of type i for batches of type i where $i = 2, 3, \ldots, k$.

The terms of transitions of the states are given in the Tables 1 and 2.

Table 1. Arrival rates

From	$(0, 0, \ldots, 0, 0, m_1, m_2, \ldots, m_{k-1}, m_k)$
To	$(0, 0, \ldots, 0, 0, m_1', m_2', \ldots, m_{k-1}', m_k')$
Description	m_i, m_i' where $i = 1, 2, \ldots, k$ $0 \leq m_i < c_i$ and $m_i' = m_i + 1$
Arrival rate	λP_i where $i = 1, 2, \ldots, k$
From	$(n_1, n_2, \ldots, n_{k-1}, n_k, m_1, m_2, \ldots, m_{k-1}, m_k)$
To	$(n_1', n_2', \ldots, n_{k-1}', n_k', m_1', m_2', \ldots, m_{k-1}', m_k')$
Description	m_i, m_i' where $i = 1, 2, \ldots, k$ $n_i, n_i' i = 1, 2, \ldots, k$ 1) $n_i = 0$ and $m_i \neq c_i$ and $m_i' = m_i + 1$ 2) $n_i = 0$ and $m_i = c_i$ and $n_i' = 1$ 3) $0 < n_i < w_i$ and $m_i = c_i$ and $n_i' = n_i + 1, i \neq 1$. 4) $0 < n_1$ and $m_1 = c_1$ and $n_1' = n_1 + 1, i = 1$.
Arrival rate	λP_i where $i = 1, 2, \ldots, k$

Table 2. Departure rates

From	$(0,0,\ldots,0,0,m_1,m_2,\ldots,m_{k-1},m_k)$
To	$(0,0,\ldots,0,0,m_1^{'},m_2^{'},\ldots,m_{k-1}^{'},m_k^{'})$
Description	$m_i, m_i^{'}$ where $i=1,2,\ldots,k$ $0 < m_i \le c_i$ and $m_i^{'} = m_i - 1$
Departure rate	μ_i where $i=1,2,\ldots,k$.
From	$(n_1,n_2,\ldots,n_{k-1},n_k,m_1,m_2,\ldots,m_{k-1},m_k)$
To	$(n_1^{'},n_2^{'},\ldots,n_{k-1}^{'},n_k,m_1^{'},m_2^{'},\ldots,m_{k-1}^{'},m_k^{'})$
Description	$m_i, m_i^{'}$ where $i=1,2,\ldots,k$ $n_i, n_i^{'} i = 1,2,\ldots,k$ 1) $1 \le n_i \le w_i$, $i \ne 1$ and $n_i^{'} = n_i - 1$ 2) $1 \le n_1$, $i=1$ and $n_i^{'} = n_i - 1$ 3) $n_i = 0$ and $0 < m_i \le c_i$ and $n_i^{'} = m_i - 1$.
Departure rate	μ_i where $i=1,2,\ldots,k$

The infinitesimal generator $Q^{'}$ of the level Independent Qusi-Birth-Death (LIQBD) process with state space is of the form

$$Q^{'} = \begin{pmatrix} A_{00} & A_{01} & A_{02} & \cdots & A_{0(k-1)} & A_{0k} & O & O & \cdots \\ A_{10} & A_{11} & A_{12} & \cdots & A_{1(k-1)} & A_{1k} & O & O & \cdots \\ A_{20} & A_{21} & A_{22} & \cdots & A_{2(k-1)} & A_{2k} & O & O & \cdots \\ \vdots & \vdots & \vdots & \cdots & \vdots & \vdots & \vdots & \vdots & \cdots \\ A_{(k-1)0} & A_{(k-1)1} & A_{(k-1)2} & \cdots & A_{(k-1)(k-1)} & A_{(k-1)k} & O & O & \cdots \\ A_{k0} & A_{k1} & A_{k2} & \cdots & A_{k(k-1)} & A_1 & A_0 & O & \cdots \\ O & O & O & \cdots & O & A_2 & A_1 & A_0 & O \cdots \\ O & O & O & \cdots & O & O & A_2 & A_1 & A_0 \\ & & & & & & & \ddots & \ddots \; \ddots \end{pmatrix};$$

Now, we modify the form of $Q^{'}$ by merging of cells. We get the following structure

$$Q = \begin{pmatrix} B_{00} & B_{01} & & & \\ B_{10} & A_1 & A_0 & & \\ & A_2 & A_1 & A_0 & \\ & & A_2 & A_1 & A_0 \\ & & \ddots & \ddots & \ddots \end{pmatrix}; \text{where}$$

$$B_{00} = \begin{pmatrix} A_{00} & A_{01} & A_{02} & \cdots & A_{0(k-1)} \\ A_{10} & A_{11} & A_{12} & \cdots & A_{1(k-1)} \\ A_{20} & A_{21} & A_{22} & \cdots & A_{2(k-1)} \\ \vdots & \vdots & \vdots & \cdots & \\ A_{(k-1)0} & A_{(k-1)1} & A_{(k-1)2} & \cdots & A_{(k-1)(k-1)} \end{pmatrix};$$

$$B_{01} = \begin{pmatrix} A_{0k} \\ A_{1k} \\ A_{2(k-1)} \\ \vdots \\ A_{(k-1)k} \end{pmatrix} \text{ and } B_{10} = \begin{pmatrix} A_{k0} \; A_{k1} \; A_{k2} \ldots A_{k(k-1)} \end{pmatrix}.$$

For example, when $k = 3$,

$$Q'_{k=3} = \begin{pmatrix} A00_{k=3} & A01_{k=3} & A02_{k=3} & A03_{k=3} & O & O & \ldots \\ A10_{k=3} & A11_{k=3} & A12_{k=3} & A13_{k=3} & O & O & \ldots \\ A20_{k=3} & A21_{k=3} & A22_{k=3} & A23_{k=3} & O & O & \ldots \\ A30_{k=3} & A31_{k=3} & A32_{k=3} & A1_{k=3} & A0_{k=3} & O & \ldots \\ O & O & O & A2_{k=3} & A1_{k=3} & A0_{k=3} & \\ & & & & \ddots & \ddots & \ddots \end{pmatrix}; \text{where}$$

$A00_{k=3} = (I_{(c_1+1)} \otimes S_0) + diag((\lambda P_1 e_{1\times(c_3+1)(c_2+1)c_1}), +((c_3+1)(c_2+1))) + diag((\mu_1 e_{1\times(c_3+1)(c_2+1)c_1}), -((c_3+1)(c_2+1))) + diag([(0e_{1\times(c3+1)(c2+1)})), (-\mu_1 e_{1\times(c3+1)(c2+1)c1})])$; where $S_0 = diag(V0_{A_{00}}, ((e_{1\times c_2} \otimes V0_{A_{00}}) - \mu_2 e_{1\times(c_3+1)c_2})) + (I_{(c_2+1)} \otimes diag(\lambda P_3 e_{1\times c_3}, +1)) + (I_{(c_2+1)} \otimes diag(\mu_3 e_{1\times c_3}, -1)) + diag(\lambda P_2 e_{1\times(c_3+1)c_2}, +(c_3+1)) + diag(\mu_2 e_{1\times(c_3+1)c_2}, -(c_3+1))$; where $V0_{A_{00}} = (-\lambda, -(\lambda+\mu_3)e_{1\times c_3})$;

$A01_{k=3} = \left(I_{(c_1+1)} \otimes \left(I_{(c_2+1)} \otimes \begin{pmatrix} O_{c_3\times 1} \\ \lambda P_3 \end{pmatrix} \right), O_{A01} \right)$; where

O_{A01} is zero matrix of order $(c_3+1)(c_2+1)(c_1+1) \times (c_2+1)(c_1+1)(w_3-1)$.

$A02_{k=3} = \left(\left(I_{(c_1+1)} \otimes \begin{pmatrix} O_{(c_3+1)c_2\times(c_3+1)} \\ \lambda P_2 I_{(c_3+1)} \end{pmatrix} \right), O_{A02} \right)$; where

$O_{A_{02}}$ is zero matrix of order $(c_3+1)(c_2+1)(c_1+1) \times (c_1+1)w_3 + ((c_3+1)(c_1+1) + ((c_1+1)w_3))(w_2-1)$.

$A03_{k=3} = \begin{pmatrix} O0_{A_{03}} & O1_{A_{03}} \\ \lambda P_1 I_{(c_3+1)(c_2+1)} & O2_{A_{03}} \end{pmatrix}$; where

$O0_{A_{03}}$ is zero matrix of order $(c_3+1)(c_2+1)c_1 \times (c_3+1)(c_2+1)$;

$O1_{A_{03}}$ is zero matrix of order $(c_3+1)(c_2+1)c_1 \times ((c_2+1)w_3 + (c_3+1+w_3)w_2)$;

$O2_{A_{03}}$ is zero matrix of order $(c_3+1)(c_2+1) \times ((c_2+1)w_3 + (c_3+1+w_3)w_2)$;

$A10_{k=3} = \begin{pmatrix} I_{(c_1+1)} \otimes (\mu_3 I_{(c_2+1)} \otimes (O_{1\times c_3} \; 1)) \\ O_{(c_2+1)(c_1+1)(w_3-1)\times(c_1+1)(c_2+1)(c_3+1)} \end{pmatrix}$;

$A11_{k=3} = (I_{w_3} \otimes S_1) + diag((0e_{1\times(c_2+1)(c_1+1)(w_3-1)}), \lambda P_3 e_{1\times(c_1+1)(c_2+1)})) + diag((\lambda P_3 e_{1\times(c_2+1)(c_1+1)(w_3-1)}), +((c_2+1)(c_1+1))) + diag((\mu_3 e_{1\times(c_2+1)(c_1+1)(w_3-1)}), -((c_2+1)(c_1+1)))$; where

$S_1 = diag(V0_{A_{11}}, (e_{1\times c_1} \otimes V0_{A_{11}}) - \mu_1 e_{1\times(c_2+1)c_1}) + (I_{(c_1+1)} \otimes diag((\lambda P_2 e_{1\times c_2}), +1)) + (I_{(c_1+1)} \otimes diag((\mu_2 e_{1\times c_2}), -1)) + diag((\lambda P_1 e_{1\times(c_2+1)c_1}), (c_2+1)) + diag((\mu_1 e_{1\times(c_2+1)c_1}), -(c_2+1))$; where $V0_{A_{11}} = (-(\lambda+\mu_3), -(\lambda+\mu_3+\mu_2)e_{1\times c_2})$.

$$A12_{k=3} = \left(O0_{A_{12}} \; (I_{w_3} \otimes (I_{(c_1+1)} \otimes \begin{pmatrix} O_{c_2\times 1} \\ \lambda P_2 \end{pmatrix})) \; O1_{A_{12}} \right); \text{where}$$

$O0_{A_{12}}$ is zero matrix of order $(c_2+1)(c_1+1)w_3 \times (c_3+1)(c_1+1)$;
$O1_{A_{12}}$ is zero matrix of order $(c_2+1)(c_1+1)w_3 \times ((c_3+1)(c_1+1)+(c_1+1)w_3)(w_2-1)$.

$$A13_{k=3} = \begin{pmatrix} O0_{A_{13}} & (I_{w_3} \otimes \begin{pmatrix} O_{c_1(c_2+1)\times(c_2+1)} \\ \lambda P_1 I_{(c_2+1)} \end{pmatrix}) & O1_{A_{13}} \end{pmatrix} ; \text{where}$$

$O0_{A_{13}}$ is zero matrix of order $(c_2+1)(c_1+1)w_3 \times (c_3+1)(c_2+1)$;
$O1_{A_{13}}$ is zero matrix of order $(c_2+1)(c_1+1)w_3 \times ((c_3+1+w_3)w_2)$.

$$A20_{k=3} = \begin{pmatrix} I_{(c_1+1)} \otimes \begin{pmatrix} O_{(c_3+1)\times(c_2(c_3+1))} & \mu_2 I_{(c_3+1)} \end{pmatrix} \\ O_{A_{20}} \end{pmatrix} ; \text{where}$$

$O_{A_{20}}$ is zero matrix of order $((c_1+1)w_3 + (c_1+1)(c_3+1+w_3)(w_2-1)) \times (c_1+1)(c_2+1)(c_3+1)$.

$$A21_{k=3} = \begin{pmatrix} O0_{A_{21}} \\ I_{w_3} \otimes (\mu_2 I_{(c_1+1)} \otimes \begin{pmatrix} O_{1\times c_2} & 1 \end{pmatrix}) \\ O1_{A_{21}} \end{pmatrix} ; \text{where}$$

$O0_{A_{21}}$ is zero matrix of order $(c_3+1)(c_1+1) \times (c_2+1)(c_1+1)w_3$;
$O1_{A_{21}}$ is zero matrix of order $(c_1+1)(c_3+1+w_3)(w_2-1) \times (c_2+1)(c_1+1)w_3$;

$A22_{k=3} = I_{w_2} \otimes S_2 + diag((0e_{1\times(c_1+1)(c_3+1+w_3)(w_2-1)}, \lambda P_2 e_{1\times((c_3+1)(c_1+1)+((c_1+1)w_3))}))$ $+diag((\lambda P_2 e_{1\times((c_1+1)(c_3+1+w_3)(w_2-1))}), ((c_3+1)(c_1+1)+((c_1+1)w_3)))$ $+$ $diag((\mu_2 e_{1\times((c_1+1)(c_3+1+w_3)(w_2-1))}), -((c_3+1)(c_1+1)+((c_1+1)w_3)))$; where

$$S_2 = \begin{pmatrix} S_{20} & S_{201} \\ S_{202} & S_{22} \end{pmatrix} ; \text{where}$$

$S_{20} = diag(V0_{A_{22}} + ((e_{1\times c_1} \otimes V0_{A_{22}}) - (\mu_1 e_{1\times c_1(c_3+1)}))) + (I_{(c_1+1)} \otimes diag((\lambda P_3 e_{1\times c_3}), +1)) + (I_{(c_1+1)} \otimes diag((\mu_3 e_{1\times c_3}), -1)) + diag((\lambda p_1 e_{1\times(c_3+1)c_1}), +(c_3+1)) + diag((\mu_1 e_{1\times(c_3+1)c_1}), -(c_3+1))$; where $V0_{A_{22}} = (-(\lambda+\mu_2), -(\lambda+\mu_2+\mu_3)e_{1\times c_3})$;

$S_{201} = \begin{pmatrix} (I_{(c_1+1)} \otimes \begin{pmatrix} O_{c_3\times 1} \\ \lambda P_3 \end{pmatrix}) & O_{(c_3+1)(c_1+1)\times(c_1+1)(w_3-1)} \end{pmatrix}$;

$S_{202} = \begin{pmatrix} I_{(c_1+1)} \otimes \begin{pmatrix} O_{1\times c_3} & \mu_3 \end{pmatrix} \\ O_{(c_1+1)(w_3-1)\times(c_3+1)(c_1+1)} \end{pmatrix}$;

$S_{22} = (I_{w_3} \otimes S_{21}) + diag((0e_{1\times(c_1+1)(w_3-1)}, \lambda P_3 e_{1\times(c_1+1)}))+$ $diag((\lambda P_3 e_{1\times(c_1+1)(w_3-1)}), +(c_1+1)) + diag((\mu_3 e_{1\times(c_1+1)(w_3-1)}), -(c_1+1))$;

$S_{21} = diag(V2_{A_{22}}) + diag((\lambda P_1 e_{1\times c_1}), +1) + diag((\mu_1 e_{1\times c_1}), -1)$; where
$V2_{A_{22}} = (-(\lambda+\mu_3+\mu_2), -(\lambda+\mu_3+\mu_2+\mu_1)e_{1\times c_1})$;

$A23_{k=3} = \begin{pmatrix} O0_{A_{23}} & (I_{w_2} \otimes S_{A_{23}}) \end{pmatrix}$; where

$O0_{A_{23}}$ is zero matrix of order $(c_1+1)(c_3+1+w_3)w_2 \times ((c_3+1)(c_2+1)+(c_2+1)w_3)$;

$$S_{A_{23}} = \begin{pmatrix} \begin{pmatrix} O_{(c_3+1)c_1\times(c_3+1)} \\ \lambda P_1 I_{(c_3+1)} \end{pmatrix} & O_{(c_3+1)(c_1+1)\times w_3} \\ O_{(c_1+1)w_3\times(c_3+1)} & I_{w_3} \otimes \begin{pmatrix} O_{c_1\times 1} \\ \lambda P_1 \end{pmatrix} \end{pmatrix};$$

$$A_{30_{k=3}} = \begin{pmatrix} \begin{pmatrix} O_{(c_3+1)(c_2+1)\times(c_3+1)(c_2+1)c_1} & \mu_1 I_{(c_3+1)(c_2+1)} \end{pmatrix} \\ O_{((c_2+1)w_3+(c_3+1+w_3)w_2)\times(c_3+1)(c_2+1)(c_1+1)} \end{pmatrix};$$

$$A_{31_{k=3}} = \begin{pmatrix} O_{(c_3+1)(c_2+1)\times((c_2+1)(c_1+1)w_3)} \\ I_{w_3} \otimes \begin{pmatrix} O_{(c_2+1)\times(c_2+1)c_1} & \mu_1 I_{(c_2+1)} \end{pmatrix} \\ O_{(c_3+1+w_3)w_2\times(c_2+1)(c_1+1)w_3} \end{pmatrix};$$

$$A_{32_{k=3}} = \begin{pmatrix} O_{((c_2+1)(c_3+1)+(c_2+1)w_3)\times(c_1+1)(c_3+1+w_3)w_2} \\ I_{w_2} \otimes S_{A32} \end{pmatrix}; \text{where}$$

$$S_{A32} = \begin{pmatrix} O_{(c_3+1)\times(c_3+1)c_1} & \mu_1 I_{(c_3+1)} & O_{(c_3+1)\times(c_1+1)w_3} \\ O_{w_3\times(c_3+1)c_1} & O_{w_3\times(c_3+1)} & (I_{w_3} \otimes \begin{pmatrix} O_{1\times c_1} & \mu_1 \end{pmatrix}) \end{pmatrix};$$

$$A_{1_{k=3}} = \begin{pmatrix} S_{30} & S_{31} & S_{32} \\ S_{33} & S_{34} & S_{35} \\ S_{36} & S_{37} & S_{38} \end{pmatrix}; \text{where}$$

$S_{30} = diag((V_{0_{A_1}}, ((e_{1\times c_2} \otimes V_{0_{A_1}}) - \mu_2 e_{1\times c_2(c_3+1)}))) + (I_{(c2+1)} \otimes (diag((\lambda P_3 e_{1\times c_3}), +1))) + (I_{(c2+1)} \otimes (diag((\mu_3 e_{1\times c_3}), -1))) + diag((\lambda P_2 e_{1\times(c_3+1)c_2}), +(c_3+1)) + diag((\mu_2 e_{1\times(c_3+1)c_2}), -(c_3+1))$; where $V_{0_{A_1}} = (-(\lambda+\mu_1), -(\lambda+\mu_1+\mu_3)e_{1\times c_3})$.

$$S_{31} = I_{(c_2+1)} \otimes \begin{pmatrix} O_{c_3\times 1} \\ \lambda P_3 \end{pmatrix}; \; S_{32} = \begin{pmatrix} O_{0_{S_{32}}} & \begin{pmatrix} O_{(c_3+1)c_2\times(c_3+1)} \\ \lambda P_2 I_{(c_3+1)} \end{pmatrix} & O_{1_{S_{32}}} \end{pmatrix}; \text{where}$$

$O_{0_{S_{32}}}$ is zero matrix of order $(c_3+1)(c_2+1) \times (c_2+1)(w_3-1)$; and $O_{1_{S_{32}}}$ is zero matrix of order $(c_3+1)(c_2+1) \times (w_3 + ((c_3+1+w_3)(w_2-1)))$;

$$S_{33} = \begin{pmatrix} \mu_3 I_{(c_2+1)} \otimes \begin{pmatrix} O_{1\times c_3} & 1 \end{pmatrix} \\ O_{(c_2+1)(w_3-1)\times(c_3+1)(c_2+1)} \end{pmatrix};$$

$S_{34} = (I_{w_3} \otimes diag(V_{1_{A_1}})) + (I_{w_3} \otimes (diag([\lambda P_2 e_{1\times c_2}], +1))) + (I_{w_3} \otimes (diag([\mu_2 e_{1\times c_2}], +1))) + diag((\lambda P_3 e_{1\times(c_2+1)(w_3-1)}), +(c_2+1)) + diag((\mu_3 e_{1\times(c_2+1)(w_3-1)}), -(c_2+1)) + diag((0 e_{1\times(c_2+1)(w_3-1)}, \lambda P_3 e_{1\times(c_2+1)}))$; where

$V_{1_{A1}} = (-(\lambda+\mu_1+\mu_3), -(\lambda+\mu_1+\mu_3+\mu_2)e_{1\times c_2})$;

$$S_{35} = \begin{pmatrix} O_{(c_2+1)w_3\times(c_3+1)} & (\lambda P_2 I_{w_3} \otimes \begin{pmatrix} O_{c_2\times 1} \\ 1 \end{pmatrix}) & O_{(c_2+1)w_3\times(c_3+1+w_3)(w_2-1)} \end{pmatrix};$$

$$S_{36} = \begin{pmatrix} O_{(c_3+1)\times(c_3+1)c_2} & \mu_2 I_{(c_3+1)} \\ O_{w_3\times(c_3+1)c_2} & O_{w_3\times(c_3+1)} \\ O_{(c3+1+w3)(w2-1)\times(c_3+1)c_2} & O_{(c3+1+w3)(w2-1))\times(c3+1)} \end{pmatrix};$$

$$S_{37} = \begin{pmatrix} O_{(c_3+1)\times(c_2+1)w_3} \\ I_{w_3} \otimes \begin{pmatrix} O_{1\times c_2} & \mu_2 \end{pmatrix} \\ O_{(c_3+1+w_3)(w_2-1)\times(c_2+1)w_3} \end{pmatrix};$$

$S_{38} = diag(V_{2_{S_{38}}}) + (I_{w_2} \otimes diag((\lambda P_3 e_{1\times(c_3+w_3)}), +1)) + (I_{w_2} \otimes diag((\mu_3 e_{1\times(c_3+w_3)}), -1)) + diag((\lambda P_2 e_{1\times(c_3+1+w_3)(w_2-1)}), +(c_3+1+w_3)) + diag((\mu_2 e_{1\times(c_3+1+w_3)(w_2-1)}), -(c_3+1+w_3))$; where $V_{2_{S_{38}}} = (e_{1\times w_2} \otimes V_{0_{S_{38}}}) + (0e_{1\times(c_3+1+w_3)(w_2-1)}, \lambda P_2 e_{1\times(c_3+1+w_3)})$; where $V_{0_{S_{38}}} = (-(\lambda+\mu_1+\mu_2), -(\lambda+\mu_1+\mu_2+\mu_3)e_{1\times(c_3+w_3-1)}, (-(\lambda+\mu_1+\mu_2+\mu_3)+\lambda P_3))$;

$$A_{0_{k=3}} = \left(\lambda P_1 I_{((c_2+1)(c_3+1)+(c_2+1)w_3+(c_3+1+w_3)w_2)}\right);$$

$$A_{2_{k=3}} = \left(\mu_1 I_{((c_2+1)(c_3+1)+(c_2+1)w_3+(c_3+1+w_3)w_2)}\right).$$

3 Steady-State Analysis

3.1 Stability Condition

Theorem 1. *The Markov chain with the infinitesimal generator Q of the level Independent Qusi-Birth-Death (LIQBD) is stable if and only if*

$$\lambda P_1 < \mu_1$$

Proof. Let $A = A_2 + A_1 + A_0$. We can notice that A is an irreducible matrix. Thus, the stationary vector $\boldsymbol{\pi}$ of A exists such that

$$\boldsymbol{\pi}\ A = 0$$

$$\boldsymbol{\pi}\ e = 1.$$

The Markov chain with the infinitesimal generator Q of the level Independent Qusi-Birth-Death (LIQBD) is stable if and only if

$$\boldsymbol{\pi}\ A_0 e < \boldsymbol{\pi}\ A_2 e.$$

Let $l_1 = (n_1, 0, \ldots, 0, 0, C_1, 0, \ldots, 0, 0)$, $l_2 = (n_1, 0, \ldots, 0, 0, c_1, 0, \ldots, 0, 1)$,..., $l_z = (n_1, w_2, \ldots, w_{k-1}, w_k, c_1, c_2, \ldots, c_{k-1}, c_k)$. Recall, A_0 is a square matrix of order z.

$$A_0 = \begin{pmatrix} \lambda P_1 & 0 & 0 & \ldots & 0 \\ 0 & \lambda P_1 & 0 & \ldots & 0 \\ 0 & 0 & \lambda P_1 & 0 & \ldots \\ 0 & 0 & 0 & \ddots & \vdots \\ 0 & 0 & 0 & 0 & \lambda P_1 \end{pmatrix};$$

$$\boldsymbol{\pi}\ A_0 e = (\pi_0\ , \boldsymbol{\pi_1}\ , \boldsymbol{\pi_2}\ , \ldots,\ \boldsymbol{\pi}_{(z-1)}\ ) A_0 e;$$

$$\pi\ A_0 e = (\ \pi_0\ ,\pi_1\ ,\pi_2, \ldots,\ \pi_{(z-1)}\ ) \begin{pmatrix} \lambda P_1 & & & \\ & \lambda P_1 & & \\ & & \ddots & \\ & & & \lambda P_1 \end{pmatrix} \begin{pmatrix} 1 \\ 1 \\ \vdots \\ 1 \end{pmatrix}_{(z)} ;$$

$$= (\pi_0\ \lambda P_1,\ \pi_1\ \lambda P_1,\ \pi_2\ \lambda P_1, \ldots,\ \pi_{(z-1)}\ \lambda P_1) \begin{pmatrix} 1 \\ 1 \\ \vdots \\ 1 \end{pmatrix}_{(z)} ;$$

$$= (\pi_0\ \lambda P_1 +\ \pi_1\ \lambda P_1 +\ \pi_2\ \lambda P_1 + \cdots +\ \pi_{(z-1)}\ \lambda P_1);$$

$$= ((\ \pi_0\ + (\ \pi_1\ + (\ \pi_2\ + \cdots + (\ \pi_{(z-1)}\ )\lambda P_1;$$

$$= (\sum_{i=0}^{(z-1)} \pi_i)\ \lambda P_1$$

$$= \lambda P_1.$$

Recall, A_2 is a square matrix of order z.

$$A_2 = \begin{pmatrix} \mu_1 & 0 & 0 & \ldots & 0 \\ 0 & \mu_1 & 0 & \ldots & 0 \\ 0 & 0 & \mu_1 & 0 & \ldots \\ 0 & 0 & 0 & \ddots & \vdots \\ 0 & 0 & 0 & 0 & \mu_1 \end{pmatrix};$$

$$\pi\ A_2 e = (\pi_0,\ \pi_1\ ,\pi_2\ , \ldots,\ \pi_{(z-1)}\ )A_2 e;$$

$$\pi\ A_2 e = \ (\pi_0\ ,\pi_1\ ,\pi_2\ , \ldots, \pi_{(z-1)}\ ) \begin{pmatrix} \mu_1 & & & \\ & \mu_1 & & \\ & & \ddots & \\ & & & \mu_1 \end{pmatrix} \begin{pmatrix} 1 \\ 1 \\ \vdots \\ 1 \end{pmatrix}_{(z)} ;$$

$$= (\pi_0\ \mu_1, \pi_1\ \mu_1, \pi_2\ \mu_1, \ldots, \pi_{(z-1)}\ \mu_1) \begin{pmatrix} 1 \\ 1 \\ \vdots \\ 1 \end{pmatrix}_{(z)} ;$$

$$= (\pi_0\ \mu_1 + \pi_1\ \mu_1 + \pi_2\ \mu_1 + \cdots + \pi_{(z-1)}\ \mu_1);$$

$$= (\pi_0\ + \pi_1\ + \pi_2\ + \cdots + \pi_{(z-1)})\ \mu_1;$$

$$= (\sum_{i=0}^{(z-1)} \pi_i)\ \mu_1$$

$$= \mu_1.$$

Since $\pi A_0 e = \lambda P_1$ and $\pi A_2 e = \mu_1$, then the queueing system is stable if and only if

$$\lambda P_1 < \mu_1$$

3.2 Stationary Distribution

The stationary distribution of the system process under consideration can be obtained by solving the set of Eqs. 1 and 2.

$$\mathbf{X}Q = 0 \tag{1}$$

$$\mathbf{X}e = 1. \tag{2}$$

Let $\mathbf{X}$ be decomposed with Q as following:

$\mathbf{X} = (\mathbf{X}_0, \mathbf{X}_1, \mathbf{X}_2, \dots)$ where $\mathbf{X}_0 = (\mathbf{X}_{00}, \mathbf{X}_{0n_k}, \mathbf{X}_{0n_{(k-1)}}, \dots, \mathbf{X}_{0n_2})$; where

$\mathbf{X}_{00} = (\mathbf{X}_{00m_k}, \mathbf{X}_{00m_{k-1}}, \mathbf{X}_{00m_{k-2}}, \dots, \mathbf{X}_{00m_2}, \mathbf{X}_{00m_1})$; where

$\mathbf{X}_{00m_k} = (0, 0, \dots, 0, 0, 0, 0, 0, \dots, 0, m_k)$;

$\mathbf{X}_{00m_{k-1}} = (0, 0, \dots, 0, 0, 0, 0, 0, \dots, 0, m_{k-1}, m_k)$;

$$\vdots$$

$\mathbf{X}_{00m_2} = (0, 0, \dots, 0, 0, 0, m_2, m_3, \dots, m_k)$;

$\mathbf{X}_{00m_1} = (0, 0, \dots, 0, 0, m_1, m_2, m_3, \dots, m_k)$ where $0 \le m_i \le c_i, i = 1, 2, \dots, k$.

$\mathbf{X}_{0n_k} = (0, 0, \dots, 0, n_k, m_1, m_2, m_3, \dots, c_k)$ where $0 < n_k \le w_k$ and $0 \le m_i \le c_i$ when $i \ne k$.

$\mathbf{X}_{0n_{k-1}} = (0, 0, \dots, 0, n_{k-1}, n_k, m_1, m_2, \dots, c_{k-1}, m_k)$ where $0 < n_{k-1} \le w_{k-1}$, and $0 \le m_i \le c_i$ when $i \ne k, (k-1)$. If $m_k = c_k$ then $0 < n_k \le w_k$ and if $m_k \ne c_k$ then $n_k = 0$.

$\mathbf{X}_{0n_{k-2}} = (0, 0, \dots, 0, n_{(k-2)}, n_{(k-1)}, n_k, m_1, m_2, \dots, c_{(k-2)}, m_{(k-1)}, m_k)$ where $0 < n_{k-2} \le w_{k-2}$, and $0 \le m_i \le c_i$ when $i \ne k, (k-1), (k-2)$. If $m_{k-1} = c_{k-1}$ then $0 < n_{k-1} \le w_{k-1}$ and if $m_{k-1} \ne c_{k-1}$ then $n_{k-1} = 0$. If $m_k = c_k$ then $0 < n_k \le w_k$ and if $m_k \ne c_k$ then $n_k = 0$.

$$\vdots$$

$\mathbf{X}_{0n2} = (0, n_2, n_3, \dots, n_k, m_1, C_2, \dots, m_k)$ where $0 < n_2 \le w_2$, and If $m_i = c_i$ then $n_i = 0$ where $i = 2, 3 \dots, k$. If $m_i \ne c_i$ then $0 \le n_i \le w_i$ where $i = 2, 3 \dots, k$.

$\mathbf{X}_{n_1} = (\mathbf{X}_{n_1 0}, \mathbf{X}_{n_1 n_k}, \mathbf{X}_{n_1 n_{(k-1)}}, \mathbf{X}_{n_1 n_{(k-2)}}, \dots, \mathbf{X}_{n_1 n_3}, \mathbf{X}_{n_1 n_2})$ where $0 < n_1$ and $\mathbf{X}_{n_1 0} = (n_1, 0, \dots, 0, 0c_1, m_2, \dots, m_k)$ where $0 \le m_i \le c_i, i \ne 1$;

$\mathbf{X}_{n_1 n_k} = (n_1, 0, \dots, 0, n_k, c_1, m_2, \dots, m_k)$ where $0 < n_k \le w_k$ and $0 \le m_i \le c_i, i \ne k$;

$\mathbf{X}_{n_1 n_{(k-1)}} = (n_1, 0, \dots, 0, n_{(k-1)}, n_k, c_1, m_2, m_3, \dots, c_{(k-1)}, m_k)$ where $0 < n_{k-1} \le w_{k-1}$ and $0 \le m_i \le c_i, i \ne k, (k-1)$. If $m_k \ne c_k$ then $n_k = 0$. If $m_k = c_k$ then $0 < n_k \le w_k$.

$$\vdots$$

$\mathbf{X}_{n_1n_3} = (n_1, 0, n_3, n_4, \ldots, n_{(k-1)}, n_k, c_1, m_2, c_3, m_4, \ldots, m_{(k-1)}, m_k)$ where $0 < n_3 \le w_3$ and $0 \le m_i \le c_i, i \ne 1, 3$. If $m_i \ne c_i$ then $n_i = 0$ where $i = 4, 5, \ldots, k$. If $m_i = c_i$ then $0 < n_i \le w_i$ where $i = 4, 5, \ldots, k$.

$\mathbf{X}_{n_1n_2} = (n_1, n_2, n_3, n_4, \ldots, n_{(k-1)}, n_k, c_1, c_2, m_3, m_4, \ldots, m_{(k-1)}, m_k)$ where $0 < n_2 \le w_2$ and $0 \le m_i \le c_i, i \ne 1, 2$. If $m_i \ne c_i$ then $n_i = 0$ where $i = 3, 4, \ldots, k$. If $m_i = c_i$ then $0 < n_i \le w_i$ where $i = 3, 4, \ldots, k$.

From Eqs. 1 and 2, we can solve the system of equations

$$\mathbf{X}_0 B_{00} + \mathbf{X}_1 B_{10} = 0; \tag{3}$$

$$\mathbf{X}_0 B_{01} + \mathbf{X}_1 A_1 + \mathbf{X}_2 A_2 = 0; \tag{4}$$

$$\vdots$$

$$\mathbf{X}_{i-1} A_0 + \mathbf{X}_i A_1 + \mathbf{X}_{i+1} A_2 = 0 \text{ for } i = 2, 3, \ldots. \tag{5}$$

From Matrix analytic method, we can obtain the following equations $\mathbf{X}_{n_1} = \mathbf{X}_1 R^{n_1 - 1}$ for $n_1 = 2, 3, \ldots$. For more details, we can see Stewart [7].

4 Performance Measures

We obtain some performance measures of the system under steady state when $k = 3$ as following:

1. Expected number of T_1 in the waiting room of type 1

$$E[N_{T_1}] = \sum_{n_1=1}^{\infty} n_1 \mathbf{X}_{n_1} e.$$

2. The probability that the server room of type 1 for T_1 is idle

$$P_{0T_1} = \sum_{i=2}^{k=3} \mathbf{X}_{00m_i} e + \sum_{n_2=1}^{w_2} \mathbf{X}_{0n_{2,m_1=0}} e + \sum_{n_3=1}^{w_3} \mathbf{X}_{0n_{3,m_1=0}} e.$$

3. The probability that the server room of type 1 for T_1 is busy

$$P_{1_{T_1}} = 1 - P_{0T_1}.$$

4. Expected number of T_2 in the waiting room of type 2

$$E[N_{T_2}] = \sum_{n_2=1}^{w_2} n_2 \mathbf{X}_{0n_2} e + \sum_{n_1=1}^{\infty} \sum_{n_2=1}^{w_2} n_2 \mathbf{X}_{n_1n_2} e.$$

5. The probability that the server room of type 2 for T_2 is idle

$$P_{0T_2} = \mathbf{X}_{00m_{1,m_2=0}} e + \mathbf{X}_{00m_3} e + \sum_{n_3=1}^{w_3} \mathbf{X}_{0n_{3,m_2=0}} e + \sum_{n_1=1}^{\infty} \mathbf{X}_{0n_{1,m_2=0}} e.$$

6. The probability that the server room of type 2 for T_2 is busy

$$P_{1_{T_2}} = 1 - P_{0T_2}.$$

7. Expected number of T_3 in the waiting room of type3

$$E[N_{T_3}] = \sum_{n_3=1}^{w_3} n_3 \mathbf{X}_{0n_3} e + \sum_{n_2=1}^{w_2} \sum_{n_3=1}^{w_3} n_3 \mathbf{X}_{0n_2} e + \sum_{n_1=1}^{\infty} \sum_{n_2=1}^{w_2} \sum_{n_3=1}^{w_3} n_3 \mathbf{X}_{n_1 n_3} e.$$

8. The probability that the server room of type 3 for T_3 is idle

$$P_{0T_3} = \sum_{i=1}^{k=2} \mathbf{X}_{00m_i} e + \sum_{n_2=1}^{w_2} \mathbf{X}_{0n_{2,m_3=0}} e + \sum_{n_1=1}^{\infty} \mathbf{X}_{0n_{1,m_3=0}} e.$$

9. The probability that the server room of type 3 for T_3 is busy

$$P_{1_{T_3}} = 1 - P_{0T_3}.$$

5 Numerical Example

To illustrate the performance measures of the system numerically, we consider the system with $k = 3$. Arrivals of customers in batches of various sizes $(1, 2, 3)$ form a marked compound Poisson process; we designate the batches as T_1, T_2 and T_3. We fix arrival rate $\lambda = 3$ with $P_1 = 0.2, P_2 = 0.5$ and $P_3 = 0.3$. The service time of T_i follows exponential distribution with parameter $\mu_1 = 0.7, \mu_2 = 0.3$ and $\mu_3 = 0.5$ where $i = 1, 2, 3$. The maximum number of tables size 2 and 3 in the waiting room of type 2 and 3 for T_2 and T_3 are $w_2 = 3$ and $w_3 = 2$ respectively. The maximum number of tables size 2 and 3 in the server room of type 1, 2 and 3 for T_1, T_2 and T_3 are $c_1 = 2, c_2 = 4$ and $c_3 = 3$ respectively. Then, we find the performance measures of the system corresponding to above parameters in the Table 3.

Table 3. Some performance measures of the system with $k = 3$.

1	$E[NT_1]$	4.4082	$E[NT_2]$	2.7520	$E[NT_3]$	1.1702
2	$P_{0_{T_1}}$	0.1429	$P_{0_{T_2}}$	0.0009	$P_{0_{T_3}}$	0.0242
3	$P_{1_{T_1}}$	0.8571	$P_{1_{T_2}}$	0.9991	$P_{1_{T_3}}$	0.9758

5.1 The Effect of c_1 on $E[N_{T_1}]$ and the Effect of w_2 and w_3 on $E[N_{T_2}]$ and on $E[N_{T_3}]$, Respectively

In this section, we show the effect of the maximum number of tables size 1 in the server room of type 1 for T_1 (c_1) on the expected number of T_1 in the waiting room of type 1 ($E[N_{T_1}]$). Besides this, we present the effect of the maximum number of tables size i in the waiting room of type i for T_i (w_i) on expected number of T_i in the waiting room of type i ($E[N_{T_i}]$) where $i = 2, 3$ as following:

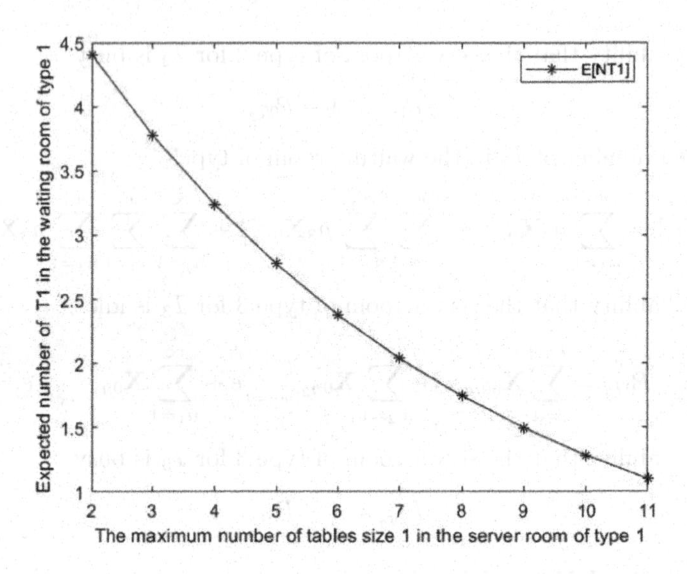

Fig. 1. Effect of c_1 on $E[N_{T_1}]$.

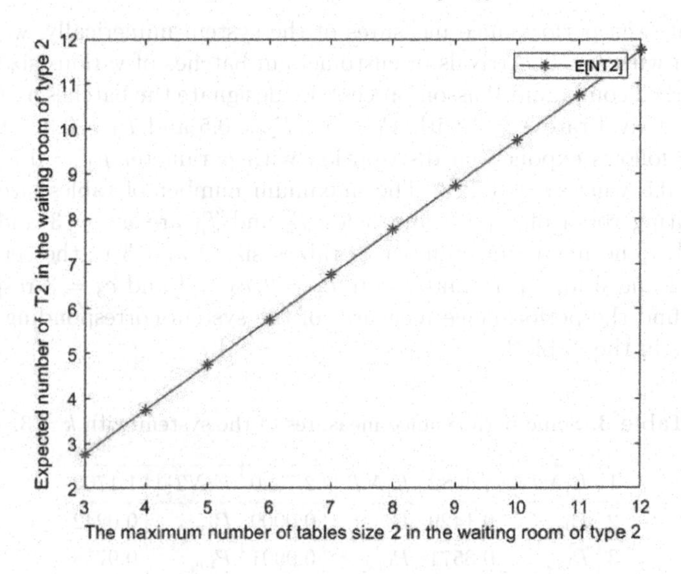

Fig. 2. Effect of w_2 on $E[N_{T_2}]$.

1. From Fig. 1, we can see that the expected number of T_1 in the waiting room of type 1 ($E[N_{T_1}]$) decreases as the maximum number of tables size 1 in the server room of type 1 for T_1 (c_1) increases.
2. In addition, from Figs. 2 and 3 we note that expected number of T_i in the waiting room of type i ($E[N_{T_i}]$) increases as the maximum number of tables size i in the waiting room of type i for T_i (w_i) increases where $i = 2, 3$.

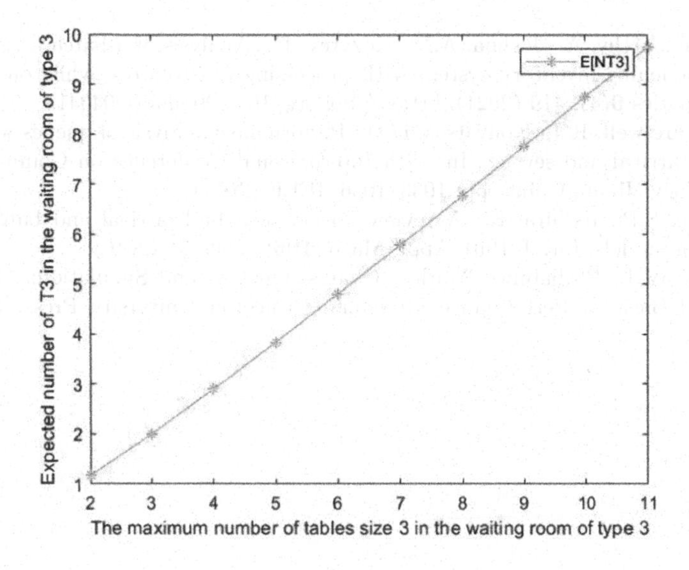

Fig. 3. Effect of w_3 on $E[N_{T_3}]$.

6 Conclusion

In this paper, we studied a queue with Marked Compound Poisson input and exponential distributed batch service. This queueing model has multi-server. Various performance measures are estimated with $k = 3$. we study the effect of the maximum number of tables size 1 in the server room of type 1 for T_1 (c_1) on the expected number of T_1 in the waiting room of type 1 ($E[N_{T_1}]$). Moreover, we study the effect of the maximum number of tables size i in the waiting room of type i for T_i (w_i) on expected number of T_i in the waiting room of type i ($E[N_{T_i}]$) where $i = 2, 3$. We find that $E[N_{T_1}]$ decreases as c_1 increases and $E[N_{T_i}]$ increases as w_i increases where $i = 2, 3$.

Acknowledgement. The first author acknowledges the Indian Council for Cultural Relations (ICCR) (Order No:2019-20/838) and Ministry of Higher Education, Research and Innovation, Order No: 2019/35 in Sultanate of Oman for their supports.

References

1. Baruah, M., Madan, K.C., Eldabi, T.: A batch arrival single server queue with server providing general service in two fluctuating modes and reneging during vacation and breakdowns. J. Probab. Stat. (2014). https://doi.org/10.1155/2014/319318
2. Chen, A., Pollett, P., Li, J., Zhang, H.: Markovian bulk-arrival and bulk-service queues with state-dependent control. Queueing Syst. **64**(3), 267–304 (2010)
3. Jayaraman, R., Matis, T.I.: Batch Arrivals and Service-Single Station Queues. Wiley Encyclopedia of Operations Research and Management Science (2010)

4. Krishnamoorthy, A., Joshua, A.N., Kozyrev, D.: Analysis of a batch arrival, batch service queuing-inventory system with processing of inventory while on vacation. Mathematics **9**(4), 419 (2021). https://doi.org/10.3390/math9040419
5. Li. W., Fretwell, R.J., Kouvatsos, D.D.: Performance analysis of queues with batch Poisson arrival and service. In: 13th International Conference on Communication Technology, Jinan, China, pp. 1033–1036. IEEE (2011)
6. Niranjan, S.P., Indhira, K.: A review on classical bulk arrival and batch service queueing models. Int. J. Pure Appl. Math. **106**(8), 45–51 (2016)
7. Stewart, W.J.: Probability, Markov Chains, Queues, and Simulation: The Mathematical Basis of Performance Modeling. Princeton University Press, Princeton (2009)

A Two Server Queueing Inventory Model with Two Types of Customers and a Dedicated Server

Nisha Mathew[1](✉), V.C. Joshua[2], and Achyutha Krishnamoorthy[2]

[1] Department of Mathematics, B.K College Amalagiri, Kottayam, India
nishamathew@cmscollege.ac.in
[2] Department of Mathematics, CMS College, Kottayam 686001, India
{vcjoshua,krishnamoorthy}@cmscollege.ac.in

Abstract. We consider a two server queueing inventory model with two types of customers. Type-I customers can form an infinite queue, while type-II customers join a finite buffer of size N. There are two servers. Server-1 is a dedicated server, giving service only to the type-I customers. Server-2 provides service to both type-I and type-II customers. Service of each customer requires an inventory item. Type-I customers are served one by one, by both servers on a FIFS basis. Type-II customers are served in batches of varying sizes by server-2. A random clock is set for type-II customers, which starts with the arrival of first type-II customer to an empty buffer. When the number of type-II customers becomes N or when the random clock realizes, server-2 starts the service of that batch of type-II customers. In such a situation, if server-2 is giving service to a type-I customer, its service is frozen and the service of that batch of type-II begins. The arrival of both types of customers are according to a stationary Marked Markovian Arrival Process. The service times of both type-I and type-II customers follow phase type distribution. The replenishment of inventory follows the (s, S) policy with positive leadtime. If the required inventory is not there at the time of service of a batch of type-II customers, local purchase is done.

Keywords: MMAP · dedicated server · phase type distribution · local purchase

1 Introduction

Inventory with positive service time were first introduced by Sigman and Simchi-Levi [1] and Melikov and Molchanov [2], independently of each other. Krishnamoorthy et al. [3] give a brief survey on inventory with positive service time. Dudin et al. [4], consider a multi-server queueing system with two types of customers, whose arrivals follow Marked Markovian arrival process. In [4], the service time distributions of different type customers have phase type distribution with different parameters. Chakravarthy et al. [5], consider a single server queueing model with (s, S) inventory system. In this paper customers are served in

V. M. Vishnevskiy et al. (Eds.): DCCN 2021, CCIS 1552, pp. 201–213, 2022.
https://doi.org/10.1007/978-3-030-97110-6_15

batches of varying size. In [6] Anbazhagan et al. consider an inventory system with two types of service and exponential service time. Krishnamoorthy et al. [7], consider a retrial queueing system with two types of customers and a single server. In this paper service time is phase-type. Krishnamoorthy and Raju [8] introduced the concept of local purchase in (s, S) inventory system. A single server (s, S) production inventory model with positive service time is discussed in [9], by Krishnamoorthy et al. Local purchase is used in this model to ensure customer satisfaction. Maqbali et al. [10], consider a queueing inventory model with one server. In [10] arrival Process is Markovian and service time is phase type. In this paper, inventories have a common life time and local purchase is used when all inventories perish. Nisha et al. [11] consider a queueing inventory system with one server and two types of customers. In the model considered in [11], arrival process is assumed to be Poisson process. The service time follows phase-type distribution and replenishment of inventory is according to (s, S) policy with positive lead time. In [12] Nisha et al. consider a queueing inventory system with one server. In this model arrival process is assumed to be Marked Markovian process and service time of both types of customers are assumed to be two different phase-type distribution. Krishnamoorthy et al. [13] consider a multiserver queueing inventory system with positive service time. Jeganathan et al. [14] consider a perishable inventory model with two stations. Here there are two dedicated servers and one flexible server, whose service rates are different with respect to stations and servers. In [15], Jeganathan et al. present a perishable stochastic inventory system with a service facility consisting of two servers and two parallel queues with jockeying. In this paper, each server has its own queue with a finite capacity and inventory is replenished as per (s, S) policy.

The motivation for the model is derived from a situation where the demands by customers that arrive at a shop are of two types. These are physical customer demands and online customer demands. The shopkeeper always tries to maintain the goodwill and so he puts two servers for service. First server is dedicated for serving physical customers. Second server serves both physical and online customers. Both the servers give service to physical customers one by one according to FIFS discipline. In order to reduce the long waiting time of online customers, online customer demands are attended by server-2 as batches. Server-2 starts its service only when the batch size reaches a fixed number or the waiting time of first online customer exceeds a particular time duration, whichever occurs first.

The remaining sections of the paper is organised as follows. Section 2, gives the mathematical description of the model. The mathematical formulation of the model is done in Sect. 3. The stability condition and stationary distribution is described in Sect. 4. In Sect. 5, we evaluate some measures of performance of the system. In Sect. 6, some numerical examples along with graphical illustrations are given. Section 7 is the conclusion of the paper.

2 Model Description

Our queueing inventory model consists of two types of customers, type-I and type-II. The arrival of both types of customers are assumed to be stationary

Marked Markovian Arrival Process (MMAP) having m phases. The matrix representation of the MMAP is (D_0, D_1, D_2). Type-I customers can form an infinite queue, while type-II customers join a finite buffer of capacity N. A random clock is also set for type-II customers. In every cycle, the clock starts when the first type-II customer arrives. Let θ be the rate of realization of the exponential clock. Type-II customers are served only when the clock expires or the number of type-II customers reaches N, whichever occurs first. Our queueing inventory system consists of two servers. Server-1 is a dedicated server, giving service to type-I customers only. Server-2 provides service to both type-I and type-II customers. Type-II customers have a preemptive priority over type-I customers. When the number of type-II customers reaches N or the clock expires, if server-2 is busy it stops service of current type-I customer and provides service to that batch of type-II customers. When the service of type-II customers is finished, server-2 returns to the service of type-I customer, whose service was frozen. The service of that type-I customer restarts from the stage where it was stopped. The service time distributions of both type-I and type-II customers are different phase type distribution. The service time of type-I customer by server-1 follows phase type distribution with representation PH$(\boldsymbol{\alpha}, W)$ with m_1 phases such that $\boldsymbol{W^0} = -W\boldsymbol{e}$. The service time of type-I customer by server-2 follows phase type distribution with representation PH$(\boldsymbol{\gamma}, U)$ with m_2 phases such that $\boldsymbol{U^0} = -U\boldsymbol{e}$. The service time of a batch of type-2 customers by server-2 follows phase type distribution with representation PH$(\boldsymbol{\tau}, T)$ with m_3 phases and $\boldsymbol{T^0} = -T\boldsymbol{e}$.

Service time is positive and the replenishment of inventory is by (s, S) policy. Lead time is assumed to be exponential with parameter β. Each customer needs one item. Service of each customer requires an inventory item. If the required inventory is not available at the time of service of a batch of type-II customers, local purchase is done. Local purchase means buying inventory locally from a nearby seller. Local purchase of inventory is made only for type-II customers and only the required inventory is purchased. Service of both types of customers are as follows: when a type-I customer arrives, if there is at least one item in inventory and both servers are idle, that customer is served by server-1. If there is at least one item in inventory and only one server is idle, that type-I customer is served by the idle server. If both the servers are busy or no inventory is available, the arriving type-I customer wait in their respective queue. A random clock is set in order to minimize the long waiting time of the type-II customer. The clock starts when the first type-II customer arrives to an empty buffer. When the number of type-II customers reaches N or when the random clock expires, server-2 provides service to that batch of type-II customers immediately. If server-2 is busy at that time, it stops service of current type-I customer and provides service to that batch of type-II customers. If required inventory is not available a local purchase is done instantaneously for that. Service of type-II customers is a batch service. The batch size vary from 1 to N. The batch size will be less than N, if the clock expires before the number of type-II customers reach N. When the service of type-II customers is finished, server-2 goes back to the service of that type-I customer whose service was frozen. Then the service of type-I customer will be resumed from the stage where it was frozen. When type-I customer enters the service facility, the inventory level drops by one unit. But when the service of type-II begins, the inventory level drops by n_2, where n_2 is the number of type-II customers present at that time. New type-II customer is not allowed to enter the buffer when server-2 is providing service to type-II customers. New type-II customer is allowed to join only after these customers leave the system. A graphical representation of the model is given below in Fig. 1

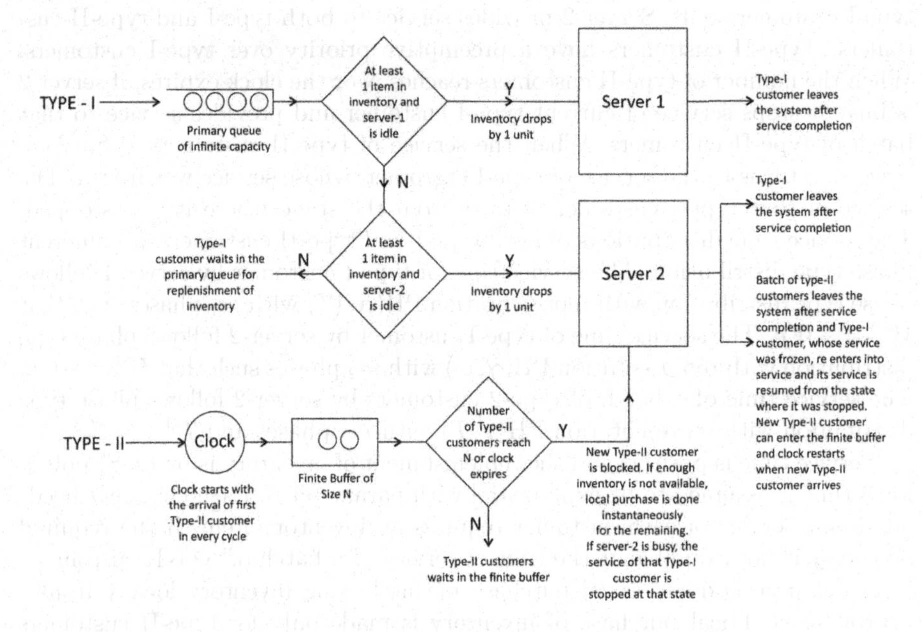

Fig. 1. Model description

3 Mathematical Formulation

Let

- $N_1(t)$ denote the number of type-I customers in the queue at time t
- $N_2(t)$ denote the number of type-II customers in the finite buffer at time t
- $I(t)$ denote the number of items in the inventory at time t
- $S_1(t)$ denote the status of server-1 at time t;

$$S_1(t) = \begin{cases} 0, & \text{if the server is idle} \\ 1, & \text{if the server is busy with a type-I customer} \end{cases}$$

- $S_2(t)$ denote the status of server-2 at time t;

$$S_2(t) = \begin{cases} 0, & \text{if the server is idle} \\ 1, & \text{if the server is busy with a type-I customer} \\ 2, & \text{if the server is busy with a batch of type-II customer} \end{cases}$$

- $C(t)$ denote the clock status at time t

$$C(t) = \begin{cases} 0, \text{if the clock is off} \\ 1, \text{if the clock is on} \end{cases}$$

- $J_1(t)$ denote the phase of the service process of server-1 at time t
- $J_2(t)$ denote the phase of the service process of server-2 at time t
- $F(t)$ represent the status of a customer whose service is frozen at time t

$$F(t) = \begin{cases} 0, & \text{if no type-I customer service is frozen at time t} \\ k, & \text{the phase of the service process of type-I customer,} \\ & \quad \text{whose service is frozen at time t} \end{cases}$$

- $A(t)$ denote the phase of the arrival process at time t

Then $\{(N_1(t), N_2(t), I(t), S_1(t), S_2(t), C(t), J_1(t), J_2(t), F(t), A(t)); t \geq 0\}$ is a continuous time Markov chain (CTMC) whose state space is described as follows. It is a Level Independent Quasi-Birth-Death(LIQBD) process. For getting a steady state solution Matrix Analytic Method is used.

The state space $\Omega = \bigcup_{n_1=0}^{\infty} l(n_1)$, where $l(n_1)$ denotes the collection of states in level n_1 and are defined as

$$l(0) = \{(0,0) \bigcup (0, n_2)\} \text{ and for } n_1 \geq 1,\ l(n_1) = \{(n_1, 0) \bigcup (n_1, n_2)\}.$$

$(0,0)$ corresponds to the state in which there is no type-I customer in queue and no type-II customer in the finite buffer and $(0, n_2)$ corresponds to the state in which there is no type-I customer in queue and n_2 type-II customer in the finite buffer. $(n_1, 0)$ corresponds to the state in which there is n_1 type-I customer in queue and no type-II customer in the finite buffer and (n_1, n_2) corresponds to the state in which there is n_1 type-I customer in queue and n_2 type-II customer in the finite buffer.

$(0,0) \quad = \quad \{(0,0,i,0,0,0,0^*,0^*,0,a) \bigcup (0,0,i,0,1,0,0^*,k,0,a) \bigcup (0,0,i,0,2,0,0^*, l,k',a) \bigcup (0,0,i,1,0,0,j,0^*,0,a) \bigcup (0,0,i,1,1,0,j,k,0,a) \bigcup (0,0,i,1,2,0,j,l,k',a) /$
$0 \leq i \leq S; j = 1,2,\ldots,m_1; k = 1,2,\ldots,m_2; l = 1,2,\ldots,m_3; k' = 0,1,2,\ldots,m_2;$
$a = 1,2,\ldots,m\}$,

$(0, n_2) = \{(0, n_2, i, 0, 0, 1, 0^*, 0^*, 0, a) \bigcup (0, n_2, i, 0, 1, 1, 0^*, k, 0, a)$
$\bigcup (0, n_2, i, 1, 0, 1, j, 0^*, 0, a) \bigcup (0, n_2, i, 1, 1, 1, j, k, 0, a) / 1 \leq n_2 \leq N-1;$
$0 \leq i \leq S; j = 1,2,\ldots,m_1; k = 1,2,\ldots,m_2; a = 1,2,\ldots,m\}$,

$(n_1, 0) = \{(n_1, 0, 0, 0, 0, 0, 0^*, 0^*, 0, a) \bigcup (n_1, 0, 0, 0, 1, 0, 0^*, k, 0, a)$
$\bigcup (n_1, 0, 0, 0, 2, 0, 0^*, l, k', a) \bigcup (n_1, 0, 0, 1, 0, 0, j, 0^*, 0, a) \bigcup (n_1, 0, 0, 1, 1, 0, j, k, 0, a)$
$\bigcup (n_1, 0, 0, 1, 2, 0, j, l, k', a) \bigcup (n_1, 0, i, 1, 1, 0, j, k, 0, a) \bigcup (n_1, 0, i, 1, 2, 0, j, l, k', a)/$
$n_1 \geq 1; 1 \leq i \leq S; j = 1, 2, \ldots, m_1; k = 1, 2, \ldots, m_2; l = 1, 2, \ldots, m_3;$
$k' = 0, 1, 2, \ldots, m_2; a = 1, 2, \ldots, m\}$,

$(n_1, n_2) = \{(n_1, n_2, 0, 0, 0, 1, 0^*, 0^*, 0, a) \bigcup (n_1, n_2, 0, 0, 1, 1, 0^*, k, 0, a)$
$\bigcup (n_1, n_2, 0, 1, 0, 1, j, 0^*, 0, a) \bigcup (n_1, n_2, 0, 1, 1, 1, j, k, 0, a) \bigcup (n_1, n_2, i, 1, 1, 1, j, k, 0, a)/$
$n_1 \geq 1; 1 \leq n_2 \leq N - 1; 1 \leq i \leq S; j = 1, 2, \ldots, m_1; k = 1, 2, \ldots, m_2; a = 1, 2, \ldots, m\}$

0^* represents phase of an idle server.

The infinitesimal generator Q of the LIQBD describing the above two server queueing inventory system is of the form

$$\begin{pmatrix} E_{00} & E_{01} & O & \cdots & & & & \\ E_{10} & E_{11} & E_{12} & O & \cdots & & & \\ E_{20} & E_{21} & E_{22} & E_{12} & O & \cdots & & \\ O & E_{31} & E_{21} & E_{22} & E_{12} & O & \cdots & \\ O & O & E_{31} & E_{21} & E_{22} & E_{12} & O & \cdots \\ O & O & O & E_{31} & E_{21} & E_{22} & E_{12} & O \\ & \ddots & \ddots & \ddots & \ddots & \ddots & & \\ & & \ddots & \ddots & \ddots & \ddots & \ddots & \end{pmatrix}$$

By merging the cells, we get the tridiagonal form given below

$$\begin{pmatrix} B_{00} & B_{01} & O & \cdots & & \\ B_{10} & A_1 & A_0 & O & \cdots & \\ O & A_2 & A_1 & A_0 & O & \cdots \\ O & O & A_2 & A_1 & A_0 & O \; \cdots \\ & \ddots & \ddots & \ddots & \ddots & \\ & & \ddots & \ddots & \ddots & \ddots \end{pmatrix}$$

where

$$B_{00} = \begin{pmatrix} E_{00} & E_{01} \\ E_{10} & E_{11} \end{pmatrix}, B_{01} = \begin{pmatrix} 0 & 0 \\ E_{12} & 0 \end{pmatrix}, B_{10} = \begin{pmatrix} E_{20} & E_{21} \\ 0 & E_{31} \end{pmatrix},$$

$$A_1 = \begin{pmatrix} E_{22} & E_{12} \\ E_{21} & E_{22} \end{pmatrix}, A_2 = \begin{pmatrix} E_{31} & E_{21} \\ 0 & E_{31} \end{pmatrix}, A_0 = \begin{pmatrix} 0 & 0 \\ E_{12} & 0 \end{pmatrix}$$

B_{00} is a square matrix of order $(M + M_1)$, B_{01} is of order $(M + M_1) \times 2M_1$, B_{10} is of order $2M_1 \times (M + M_1)$. A_0, A_1 and A_2 are square matrix of order $2M_1$.

$M = m(1 + m_1)(1 + m_2)(N + m_3)(S + 1)$ and
$M_1 = m(1 + m_1)(1 + m_2)(N + m_3) + mm_1S(m_2(m_3 + N) + m_3)$

4 Stability Condition

The Markov chain with Q as generator is positive recurrent if and only if

$$\pi A_0 e < \pi A_2 \mathbf{e}, \tag{1}$$

where the stationary vector $\boldsymbol{\pi}$ of A is obtained by solving

$$\boldsymbol{\pi} A = 0; \boldsymbol{\pi}\mathbf{e} = 1, \tag{2}$$

where the matrix A be defined as $A = A_0 + A_1 + A_2$.

4.1 Stationary Distribution

The stationary distribution of the Markov process under consideration is obtained by solving the set of equations

$$\boldsymbol{\pi} A = 0; \boldsymbol{\pi} e = 1. \tag{3}$$

Let $\mathbf{x}$ be decomposed in conformity with Q. Then
$\mathbf{x} = (\mathbf{x_0}, \mathbf{x_1}, \mathbf{x_2}, \dots)$, where $\mathbf{x_i} = (\mathbf{y_{2i}}, \mathbf{y_{2i+1}})$.
$\mathbf{y_i} = (\mathbf{y_{i0}}, \mathbf{y_{i1}}, \dots\dots, \mathbf{y_{iN-1}})$
For $j = 1, 2, \dots, N-1$, $\mathbf{y_{ij}} = (\mathbf{y_{ij0}}, \mathbf{y_{ij1}}, ..., \mathbf{y_{ijS}})$
For $k = 0, 1, ..., S$, the vectors $\mathbf{y_{ijk}} = (\mathbf{y_{ijk0}}, \mathbf{y_{ijk1}})$
For $l = 0, 1$, the vectors $\mathbf{y_{ijkl}} = (\mathbf{y_{ijkl0}}, \mathbf{y_{ijkl1}}, \mathbf{y_{ijkl2}})$
For $r = 0, 1, 2$, $\mathbf{y_{ijklr}} = (\mathbf{y_{ijklr0}}, \mathbf{y_{ijklr1}})$
For $t = 0, 1$, $\mathbf{y_{ijklrt}} = (\mathbf{y_{ijk0rt0^*}}, \mathbf{y_{ijk1rt1}}, ..., \mathbf{y_{ijk1rtm_1}})$
For $u = 0^*, 1, ..., m_1$, $\mathbf{y_{ijklrtu}} = (\mathbf{y_{ijkl0tu0^*}}, \mathbf{y_{ijkl1tu1}}, ..., \mathbf{y_{ijkl1tum_2}}, \mathbf{y_{ijkl2tu1}}, ..., \mathbf{y_{ijkl2tum_3}})$
For $v = 0^*, 1, ..., m_2, 1, ..., m_3$,
$\mathbf{y_{ijklrtuv}} = (\mathbf{y_{ijkl0tuv0}}, \mathbf{y_{ijkl1tuv0}}, \mathbf{y_{ijkl2tuv0}}, \mathbf{y_{ijkl2tuv1}}, ..., \mathbf{y_{ijkl2tuvm_2}})$
For $w = 0, 1, ..., m_2$, $\mathbf{y_{ijklrtuvw}} = (y_{ijklrtuvw1}, y_{ijklrtuvw2}, ..., y_{ijklrtuvwm})$
$y_{ijklrtuvwy}$ is the probability of being in state $(i, j, k, l, r, t, u, v, w, y)$ for $i \geq 0; j = 1, 2, \dots, N-1; 0 \leq k \leq S; l = 0, 1; r = 0, 1, 2; t = 0, 1; u = 0^*, 1, ..., m_1; v = 0^*, 1, ..., m_2, 1, ..., m_3; w = 0, 1, ..., m_2$ and $y = 1, ..., m$
From $\mathbf{x}Q = 0$, we get the following equations:

$$\mathbf{x_0} B_{00} + \mathbf{x_1} B_{10} = 0 \tag{4}$$

$$\mathbf{x_0} B_{01} + \mathbf{x_1} A_1 + \mathbf{x_2} A_2 = 0 \tag{5}$$

$$\mathbf{x_1} A_0 + \mathbf{x_2} A_1 + \mathbf{x_3} A_2 = 0 \tag{6}$$

$$\mathbf{x_{i-1}} A_0 + \mathbf{x_i} A_1 + \mathbf{x_{i+1}} A_2 = 0, i = 2, 3, .. \tag{7}$$

There exists a constant matrix R such that

$$\mathbf{x_i} = \mathbf{x_{i-1}} R, i = 2, 3, ... \tag{8}$$

The sub vectors $\mathbf{x_i}$ are geometrically related by the equation

$$\mathbf{x_i} = \mathbf{x_1} R^{i-1}, i = 2, 3, ... \tag{9}$$

R can be obtained from the matrix quadratic equation

$$R^2 A_2 + R A_1 + A_0 = O \tag{10}$$

5 Performance Measures

Some measures of performance of the system are evaluated as follows.

1. Expected number of type-I customers in the queue

$$E[N_1] = \sum_{i=0}^{\infty} i\mathbf{y_i e} \tag{11}$$

2. Expected number of type-II customers in the finite buffer

$$E[N_2] = \sum_{i=0}^{\infty}\sum_{j=0}^{N-1} j\mathbf{y_{ij} e} \tag{12}$$

3. Expected number of items in the inventory

$$E[I] = \sum_{i=0}^{\infty}\sum_{j=0}^{N-1}\sum_{k=0}^{S} k\mathbf{y_{ijk} e} \tag{13}$$

4. Probability that server-1 is idle

$$b_0 = \sum_{i=1}^{\infty}\sum_{j=0}^{N-1} \mathbf{y_{ij00} e} + \sum_{j=0}^{N-1}\sum_{k=0}^{S} \mathbf{y_{0jk0} e} \tag{14}$$

5. Probability that server-1 is busy with a type-I customer

$$b_1 = \sum_{i=0}^{\infty}\sum_{j=0}^{N-1}\sum_{k=0}^{S} \mathbf{y_{ijk1} e} \tag{15}$$

6. Probability that the server-2 is idle

$$b_2 = \sum_{i=1}^{\infty}\sum_{j=0}^{N-1}\sum_{l=0}^{1} \mathbf{y_{ij0l0} e} + \sum_{j=0}^{N-1}\sum_{k=0}^{S}\sum_{l=0}^{1} \mathbf{y_{0jkl0} e} \tag{16}$$

7. Probability that the server-2 is busy with a type-I customer

$$b_3 = \sum_{i=0}^{\infty}\sum_{j=0}^{N-1}\sum_{k=0}^{S}\sum_{l=0}^{1} \mathbf{y_{ijkl1} e} \tag{17}$$

8. Probability that the server-2 is busy with a batch of type-II customers

$$b_4 = \sum_{i=0}^{\infty}\sum_{k=0}^{S}\sum_{l=0}^{1} \mathbf{y_{i0kl2} e} \tag{18}$$

9. Probability that the clock status is on

$$c_1 = \sum_{i=0}^{\infty}\sum_{j=1}^{N-1}\sum_{k=0}^{S}\sum_{l=0}^{1}\sum_{r=0}^{1}\mathbf{y_{ijklr1}e} \quad (19)$$

10. The probability that a type-II customer is blocked from entering the system:

$$p_b = \sum_{i=0}^{\infty}\sum_{k=0}^{S}\sum_{l=0}^{1}\mathbf{y_{i0kl2}e} \quad (20)$$

11. Expected rate at which inventory replenishment occurs

$$E_R = \sum_{i=0}^{\infty}\sum_{j=0}^{N}\sum_{k=0}^{s}\beta\mathbf{y_{ijk}e} \quad (21)$$

12. Expected number of type-I customers waiting in the system due to lack of inventory

$$E[W] = \sum_{i=1}^{\infty}\sum_{j=0}^{N-1} i\mathbf{y_{ij0}e} \quad (22)$$

6 Numerical Examples

In this section, we give some numerical illustrations of variation in the measures of performance with regard to variation in values of the parameters. Here the MMAP describing the arrival is represented by (D_0, D_1, D_2).The following values are kept fixed:

$$\alpha = (0.2000\ 0.4000\ 0.4000)\,;$$

$$W = \begin{pmatrix} -6.0000 & 3.0000 & 2.0000 \\ 1.0000 & -6.0000 & 2.0000 \\ 2.5000 & 1.5000 & -8.0000 \end{pmatrix}$$

$$\gamma = (0.3000\ 0.7000)\,; U = \begin{pmatrix} -7 & 2 \\ 4 & -12 \end{pmatrix}$$

$$\tau = (.2\ .4\ .1\ .3)\,; T = \begin{pmatrix} -10 & 3 & 4 & 2 \\ 5 & -15 & 1 & 6 \\ 2 & 4 & -14 & 3 \\ 5 & 3 & 2 & -12 \end{pmatrix}$$

$$W_0 = \begin{pmatrix} 1 \\ 3 \\ 4 \end{pmatrix} U_0 = \begin{pmatrix} 5 \\ 8 \end{pmatrix} T_0 = \begin{pmatrix} 1 \\ 3 \\ 5 \\ 2 \end{pmatrix}$$

$$D_0 = \begin{pmatrix} -7.3000 & 0.5000 \\ 0.8000 & -5.1000 \end{pmatrix}, D_1 = \begin{pmatrix} 1.5000 & 0.6000 \\ 1.1000 & 0.2000 \end{pmatrix}, D_2 = \begin{pmatrix} 2.2000 & 2.5000 \\ 1.9000 & 1.1000 \end{pmatrix}$$

6.1 Effect of θ on Various Measures of Performance

We fix $m = 2$; $m_1 = 3$; $m_2 = 2$; $m_3 = 4$; $s = 4$; $S = 9$; $\beta = 5$; $N = 5$;

Table 1 shows the variation in various measures of performance for different values of θ

Table 1. Effect of θ on some measures of performance

θ	$E[N_1]$	$E[N_2]$	$E[W]$	b_0	b_1	b_2	b_3	$b_4 = p_b$
3	0.1236	0.5661	0.0036	0.5708	0.4292	0.4886	0.0841	0.4273
4	0.1319	0.4443	0.0029	0.5675	0.4325	0.4599	0.0827	0.4574
5	0.1384	0.3605	0.0025	0.565	0.435	0.4385	0.0816	0.4799
6	0.1437	0.3007	0.0022	0.5629	0.4371	0.4222	0.0807	0.4970
7	0.1481	0.2563	0.002	0.5613	0.4387	0.4094	0.08	0.5105
8	0.1517	0.2225	0.0019	0.56	0.44	0.3992	0.0795	0.5214
9	0.1548	0.1961	0.0018	0.5589	0.4411	0.3908	0.079	0.5302
10	0.1575	0.1749	0.0017	0.5579	0.4421	0.3838	0.0786	0.5376
11	0.1598	0.1576	0.0016	0.5571	0.4429	0.3779	0.0782	0.5439

The expected number of type-I customers in the system increases as θ increases. But expected number of type-II customers in the finite buffer decreases as θ increases. This happens because as θ increases, the chances of getting service to a batch of type-II customers increases and so server-2 will be serving type-II customers most of the time. As a result, the expected number of type-I customers increases and the probability of server-1 being idle, decreases. As the expected number of type-I customers increases the probability that the server-1 is busy with a type-I customer increases because server-1 is a dedicated server for type-I customer. With an increase in values of θ, the probability that the server-2 is serving a batch of type-II customers, increases. As a result, both the probabilities of server-2 being idle and server-2 being busy with a type-I customer, decreases. The blocking probability of the type-II customers increases with an increase in θ. It happens because as θ increases, server-2 is busy with type-II customers more time and during that time the arriving type-II customers are blocked from entering the system.

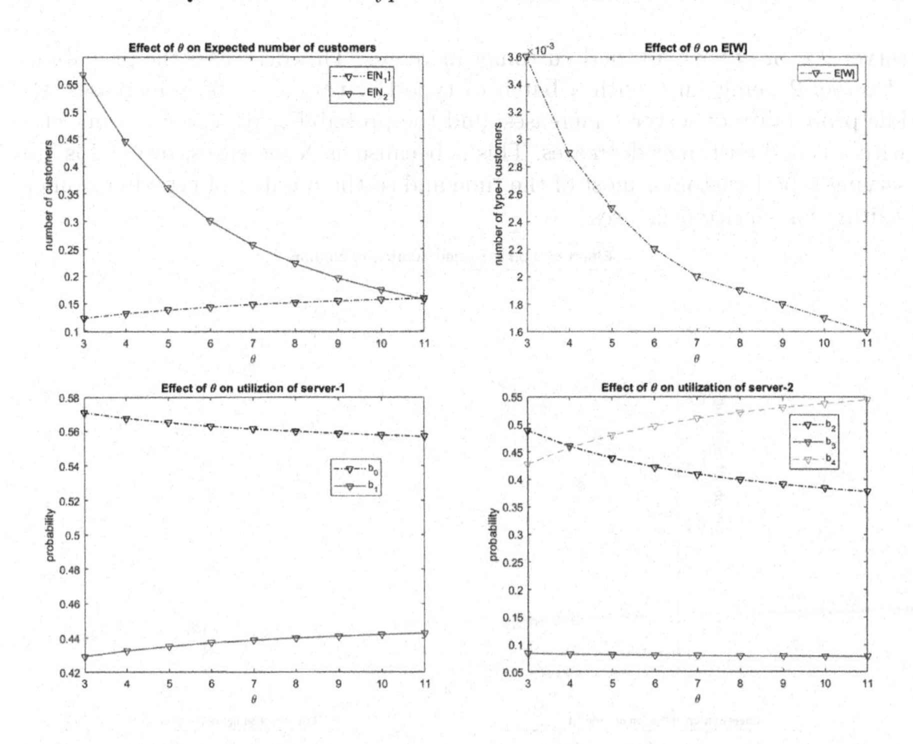

6.2 Effect of N on Various Measures of Performance

We fix $m = 2$; $m_1 = 3$; $m_2 = 2$; $m_3 = 4$; $s = 4$; $S = 9$; $\beta = 5$; $\theta = 3$;

Table 2 indicates the variation in various measures of performance for different values off N.

Table 2. Effect of N on some measures of performance

N	$E[N_1]$	$E[N_2]$	$E[W]$	b_0	b_1	b_2	b_3	b_4
3	0.132	0.3424	0.002	0.567	0.433	0.456	0.0825	0.4615
4	0.126	0.4718	0.0027	0.5695	0.4305	0.4775	0.0835	0.439
5	0.1236	0.5661	0.0036	0.5708	0.4292	0.4886	0.0841	0.4273
6	0.1224	0.632	0.0041	0.5716	0.4284	0.4946	0.0844	0.4209
7	0.1219	0.6765	0.0045	0.572	0.428	0.498	0.0846	0.4174

As the maximum batch size N of type-II customers increases, the expected number of type-I customers in the system decreases and expected number of type-II customers in the finite buffer increases. This is because of the fact that as N increases, the chances of getting service to a batch of type-II customers decreases and so server-2 will be serving type-I customers most of the time. So both the probabilities of server-2 being idle and the probability that the server-2 is busy with a type-I customer increases. This decreases the probability of server-2 being busy with a batch of type-II customers. As N increases, the idle probability of server-1 increases, and the probability of server-1 being busy with a type-I customer decreases. This is because as N increases, server-2 is also serving type-I customer most of the time and so the number of type-I customers waiting for service decreases.

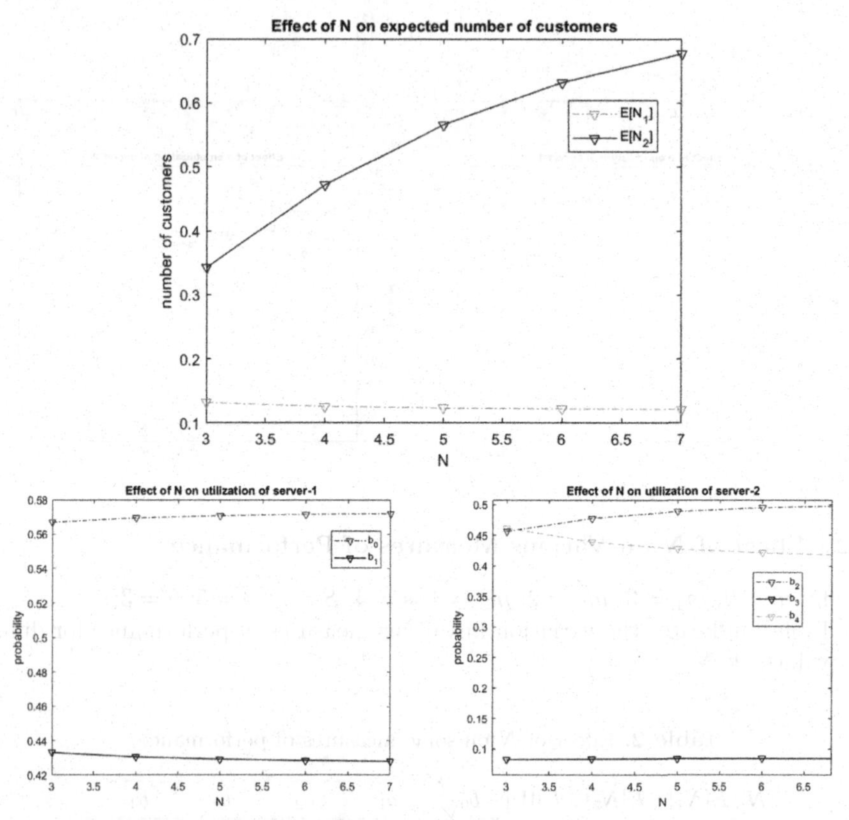

7 Conclusion

We considered a queueing inventory model with two servers and two types of customers. Server-1 is a dedicated server, giving service to type-I customers only. Service to type-I customers are provided by both servers and service to type-II customers is only by server-2. Various performance measures are evaluated at steady state conditions.

References

1. Sigman, K., Simchi-Levi, D.: Light traffic heutrestic for an M/G/1 queue with limited inventory. Ann. Oper. Res. **40**, 371–380 (1992)

2. Melikov, A.Z., Molchanov, A.A.: Stock optimization in transportation/storage systems. Cybern. Syst. Anal. **28**(3), 484–487 (1992)
3. Krishnamoorthy, A., Shajin, D., Narayanan, V.C.: Inventory with positive service time: a survey. In: Advanced Trends in Queueing Theory: Series of Books "Mathematics and Statistics", Sciences. ISTE & J. Wiley, London (2019)
4. Kim, C., Dudin, A., Dudin, S., Dudina, O.: Analysis of an MMAP/PH1, PH2/N/∞ queueing system operating in a random environment. Int. J. Appl. Math. Comput. Sci. **24**, 485–501 (2014)
5. Chakravarthy, S.R., Maity, A., Gupta, U.C.: An (s, S) inventory in a queueing system with batch service facility. Ann. Oper. Res. **258**, 263–283 (2015)
6. Anbazhagan, N., Vigneshwaran, B., Jeganathan, K.: Stochastic inventory system with two types of services. Int. J. Adv. Appl. Math. Mech. **2**(1), 120–127 (2014)
7. Krishnamoorthy, A., Joshua, V.C., Mathew, A.P.: A retrial queueing system with abandonment and search for priority customers. In: Vishnevskiy, V.M., Samouylov, K.E., Kozyrev, D.V. (eds.) DCCN 2017. CCIS, vol. 700, pp. 98–107. Springer, Cham (2017). https://doi.org/10.1007/978-3-319-66836-9_9
8. Krishnamoorthy, A., Raju, N.: N-Policy for (s, S) perishable inventory system with positive lead time. Korean J. Comput. Appl. Math. **5**(1), 253–261 (1998)
9. Krishnamoorthy, A., Varghese, R., Lakshmy, B.: Production inventory system with positive service time under local purchase. In: Dudin, A., Nazarov, A., Moiseev, A. (eds.) ITMM 2019. CCIS, vol. 1109, pp. 243–256. Springer, Cham (2019). https://doi.org/10.1007/978-3-030-33388-1_20
10. AL Maqbali, K.A.K., Joshua, V.C., Krishnamoorthy, A.: On a single server queueing inventory system with common life time for inventoried items. In: Dudin, A., Nazarov, A., Moiseev, A. (eds.) ITMM 2020. CCIS, vol. 1391, pp. 186–197. Springer, Cham (2021). https://doi.org/10.1007/978-3-030-72247-0_14
11. Mathew, N., Joshua, V.C., Krishnamoorthy, A.: A queueing inventory system with two channels of service. In: Vishnevskiy, V.M., Samouylov, K.E., Kozyrev, D.V. (eds.) DCCN 2020. LNCS, vol. 12563, pp. 604–616. Springer, Cham (2020). https://doi.org/10.1007/978-3-030-66471-8_46
12. Mathew, N., Joshua, V.C., Krishnamoorthy, A.: On an MMAP/(PH, PH)/1/(∞, N) queueing-inventory system. In: Dudin, A., Nazarov, A., Moiseev, A. (eds.) ITMM 2020. CCIS, vol. 1391, pp. 363–377. Springer, Cham (2021). https://doi.org/10.1007/978-3-030-72247-0_27
13. Krishnamoorthy, A., Manikandan, R., Shajin, D.: Analysis of a multiserver queueing-inventory system. Adv. Oper. Res. **2015** (2015). https://doi.org/10.1155/2015/747328
14. Jeganathan, K., Melikov, A.Z., Padmasekaran, S., Kingsly, S.J., Lakshmi, K.P.: A stochastic inventory model with two queues and a flexible server. Int. J. Appl. Comput. Math. **5**(1), 1–27 (2019). https://doi.org/10.1007/s40819-019-0605-3
15. Jeganathan, K., Sumathi, J., Mahalakshmi, G.: Markovian inventory model with two parallel queues, jockeying and impatient customers. Yugoslav J. Oper. Res. **26**(4), 467–506 (2016)

On Convergence of Tabu-Enhanced Quantum Annealing Algorithm

A. S. Rumyantsev[1,2(✉)], D. Pastorello[3,4], E. Blanzieri[3,4,5], and V. Cavecchia[5]

[1] Institute of Applied Mathematical Research, Karelian Research Centre of RAS, Pushkinskaya Street 11, 185000 Petrozavodsk, Republic of Karelia, Russia
ar0@krc.karelia.ru

[2] Petrozavodsk State University, 33 Lenina Pr., Petrozavodsk, Russia

[3] Department of Information Engineering and Computer Science, University of Trento, via Sommarive 9, 38123 Povo, Trento, Italy
{d.pastorello,enrico.blanzieri}@unitn.it

[4] Trento Institute of Fundamental Physics and Applications, via Sommarive 14, 38123 Povo, Trento, Italy

[5] Institute of Materials for Electronics and Magnetism (CNR), via alla Cascata 56/c, 38123 Povo, Trento, Italy
valter.cavecchia@unitn.it

Abstract. The convergence of a recently proposed tabu-enhanced quantum annealing algorithm depends critically on the finiteness of memory of the related stochastic process. We discuss the background of quantum annealing and the convergence issues of the tabu-enhanced algorithm. Given the details of the tabu data structure, the so-called tabu matrix, we consider the sequences of solutions that result in a tabu matrix collisions. As such, convergence of the algorithm is related to the problem of studying the so-called matrix kernel, which we investigate and give an example of such a collision as well.

Keywords: Quantum Annealing · Tabu Search · Ising Model

1 Introduction

Quantum computing is an emerging technology of computing systems design, which allows to use the advanced physical effects to overcome the traditional limitations of sequential/parallel computing in semiconductor systems. In particular, it allows to utilize quantum tunneling effect and quantum entanglement to dramatically increase the speed [7] or decrease the dimension [1] of the optimization problem being solved. The above effects, however, come within a limited class of tasks suitable for computing systems of this type, with quantum annealing and quantum Fourier transform being among the most actively studied fields (we refer the reader to [8] for a comprehensive review of the active fields of research in quantum computing). As such, a trending research subject

V. M. Vishnevskiy et al. (Eds.): DCCN 2021, CCIS 1552, pp. 214–219, 2022.
https://doi.org/10.1007/978-3-030-97110-6_16

is the study of advanced algorithms and computational problems that can utilize the quantum computing environment, both in classical and in hybrid quantum-classical computing environment [2], where the latter corresponds to utilization of the quantum computer as a co-processor when a classical semiconductor system is used for the computations management.

Simulated annealing is a well-known local search optimization heuristics used to obtain an approximate solution on a huge parameter space, where explicit solution can not be obtained in a reasonable time [5]. In brief, such an algorithm allows to build up a sequence of random trajectories in the parameter space, and such trajectories are dependent on a single parameter known as the temperature. The so-called cooling scheme, which is a sequence of conditions and temperature levels modulating acceptance of the non-optimal candidates in a neighborhood of the currently optimal solution, is to be chosen in advance. Such a scheme resembles the known annealing physical process in the metal production. Compared to that, quantum annealing uses quantum tunneling in the solution space, instead of accepting non-optimal solution at the neighborhood of the currently optimal candidate. Both algorithms target to obtain the optimal state of the so-called Ising model widely used in applications related to distributed networks, such as the sensor network responsiveness modeling [11].

Convergence of the algorithm to the optimal solution is a necessary step in the analysis of any heuristics. In the case of simulated annealing, the proof is usually done in one of the two possible directions: either the steady state of a Markov chain which models the trajectory of the random search is reached at a single temperature (that is, the algorithm stays at each temperature for sufficiently long period), or the sample path is considered to be derived from a sequence of non-homogeneous Markov chains [5]. In general case, the same is applicable to the so-called Generalized Hill Climbing class of algorithms, of which the simulated annealing is a representative [6]. In a recent paper, the convergence of quantum annealing algorithm by means of irreducibility of the corresponding Markov chain was shown [9]. In the present paper, we discuss the ways of demonstrating the convergence in the case quantum annealing is combined with the tabu search, which in general disallows to apply Markov chain convergence results directly due to virtually unlimited memory of the previous states, and thus it is important to reason the finiteness of the memory of corresponding tabu search mechanism [3–5]. We address this issue by studying the collisions in the object storing the tabu states.

The structure of the paper is as follows. In Sect. 2 we give the necessary details on the tabu enhanced quantum annealing algorithm. In Sect. 3 we describe an approach to the formal study of convergence of the proposed algorithm based on the collisions in the tabu matrix. We finalize the paper with a conclusion and discussion of further research directions.

2 Tabu Enhanced Quantum Annealing

In this section we give a brief description of the quantum annealing algorithm applied to the so-called Quantum Unconstrained Binary Optimization (QUBO) problem and enhanced with tabu search [9], as seen from the perspective of application on a quantum annealer machine such as D-Wave. Such a machine is capable of solving optimization problems encoded as the Ising model, that is, minimization of the cost function

$$\mathsf{E}(\boldsymbol{\Theta}, \boldsymbol{z}) = \sum_{i \in V} \theta_i z_i + \sum_{(i,j) \in E} \theta_{ij} z_i z_j, \quad \boldsymbol{z} = (z_1, ..., z_n) \in \{-1, 1\}^n. \tag{1}$$

The matrix $\boldsymbol{\Theta}$ holds the (real) coefficients θ_i known as individual fields, and θ_{ij} known as interaction (coupling) between the variables z_i corresponding to the measurements of n qubits on a quantum machine. The graph $\langle V, E \rangle$, with $|V| = n$ corresponds to physical architecture of the machine, where E is the edge set between the vertices V, having an adjacency matrix $\boldsymbol{A}$. However, to solve a general optimization problem on such a machine, one needs to obtain the correct *encoding* of the optimization function in terms of the cost function (1).

Consider the following QUBO problem

$$f(\boldsymbol{z}) = \boldsymbol{z}^T \boldsymbol{Q} \boldsymbol{z} \to \min, \quad \boldsymbol{z} \in \{-1, 1\}^n, \tag{2}$$

where $\boldsymbol{z}$ is the n-dimensional vector of ± 1 and $\boldsymbol{Q}$ is a real symmetric matrix of size $n \geq 1$. In order to solve QUBO problem (2) by applying the optimization procedure on a quantum annealer, a mapping of the variables is performed:

$$\mathsf{E}(\boldsymbol{P}^T \boldsymbol{Q} \boldsymbol{P} \circ \boldsymbol{A}, \boldsymbol{P} \boldsymbol{z}),$$

where $\boldsymbol{P}$ is a permutation matrix, and $\circ$ is the Hadamard (componentwise) product. This means that elements of $\boldsymbol{Q}$ are mapped to the coefficients of $\boldsymbol{\Theta}$ if there are corresponding edges in the set E, possibly after permutation. After performing a solution, the result is read from the quantum state, and transformed to the original state space.

However, since the best mapping $\boldsymbol{P}$ is not known in advance, the optimization can be done iteratively, e.g. by using heuristic approach known as tabu search. In such a case, the solution $\boldsymbol{z}^*$ of the original optimization problem (2) is obtained simultaneously with the best mapping $\boldsymbol{P}^*$ by penalizing the relatively worst solutions in favor of currently best known candidate. In particular, due to the problem structure, the following approach is suggested [9]: obtain the minimum of

$$\mathsf{E}(\boldsymbol{P}^T (\boldsymbol{Q} + \lambda_k \boldsymbol{S}_k) \boldsymbol{P} \circ \boldsymbol{A}, \boldsymbol{P} \boldsymbol{z}),$$

where $\boldsymbol{S}_k$ is the so-called tabu matrix obtained after k iterations and λ_k is the real coefficient encoding the penalty strictness. The matrix $\boldsymbol{S}_k$ is constructed as the sum of outer products of the k already visited (worst) solutions $\{\boldsymbol{z}_{(j)}\}_{j \leq k}$,

$$\boldsymbol{S}_k = \sum_{j=1}^{k} \left[\boldsymbol{z}_{(j)} \boldsymbol{z}_{(j)}^T - \boldsymbol{I} + \operatorname{diag} \boldsymbol{z}_{(j)} \right], \tag{3}$$

where $\boldsymbol{I}$ is the identity matrix of size n, and diag constructs a diagonal matrix from a vector.

Note that such an iterative approach is valid not only for QUBO, but for more general class of functions [10]. Since the scheme is iteratively applied, the convergence of the procedure to the optimal solution $\boldsymbol{z}^*$ is usually proved by considering the convergence of the sequence of steady-state probabilities of non-homogeneous Markov chains modeling the sequences of candidate solutions obtained. However, due to (3), proving the Markovian nature of such a process is problematic, since the tabu matrix holds virtually unlimited memory of previous states. Thus, it is important to study the collisions in the tabu matrix $\boldsymbol{S}_k$. We address this issue in the next section.

3 Collisions in Tabu Matrix

Let $\boldsymbol{a}, \boldsymbol{b} \in 2^{2^n}$ be the binary sequences of length 2^n encoding two independent *trajectories* of the iterative approach after some iterations. For each such a sequence $\boldsymbol{a} = (a_1, \ldots, a_{2^n})$, construct the matrix $\boldsymbol{S}_{\boldsymbol{a}}$ using (3) in the following way:

$$\boldsymbol{S}_{\boldsymbol{a}} = \sum_{i=1}^{2^n} a_i(\boldsymbol{z}_i\boldsymbol{z}_i^T - \boldsymbol{I} + \mathrm{diag}\boldsymbol{z}_i), \tag{4}$$

where $\boldsymbol{z}_i \in \{-1, 1\}^n$ is one of the possible solutions in the state space numbered lexicographically. Then the *collision* in the tabu matrix is the solution of a linear system $\boldsymbol{S}_{\boldsymbol{a}} - \boldsymbol{S}_{\boldsymbol{b}} = \mathbb{O}$, which is equivalent to

$$\sum_{i=1}^{2^n} x_i(\boldsymbol{z}_i\boldsymbol{z}_i^T - \boldsymbol{I} + \mathrm{diag}\boldsymbol{z}_i) = \mathbb{O}, \tag{5}$$

where $\mathbb{O}$ is the square zero matrix of size n and $x_i := a_i - b_i$ are the variables. Due to symmetric nature of the tabu matrix, there are essentially $n(n+1)/2$ equations and 2^n variables x_i, $i = 1, \ldots, 2^n$. It is easy to construct rowwise a $2^n \times n(n+1)/2$ matrix $\boldsymbol{M}$ containing at ith column the individual elements of the matrix $\boldsymbol{z}_i\boldsymbol{z}_i^T - \boldsymbol{I} + \mathrm{diag}\boldsymbol{z}_i$, and thus the solution of the system (5) is a vector in the matrix kernel of $\boldsymbol{M}$. In particular, this means that two sequences of states $\boldsymbol{a}$ and $\boldsymbol{b}$ produce the same tabu matrix after a non-zero number of iterations of the quantum annealing algorithm. Restricting $\boldsymbol{x}$ to a binary vector, it follows from (5) that the corresponding sequence of states produces a zero tabu matrix $\boldsymbol{S}_{\boldsymbol{x}}$ after a nonzero number of iterations.

As an example, take $n = 4$ and consider the vectors $\boldsymbol{z}_i$ ordered lexicographically, starting from the vector $\boldsymbol{z}_1 = (-1\ -1\ -1\ 1)$ and $\boldsymbol{z}_2 = (-1\ -1\ 1\ -1)$, ending up with $\boldsymbol{z}_{16} = (-1\ \ -1\ \ -1\ \ -1)$. Thus the following sequence $\boldsymbol{x} = (1101001100101100)$ solves (5). As such, the tabu matrix with positive probability returns to zero state $\mathbb{O}$, and thus, the memory of the sequence of iterations is finite, which opens a possibility to use the non-homogeneous Markov chain approach to convergence analysis of the tabu-enhanced quantum annealing algorithm.

4 Conclusion and Discussion

In this paper we presented an approach to the formal proof of convergence of the tabu enhanced quantum annealing algorithm. However, the speed of convergence and ways to improve the efficiency of the algorithm are to be studied separately. Among the possible ways to continue this research is in the direction of the tabu matrix parametrization so as to balance the depth of dependency vs. the speed of convergence of the optimization algorithm. Another possibility is to study the convergence problem within a more general class of hybrid quantum-classical optimization algorithms with penalty-type restrictions. However, we leave all these for future research.

Acknowledgements. The publication has been prepared with the support of Russian Science Foundation according to the research project No.21-71-10135 https://rscf.ru/en/project/21-71-10135/. The work has been supported by the Q@TN consortium.

References

1. Elliott, T.J., Yang, C., Binder, F.C., Garner, A.J.P., Thompson, J., Gu, M.: Extreme dimensionality reduction with quantum modeling. Phys. Rev. Lett. **125**(26),(2020). https://doi.org/10.1103/PhysRevLett.125.260501. place: College Pk Publisher: Amer Physical Soc WOS:000600849400001
2. Endo, S., Cai, Z., Benjamin, S.C., Yuan, X.: Hybrid quantum-classical algorithms and quantum error mitigation. J. Phys. Soc. Jpn. **90**(3), 032001 (2021). https://doi.org/10.7566/JPSJ.90.032001
3. Faigle, U., Kern, W.: Some convergence results for probabilistic Tabu search. ORSA J. Comput. **4**(1), 32–37 (1992). https://doi.org/10.1287/ijoc.4.1.32, http://pubsonline.informs.org/doi/abs/10.1287/ijoc.4.1.32
4. Glover, F.: Tabu search and finite convergence. Discrete Appl. Math. 34 (2002). https://doi.org/10.1016/S0166-218X(01)00263-3
5. Henderson, D., Jacobson, S.H., Johnson, A.W.: The theory and practice of simulated annealing. In: Glover, F., Kochenberger, G.A. (eds.) Handbook of Metaheuristics, pp. 287–319. Springer, Boston (2003). https://doi.org/10.1007/0-306-48056-5_10
6. Johnson, A., Jacobson, S.: On the convergence of generalized hill climbing algorithms. Discrete Appl. Math. **119**(1-2), 37–57 (2002). https://doi.org/10.1016/S0166-218X(01)00264-5, https://linkinghub.elsevier.com/retrieve/pii/S0166218X01002645
7. Nielsen, M.A., Chuang, I.L.: Quantum computation and Quantum Information. Cambridge University Press, Cambridge 10th anniversary ed edn. (2010)
8. Nimbe, P., Weyori, B.A., Adekoya, A.F.: Models in quantum computing: a systematic review. Quantum Inf. Process. **20**(2), 80 (2021). https://doi.org/10.1007/11128-021-03021-3, http://link.springer.com/10.1007/s11128-021-03021-3
9. Pastorello, D., Blanzieri, E.: Quantum annealing learning search for solving qubo problems. Quantum Inf. Process. **18**(10), 303 (2019)

10. Pastorello, D., Blanzieri, E., Cavecchia, V.: Learning adiabatic quantum algorithms over optimization problems. Quantum Mach. Intell. **3**(1), 1–19 (2021). https://doi.org/10.1007/s42484-020-00030-w
11. Paszkiewicz, A., Wegrzyn, J.: Responsiveness of the sensor network to alarm events based on the potts model. Sensors **20**(23) (2020). https://doi.org/10.3390/s20236979, https://www.mdpi.com/1424-8220/20/23/6979

Semi-markov Resource Flow as a Bit-Level Model of Traffic

Anatoly Nazarov[1], Alexander Moiseev[1], Ivan Lapatin[1], Svetlana Paul[1(✉)], Olga Lizyura[1], Pavel Pristupa[1], Xi Peng[2], Li Chen[2], and Bo Bai[2]

[1] Institute of Applied Mathematics and Computer Science, National Research Tomsk State University, 36 Lenina Avenue, Tomsk 634050, Russia

[2] Theory Lab, Central Research Center, 2012 Labs, Huawei Tech. Investment Co., Ltd, 8/F, Bio-informatics Center, No. 2 Science Park West Avenue, Hong Kong Science Park, Pak Shek Kok, Shatin, N.T., Hong Kong

{pancy.pengxi,chen.li7}@huawei.com

Abstract. In this paper, we consider semi-Markov flow as a bit-level model of traffic. Each request of the flow brings some arbitrary distributed amount of information to the system. The current paper aims to investigate the amount of information received in semi-Markov flow during time unit. We use the asymptotic analysis method under the limit condition of growing time of observation to derive the limiting probability distribution of the amount of information received in the flow and build the approximation of its prelimit distribution function.

Keywords: semi-Markov flow · asymptotic analysis · Gaussian approximation · traffic modeling

1 Introduction

In telecommunication systems, the models of arrivals usually capture the structure of traffic from a packet-level point of view. Despite the interest in traffic models, few studies take into account packet length. Traffic modeling is focused on capturing such properties of telecommunication flows as burstiness, self-similarity and long-range dependence [5,13–15].

The idea of modeling arrivals together with the size of packets described in paper [4]. Authors use batch Markovian arrival process (BMAP) to model packet size as a size of the batch. In paper [12], authors build the model of traffic based on discrete-time BMAP model using two counting processes: the number of arriving packets and the number of bytes in those packets. Both processes in the model are affected by the state of the underlying Markov chain. More ideas of using packet size in traffic modeling are described in [3]. In some cases, for example, in papers [10,11], the model cannot be investigated when the input process describe only the number of received packets.

Resource flows are applicable in such area of research as queueing systems with random resource requirements. In such systems, each request of the flow

V. M. Vishnevskiy et al. (Eds.): DCCN 2021, CCIS 1552, pp. 220–232, 2022.
https://doi.org/10.1007/978-3-030-97110-6_17

has some random requirement on the resources [1,2,6]. Similar resource systems are described in papers [7,8].

We propose semi-Markov flow as a model of bit-level traffic, which allows us to take into account the length of packets in telecommunication systems. In our model, packets arrivals are driven by the semi-Markov process and the lengths of packets follow the arbitrary distribution. To research the model, we use the asymptotic analysis method under the limit condition of the growing time of the flow observation. We build a Gaussian approximation of the cumulative distribution function of the amount of information received in the flow.

We have organized the paper as follows. In Sect. 2, we present a mathematical model of semi-Markov flow. Section 3 is devoted to the derivation of the balance equation for the probability distribution of the process describing the amount of information received in the flow. In Sect. 4, we investigate the model using the asymptotic analysis method under the limit condition of growing time and build a Gaussian approximation. In Sect. 5, we show the numerical experiments and the area of applicability of the approximation. Section 6 is dedicated to the concluding remarks.

2 Mathematical Model of Semi-markov Flow

Semi-Markov flow is determined by semi-Markov matrix $\mathbf{A}(x)$. Elements $A_{k\nu}(x)$ of the matrix has the following from:

$$A_{k\nu}(x) = P\{\xi(n+1) = \nu, \tau(n+1) < x|\xi(n) = k\}. \tag{1}$$

We also take into account that

$$\mathbf{P} = \mathbf{A}(\infty), \tag{2}$$

where $\mathbf{P}$ is the transition matrix of embedded Markov chain $\xi(n)$ at the moments of state changes of the semi-Markov process. Moments t_n of arrivals in semi-Markov flow we determine as follows:

$$t_{n+1} = t_n + \tau(n+1).$$

Further, we use semi-Markov process $k(t)$, which is defined by equality

$$k(t) = \xi(n+1), \text{if } t_n < t \leq t_{n+1} = t_n + \tau(n+1). \tag{3}$$

Each request of the flow brings some random amount of information with arbitrary distribution given by cumulative distribution function $B(x)$.

We denote $S(t)$ as the amount of information received in semi-Markov flow during time t. The problem is to derive the probability distribution of process $S(t)$.

We also denote $z(t)$ as the residual time of next arrival in the flow and consider three-dimensional process $\{k(t), S(t), z(t)\}$.

3 Balance Equation for the Probability Distribution of the Flow State

Three-dimensional process $\{k(t), S(t), z(t)\}$ is Markovian. Thus, we consider the function

$$P_k(s, z, t) = P\{k(t) = k, S(t) < s, z(t) < z\}$$

and derive balance equation

$$\frac{\partial P_k(s,z,t)}{\partial t} = \frac{\partial P_k(s,z,t)}{\partial z} - \frac{\partial P_k(s,0,t)}{\partial z} + \sum_{\nu=1}^{K} \int_0^s \frac{\partial P_\nu(s-x,0,t)}{\partial z} dB(x) A_{\nu k}(z), \tag{4}$$

where $\frac{\partial P_k(s,0,t)}{\partial z} = \left.\frac{\partial P_k(s,z,t)}{\partial z}\right|_{z=0}$.

We introduce partial characteristic functions

$$H_k(u, z, t) = \int_0^\infty e^{jus} d_s P_k(s, z, t)$$

and denote vector characteristic function

$$\mathbf{H}(u, z, t) = \{H_1(u, z, t), H_2(u, z, t), ..., H_K(u, z, t)\},$$

identity matrix $\mathbf{I}$ and vector of ones $\mathbf{e}$. After that, we rewrite Eq. (4) together with additional equation obtained taking the limit by $z \to \infty$

$$\frac{\partial \mathbf{H}(u,z,t)}{\partial t} = \frac{\partial \mathbf{H}(u,z,t)}{\partial z} - \frac{\partial \mathbf{H}(u,0,t)}{\partial z}\{\mathbf{I} - \mathbf{A}(z)B^*(u)\},$$

$$\frac{\partial \mathbf{H}(u,t)}{\partial t}\mathbf{e} = \frac{\partial \mathbf{H}(u,0,t)}{\partial z}\{B^*(u) - 1\}\mathbf{e}, \tag{5}$$

where $B^*(u) = \int_0^\infty e^{jux} dB(x)$ is the characteristic function of the amount of information in one request of the semi-Markov flow and $\mathbf{H}(u,t) = \mathbf{H}(u,\infty,t)$.

We cannot solve system (5) directly. Thus, we use asymptotic analysis method to investigate the amount of information received in the flow per time unit.

4 Asymptotic Probability Distribution

We introduce the equality $t = \tau T$, where $\tau \geq 0$ and T is an infinite parameter, as the limit condition of growing time. Solving system (5) in the limit by $T \to \infty$, we formulate the following theorem.

Theorem 1. *For characteristic function $H(u,t) = \mathbb{E}e^{juS(t)} = \mathbf{H}(u,t)\mathbf{e}$ in the limit condition of growing time, the following equality holds:*

$$\lim_{t\to\infty}\left\{H(u,t) - \exp\left(ju\kappa_1 t + \frac{(ju)^2}{2}\kappa_2 t\right)\right\} = 0, \tag{6}$$

where

$$\kappa_1 = \frac{b_1}{\mathbf{r}\mathbf{A}_1\mathbf{e}}, \tag{7}$$

$$\kappa_2 = \frac{b_2}{\mathbf{r}\mathbf{A}_1\mathbf{e}} + 2b_1\mathbf{g}'(0)\mathbf{e}. \tag{8}$$

Here b_1 and b_2 are the first and second raw moments of distribution function $B(x)$, matrices $\mathbf{A}_1$ and $\mathbf{A}_2$ are determined by formulas

$$\mathbf{A}_1 = \int_0^\infty (\mathbf{P} - \mathbf{A}(x))dx,$$

$$\mathbf{A}_2 = \int_0^\infty x^2 d\mathbf{A}(x).$$

Vector $\mathbf{g}'(0)$ is the solution of the inhomogeneous system of equations

$$\mathbf{g}'(0)(\mathbf{I} - \mathbf{P}) = \kappa_1(\mathbf{r} - \mathbf{R}),$$

$$\mathbf{g}'(0)\mathbf{A}_1\mathbf{e} = \frac{b_1}{2}\frac{\mathbf{r}\mathbf{A}_1\mathbf{e}}{(\mathbf{r}\mathbf{A}_2\mathbf{e})^2} - b_1.$$

Vector $\mathbf{r}$ is the steady state probability distribution of embedded Markov chain $\xi(n)$, which is the solution of the system

$$\mathbf{r} = \mathbf{r}\mathbf{P},$$

$$\mathbf{r}\mathbf{e} = 1.$$

Vector $\mathbf{R}$ is the steady-state probability distribution of semi-Markov process $k(t)$, which is given by formula

$$\mathbf{R} = \frac{\mathbf{r}\mathbf{A}_1}{\mathbf{r}\mathbf{A}_1\mathbf{e}}.$$

Proof. In system (5), we denote $\frac{1}{T} = \varepsilon$ and make the following substitutions:

$$\tau = \varepsilon t,\ u = \varepsilon w,\ \mathbf{H}(u,z,t) = \mathbf{F}(w,z,\tau,\varepsilon). \tag{9}$$

We obtain

$$\varepsilon\frac{\partial \mathbf{F}(w,z,\tau,\varepsilon)}{\partial \tau} - \frac{\partial \mathbf{F}(w,z,\tau,\varepsilon)}{\partial z} = \frac{\partial \mathbf{F}(w,0,\tau,\varepsilon)}{\partial z}\left\{\mathbf{A}(z)B^*(\varepsilon w) - \mathbf{I}\right\},$$

$$\varepsilon \frac{\partial \mathbf{F}(w,z,\tau,\varepsilon)}{\partial \tau}\mathbf{e} = \frac{\partial \mathbf{F}(w,z,\tau,\varepsilon)}{\partial z}\{B^*(\varepsilon w) - 1\}\,\mathbf{e}. \tag{10}$$

After that, we take the limit by $\varepsilon \to 0$ in the first equation of system (10) taking into account that $B^*(0) = 1$, which yields

$$\frac{\partial \mathbf{F}(w,z,\tau)}{\partial z} = \frac{\partial \mathbf{F}(w,0,\tau)}{\partial z}\{\mathbf{I} - \mathbf{A}(z)\}.$$

The idea of the asymptotic analysis method, which is outlined in paper [9], is to present the solution of the last equation in the following form:

$$\mathbf{F}(w,z,\tau) = \Phi(w,\tau)\mathbf{R}(z), \tag{11}$$

where $\mathbf{R}(z)$ is the steady-state distribution of two-dimensional process $\{k(t),\ z(t)\}$, which satisfies the equality

$$\mathbf{R}(z) = \mathbf{R}'(0)\int_0^z (\mathbf{P} - \mathbf{A}(x))dx.$$

Here

$$\mathbf{R}'(0) = \frac{\mathbf{r}}{\mathbf{r}\mathbf{A}_1\mathbf{e}},$$

matrix $\mathbf{A}_1$ is given by

$$\mathbf{A}_1 = \int_0^\infty (\mathbf{P} - \mathbf{A}(x))dx,$$

vector $\mathbf{r}$ is the steady-state distribution of the embedded Markov chain, which is the solution of the system

$$\mathbf{r} = \mathbf{r}\mathbf{P},$$

$$\mathbf{r}\mathbf{e} = 1. \tag{12}$$

Consider the second equation of system (10), making the decomposition of $B^*(\varepsilon w)$ into the Taylor series up to $O(\varepsilon^2)$:

$$\varepsilon \frac{\partial \mathbf{F}(w,\tau,\varepsilon)}{\partial \tau}\mathbf{e} = jw\varepsilon b_1 \frac{\partial \mathbf{F}(w,0,\tau,\varepsilon)}{\partial z}\mathbf{e} + O(\varepsilon^2),$$

where b_1 is the mean packet length. Substituting the solution (11) into the last equation, we take the limit by $\varepsilon \to 0$ and obtain

$$\frac{\partial \Phi(w,\tau)}{\partial \tau} = jwb_1\Phi(w,\tau)\mathbf{R}'(0)\mathbf{e}.$$

It is easy to see that the solution of the last equation is given by

$$\Phi(w,\tau) = e^{jw\kappa_1\tau}.$$

Here κ_1 has the following form:

$$\kappa_1 = \frac{b_1}{\mathbf{r}\mathbf{A}_1\mathbf{e}},$$

which coincides with (7).

Making substitutions $w = \dfrac{u}{\varepsilon}$ and $\tau = \varepsilon t$ reverse to (9), we obtain the equality

$$e^{jw\kappa_1\tau} = e^{ju\kappa_1 t}.$$

For the more detailed analysis, we make the following substitution in system (5):

$$\mathbf{H}(u,z,t) = e^{ju\kappa_1 t}\mathbf{H}_1(u,z,t). \tag{13}$$

Substituting (13) into system (5), we obtain the system of equations for characteristic function $\mathbf{H}_1(u,z,t)$:

$$\frac{\partial \mathbf{H}_1(u,z,t)}{\partial t} + ju\kappa_1\mathbf{H}_1(u,z,t) = \frac{\partial \mathbf{H}_1(u,z,t)}{\partial z} + \frac{\partial \mathbf{H}_1(u,z,t)}{\partial z}\left\{\mathbf{A}(z)B^*(u) - \mathbf{I}\right\},$$
$$\frac{\partial \mathbf{H}_1(u,z,t)}{\partial t}\mathbf{e} + ju\kappa_1\mathbf{H}_1(u,z,t)\mathbf{e} = \frac{\partial \mathbf{H}_1(u,z,t)}{\partial z}\left\{B^*(u) - 1\right\}\mathbf{e}. \tag{14}$$

We denote $\dfrac{1}{T} = \varepsilon^2$ and make the following substitutions in system (14):

$$\tau = \varepsilon^2 t, u = \varepsilon w, \mathbf{H}_1(u,z,t) = \mathbf{F}_1(w,z,\tau,\varepsilon). \tag{15}$$

We obtain the system of equations

$$\varepsilon^2\frac{\partial \mathbf{F}_1(w,z,\tau,\varepsilon)}{\partial \tau} + j\varepsilon w\kappa_1\mathbf{F}_1(w,z,\tau,\varepsilon)$$
$$= \frac{\partial \mathbf{F}_1(w,z,\tau,\varepsilon)}{\partial z} - \frac{\partial \mathbf{F}_1(w,0,\tau,\varepsilon)}{\partial z}\left\{\mathbf{I} - \mathbf{A}(z)B^*(\varepsilon w)\right\},$$
$$\varepsilon^2\frac{\partial \mathbf{F}_1(w,\tau,\varepsilon)}{\partial \tau}\mathbf{e} + j\varepsilon w\kappa_1\mathbf{F}_1(w,\tau,\varepsilon)\mathbf{e} = \frac{\partial \mathbf{F}_1(w,0,\tau,\varepsilon)}{\partial z}\left\{B^*(\varepsilon w) - 1\right\}\mathbf{e}. \tag{16}$$

We will seek the solution of system (16) in the following form:

$$\mathbf{F}_1(w,z,\tau,\varepsilon) = \Phi(w,\tau)\left\{\mathbf{R}(z) + j\varepsilon w\mathbf{f}(z)\right\} + O(\varepsilon^2), \tag{17}$$

which we substitute into (16):

$$j\varepsilon w\kappa_1\mathbf{R}(z) = \mathbf{R}'(z) + j\varepsilon w\mathbf{f}'(z) - \mathbf{R}'(0)\left\{\mathbf{I} - \mathbf{A}(z)(1 + j\varepsilon w b_1)\right\}$$
$$- j\varepsilon w\mathbf{f}'(0)\left\{\mathbf{I} - \mathbf{A}(z)\right\} + O(\varepsilon^2).$$

After that, we present the last equation as follows:

$$\mathbf{f}'(z) - \mathbf{f}'(0)\left\{\mathbf{I} - \mathbf{A}(z)\right\} = \kappa_1\left[\mathbf{R}(z) - \mathbf{r}\mathbf{A}(z)\right]. \tag{18}$$

According to the superposition principle, we present the solution of Eq. (18) as the sum:

$$\mathbf{f}(z) = C\mathbf{R}(z) + \mathbf{g}(z), \tag{19}$$

which we substitute into (18):

$$\mathbf{g}'(z) - \mathbf{g}'(0)\left\{\mathbf{I} - \mathbf{A}(z)\right\} = \kappa_1\left[\mathbf{R}(z) - \mathbf{r}\mathbf{A}(z)\right]. \tag{20}$$

Since $\mathbf{g}(z)$ by virtue of (19) is a particular solution of (18), then we assume that it satisfies the additional condition $\mathbf{g}(\infty)\mathbf{e} = 0$. We take the limit by $z \to \infty$ in Eq. (20) and obtain

$$\mathbf{g}(\infty) = \int_0^\infty \mathbf{g}'(0)\left\{\mathbf{I} - \mathbf{A}(z)\right\}dz - \kappa_1\int_0^\infty (\mathbf{r}\mathbf{A}(z) - \mathbf{R}(z))dz.$$

For the improper integral, we set the integrand as $z \to \infty$ equal to zero:

$$\mathbf{g}'(0)(\mathbf{I} - \mathbf{P}) - \kappa_1(\mathbf{r} - \mathbf{R}) = 0, \tag{21}$$

where $\mathbf{R} = \mathbf{R}(\infty)$ is the vector of steady-state distribution of semi-Markov process $k(t)$, which satisfies the system of equations

$$\mathbf{R} = \frac{\mathbf{r}\mathbf{A}_1}{\mathbf{r}\mathbf{A}_1\mathbf{e}},$$
$$\mathbf{R}\mathbf{e} = 1. \tag{22}$$

Taking back to Eq. (21), we represent it as follows:

$$\mathbf{g}'(0)(\mathbf{I} - \mathbf{P}) = \kappa_1(\mathbf{r} - \mathbf{R}).$$

The obtained system of linear algebraic equations has unlimited number of solutions. Thus, we apply the additional condition, which we derive from the equality

$$0 = \mathbf{g}(\infty)\mathbf{e} = \int_0^\infty \left\{\mathbf{g}'(0)(\mathbf{I} - \mathbf{A}(z)) - \kappa_1(\mathbf{r}\mathbf{A}(z) - \mathbf{R}(z))\right\}dz\ \mathbf{e}.$$

Taking (21) into account, we can transform the last equality:

$$0 = \mathbf{g}(\infty)\mathbf{e} = \int_0^\infty \left\{\mathbf{g}'(0)(\mathbf{I} - \mathbf{A}(z)) + \kappa_1\mathbf{r}(\mathbf{P} - \mathbf{A}(z)) + \kappa_1(\mathbf{R}(z) - \mathbf{R})\right\}dz\mathbf{e}$$

$$= \mathbf{g}'(0)\int_0^\infty (\mathbf{P} - \mathbf{A}(x))dx\mathbf{e} + \kappa_1\int_0^\infty (\mathbf{R}(x) - \mathbf{R})dx\mathbf{e} + \kappa_1\mathbf{r}\int_0^\infty (\mathbf{P} - \mathbf{A}(x))dx\mathbf{e}$$

$$= \mathbf{g}'(0)\mathbf{A}_1\mathbf{e} - \kappa_1\int_0^\infty (\mathbf{R} - \mathbf{R}(x))dx\mathbf{e} + b_1.$$

Here the integral can be transformed as follows:

$$\int_0^\infty (\mathbf{R} - \mathbf{R}(x))dx = (\mathbf{R} - \mathbf{R}(x))x\Big|_0^\infty + \int_0^\infty x d\mathbf{R}(x)$$

$$= \mathbf{R}'(0)\int_0^\infty x(\mathbf{I} - \mathbf{A}(x))dx = \mathbf{R}'(0)\int_0^\infty (\mathbf{I} - \mathbf{A}(x))d\frac{x^2}{2}$$

$$= \mathbf{R}'(0)\left\{(\mathbf{I} - \mathbf{A}(x))\frac{x^2}{2}\Big|_0^\infty + \int_0^\infty \frac{x^2}{2}d\mathbf{A}(x)\right\} = \frac{1}{2}\mathbf{R}'(0)\mathbf{A}_2 = \frac{\mathbf{r}\mathbf{A}_2}{\mathbf{r}\mathbf{A}_2\mathbf{e}}.$$

Here matrix $\mathbf{A}_2$ is given by

$$\mathbf{A}_2 = \int_0^\infty \frac{x^2}{2}d\mathbf{A}(x).$$

Finally, we have the system of linear algebraic equations with a solution

$$\mathbf{g}'(0)(\mathbf{I} - \mathbf{P}) = \kappa_1(\mathbf{r} - \mathbf{R}),$$

$$\mathbf{g}'(0)\mathbf{A}_1\mathbf{e} = \frac{b_1}{2}\frac{\mathbf{r}\mathbf{A}_2\mathbf{e}}{(\mathbf{r}\mathbf{A}_1\mathbf{e})^2} - b_1. \tag{23}$$

After that, we consider the second equation of system (16), in which we substitute decomposition (17):

$$\varepsilon^2\frac{\partial\Phi(w,\tau)}{\partial\tau} + jw\varepsilon\kappa_1\Phi(w,\tau)(1 + jw\varepsilon C)$$

$$= \Phi(w,\tau)\left\{\mathbf{R}'(0)\left[jw\varepsilon b_1 + \frac{(jw\varepsilon)^2}{2}b_2\right] - jw\varepsilon\mathbf{f}'(0)(-jw\varepsilon b_1)\right\}\mathbf{e} + O(\varepsilon^3).$$

By simple transformations, we obtain

$$\frac{\partial\Phi(w,\tau)}{\partial\tau} + (jw)^2\kappa_1\Phi(w,\tau)C = \Phi(w,\tau)\left\{\frac{(jw)^2}{2}\mathbf{R}'(0)b_2 + (jw)^2 b_1\mathbf{f}'(0)\right\}\mathbf{e}.$$

By the virtue of (19), we can write

$$\frac{\partial\Phi(w,\tau)}{\partial\tau} + (jw)^2\kappa_1\Phi(w,\tau)C$$

$$= \Phi(w,\tau)\left\{\frac{(jw)^2}{2}\mathbf{R}'(0)b_2 + (jw)^2 b_1(C\mathbf{R}'(0) + \mathbf{g}'(0))\right\}\mathbf{e},$$

from which we obtain

$$\frac{\partial\Phi(w,\tau)}{\partial\tau} = \Phi(w,\tau)\frac{(jw)^2}{2}\left\{\frac{b_2}{\mathbf{r}\mathbf{A}_1\mathbf{e}} + 2b_1\mathbf{g}'(0)\mathbf{e}\right\}.$$

Denoting

$$\kappa_2 = \frac{b_2}{\mathbf{r}\mathbf{A}_1\mathbf{e}} + 2b_1\mathbf{g}'(0)\mathbf{e}, \tag{24}$$

which coincides with (8), we derive the solution of differential equation above

$$\Phi(w,\tau) = \exp\left\{\frac{(jw)^2}{2}\kappa_2\tau\right\}.$$

From substitutions (15), we make the reverse substitutions

$$w = \frac{u}{\varepsilon},\ \tau = \varepsilon^2 t,$$

which yields

$$\Phi(w,\tau) = \exp\left\{\frac{(jw)^2}{2}\kappa_2\tau\right\} = \exp\left\{\frac{(ju)^2}{2\varepsilon^2}\kappa_2\varepsilon^2 t\right\} = \exp\left\{\frac{(ju)^2}{2}\kappa_2 t\right\}.$$

Finally, in (13), we set $z \to \infty$ and obtain the asymptotic characteristic function

$$h_1(u,t) = \exp\{ju\kappa_1 t\}\exp\left\{\frac{(ju)^2}{2}\kappa_2 t\right\} = \exp\left\{ju\kappa_1 t + \frac{(ju)^2}{2}\kappa_2 t\right\}.$$

As we can see, the distribution of the amount of information received in semi-Markov flow is asymptotically Gaussian with mean $\kappa_1 t$ and variance $\kappa_2 t$.

We note that by setting $b_1 = 1$ and $b_2 = 1$, we obtain the case when the amount of information in a packet is deterministic and equal to one. Thus, the obtained result is valid for the number of packet arrivals in the flow.

Since Gaussian distribution allows negative values, we propose the following approximation for distribution function of the amount of information received in the flow during time t:

$$F_{Approx}(x,t) = \frac{G(x,t) - G(0,t)}{1 - G(0,t)}, \tag{25}$$

where $G(x,t)$ is the Gaussian distribution function with mean $\kappa_1 t$ and variance $\kappa_2 t$.

5 Numerical Example

We set semi-Markov matrix as follows:

$$\mathbf{A}(x) = \mathbf{P} \circ \mathbf{G}(x),$$

where $\mathbf{P}$ is the transition matrix of the embedded Markov chain $\xi(n)$ and $\mathbf{G}(x)$ is the matrix of conditional distributions of the process $\tau(n)$, operation $\circ$ is Hadamard product of matrices.

Matrix $\mathbf{P}$ is given by

$$\mathbf{P} = \begin{bmatrix} 0.95 & 0.05 \\ 0.8 & 0.2 \end{bmatrix}.$$

The elements of matrix $\mathbf{G}(x)$ are gamma distribution functions with shape parameters $\alpha_{11} = 0.005$, $\alpha_{12} = 0.01$, $\alpha_{21} = 0.1$, $\alpha_{22} = 1$ and scale parameter $\beta = 1$. We assume that the amount of information in one packet is deterministic and equals to $b_1 = 1.5$.

Figures 1, 2, 3 show the distribution function of the amount of information received in semi-Markov flow via simulation (solid line) compared with asymptotic results (dash line) for $t = 20$, $t = 50$ and $t = 75$.

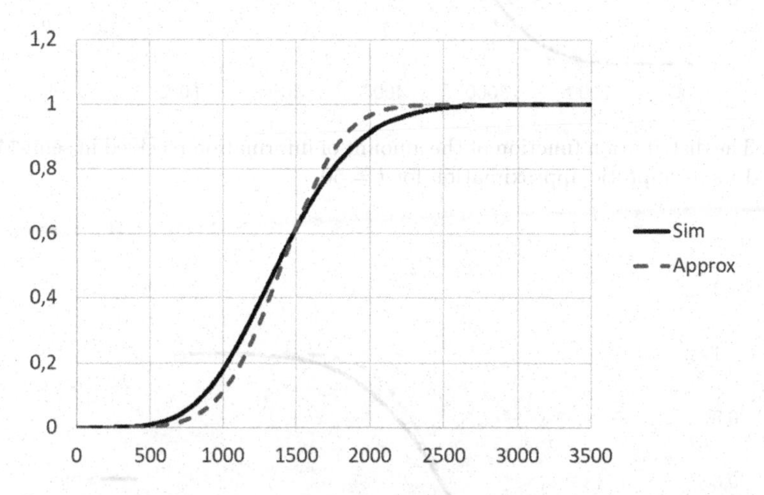

Fig. 1. The distribution function of the amount of information received in semi-Markov flow and its asymptotic approximation for $t = 20$

Table contains the values of Kolmogorov distance

$$\Delta = \max_{0 \leq x < \infty} \left| F_{Sim}(x,t) - F_{Approx}(x,t) \right|$$

between empirical distribution function obtained via simulation $F_{Sim}(x,t)$ and asymptotic cumulative distribution function $F_{Approx}(x,t)$ of the amount of information received in the flow during time t given by (25) (Table 1).

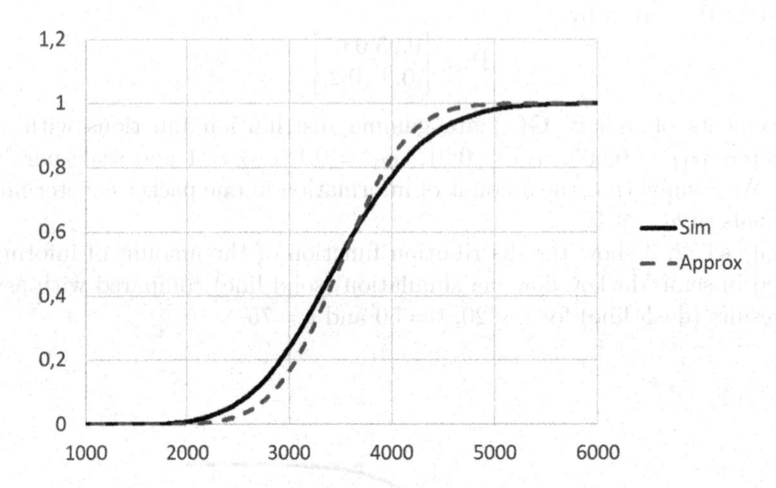

Fig. 2. The distribution function of the amount of information received in semi-Markov flow and its asymptotic approximation for $t = 50$

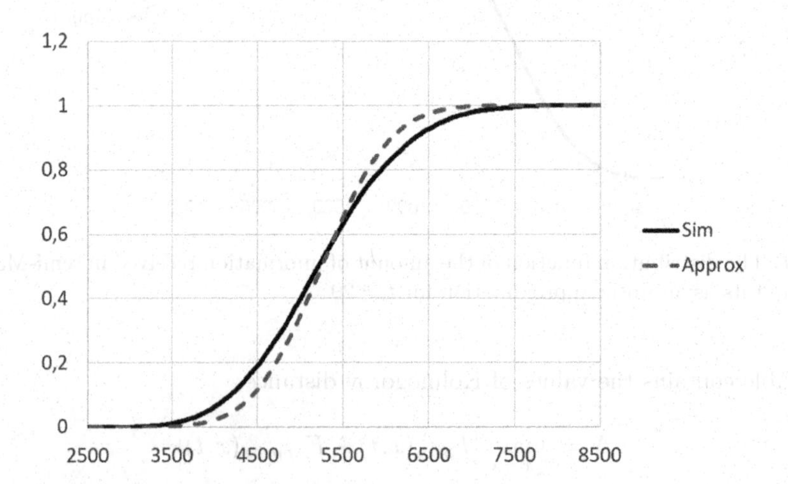

Fig. 3. The distribution function of the amount of information received in semi-Markov flow and its asymptotic approximation for $t = 75$

Table 1. Kolmogorov distance between empirical distribution function of the amount of information in the buffer and its asymptotic approximation

	$t = 10$	$t = 20$	$t = 50$	$t = 75$	$t = 100$
Δ	0.0817	0.0768	0.0766	0.0758	0.0752

6 Conclusion

We have considered the bit-level traffic model in form of semi-Markov flow. For the amount of information received in the flow, we have obtained the limiting probability distribution under the limit condition of growing time of observation. We have derived the explicit formula for the mean and variance of Gaussian distribution. Since the distribution of the packet length in the model is arbitrary, the results are applicable for the number of packets arrivals when we set the size of each packet is equal to one.

References

1. Galileyskaya, A., Lisovskaya, E., Fedorova, E.: Resource queueing system with the requirements copying at the second phase. In: Vishnevskiy, V.M., Samouylov, K.E., Kozyrev, D.V. (eds.) DCCN 2019. CCIS, vol. 1141, pp. 352–363. Springer, Cham (2019). https://doi.org/10.1007/978-3-030-36625-4_28
2. Galileyskaya, A., Lisovskaya, E., Pagano, M.: On the total amount of the occupied resources in the multi-resource QS with renewal arrival process. In: Dudin, A., Nazarov, A., Moiseev, A. (eds.) ITMM 2019. CCIS, vol. 1109, pp. 257–269. Springer, Cham (2019). https://doi.org/10.1007/978-3-030-33388-1_21
3. Gao, J., Rubin, I.: Multifractal analysis and modeling of long-range-dependent traffic. In: 1999 IEEE International Conference on Communications (Cat. No. 99CH36311), vol. 1, pp. 382–386. IEEE (1999)
4. Klemm, A., Lindemann, C., Lohmann, M.: Modeling IP traffic using the batch Markovian arrival process. Perform. Eval. **54**(2), 149–173 (2003)
5. Li, M.: Long-range dependence and self-similarity of teletraffic with different protocols at the large time scale of day in the duration of 12 years: Autocorrelation modeling. Physica Scripta **95**(6), 065222 (2020)
6. Lisovskaya, E., Moiseeva, S., Pagano, M.: Multiclass GI/GI/∞ queueing systems with random resource requirements. In: Dudin, A., Nazarov, A., Moiseev, A. (eds.) ITMM/WRQ -2018. CCIS, vol. 912, pp. 129–142. Springer, Cham (2018). https://doi.org/10.1007/978-3-319-97595-5_11
7. Naumov, V., Samouylov, K., Yarkina, N., Sopin, E., Andreev, S., Samuylov, A.: Lte performance analysis using queuing systems with finite resources and random requirements. In: 2015 7th International Congress on Ultra Modern Telecommunications and Control Systems and Workshops (ICUMT), pp. 100–103. IEEE (2015)
8. Naumov, V.A., Samuilov, K.E., Samuilov, A.K.: On the total amount of resources occupied by serviced customers. Autom. Remote Control **77**(8), 1419–1427 (2016). https://doi.org/10.1134/S0005117916080087
9. Nazarov, A., et al.: Multi-level MMPP as a model of fractal traffic. In: Dudin, A., Nazarov, A., Moiseev, A. (eds.) ITMM 2020. CCIS, vol. 1391, pp. 61–77. Springer, Cham (2021). https://doi.org/10.1007/978-3-030-72247-0_5
10. Nazarov, A., et al.: Mathematical model of scheduler with semi-markov input and bandwidth sharing discipline. In: 2021 International Conference on Information Technology (ICIT), pp. 494–498. IEEE (2021)
11. Peng, X., Bai, B., Zhang, G., Lan, Y., Qi, H., Towsley, D.: Bit-level power-law queueing theory with applications in lte networks. In: 2018 IEEE Global Communications Conference (GLOBECOM), pp. 1–6. IEEE (2018)

12. Salvador, P., Pacheco, A., Valadas, R.: Modeling IP traffic: joint characterization of packet arrivals and packet sizes using BMAPs. Comput. Netw. **44**(3), 335–352 (2004)
13. Willinger, W., Taqqu, M.S., Leland, W.E., Wilson, D.V., et al.: Self-similarity in high-speed packet traffic: analysis and modeling of ethernet traffic measurements. Stat. Sci. **10**(1), 67–85 (1995)
14. Yang, T., Zhao, R., Zhang, W., Yang, Q.: On the modeling and analysis of communication traffic in intelligent electric power substations. IEEE Trans. Power Delivery **32**(3), 1329–1338 (2016)
15. Yang, X., Petropulu, A.P.: The extended alternating fractal renewal process for modeling traffic in high-speed communication networks. IEEE Trans. Signal Process. **49**(7), 1349–1363 (2001)

Asymptotic Diffusion Analysis of an Retrial Queueing System M/M/1 with Impatient Calls

Elena Danilyuk(✉), Svetlana Moiseeva, and Anatoly Nazarov

National Research Tomsk State University, Lenina Avenue, 36, Tomsk, Russia

Abstract. In the paper, the retrial queueing system of M/M/1 type with input Poison flow of events and impatient calls is considered. The service time, delay time of calls in the orbit and the impatience time of calls in the orbit have exponential distribution. Asymptotic diffusion analysis method is proposed for the solving problem of finding distribution of the number of calls in the orbit under a long delay of calls in orbit and long time patience of calls in the orbit condition. Numerical results confirm that the probability distribution of the number of calls in the orbit is Gaussian. The range of applicability of the obtained results is given for different values of system parameters. The RQ-system under consideration is also investigated by the asymptotic analysis method. The results of the comparison of the two methods are presented.

Keywords: Retrial queueing system · Impatient calls · Asymptotic diffusion analysis

1 Introduction

At the present time retrial queueing systems (RQ-systems) research is in the demand as evidenced by numerous papers in this area and grants support. The systems as mathematical models are very suitable for modern telecommunication systems, networks, mobile networks describing. Along with the construction of mathematical models of RQ-systems, new methods of their study are being developed. A fairly new method is asymptotic diffusion analysis method, as a modification of the asymptotic analysis method. Both of them are suggested by the Tomsk research school, and there are interesting works [1–4], in which the asymptotic diffusion analysis method is used.

The present paper is devoted to study of the retrial queueing system of M/M/1 with impatient calls by the asymptotic diffusion analysis method.

The general information about mathematical model of the retrial queueing system discussed in the paper and the problem statement are presented in the Sect. 2. In the Sect. 3 the detailed derivation of the model and the system of

The reported study was funded by the RFBR and Tomsk region according to the research project No.19-41-703002.

V. M. Vishnevskiy et al. (Eds.): DCCN 2021, CCIS 1552, pp. 233–246, 2022.
https://doi.org/10.1007/978-3-030-97110-6_18

Kolmogorov equations for the stationary state probabilities are cited. The Sect. 4 consists of the decision of the problem under study by the asymptotic analysis method. This part of the present study is obtained to compare the asymptotic analysis method and the asymptotic diffusion analysis method. The last one for the decision of the problem is given in Sect. 5. Some numerical results, graphs, that proved the theoretical results, are performed in the Sect. 6. Section 7 concludes the paper.

2 Mathematical Model

We consider an retrial queueing system with one server and Poisson arrival process with intensity λ. An arriving call (or customer) that has found the service device free takes it for the service for a random time distributed exponentially with parameter μ. If the device is busy, calls that arrive go into the orbit. On the orbit, each call, independently of others, waits for a random time whose duration has an exponential distribution with parameter σ, and then again accesses the device with a second attempt to obtain servicing. If the device is free, the call from orbit occupies it for random servicing time. If the device is busy, call immediately goes into the orbit and wait once more random time. Moreover, a call from the orbit leaves the system after exponential distributed time with parameter α, demonstrating the "impatience" property. Figure 1 shows a model of the RQ-system $M/M/1$ with impatient calls.

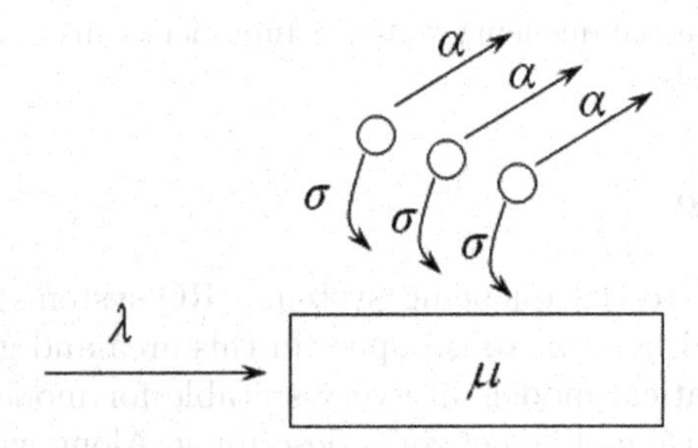

Fig. 1. RQ-system $M/M/1$ with impatient calls

The problem is to find the stationary distribution of the number of calls in the orbit. This problem has been solved in [5] by the asymptotic analysis method under a long time patience of calls in the orbit condition. In the present paper we use asymptotic diffusion analysis under a long delay of calls in orbit and long time patience of calls in the orbit condition to study the stationary distribution $P(i)$ of the number of calls in the orbit (when $\sigma \to \varepsilon$, $\alpha \to \varepsilon$).

3 Process of the System States: System of Kolmogorov Differential Equations in Terms of Partial Characteristic Functions

Let us consider Markovian process $\{k(t), i(t)\}$ determined states of the considered RQ-system where $i(t)$ is the number of calls in the orbit at the moment t, $i(t) = 0, 1, 2, 3, \ldots$, and $k(t)$ defines device state at the moment t and takes one of the following values

$$k(t) = \begin{cases} 0, & \textit{if server is free at the moment } t; \\ 1, & \textit{if server is busy at the moment } t. \end{cases}$$

Denote as $P_0(i,t) = P\{k(t) = 0, i(t) = i\}$ and $P_1(i,t) = P\{k(t) = 1, i(t) = i\}$ the probability that, at the moment t, there are i calls in the orbit, $i = 0, 1, 2, \ldots$, and the service device is free or the server is busy respectively.

Introduce the partial characteristic functions

$$\begin{aligned} H_k(u) &= \sum_{i=0}^{\infty} e^{jui} P_k(i,t), \\ H_k(0) &= \sum_{i=0}^{\infty} P_k(i) = R_k, \end{aligned} \tag{1}$$

where $j = \sqrt{-1}$, $k = 0, 1$, $i(t) = 0, 1, 2, \ldots$, and R_k have the meaning of the stationary probability distribution of the $k(t)$ values. It is obvious that $H(u) = H_0(u) + H_1(u)$.

To obtain the probability distribution $P_0(i,t), P_1(i,t)$ for the states of the retrial queue M/M/1 with impatient calls in the orbit, we construct a system of Kolmogorov differential equations (2) [5]

$$\begin{cases} \dfrac{\partial P_0(i,t)}{\partial t} = -(\lambda + i\sigma + i\alpha) P_0(i,t) + (i+1)\alpha P_0(i+1,t) + \mu P_1(i,t), \\ \dfrac{\partial P_1(i,t)}{\partial t} = -(\lambda + i\alpha + \mu) P_1(i,t) + (i+1)\alpha P_1(i+1,t) + \lambda P_1(i-1,t) \\ +\lambda P_0(i,t) + (i+1)\sigma P_0(i+1,t), \end{cases} \tag{2}$$

$i = 0, 1, 2, \ldots$

and write it in terms of partial characteristic functions (1)

$$\begin{cases} \dfrac{\partial H_0(u,t)}{\partial t} = -\lambda H_0(u,t) + \mu H_1(u,t) + j\left(\sigma + \alpha\left(1 - e^{-ju}\right)\right) \dfrac{\partial H_0(u,t)}{\partial u}, \\ \dfrac{\partial H_1(u,t)}{\partial t} = \lambda H_0(u,t) - \mu H_1(u,t) - \lambda\left(1 - e^{ju}\right) H_1(u,t) \\ -j\sigma e^{-ju} \dfrac{\partial H_0(u,t)}{\partial u} + j\alpha\left(1 - e^{-ju}\right) \dfrac{\partial H_1(u,t)}{\partial u}. \end{cases} \tag{3}$$

In adding the first equation by the second equation of (3) we get (4)

$$\frac{\partial H(u,t)}{\partial t} = \left(1 - e^{-ju}\right)\left(\lambda e^{ju} H_1(u,t) + j(\sigma+\alpha)\frac{\partial H_0(u,t)}{\partial u} + j\alpha \frac{\partial H_1(u,t)}{\partial u}\right), \tag{4}$$

where $H(u,t) = H_0(u,t) + H_1(u,t)$.

Since calls are "impatient," the considered system has a stationary mode for any values of λ and μ.

Let in (2) $\lim_{t\to\infty} P_k(i,t) = P_k(i)$, $k = 0, 1$, and then we write system (2) in the form

$$\begin{cases} -(\lambda + i\sigma + i\alpha)\, P_0(i) + (i+1)\alpha P_0(i+1) + \mu P_1(i) = 0, \\ -(\lambda + i\alpha + \mu)\, P_1(i) + (i+1)\alpha P_1(i+1) + \lambda P_1(i-1) \\ +\lambda P_0(i) + (i+1)\sigma P_0(i+1) = 0, \\ \sum\limits_{i=0}^{\infty} (P_0(i) + P_1(i)) = 1, \end{cases} \tag{5}$$

$i = 0, 1, 2, ...$

In (5), we have a system of difference equations of infinite dimension with variable coefficients, which would be very hard to solve by mathematical methods. Therefore, to solve it we propose the follow approaches: asymptotic analysis method, asymptotic diffusion analysis method, and numerical method. The numerical algorithm for finding the final probabilities is based on truncating the system dimension (5); to do that, we represent system (5) for $i = 0, 1, 2, ..., N$ as

$$PS = B, \tag{6}$$

where the row vector P of dimension $2\,(N+1)$ is the desired stationary probability distribution of the number of calls in orbit for each state of the device $k = \{0; 1\}$

$$P = \left(P\,(0)\;\, P\,(1)\right),$$

and $P(0)$, $P(1)$ is a row vector with elements $P(0,i)$, $P(1,i)$, $i = 0, 1, 2, ..., N$, respectively. Matrix S of dimension $2\,(N+1)\times 2\,(N+1)+1$ is represented in block form as

$$S = \begin{pmatrix} S_{11} & S_{12} & S_{13} \\ S_{21} & S_{22} & S_{23} \end{pmatrix},$$

where $S_{11} = ||s_{ij}^{11}||_1^{N+1}$, $S_{12} = ||s_{ij}^{12}||_1^{N+1}$, $S_{21} = ||s_{ij}^{21}||_1^{N+1}$, $S_{22} = ||s_{ij}^{22}||_1^{N+1}$ are sparse matrices whose nonzero elements are defined as $s_{ii}^{11} = -(\lambda + (i-1)(\sigma+\alpha))$, $s_{i+1,i}^{11} = i\alpha$, $s_{ii}^{12} = \lambda$, $s_{i+1,i}^{12} = (i-1)\sigma$, $s_{ii}^{21} = \mu$, $s_{ii}^{22} = -(\lambda + \mu + (i-1)\alpha)$, $s_{i+1,i}^{22} = i\alpha$, $s_{i,i+1}^{22} = \lambda$.

Blocks S_{13}, S_{23} are unit vector columns of dimension $(N+1)$, row vector $B = ||b_i||$ of dimension $2(N+1)+1$ is a row of free coefficients with elements $b_i = 0$ $(i = 0, 1, 2, ..., N-1)$, $b_N = 1$.

We solve (6) with the help of a numerical algorithm using the Mathcad software suite. We choose N to be so large that probabilities $P(0,N)$, $P(1,N)$ are equal to the machine zero. The obtained results give us so called pre-limit probabilities distribution (we get that distribution of the number of calls in orbit is normal) that we will use for comparing with probabilities distributions produced by the asymptotic diffusion analysis method and asymptotic analysis method under the same asymptotic conditions.

4 Asymptotic Analysis Method

We propose to find the solution of the system of (5) by the method of asymptotic analysis under the assumption that there is a long delay between calls from the orbit and high "patience" of calls and summarize the results in two stage.

Using (1), $\frac{\partial H_k(u)}{\partial u} = j\sum_{i=0}^{\infty} ie^{jui}P_k(i)$, $k = 0, 1$, we can write the system (5) as

$$\begin{cases} -\lambda H_0(u) + j\left(\sigma + \alpha\left(1 - e^{-ju}\right)\right) H_0'(u) + \mu H_1(u) = 0, \\ \lambda H_0(u) - j\sigma e^{-ju} H_0'(u) + \left(\lambda\left(e^{ju} - 1\right) - \mu\right) H_1(u) \\ +j\alpha\left(1 - e^{-ju}\right) H_1'(u) = 0, \\ \lambda e^{ju} H_1(u) + j\left(\sigma + \alpha\right) H_0'(u) + j\alpha H_1'(u) = 0, \end{cases} \tag{7}$$

where the third equation is the sum of the sum of the first two from (7).

4.1 Finding First-order Asymptotics

In the (7) we make the substitutions $\sigma = \varepsilon$, $\alpha = q\varepsilon$, $u = \varepsilon w$, $H_k(u) = F_0(w, \varepsilon)$, $k = 0, 1$, where ε is infinitesimal value ($\varepsilon \to 0$).

Since $H_k'(u) = \frac{1}{\varepsilon}\frac{\partial F_k(w,\varepsilon)}{\partial w}$, $k = 0, 1$, the equations system (7) can be written as

$$\begin{cases} -\lambda F_0(w,\varepsilon) + j(1 + qjw\varepsilon)\frac{\partial F_0(w,\varepsilon)}{\partial w} + \mu F_1(w,\varepsilon) = O(\varepsilon^2), \\ -\lambda F_0(w,\varepsilon) - j(1 - jw\varepsilon)\frac{\partial F_0(w,\varepsilon)}{\partial w} + \left(\lambda jw\varepsilon - \mu\right) F_1(w,\varepsilon) \\ +jqjw\varepsilon\frac{\partial F_1(w,\varepsilon)}{\partial w} = O(\varepsilon^2), \\ \lambda(1 + jw\varepsilon)F_1(w,\varepsilon) + j(1+q)\frac{\partial F_0(w,\varepsilon)}{\partial w} + jq\frac{\partial F_1(w,\varepsilon)}{\partial w} = O(\varepsilon^2). \end{cases} \tag{8}$$

The transformation of equations of (8) under $\varepsilon \to 0$ with $F_k(w) = \lim_{\varepsilon\to 0} F_k(w,\varepsilon)$, $k = 0, 1$, leads to equations system as follows

$$\begin{cases} -\lambda F_0(w) + jF_1'(w) + \mu F_1(w) = 0, \\ \lambda F_0(w) - jF_1'(w) - \mu F_1(w) = 0, \\ \lambda F_1(w) + j(1+q)F_0'(w) + jqF_1'(w) = 0. \end{cases} \tag{9}$$

We suggest to find the Eq. (9) solution $F_k(w)$, $k = 0, 1$, in the form

$$F_k(w) = R_k\Phi(w), \quad k = 0, 1, \tag{10}$$

where $R_k = H_k(0)$, $k = 0, 1$ (1).

Substituting (10) in (9) we have solution of the (9)

$$\Phi(w) = \exp\left\{jG_1 w\right\}, \tag{11}$$

where

$$G_1 = \frac{\mu R_1 - \lambda R_0}{R_0} \quad or \quad G_1 = \frac{\lambda R_1}{q + R_0},$$

and R_0 is the root of the Eq. (12), $R_1 = 1 - R_0$,

$$\mu R_0^2 + (\lambda - \mu + q(\lambda + \mu)) - \mu q = 0. \tag{12}$$

Pre-limit characteristic function $h(u)$ is approximately equal to

$$H(u) = H_0(u) + H_1(u) \approx F_0\left(\frac{u}{\varepsilon}\right) + F_1\left(\frac{u}{\varepsilon}\right) = h_1(u).$$

So, the first-order asymptotic characteristic function $h_1(u)$ of the probability distribution of the number of calls in the orbit under the assumption of a long delay of calls in orbit and their high "patience" can be presented as

$$h_1(u) = \exp\left\{\frac{G_1}{\sigma} ju\right\}. \tag{13}$$

4.2 Finding Second-order Asymptotics

In the (7) with (13) we let

$$H_k(u) = \exp\left\{\frac{G_1}{\sigma} ju\right\} H_k^{(2)}(u), \quad k = 0, 1. \tag{14}$$

Let $\sigma = \varepsilon^2, \alpha = q\varepsilon^2$, $u = \varepsilon w$, $H_k^{(2)}(u) = F_k^{(2)}(w, \varepsilon)$, $k = 0, 1$, where ε is an infinitesimal, then (7) with some transformations can be rewritten as

$$\begin{cases} -(\lambda + (1 + jw\varepsilon)G_1) F_0^{(2)}(w,\varepsilon) + j\varepsilon \dfrac{\partial F_0^{(2)}(w,\varepsilon)}{\partial w} + \mu F_1^{(2)}(w,\varepsilon) = o(\varepsilon^2), \\ (\lambda + (1 - jw\varepsilon)G_1) F_0^{(2)}(w,\varepsilon) - j\varepsilon \dfrac{\partial F_0^{(2)}(w,\varepsilon)}{\partial w} \\ + (\lambda jw\varepsilon - \mu - qG_1 jw\varepsilon)) F_1^{(2)}(w,\varepsilon) = o(\varepsilon^2), \\ -j(1+q)G_1 F_0^{(2)}(w,\varepsilon) + j(1+q)\varepsilon \dfrac{\partial F_0^{(2)}(w,\varepsilon)}{\partial w} \\ + (\lambda(1 + jw\varepsilon) - qG_1) F_1^{(2)}(w,\varepsilon) + jq\varepsilon \dfrac{\partial F_1^{(2)}(w,\varepsilon)}{\partial w} = o(\varepsilon^2). \end{cases} \tag{15}$$

When $\varepsilon \to 0$ in (15) and $\lim_{\varepsilon \to 0} F_k^{(2)}(w, \varepsilon) = F_k^{(2)}(w)$, $k = 0, 1$, we get

$$\begin{cases} -(\lambda + G_1) F_0^{(2)}(w) + \mu F_1^{(2)}(w) = 0, \\ (\lambda + G_1) F_0^{(2)}(w) - \mu F_1^{(2)}(w) = 0, \\ (1+q)G_1 F_0^{(2)}(w) + (\lambda - qG_1) F_1^{(2)}(w) = 0. \end{cases} \tag{16}$$

The solution of equations system (15) has the following form

$$\begin{cases} F_k^{(2)}(w,\varepsilon) = (R_k + jw\varepsilon f_k)\Phi^{(2)}(w) + O(\varepsilon^2), \quad k = 0,1, \\ R_0 + R_1 = 1, \end{cases} \tag{17}$$

where R_0, R_1 are defined above, f_0, f_1 are constants, and function $\Phi^{(2)}(w)$ is to be determined.

Substituting (17) into (15) and taking into account (12), (16) we have $\Phi^{(2)}(w)$ in (18) under $\varepsilon \to 0$

$$\Phi^{(2)}(w) = \exp\left\{ G_2 \frac{(jw)^2}{2} \right\}, \tag{18}$$

where

$$G_2 = \frac{\mu f_1 - (\lambda + f_0)G_1 - qG_1R_0}{R_0}, \quad or$$
$$G_2 = \frac{(\lambda - qG_1)f_1 - (1+q)G_1f_0 + \lambda R_1}{q + R_0},$$

and R_0, R_1, G_1 are determined above, $f_0 + f_1 = 0$.

Taking into account (2), (14), (17), (18), the characteristic function $H(u) = H_0(u) + H_1(u)$, provided that the calls in orbit have long delays and the "patience" is high, is a Gaussian

$$H(u) = H_0(u) + H_1(u) \approx \sum_{k=0}^{1} R_k \exp\left\{ \frac{G_1}{\sigma} ju + \frac{G_2}{\sigma} \frac{(ju)^2}{2} \right\} = h_2(u). \tag{19}$$

5 Asymptotic Diffusion Analysis Method

We use the system (3) and Eq. (4) for diffusion approximation in three stages: 1) obtaining the drift (**transfer**) coefficient; 2) centering the process and obtaining the diffusion coefficient; 3) diffusion approximation.

5.1 Obtaining the Drift (Transfer) Coefficient

In the system (3) and Eq. (4), we make the substitutions $\sigma = \varepsilon$, $\alpha = q\varepsilon$, $u = \varepsilon w$, $\tau = \varepsilon t$, $H_k(u,t) = F_k(w,\varepsilon,\tau)$, $k = 0,1$, where ε is infinitesimal value,

$$\begin{cases} \varepsilon\frac{\partial F_0(w,\varepsilon,\tau)}{\partial \tau} = -\lambda F_0(w,\varepsilon,\tau) + \mu F_1(w,\varepsilon,\tau) + j\left(1+q-qe^{-jw\varepsilon}\right)\frac{\partial F_0(w,\varepsilon,\tau)}{\partial w}, \\ \varepsilon\frac{\partial F_1(w,\varepsilon,\tau)}{\partial \tau} = \lambda F_0(w,\varepsilon,\tau) - \mu F_1(w,\varepsilon,\tau) - \lambda\left(1 - e^{jw\varepsilon}\right)F_1(w,\varepsilon,\tau) \\ -je^{-jw\varepsilon}\frac{\partial F_0(w,\varepsilon,\tau)}{\partial w} + jq\left(1 - e^{-jw\varepsilon}\right)\frac{\partial F_1(w,\varepsilon,\tau)}{\partial w}, \\ \varepsilon\frac{\partial F(w,\varepsilon,\tau)}{\partial \tau} = \left(e^{jw\varepsilon} - 1\right)\left(\lambda F_1(w,\varepsilon,\tau) + j(1+q)e^{-jw\varepsilon}\frac{\partial F_0(w,\varepsilon,\tau)}{\partial w} \right. \\ \left. +jqe^{-jw\varepsilon}\frac{\partial F_1(w,\varepsilon,\tau)}{\partial w}\right). \end{cases} \tag{20}$$

Transform the equations of (20) under $\varepsilon \to 0$ with $F_k(w,\tau) = \lim_{\varepsilon\to 0} F_k(w,\varepsilon,\tau)$, $k = 0, 1$, and find their solution $F_k(w,\tau)$, $k = 0, 1$, in the form

$$F_k(w,\tau) = R_k \exp\{jwx(\tau)\}, \quad k = 0, 1, \tag{21}$$

where $R_k = H_k(0)$, $k = 0, 1$, $x(\tau)$ - unknown function of time τ.

Substituting (21) in (20) we get the following

$$\begin{cases} R_0 = R_0(x(\tau)) = \dfrac{\mu}{\lambda+\mu+x(\tau)}, \\ R_1 = R_1(x(\tau)) = \dfrac{\lambda+x(\tau)}{\lambda+\mu+x(\tau)}, \\ x'(\tau) = a(x(\tau)) = \lambda - qx(\tau) - (\lambda + x(\tau))\, R_0(x(\tau)). \end{cases} \tag{22}$$

5.2 Centering the Process and Obtaining the Diffusion Coefficient

In (3) and (4) we let

$$H_k(u,t) = \exp\left\{\frac{ju}{\sigma}x(\sigma t)\right\} H_k^{(2)}(u,t), \quad k = 0, 1, \tag{23}$$

and get the system of equations as follows

$$\begin{cases} \dfrac{\partial H_0^{(2)}(u,t)}{\partial t} + jux'(\sigma t)H_0^{(2)}(u,t) = -\left(\lambda + \dfrac{\sigma+\alpha\left(1-e^{-ju}\right)}{\sigma}x(\sigma t)\right) H_0^{(2)}(u,t) \\ +\mu H_1^{(2)}(u,t) + j\left(\sigma + \alpha\left(1-e^{-ju}\right)\right) \dfrac{\partial H_0^{(2)}(u,t)}{\partial u}, \\ \dfrac{\partial H_1^{(2)}(u,t)}{\partial t} + jux'(\sigma t)H_1^{(2)}(u,t) = \left(\lambda + e^{-ju}x(\sigma t)\right) H_0^{(2)}(u,t) \\ + \left(\lambda e^{ju} - (\lambda+\mu) - \dfrac{\alpha\left(1-e^{-ju}\right)}{\sigma}x(\sigma t)\right) H_1^{(2)}(u,t) \\ -j\sigma e^{-ju}\dfrac{\partial H_0^{(2)}(u,t)}{\partial u} + j\alpha\left(1-e^{-ju}\right)\dfrac{\partial H_1^{(2)}(u,t)}{\partial u}, \\ \dfrac{\partial H^{(2)}(u,t)}{\partial t} + jux'(\sigma t)H^{(2)}(u,t) = \left(e^{ju}-1\right) \\ \times\left\{\dfrac{-(\sigma+\alpha)\,e^{-ju}}{\sigma}x(\sigma t)H_0^{(2)}(u,t) + \left(\lambda - \dfrac{\alpha e^{-ju}}{\sigma}x(\sigma t)\right) H_1^{(2)}(u,t) \right. \\ \left. +j(\sigma+\alpha)\,e^{-ju}\dfrac{\partial H_0^{(2)}(u,t)}{\partial u} + j\alpha e^{-ju}\dfrac{\partial H_1^{(2)}(u,t)}{\partial u}\right\} \end{cases} \tag{24}$$

In the system (24) we make the substitutions $\sigma = \varepsilon^2$, $\alpha = q\varepsilon^2$, $u = \varepsilon w$, $\tau = \varepsilon^2 t$, $H_k^{(2)}(u,t) = F_k^{(2)}(w,\varepsilon,\tau)$, $k = 0,1$, to obtain the system below

$$\begin{cases} \varepsilon^2 \dfrac{\partial F_0^{(2)}(w,\varepsilon,\tau)}{\partial \tau} + jw\varepsilon a(x(\tau)) F_0^{(2)}(w,\varepsilon,\tau) = \mu F_1^{(2)}(w,\varepsilon,\tau) - [\lambda + x(\tau) \\ +q\left(1-e^{-jw\varepsilon}\right)x(\tau)\big] F_0^{(2)}(w,\varepsilon,\tau) + j\varepsilon\left[1+q\left(1-e^{-jw\varepsilon}\right)\right] \dfrac{\partial F_0^{(2)}(w,\varepsilon,\tau)}{\partial w}, \\ \varepsilon^2 \dfrac{\partial F_1^{(2)}(w,\varepsilon,\tau)}{\partial \tau} + jw\varepsilon a(x(\tau)) F_1^{(2)}(w,\varepsilon,\tau) = \left(\lambda + e^{-jw\varepsilon} x(\tau)\right) F_0^{(2)}(w,\varepsilon,\tau) \\ + \left[\lambda e^{jw\varepsilon} - (\lambda+\mu) - q\left(1-e^{-jw\varepsilon}\right) x(\tau)\right] F_1^{(2)}(w,\varepsilon,\tau) \\ -j\varepsilon e^{-jw\varepsilon} \dfrac{\partial F_0^{(2)}(w,\varepsilon,\tau)}{\partial w} + jq\varepsilon\left(1-e^{-jw\varepsilon}\right) \dfrac{\partial F_1^{(2)}(w,\varepsilon,\tau)}{\partial w}, \\ \varepsilon^2 \dfrac{\partial F^{(2)}(w,\varepsilon,\tau)}{\partial \tau} + jw\varepsilon a(x(\tau)) F^{(2)}(w,\varepsilon,\tau) = \left(e^{jw\varepsilon} - 1\right) \\ \Big\{-(1+q)\, e^{-jw\varepsilon} x(\tau) F_0^{(2)}(w,\varepsilon,\tau) + \left[\lambda - q e^{-jw\varepsilon} x(\tau)\right] F_1^{(2)}(w,\varepsilon,\tau) \\ +j\varepsilon(1+q)\, e^{-jw\varepsilon} \dfrac{\partial F_0^{(2)}(w,\varepsilon,\tau)}{\partial w} + jq\varepsilon e^{-jw\varepsilon} \dfrac{\partial F_1^{(2)}(w,\varepsilon,\tau)}{\partial w} \Big\}. \end{cases} \tag{25}$$

The equations of the (25) with some transformations can be rewritten as

$$\begin{cases} jw\varepsilon a(x(\tau)) F_0^{(2)}(w,\varepsilon,\tau) = \mu F_1^{(2)}(w,\varepsilon,\tau) \\ -[\lambda + x(\tau) + qjw\varepsilon x(\tau)] F_0^{(2)}(w,\varepsilon,\tau) + j\varepsilon \dfrac{\partial F_0^{(2)}(w,\varepsilon,\tau)}{\partial w} + O(\varepsilon^2), \\ jw\varepsilon a(x(\tau)) F_1^{(2)}(w,\varepsilon,\tau) = [\lambda + (1-jw\varepsilon)\, x(\tau)] F_0^{(2)}(w,\varepsilon,\tau) \\ + [\lambda jw\varepsilon - \mu - qjw\varepsilon x(\tau)] F_1^{(2)}(w,\varepsilon,\tau) - j\varepsilon \dfrac{\partial F_0^{(2)}(w,\varepsilon,\tau)}{\partial w} + O(\varepsilon^2), \\ \varepsilon^2 \dfrac{\partial F^{(2)}(w,\varepsilon,\tau)}{\partial \tau} + jw\varepsilon a(x(\tau)) F^{(2)}(w,\varepsilon,\tau) = \left(jw\varepsilon + \dfrac{(jw\varepsilon)^2}{2}\right) \\ \Big\{-(1+q)(1-jw\varepsilon)\, x(\tau) F_0^{(2)}(w,\varepsilon,\tau) + [\lambda - q(1-jw\varepsilon)\, x(\tau)] F_1^{(2)}(w,\varepsilon,\tau) \\ +j\varepsilon(1+q) \dfrac{\partial F_0^{(2)}(w,\varepsilon,\tau)}{\partial w} + jq\varepsilon \dfrac{\partial F_1^{(2)}(w,\varepsilon,\tau)}{\partial w} \Big\} + O(\varepsilon^3). \end{cases} \tag{26}$$

The solution of equations system (26) has the following form

$$\begin{cases} F_k^{(2)}(w,\varepsilon,\tau) = \Phi(w,\tau)\,(R_k + jw\varepsilon f_k) + O(\varepsilon^2), \quad k = 0,1, \\ R_0 + R_1 = 1, \end{cases} \tag{27}$$

where $R_k = R_k(x(\tau))$, $k = 0,1$, are defined above, f_0, f_1, $(f_0 + f_1 = f)$, are constants, and $\Phi(w,\tau)$ is determined function.

Using (22) and (27) in (26) after transformations we can get

$$\begin{cases} -[\lambda + x(\tau)]\, f_0 + \mu f_1 = [a(x(\tau)) + qx(\tau)]\, R_0 - R_0 \dfrac{\partial \Phi(w,\tau)/\partial w}{w\Phi(w,\tau)}, \\ [\lambda + x(\tau)]\, f_0 - \mu f_1 = [a(x(\tau)) - \lambda + qx(\tau)]\, R_1 + x(\tau) R_0 + R_0 \dfrac{\partial \Phi(w,\tau)/\partial w}{w\Phi(w,\tau)}, \\ \dfrac{\partial \Phi(w,\tau)}{\partial \tau} = (jw)^2 \Phi(w,\tau) \{-a(x(\tau)) f + qx(\tau) R_1 + [\lambda - qx(\tau)]\, f_1 \\ -(1+q)\, x(\tau) f_0 + (1+q)\, x(\tau) R_0\} - w(1+q) R_0 \dfrac{\partial \Phi(w,\tau)}{\partial w} \\ -wqR_1 \dfrac{\partial \Phi(w,\tau)}{\partial w} + \dfrac{(jw)^2}{2} a(x(\tau)) \Phi(w,\tau). \end{cases} \tag{28}$$

The solution of system (28) has the form

$$f_k = CR_k + g_k - \varphi_k \frac{\partial \Phi(w,\tau)/\partial w}{w\Phi(w,\tau)}, \quad k = 0, 1, \tag{29}$$

and after substitution (29) in the first and the second equations of the (28) we obtain the equations systems (30), (31) for the φ_k and g_k, $k = 0, 1$, respectively

$$\begin{cases} -[\lambda + x(\tau)]\, g_0 + \mu g_1 = [a(x(\tau)) + qx(\tau)]\, R_0, \\ [\lambda + x(\tau)]\, g_0 - \mu g_1 = [a(x(\tau)) - \lambda + qx(\tau)]\, R_1 + x(\tau) R_0, \end{cases} \tag{30}$$

$$\begin{cases} [\lambda + x(\tau)]\, \varphi_0 - \mu \varphi_1 = -R_0, \\ -[\lambda + x(\tau)]\, \varphi_0 + \mu \varphi_1 = R_0. \end{cases} \tag{31}$$

Equations (22) and additional condition $g_0 + g_1 = 0$ for the (30) lead us to (32)

$$\begin{cases} \varphi_k = \varphi_k(x(\tau)) = \dfrac{\partial R_k(x(\tau))}{\partial x(\tau)}, \quad \varphi_0 + \varphi_1 = 0, \quad k = 0, 1, \\ g_0 = g_0(x(\tau)) = -\dfrac{a(x(\tau)) + qx(\tau)}{\lambda + \mu + x(\tau)} R_0(x(\tau)), \quad g_1 = -g_0. \end{cases} \tag{32}$$

The third equation of the (28) with (22), (29), (32) can be rewritten as

$$\frac{\partial \Phi(w,\tau)}{\partial \tau} = a'(x(\tau)) w \frac{\partial \Phi(w,\tau)}{\partial w} + b(x(\tau)) \frac{(jw)^2}{2} \Phi(w,\tau), \tag{33}$$

where

$$b(x(\tau)) = a(x(\tau)) + 2\Big(qx(\tau) R_1(x(\tau)) + (1+q) x(\tau) R_0(x(\tau)) + [\lambda + x(\tau)]\, g_1\Big). \tag{34}$$

5.3 Diffusion Approximation

Using (33) and (1), (23), (26) we can get the Fokker-Plank equation for the probability density of an diffusion process $y(\tau)$ with drift (transfer) coefficient $a'(x(\tau))y(\tau)$ and diffusion coefficient $b(x(\tau))$

$$\frac{\partial P(y(\tau),\tau)}{\partial \tau} = -a'(x(\tau)) \frac{\partial \{y(\tau) P(y(\tau),\tau)\}}{\partial y(\tau)} + \frac{b(x(\tau))}{2} \frac{\partial^2 P(y(\tau),\tau)}{\partial y^2(\tau)}, \tag{35}$$

and the process $y(\tau)$ is the solution of the stochastic differential Eq. (36)

$$dy(\tau) = a'(x(\tau))y(\tau)d\tau + \sqrt{b(x(\tau))}d\omega(\tau), \tag{36}$$

where $\omega(\tau)$ is the Wiener process.

Introduce diffusion process $z(\tau) = x(\tau) + \varepsilon y(\tau)$ and write the stochastic differential Eq. (37) for $z(\tau)$

$$dz(\tau) = a(z(\tau))d\tau + \varepsilon\sqrt{b(z(\tau))}d\omega(\tau). \tag{37}$$

Denote the probability density of the $z(\tau)$ as $\Pi(z(\tau),\tau) = \dfrac{\partial P\{z(\tau) < z\}}{\partial z}$ and he Fokker-Plank equation for it can be written as follows

$$\frac{\partial \Pi(z(\tau),\tau)}{\partial \tau} = -\frac{\partial\{a(z(\tau))\Pi(z(\tau),\tau)\}}{\partial z(\tau)} + \frac{\varepsilon^2}{2}\frac{\partial^2\{b(z(\tau))\Pi(z(\tau),\tau)\}}{\partial z^2(\tau)}. \tag{38}$$

The solution of the Eq. (38) for stationary probability distribution of the process $z(\tau)$ has the form

$$\Pi(z) = \frac{C}{b(z)}\exp\left\{\frac{2}{\sigma}\int_0^z \frac{a(x)}{b(x)}dx\right\}, \quad C - constant. \tag{39}$$

Finally, based on the (39) in (40) we get diffusion approximation $P_{ADA}(i)$ for the stationary distribution $P(i)$ of the number of calls in the orbit

$$P_{ADA}(i) = \frac{\Pi(i\sigma)}{\sum\limits_{k=0}^{\infty}\Pi(ki)}. \tag{40}$$

6 Numerical Results

In this section we give several numerical examples. Preliminary calculations suggest that theoretical results are consistent with simulation ones for the asymptotic analysis method. According to the asymptotic diffusion analysis method it is fashionable to conclude that the number of calls in the orbit is asymptotically normal [6].

To compare the pre-limit probability distribution of the number of calls in the orbit of considered queueing system $P(i)$ calculated via matrix method and its approximation $P_{AA}(i)$ and $P_{ADA}(i)$ constructed by using the asymptotic analysis method and asymptotic diffusion analysis method respectively for different values of the system parameters we use Kolmogorov distance Δ between respective distribution functions

$$\begin{cases} \Delta_{AA} = \max\limits_{n\geq 0}\sum\limits_{i=0}^{n}|P(i) - P_{AA}(i)|, \ \textit{for asymptotic analysis}, \\ \Delta_{ADA} = \max\limits_{n\geq 0}\sum\limits_{i=0}^{n}|P(i) - P_{ADA}(i)|, \ \textit{for asymptotic diffusion analysis}. \end{cases} \tag{41}$$

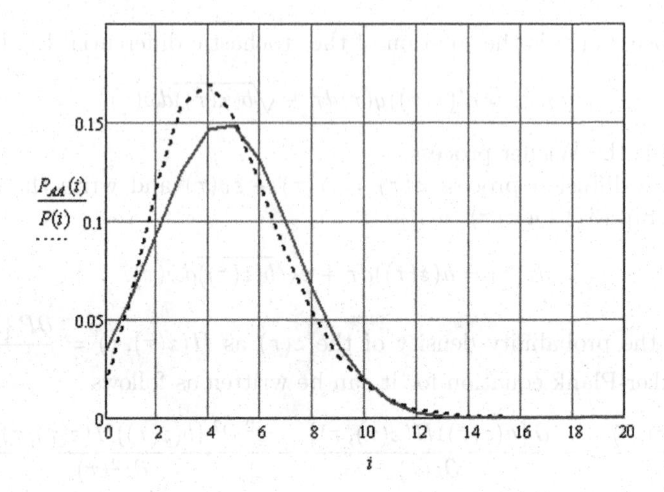

Fig. 2. Comparison of the asymptotic (solid line) and the pre-limit (dashed line) distributions for $\sigma = 0.01$, $\lambda = 0.4$, $\mu = 1$, $q = 2$, $\Delta = 0.069$.

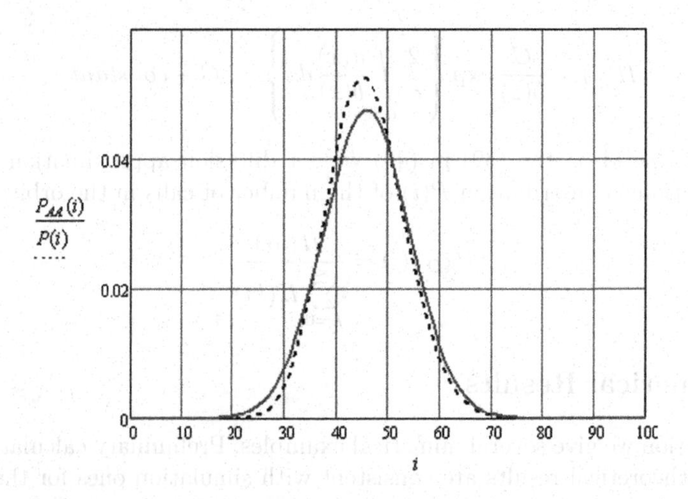

Fig. 3. Comparison of the asymptotic (solid line) and the pre-limit (dashed line) distributions for $\sigma = 0.001$, $\lambda = 0.4$, $\mu = 1$, $q = 2$, $\Delta = 0.028$.

The comparison of the pre-limit distribution and distribution obtained with the asymptotic analysis method is shown in Figs. 2, 3.

The comparison of the pre-limit distribution and distribution obtained with the asymptotic diffusion analysis method is shown in Figs. 4, 5.

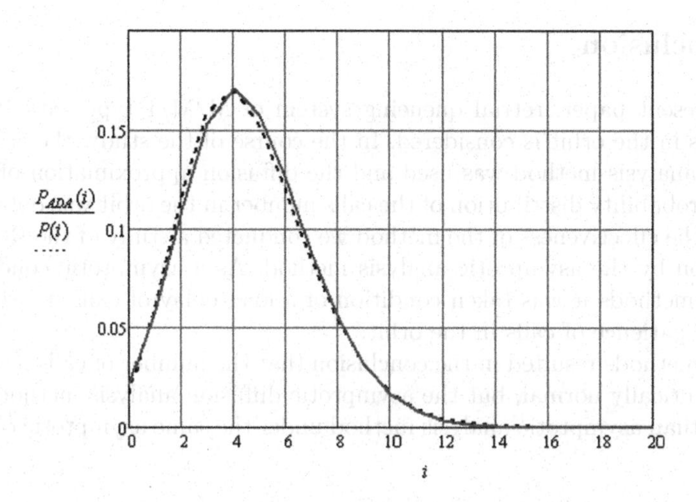

Fig. 4. Comparison of the asymptotic (solid line) and the pre-limit (dashed line) distributions for $\sigma = 0.01$, $\lambda = 0.4$, $\mu = 1$, $q = 2$, $H = 1$, $\Delta = 0.012$.

The Table 1 shows that the value of the Kolmogorov distance decreases with the growth of the system load λ/μ (if $\lambda/\mu \leq 1$) and with the increase in delay time of orders in orbit σ. The case when $\lambda/\mu > 1$ needs to be investigated further.

Table 1. Kolmogorov distance.

σ	$\lambda/\mu = 0.2$		$\lambda/\mu = 0.4$		$\lambda/\mu = 0.6$		$\lambda/\mu = 0.8$		$\lambda/\mu = 1$	
	Δ_{AA}	Δ_{ADA}	Δ_{AA}	Δ_{ADA}	Δ_{AA}	Δ_{ADA}	Δ_{AA}	Δ_{ADA}	Δ_{AA}	Δ_{ADA}
0, 01	0,175	0,036	0,069	0,012	0,033	0,006	0,022	0,004	0,017	0,003
0, 001	0,041	0,004	0,028	0,003	0,019	0,0018	0,012	0,0013	0,008	0,0009

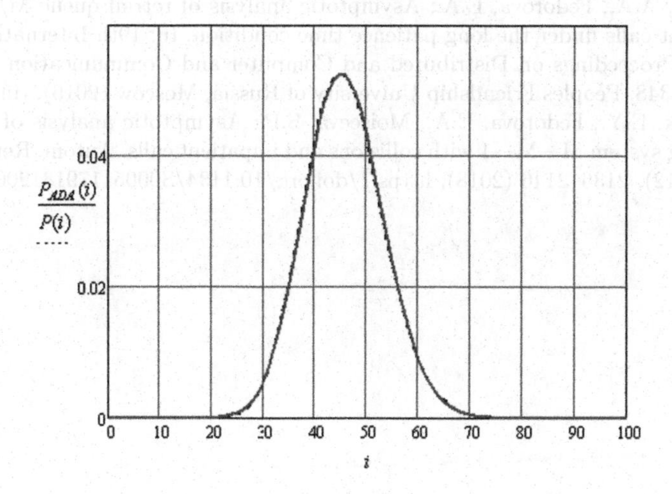

Fig. 5. Comparison of the asymptotic (solid line) and the pre-limit (dashed line) distributions for $\sigma = 0.001$, $\lambda = 0.4$, $\mu = 1$, $q = 2$, $H = 1$, $\Delta = 0.003$.

7 Conclusion

In the present paper, retrial queueing system of M/M/1 type with impatient customers in the orbit is considered. In the course of the study, the asymptotic diffusion analysis method was used and the diffusion approximation of the stationary probability distribution of the calls number in the orbit was obtained. To evaluate the effectiveness of the method we conducted a study of the RQ-system in question by the asymptotic analysis method. As a asymptotic condition for the both methods it was taken condition of a long delay of calls in orbit and a long time patience of calls in the orbit.

Both methods resulted in the conclusion that the number of calls in the orbit is asymptotically normal, but the asymptotic diffusion analysis method is more accurate than asymptotic analysis method under the same asymptotic condition.

References

1. Nazarov, A.A., Paul, S.V., Lizyura, O.D.: Asymptotic diffusion analysis of retrial queue M/M/1/1 with outgoing calls. In: 22th International Conference Proceedings on Distributed and Computer and Communication Networks, pp. 148–155. Peoples Friendship University of Russia, Moscow (2019). (in Russian)
2. Nazarov, A., Phung-Duc, T., Paul, S., Lizyura, O.: Asymptotic-diffusion analysis of retrial queue with two-way communication and renewal input. In: Proceedings of The 5th International Conference on Stochastic Methods, pp. 339–345. Peoples Friendship University of Russia, Moscow (2020)
3. Nazarov, A. A., Phung-Duc, T., Izmailova, Ya. Ye.: Asymptotic-diffusion analysis of multiserver retrial queueing system with priority customers. In: Proceedings of the XIX International Conference named after A. F. Terpugov on Information Technologies and Mathematical Modelling, pp. 88–98. NTL, Tomsk (2021)
4. Moiseev, A., Nazarov, A., Paul, S.: Asymptotic diffusion analysis of multi-server retrial queue with hyperexponential service. Mathematics **8**(4), 531 (2020)
5. Nazarov, A.A., Fedorova, E.A.: Asymptotic analysis of retrial queue M/M/1 with impatient calls under the long patience time condition. In: 19th International Conference Proceedings on Distributed and Computer and Communication Networks, pp. 342–348. Peoples Friendship University of Russia, Moscow (2016). (in Russian)
6. Danilyuk, E.Y., Fedorova, E.A., Moiseeva, S.P.: Asymptotic analysis of an retrial queueing system M—M—1 with collisions and impatient calls. Autom. Remote Control **79**(12), 2136–2146 (2018). https://doi.org/10.1134/S0005117918120044

Sufficient Stability Conditions for a Multi-orbit Retrial System with General Retrials Under Classical Retrial Policy

Ruslana Nekrasova[1,2(✉)]

[1] Institite of Applied Mathematical Research KarRC RAS, Petrozavodsk 185000, Russia
[2] Petrozavodsk State University, Petrozavodsk 185000, Russia

Abstract. We consider a single server multi-class retrial system. The arrival customer, who meets the server busy, joins the corresponding orbit and then retries to capture the server. The model obeys to the classical retrial policy: the total rate of orbit customers depends on their number. Retrial times are assumed to be generally distributed, and that makes the analysis much more complicated.

We use the previous results for the systems with exponential retrials and regenerative approach to establish the sufficient stability conditions to the model under consideration. The key element of the proof relies on Lorden's inequality, which is a the significant result from the renewal theory.

Keywords: Retrial model · Classical retrial policy · Stability analysis · Renewal theory · Lorden's inequality · Regenerative approach

1 Introduction

The paper deals with a multi-class retrial queue under classical retrial policy. The customers arrive according to the renewal input. The model admits a number infinite capacity *orbits* associated with the corresponding class. If arrival is blocked, it joins the orbit and then after generally distributed class-dependent retrial time attacks the server again. The system obeys to the classical retrial policy, thus the total retrial rate grows proportionally to the number of orbit customers.

To motivate the presented research, we touch the applicability of the retrial systems. Retrial queues were successfully used in simulation of widespread multiple access systems like call centers [1], telephone networks [2], cellular mobile

Supported by Russian Science Foundation according to the research project No.21-71-10135 https://rscf.ru/en/project/21-71-10135/.

V. M. Vishnevskiy et al. (Eds.): DCCN 2021, CCIS 1552, pp. 247–258, 2022.
https://doi.org/10.1007/978-3-030-97110-6_19

networks, etc. Moreover such systems are applicable in modeling of a large number of a modern objects like wireless telecommunication systems or multi-access protocols, where blocked data packets are sent again after some waiting period.

Retrial queuing models are widely studied in the literature. It is worth mention the basic books [3,4] and, for instance, quite recent survey papers [5,6].

Obliviously that most of stability results were obtained for more simple retrial single-class models with exponential retrials, where authors may obtain explicit statements for stability conditions or steady-state performance measures (see, for instance [7,8]). The analysis of more general case with an arbitrary distribution of retrial times is a challenging problem.

One of the first stability results related model with non-exponential distribution of inter-retrial times is presented in paper [9], where authors considered single-class model. In a later paper [10] the analysis was extended to a multi-server single-class system, the research was based on the regenerative approach. The similar model was considered in [11], where the analysis relayed on the fluid limit approach.

Our goal in this paper is to establish sufficient stability conditions for a single server model under consideration. Namely, we expand the previous analysis, obtained for particular cases of multi-class retrial models [15,16]. The research is based on regenerative approach. We present just the main steps of the proof, which are focused on the application for general retrials and rely on the results from renewal theory, namely, Lorden's inequality.

The paper is organized as follows. Section 2 contains a detailed description of multi-class retrial system. Then, in a basic Sect. 3, we obtain sufficient stability conditions for the presented model. In Sect. 4 the simulation results for the system with Pareto distribution of retrial time and non zero initial conditions are presented. Section 5 concludes the talk.

2 Description of the Model

We define a multi-class retrial queue with a single server. Incoming customers arrive at instants $\{t_n,\, n \geq 1\}$ according to renewal input. Assume (for the simplicity) $t_0 = 0$ and define $\tau_n = t_n - t_{n-1}$. Thus the sequence of inter-arrival times $\{\tau_n,\, n \geq 1\}$ is iid, we denote its generic element by τ.

The system admits $K \geq 1$ classes of customers. Assume that class-i customer, where $i = 1, \dots, K$, arrives according to *Poisson input* with a marginal rate $\lambda_i > 0$. Thus we obtain the total input rate as follows

$$\lambda = \lambda_1 + \cdots + \lambda_K.$$

Note that $\mathsf{E}\tau = 1/\lambda$ and the summary input is Poisson as a superposition of K Poisson streams. On the other hand, we can assume that class-i customer arrives with the probability p_i, defined in some given distribution

$$\mathbf{p} = (p_1, \dots, p_K),$$

and $\lambda_i = p_i \lambda$.

Service times form an independent sequence, which elements have class-dependent distributions. Define the generic service time corresponding to class i by $S^{(i)}$. Then denote the marginal load coefficients by $\rho_i = \lambda_i \mathsf{E}S^{(i)}$, and the total load coefficient as follows

$$\rho = \lambda\big(p_1 \mathsf{E}S^{(1)} + \cdots + p_K \mathsf{E}S^{(K)}\big).$$

Class-i arrival, who meets the server busy, joins to the corresponding infinite capacity buffer so-called *orbit* and then after a random time, distributed as $\xi^{(i)}$, makes attempts to capture the server again. If the server is still busy, the orbital or secondary customer, returns to the orbit immediately, waits again for a random time and then makes new attempts. (From this point of view, the customers from the orbits are called secondary, while the arrivals from input stream are called primary.) The i-th orbit rate is defined as follows

$$\gamma_i := 1/\mathsf{E}\xi^{(i)}, \quad i = 1, \dots, K.$$

The model under consideration obeys to *classical retrial policy.* Thus, all the secondary customers make independent attempts and the total (actual) orbit rate grows proportionally to sum orbit size. In case $\gamma_i = 0$ we obtain classical loss system, while if $\gamma_i = \infty$, the orbit customer immediately captures the server, as in becomes empty, and the model is equivalent to the infinite-buffer queuing system. One more significant feature of a model under consideration is *non-exponential retrials*: class-i random orbit waiting times $\xi^{(i)}$, $i = 1, \dots, K$ are generally distributed, which makes the analysis much more complicated.

Note that retrial policy admits the case when the server is idle, while the system is not empty (some customers wait in orbits). Thus the model under consideration is an example of *non-conservative* queues.

Define the number of customers on the i-th orbit just before time instant t by $N^{(i)}(t)$, thus the summary orbit is obtained as follows

$$N(t) := N^{(1)}(t) + \cdots + N^{(K)}(t), \quad t \geq 0.$$

We deal with a single-server model with no buffer. Thus the system becomes overloaded just in case the number of secondary customer increases. That means, the only reason of instability is the infinite the growth of orbit size $N(t)$, as $t \to \infty$.

Our goal is to find the conditions, which guarantee the stable summary orbit. The basic stability analysis is relied on results from regeneration theory, well-presented in [12–14].

3 Stability Analysis

In this section we present the sufficient stability conditions for the system under consideration. The proof relies on the regenerative approach. Note that regenerative method of stability analysis is well-presented, at instance, in [12, 14].

To present the regenerations we consider the process $\nu(t) \in \{0, 1\}$, which indicates the server state (idle/busy), and the basic process

$$X(t) = \nu(t) + N(t),$$

associated with the total number of customers. Next, we consider zero initial state $X(t_0) = 0$, $t_0 = 0$ and define

$$T_k = \min_n\{t_n > T_{k-1} : X(t_n^-) = 0\}, \quad k \geq 1, T_0 = 0.$$

Note that the sequence $\{T_k, k \geq 1\}$ represents instants, when new arrivals join to the totally empty systems. In such instants the system starts over in stochastic sense or *regenerates*. From this point of view the process $X(t)$ is called regenerative. Moreover the sequence of regenerative cycle lengths $\{T_k - T_{k-1}\}$ is iid, we denote the generic length by T. The process $\{X(t)\}$ is called *positive recurrent* if $\mathsf{E}T < \infty$.

In case T is non-lattice (which holds for the models with Poisson input) and if $X(t)$ is positive recurrent, there exists the stationary distribution for the process $X(t)$, as $t \to \infty$. The existence of stationary distribution implies the stability of the model under consideration. Note that under the term "stability" we actually mean positive recurrence of the basic regenerative processes. Thus following the regeneration arguments, for establishing the stability, it is enough to show that $\mathsf{E}T < \infty$. Next we define by

$$T(t) = \min_n\{T_n - t : T_n - t > 0\}, \quad n \geq 1.$$

the remaining at the instant t regeneration time. According to the results from regeneration theory, $T(t)$ does not converge to infinity in distribution if and only if $\mathsf{E}T < \infty$. Thus for the positive recurrence of the basic regenerative process (and as a consequence for the stability of the system under consideration) it is enough to show that

$$T(t) \not\Rightarrow \infty, \quad t \to \infty. \tag{1}$$

Actually the result (1) defines that the process $X(t)$, starting at the arbitrary moment t, with a positive will achieve the regeneration point in a finite time.

Note that such an approach was successfully applied for stability analysis of a multi-class and multi-server retrial system with exponential retrials in [15] and developed for the single-server system with Poisson input, where retrial waiting time $\xi^{(i)}$ belongs to the special subclass of *New Better than Used* (NBU) distributions (see [16]). An arbitrary random variable $\xi \geq 0$ is called NBU if, for each x, $y \geq 0$

$$\mathsf{P}(\xi > x + y | \xi > y) \geq \mathsf{P}(\xi > x).$$

Next we present the sufficient stability conditions for a multi-class single server retrial system with non-exponential distribution of retrial times.

Theorem 1. *Consider a single-server* K*-class retrial queuing system with zero initial state and assume*

$$\rho < 1, \tag{2}$$

$$\mathsf{P}(\tau > x) > 0, \quad \forall x \geq 0, \tag{3}$$

$$\mathsf{E}(\xi^{(i)})^2 < \infty, \quad i = 1, \ldots, K. \tag{4}$$

Then the system is stable.

Proof. Note that new results are focused on the application for general retrials case. Thus we give details of the proof, related to the general distribution of $\xi^{(i)}$, as the rest steps follow the analysis for the particular case of exponential retrials, presented in [15].

Denote by Δ_n the sum idle period in $[t_n, t_{n+1})$. The condition $\rho < 1$ implies $\mathsf{E}\Delta_n \not\to 0$, see [16].

Next define the summary orbit size just before t_n and the total number of departures in $[t_n, t_{n+1})$ by N_n and D_n, respectively. Then for some arbitrary constants d, $d_0 > 0$ we present mean idle period as follows

$$\mathsf{E}\Delta_n = \mathsf{E}[\Delta_n, N_n \leq d + d_0] + \mathsf{E}[\Delta_n, N_n > d + d_0, D_n > d_0] \tag{5}$$

$$+ \mathsf{E}[\Delta_n, N_n > d + d_0, D_n \leq d_0]. \tag{6}$$

From [15,16] and independently on distribution of $\xi^{(i)}$ we obtain the upper bounds for the first summands in (5) as follows

$$\mathsf{E}[\Delta_n, N_n \leq d + d_0] \leq \mathsf{E}\tau\mathsf{P}(N_n < d + d_0), \tag{7}$$

$$\mathsf{E}[\Delta_n, N_n > d + d_0, D_n > d_0] \leq a\mathsf{P}(M(a) > d_0) + \mathsf{E}[\tau, \tau > a], \tag{8}$$

where $a > 0$ is an arbitrary constant and $\{M(t)\}$ defines a zero-delayed renewal process, built on intervals, stochastically equivalent to $\min_{i=1,\ldots,K} S^{(i)}$.

Our goal is to obtain the upper bound of (6). Namely, we explore a mean idle period Δ_n in case the number of departures D_n is not greater, than d_0 and the summary orbit just before the instant t_n is lower bounded by $d + d_0$. Thus up to the next arrival the summary orbit contains at least d customers: $N_{n+1} > d$. That means, the retrial attempts at least of d orbit customers are unsuccessful at τ_n. Define the set of numbers for such customers by $\mathcal{C}$. Then denote by $A_1 < A_2 < \cdots < A_d \leq t_n$ the arrival instants of customers from $\mathcal{C}$.

Consider $c_0 = \max(1, \lfloor d/2 \rfloor)$ and $c_1 = d - c_0$ and divide $\mathcal{C}$ for two sets: $\mathcal{C}_0$ the numbers of customers, arrived at instants $A_1, \ldots, A_{c_0}$, and $\mathcal{C}_1$ the numbers of customers, arrived at instants $A_{c_0+1}, \ldots, A_d$.

Next, denote by t_n^* the first *departure* instant after t_n. (Note that if $t_n^* > t_{n+1}$, then $\Delta_n = 0$ with probability 1.) Thus assume $t_n^* < t_{n+1}$. Then define by $\mathcal{T}(t_n^*)$ an interval since t_n^* up to the next retrial. Namely, $\mathcal{T}(t_n^*)$ coincides with the first idle period in $[t_n, t_{n+1})$, note $t_n^* + \mathcal{T}(t_n^*) < t_{n+1}$.

Define by $\mathcal{T}_{c_0}(t_n^*)$ the remaining retrial time for the costumers from the set $\mathcal{C}_0$. Recall $N_{n+1} > d$ and the assumption that $\mathcal{C}_0$ contains only customers, that would not capture the server before t_{n+1}. Thus

1. $t_n^* + \mathcal{T}_{c_0}(t_n^*) \geq t_{n+1}$ or
2. $t_n^* + \mathcal{T}_{c_0}(t_n^*) < t_{n+1}$ and the server is busy at instant $t_n^* + \mathcal{T}_{c_0}(t_n^*)$.

Hence

$$\mathcal{T}(t_n^*) \leq \mathcal{T}_d(t_n^*). \tag{9}$$

The relation of remaining retrial times for successful and unsuccessful attempts for the case $t_n^* + \mathcal{T}_{c_0}(t_n^*) < t_{n+1}$ is illustrated on Fig. 1. Note that the server is busy at instant $t_n^* + \mathcal{T}_{c_0}(t_n^*)$, while there could be idle intervals in $\left(t_n^* + \mathcal{T}(t_n^*),\, t_n^* + \mathcal{T}_{c_0}(t_n^*)\right)$.

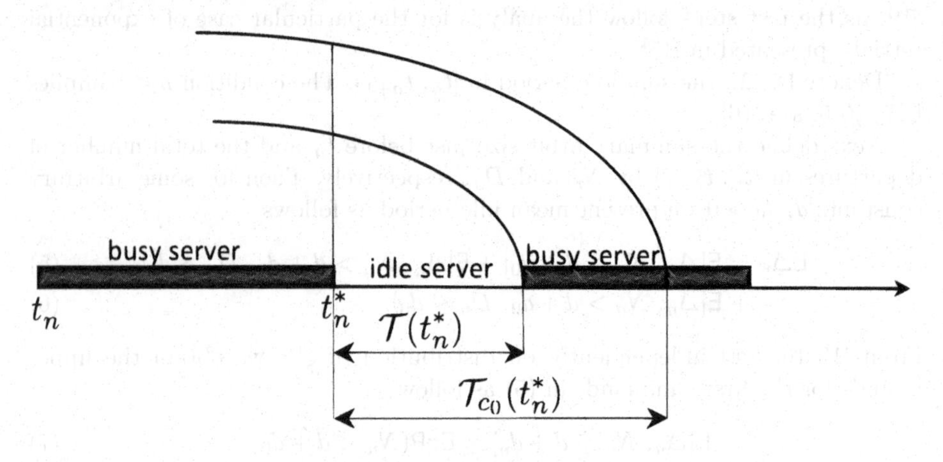

Fig. 1. Remaining retrial times

Then for $t \geq A_{c_0+1}$ we construct a set of *renewal processes* $\Lambda_j(t)$, $j = 1, \ldots, c_0$, associated with the number of unsuccessful attempts for the j-th orbit customer from $\mathcal{C}_0$. Note that all the customers in $\mathcal{C}_0$ had already been in the system before the moment A_{c_0+1}. Consider the j-th orbit customer belongs to class $i_j \in \{1, \ldots, K\}$, thus inter-renewal times of a process $\Lambda_j(t)$ are stochastically equivalent to $\xi^{(i_j)}$. Next construct $\mathbf{B}_j(t_n^*)$ – remaining time from t_n^* up to the next renewal in a process $\Lambda_j(t)$ (the next after the instant t_n^* attempt of the j-th orbit customer from the set $\mathcal{C}_0$). Namely, $\mathbf{B}_j(t_n^*)$ coincides with a remaining retrial time of the corresponding class $i_j \in \{1, \ldots, K\}$. Then

$$\mathsf{P}\big(\mathcal{T}(t_n^*) > x\big) \leq \mathsf{P}\big(\mathcal{T}_{c_0}(t_n^*) > x\big) = \Pi_{j=1}^{c_0} \mathsf{P}(\mathbf{B}_j(t_n^*) > x) \leq \big(\max_j \mathsf{P}(\mathbf{B}_j(t_n^*) > x)\big)^{c_0}.$$

Consider for simplicity

$$\beta(t_n^*) := \mathbf{B}_j(t_n^*) : \mathsf{P}(\beta(t_n^*) > x) = \max_{j=1,\ldots,c_0} \mathsf{P}(\mathbf{B}_j(t_n^*) > x),$$

and define by ξ the generic renewal time of the j-th renewal process, which corresponds to the maximal value of $\mathsf{P}(\mathbf{B}_j(t_n^*) > x)$. Note that ξ is stochastically equivalent to the retrial time $\xi^{(i_j)}$ of corresponding class $i_j \in \{1, \dots, K\}$, and the number i_j depends on t_n^*. The instant t_n^* is random with distribution $F_{t_n^*}$ and depends on the service time, while $\mathcal{T}(t_n^*)$ depends on the number of orbit customers. Thus the mean for the first idle period is defined as follows

$$\mathsf{E}\mathcal{T}(t_n^*) = \int_{u \in \tau_n} \mathsf{E}\mathcal{T}(u) dF_{t_n^*}(u). \tag{10}$$

Next from (9) for all deterministic $u \in \tau_n$

$$\mathsf{E}\mathcal{T}(u) = \int_0^\infty \mathsf{P}(\mathcal{T}(u) > x) dx \le \int_0^\infty \big(\mathsf{P}(\beta(u) > x)\big)^{c_0} dx. \tag{11}$$

Because β defines remaining renewal time for a corresponding renewal process, then by **Lorden's inequality** (see [12]) :

$$\int_0^\infty \mathsf{P}(\beta(u) > x) dx \equiv \mathsf{E}\beta(u) \le \frac{\mathsf{E}\xi^2}{\mathsf{E}\xi}.$$

By condition of the theorem $\mathsf{E}\big[\xi^{(i)}\big]^2 < \infty$ for all $i = 1, \dots, K$, then $\mathsf{E}\xi^2/\mathsf{E}\xi < \infty$. Thus $\mathsf{P}(\beta(u) > x)$ is integrable with respect to x. Hence $\Big(\mathsf{P}(\beta(u) > x)\Big)^{c_0}$ is dominated by integrable function and we can apply dominance convergence (Lebesgue) as follows:

$$\lim_{d \to \infty} \mathsf{E}\mathcal{T}(u) \le \lim_{d \to \infty} \int_0^\infty \Big(\mathsf{P}(\beta(u) > x)\Big)^{\lfloor d/2 \rfloor} dx = 0.$$

Taking into account (10), we obtain

$$\mathsf{E}\mathcal{T}(t_n^*) \to 0, \quad d \to \infty. \tag{12}$$

Note that on the event $\{N_n > d + d_0, D_n \le d_0\}$ the system admits not more than $(d_0 + 1)$ idle periods in τ_n, while the orbit is not less than d. By the same arguments, as in (12), we can obtain that each mean idle period in τ_n goes to zero with a growth of d. Define by $\mathbf{T}_n(d)$ the longest mean idle period in τ_n. Thus

$$\mathsf{E}[\Delta_n, N_n > d + d_0, D_n \le d_0] \le (d_0 + 1)\mathbf{T}_n(d). \tag{13}$$

Next, taking into account bounds (7), (8) and (13), for all $\varepsilon > 0$ we chose appropriate values of the constants $a = a(\varepsilon)$, $d_0 = d_0(a)$, $d = d(d_0)$ such, that

$$\mathsf{E}[\tau, \tau > a] + a\mathsf{P}(M(a) > d_0) + (d_0 + 1)\mathbf{T}_n(d) < \varepsilon/2.$$

Now assume that the summary orbit infinitely grows: $N_n \Rightarrow \infty$, as $n \to \infty$. The assumption implies that there exist such a number n_1, that

$$\mathsf{E}\tau\mathsf{P}(N_n < d + d_0) \le \varepsilon/2$$

for all $n \ge n_1$. Thus, we obtain that $\mathsf{E}\Delta_n < \varepsilon$, $n \ge n_1$, which leads to the contradiction. Hence, the orbit is tight: $N_n \not\Rightarrow \infty$. Next, using the condition $\mathsf{P}(\tau > x) > 0$ and regenerative approach, exactly as in [16], we are able to show that the system under consideration is stable (positive recurrent). Note that the demand of zero initial state is used in regenerative method.

Note that in paper [15], which deals with exponential-retrial model, the authors demands the fulfilness of a weaker condition

$$\max_{1\le i\le K} \mathsf{P}(\tau > S^{(i)}) > 0$$

instead of $\mathsf{P}(\tau > x) > 0$.

4 Simulation

In a recent work [17] the simulation results illustrated that at least in case $\mathsf{E}(\xi^{(i)})^2 < \infty$, the condition $\rho < 1$ defines the stable orbit and the demand of unbounded inter-arrival times is rather technical. In this section we explore the behavior of the model with non-zero initial conditions.

Consider two-class retrial model with Poisson input and exponential service times. We set three cases of load coefficients to simulate the behavior of the model under consideration. The explicit values of input parameters are presented in Table 1. Note that in all presented cases the first class arrivals are more intensive then the second class: $\lambda_1 > \lambda_2$. Moreover in the first two configurations the condition $\rho < 1$ holds true, while in the third case is violated.

Table 1. Load configurations

ρ	ρ_1	λ_1	μ_1	ρ_2	λ_2	μ_2
0.75	0.5	0.7	1.4	0.25	0.3	1.2
0.90	0.6	1.5	2.5	0.3	0.9	3.0
1.10	0.7	3.5	5.0	0.4	2.0	5.0

Next assume Pareto distribution of inter retrial time. Namely

$$\mathsf{P}(\xi^{(i)} > x) = \left(x_i/x\right)^{\alpha_i}, \quad x \ge x_i,\ i = 1, 2,$$

where$x_i > 0$ defines the corresponding scale parameter and $\alpha_i > 0$. Thus

$$\gamma_i = \frac{\alpha_i - 1}{x_i \alpha_i}, \quad \mathsf{E}\left(\xi^{(i)}\right)^2 = x_i^2 \frac{\alpha_i}{\alpha_i - 2}.$$

We chose $\alpha_i > 2$ to provide the second moment finite. The values of orbit rates used in simulation are presented in Table 2.

Table 2. Orbit configurations

Case	x_1	α_1	x_2	α_2	γ_1	γ_2
1.	0.20	3.0	2.00	4.0	3.333	0.375
2.	2.00	4.0	0.20	3.0	0.375	3.333
3.	0.50	3.0	1.00	2.5	1.333	0.600
4.	1.00	2.5	0.50	3.0	0.600	1.333
5.	0.75	4.0	0.75	4.0	1.000	1.000

In case 1 the first orbit is much more intensive then the second one: $\gamma_1 >> \gamma_2$, case 2 is the opposite: $\gamma_1 << \gamma_2$. The first orbit rate in case 3 is greater than the second orbit rate, but the difference is not as significant as in case 1: $\gamma_1 > \gamma_2$, the case 4 is opposite to case 3. In case 5 orbit rates are equal.

We simulate the mean orbits behavior $\hat{N}_n^{(i)}$ for n arrivals among m independent replications and under non-zero initial conditions: $\hat{N}_1^{(1)} := 100$, $\hat{N}_1^{(2)} = 100$.

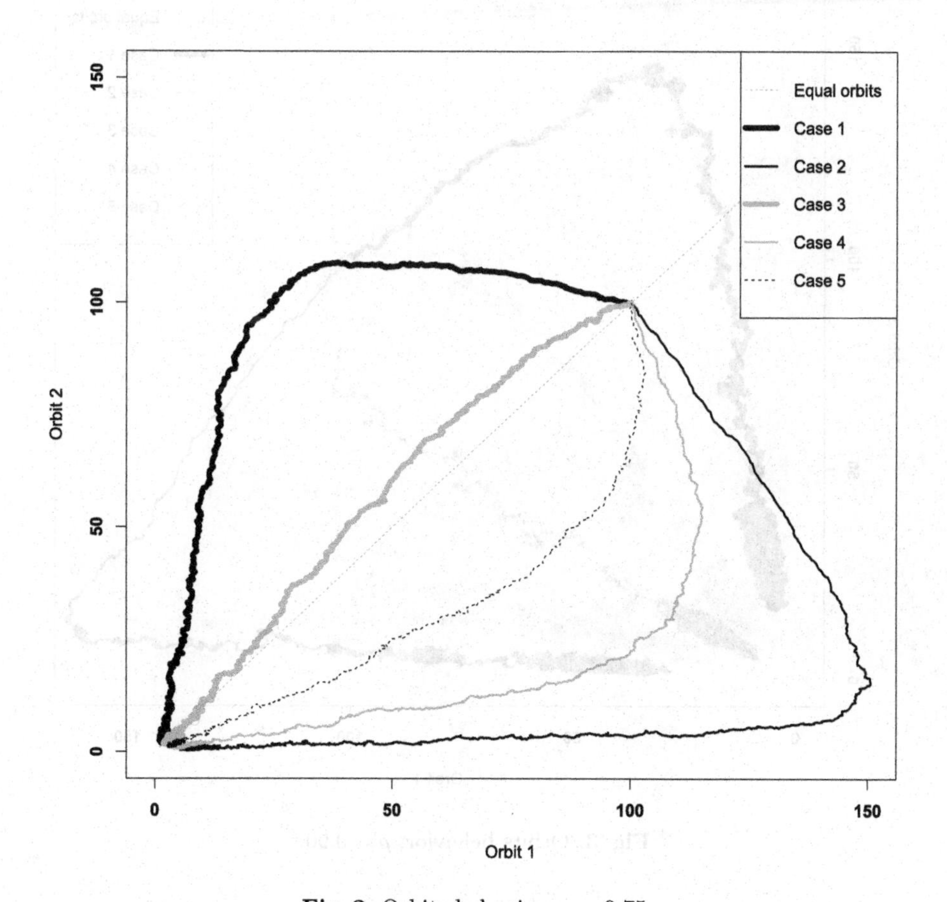

Fig. 2. Orbits behavior, $\rho = 0.75$

The results for different values of load coefficient are illustrated in $\hat{N}^{(1)} \times \hat{N}^{(2)}$ plots and presented in Fig. 2, 3 and 4.

Figure 2 shows the results for "light" load case when $\rho = 0.75$. We simulation was based on $m = 100$ sample paths and $n = 5000$ arrivals. Note that dash grey is extra and corresponds to the case $\hat{N}_n^{(1)} = \hat{N}_n^{(2)}$. In all five cases (see Table 2) both orbits decreases close to zero, that corresponds to the stable behavior of the system. Note that in cases 1 and 3 (thick black and thick grey curves respectively) the second orbit dominates the first orbit (the curves are higher than grey dash line). That phenomenon is explained by the rates relation: $\gamma_2 < \gamma_1$. Thus the first orbit is unloaded faster. By the same season the curves corresponding to cases 2 and 4 (thin black and thin grey lines respectively) are lower than the line $\hat{N}_n^{(1)} = \hat{N}_n^{(2)}$: the first orbit dominates because its retrials is less intensive. Note that in equal rate case (black dash curve) the first orbit also dominates. That is explained by input rate relation: the first class arrivals are more intensive, $\lambda_1 > \lambda_2$.

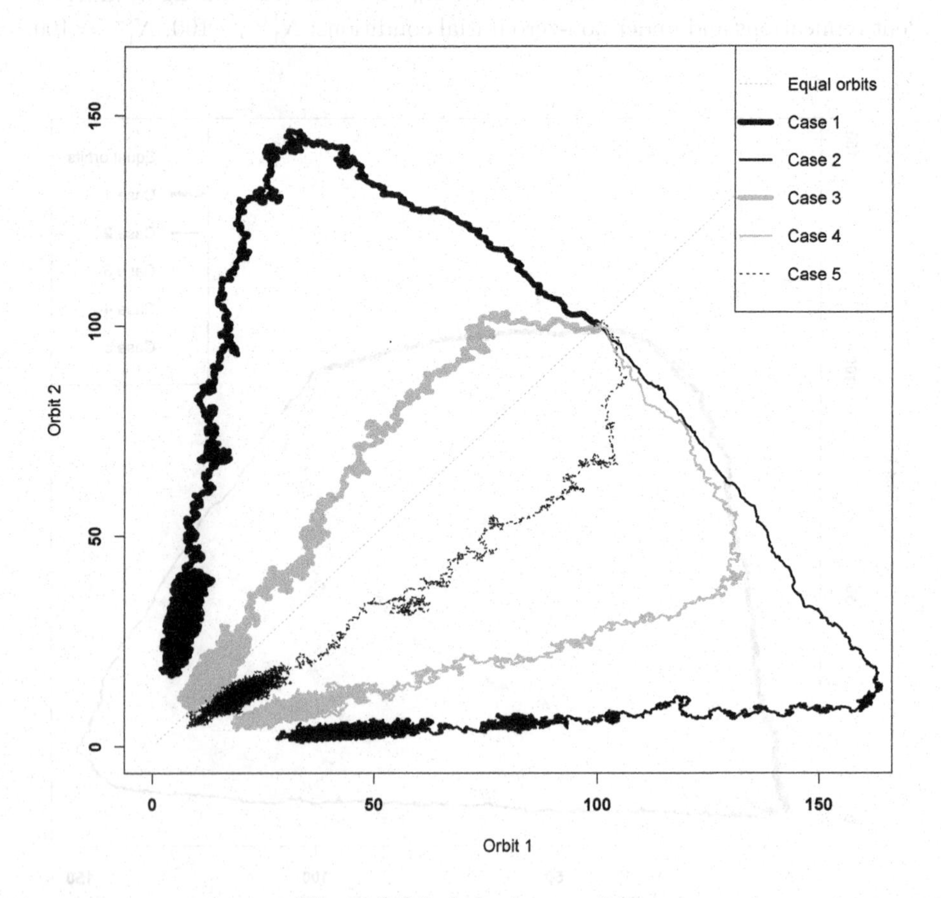

Fig. 3. Orbits behavior, $\rho = 0.90$

Figure 3 illustrates the results for $\rho = 0.90$. We simulation was based on $m = 20$ sample paths and $n = 10000$ arrivals. As in previous example for $\rho = 0.75$ both orbits in all rate configurations are stable. Note that in case 3: $\lambda_1 = 1.5$, $\gamma_1 = 1.333$, $\gamma_2 = 0.600$. Thus $\lambda_1 > \gamma_1$, the first class primary customers are more intensive than the first class retrials, moreover the first class primary customers are also more intensive than the second class arrivals ($\lambda_1 > \lambda_2$). Simulations show that in such a configuration the second orbit still dominates the first one (thick grey line). Thus the orbit rate relation is more significant for the system behavior than the input rates, and we can manage the load in whole system shifting the orbit parameters, at instance to redistribute the retrial attempts in case one of the classes has too intensive arrivals.

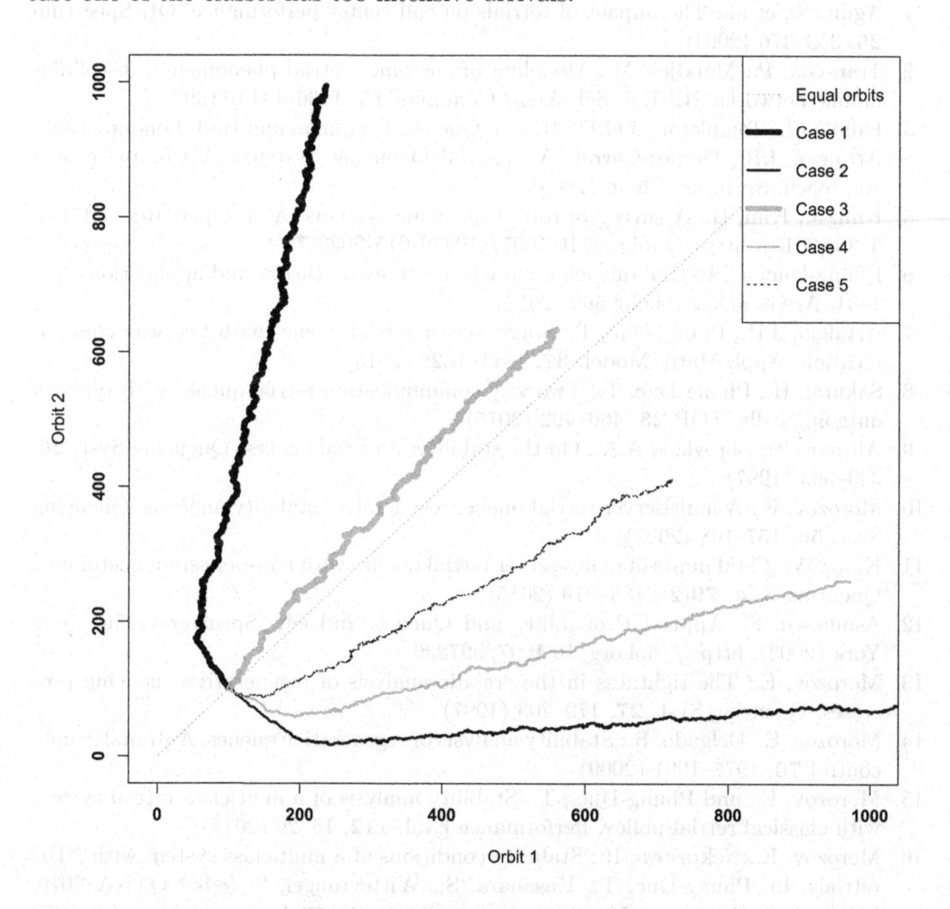

Fig. 4. Orbits behavior, $\rho = 1.10$

Figure 4 corresponds to the case $\rho = 1.10$. Thus stability condition from Theorem 1 is violated. The experiments for the considered rate configurations show the grow of both orbits. Thus we can assume that the condition $\rho < 1$ is not sensitive to the initial conditions.

5 Conclusion

We considered a single server multi-class retrial queue under classical retrial policy. Relying on previous analysis for particular distributions of retrial times $\xi^{(i)}$, we obtained that the condition $\rho < 1$ indeed guarantees the stability, at least if inter-arrival times are unbounded and additional moment properties of $\xi^{(i)}$ holds true.

References

1. Aguir, S., et al.: The impact of retrials on call center performance. OR Spectrum **26**, 353–376 (2004)
2. Tran-Gia, P., Mandjes, M.: Modeling of customer retrial phenomenon in cellular mobile networks. IEEE J. Sel. Areas Commun. **15**, 1406–1414 (1997)
3. Falin, G.I., Templeton, J.G.D.: Retrial Queues. Chapman and Hall, London (1997)
4. Artalejo, J.R., Gomez-Corral, A.: Retrial Queueing Systems: A Computational Approach. Springer, Cham (2008)
5. Kim, J., Kim, B.: A survey of retrial queueing systems. Ann. Oper. Res. **247**(1), 3–36 (2015). https://doi.org/10.1007/s10479-015-2038-7
6. Phung-Duc, T.: Retrial queueing models: a survey on theory and applications, pp. 1–31. ArXiv arXiv:1906.09560 (2019)
7. Artalejo, J.R., Phung-Duc, T.: Single server retrial queues with two way communication. Appl. Math. Model. **37**, 1811–1822 (2013)
8. Sakurai, H., Phung-Duc, T.: Two-way communication retrial queues with types of outgoing calls. TOP **23**, 466–492 (2015)
9. Altman, E., Borovkov, A.A.: On the stability of retrial queues. Queueing Syst. **26**, 343–363 (1997)
10. Morozov, E.: A multiserver retrial queue: regenerative stability analysis. Queueing Syst. **56**, 157–168 (2007)
11. Kang, W.: Fluid limits of many-server retrial queues with non-persistent customers. Queueing Syst. **79**(2), 183–219 (2015)
12. Asmussen, S.: Applied Probability and Queues. 2nd edt. Springer-Verlag, New York (2003). https://doi.org/10.1007/b97236
13. Morozov, E.: The tightness in the ergodic analysis of regenerative queueing processes. Queueing Syst. **27**, 179–203 (1997)
14. Morozov, E., Delgado, R.: Stability analysis of regenerative queues. Autom. Remote control **70**, 1977–1991 (2009)
15. Morozov, E., and Phung-Duc., T.: Stability analysis of a multiclass retrial system with classical retrial policy. Performance Eval. **112**, 15–26 (2017)
16. Morozov, E., Nekrasova, R.: Stability conditions of a multiclass system with NBU retrials. In: Phung-Duc, T., Kasahara, S., Wittevrongel, S. (eds.) QTNA 2019. LNCS, vol. 11688, pp. 51–63. Springer, Cham (2019). https://doi.org/10.1007/978-3-030-27181-7_4
17. Nekrasova, R.: Stability analysis of a multi-class retrial queue with general retrials and classical retrial policy. In: 28th Conference of Open Innovations Association (FRUCT), pp. 328–333 (2021)

Analysis of the Probabilistic and Cost Characteristics of the Queueing Network with a Control Queue and Quarantine in Systems and Negative Requests by Means of Successive Approximations

Katsiaryna Kosarava(✉) and Dmitry Kopats

Yanka Kupala State University of Grodno, 230012 Grodno, Belarus

Abstract. In paper we sudy a queuing network with positive and negative requests, consisting of systems with control queue and quarantine node. The application of such a network as a stochastic model of a computer network with antivirus is described. A system of difference-differential equations (DDE) for possible states of described queueing network is derived. In a similar way, a system of DDE for the expected incomes of the network's systems, in the case when the incomes from the network's transitions between states depend on these states, was constructed. To solve the obtained DDE systems, the method of successive approximations, combined with the method of series, was used. It is shown that the probabilities of the network's states and the expected incomes of the network's systems can be represented in the form of converging power series. Recurrence relations for calculating the coefficients of these series are given. The use of this technique is illustrated by the example of finding the state probabilities of the studied queueing network.

Keywords: G-network · Computer network · Negative arrivals · Control queue · Quarantine node · Successive approximations

1 Introduction

Consider a computer network consisting of computers with antivirus software installed. In general, antivirus software perfom 3 basic function: detecting malicious codes in the system, removing them by destroying or isolating them, take preventive measure [1]. For each of the three types of software, its own part of the RAM(Random Access Memory) and CPU(Central Processing Unit) is allocated. The RAM stores the codes of files active in this session. Antivirus software monitors and checks the contents of the computer's RAM command blocks for viruses [2]. If the check is successful, the file is passed to the queue for program execution. But there is a possibility that the antivirus software might

V. M. Vishnevskiy et al. (Eds.): DCCN 2021, CCIS 1552, pp. 259–271, 2022.
https://doi.org/10.1007/978-3-030-97110-6_20

not recognize the virus and infect the user's computer, for example, due to an untimely update of the antivirus signature database. We will assume that there can be 2 categories of unidentified viruses: 1) resident, which do not affect the file queue for processing and are attached to the Personal computer (PC), but during file processing they can infect an executable file, and 2) viruses that make it impossible to execute the file, pending processing. After "processing" the file can be transferred over the network to another computer for further processing, or transferred to the computer's hard drive for storage and waiting for a subsequent call. Files downloaded to a computer from the network are automatically scanned by antivirus software for viruses, and if this option is disabled, the user receives a warning about the dangers of this file. A file that is declared infected after being scanned for viruses is quarantined on the computer [3]. Quarantine consists of 2 components: 1) storage of infected files that will not be executed by the processor and 2) software that "cures" malicious files. We will assume that 3 categories of files can be quarantined: 1) mistakenly recognized as malicious, the user has the ability to manually remove them from the quarantine; 2) files containing a virus code which (while in quarantine) can be "cured" of this code and continue execution on the user's computer; 3) viruses that cannot be neutralized and must be removed from the computer's memory. The extracted file is returned back to the location on the computer where it was extracted from and placed in RAM for loading and subsequent execution.

In [4] a stochastic model of a computer network consisting of systems with a control queue and one quarantine node in a stationary mode is investigated. To simulate the described computer network with an antivirus, in current work we propose to use a G-network with positive and negative requests, consisting of single-channel queueing systems (QS) with a control queue and quarantine node. In this research we don't impose strict restrictions on behavior of a negative request introduced by Gelenbe [5]: after the destroying of one positive request, negative request can either leave the network or go to the quarantine queue of another system. Recently, G-networks have been widely used in a number of applications related to the simulation of attacks in computer networks (attacks on smart technologies, DoS attacks) [6–8], modelling of Intrusion Detection Systems [9], customers resets [10,11], optimization of supply chains [12] and solving deep learning problems [13,14].

We also aim to predict the expected incomes and losses of such a network associated with potential threats of virus infection, loss of valuable information, drop in user productivity and additional security risks. Finding incomes in queuing networks, which are stochastic models of various objects in economics, technology and production, was previously studied in works [15,16].

2 Model Description

Let us consider a G-network with n QS. Each QS S_i has external arrivals of positive and negative requests which occur according to mutually independent Poisson processes, with rates λ_{0i}^{+}, λ_{0i}^{-} respectively, $i = \overline{1, n}$. For described network

λ_{0i}^{+}, λ_{0i}^{-} mean the number of files that are not dangerous for the PC and files that pose a threat to it respectively, which entered the computer's RAM from its hard disk or from outside the network. The request initially received by the i-th QS enters the control queue, where it is checked for standardness, i.e. for the presence of a virus. The verification time of a request for standardness in the i-th QS has an exponential distribution with the parameter $\mu_i^{(v)}$, $i = \overline{1,n}$. After verification in the i-th QS a positive request is recognized as such with probability p_i^{+} and enters the QS for servicing and with a probability $(1-p_i^{+})$ it is recognized as negative and redirected to quarantine for treatment. A negative request after verification in the control queue of the i-th QS with probability p_i^{-} is recognized as such and redirected to the quarantine queue for treatment, and with probability $(1-p_i^{-})$ it can be mistakenly recognized as positive (for example, due to a failure to update the antivirus databases) and sent to a processing queue, where it immediately destroys the positive request. In this paper, we investigate the model under the assumption that a negative request destroys one positive request if the QS is not empty, and leaves the system without having any impact on it, otherwise. In the physical model, this may correspond, for example, to the fact that a memory resident virus overwrites a copy of itself into a piece of computer memory, regardless of what was in that location. The files are no longer readable, and it is completely impossible to restore them using special programs. The resident copy of the virus remains active and infects newly created files. Let the service time of requets in the i-th QS has an exponential distributed function (d.f.) with the parameter μ_i, $i = \overline{1,n}$. A positive request after being served in the i-th QS can make several transitions: 1) with probability p_{ij}^{+} it goes to the j-th QS control queue as a positive request, 2) with probability p_{ij}^{-} it goes to S_j as a negative one, infected during the service with resident viruses and 3) with probability $p_{i0} = 1 - \sum_{j=1}^{n}(p_{ij}^{+} + p_{ij}^{-})$ it leaves the network, $i, j = \overline{1,n}$.

Let us describe the behavior of a quarantine: requests recognized as non-standard are placed in the quarantine queue for treatment. Physically, the quarantine queue is a folder of files placed in quarantine by the antivirus. Suppose that the treatment time in the quarantine queue of the QS S_i has an exponential d.f. with a parameter $\mu_i^{(c)}$, $i = \overline{1,n}$. If the treatment was successful, then the request with probability $p_i^{(s)}$ is returned to the i-th QS for servicing, otherwise the request (infected file) with probability $(1 - p_i^{(s)})$ turns out to be a virus and is removed, i.e. leaves the network, $i = \overline{1,n}$. In this description of the quarantine, we assume that the virus cannot trick it during treatment.

The structure of the QS of the described stochastic model is represented on Fig. 1.

3 Network's State Probabilities

The state vector of the described network has the form

$$\left(\overrightarrow{k}, \overrightarrow{l}, t\right) = \left(\overrightarrow{k_1}, \overrightarrow{k_2}, \dots \overrightarrow{k_n}, \overrightarrow{l_1}, \overrightarrow{l_2}, \dots \overrightarrow{l_n}, t\right), \tag{1}$$

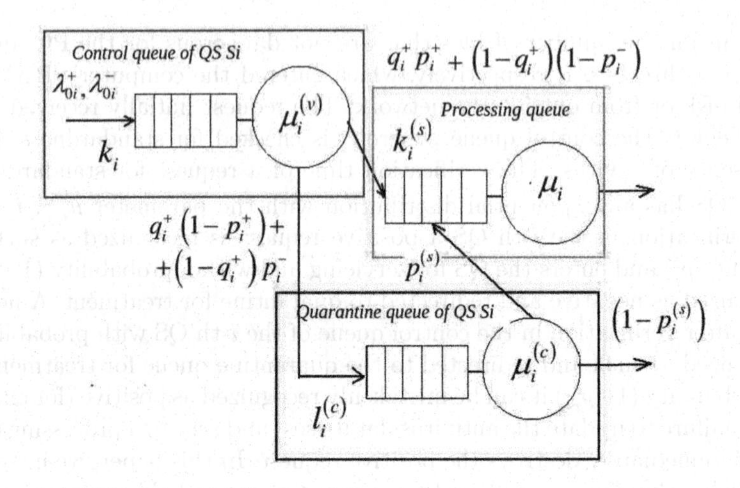

Fig. 1. Queueing system with control quaeue and quarantine.

where $(\overrightarrow{k_i}, \overrightarrow{l_i}, t) = (k_i^{(p)}, k_i^{(s)}, l_i^{(n)}, l_i^{(c)}, t)$, $k_i^{(p)}$ and $l_i^{(n)}$ - the number of positive and negative requests in the control queue of QS S_i, respectively; $k_i^{(s)}$ is the number of requests for service in i-th queue; $l_i^{(c)}$ - the number of the requests in the quarantine, $i = \overline{1,n}$. Let requests are choosen from the control queue randomly, then we estimate the probability to choose a positive request as

$$q_i^+ = \frac{\boldsymbol{E}\left[k_i^{(p)}\right]}{\boldsymbol{E}\left[k_i^{(p)} + l_i^{(n)}\right]}, i = \overline{1,n}.$$

Let us find the possible transitions of our Markov process in the state $(\overrightarrow{k}, \overrightarrow{l}, t + \Delta t)$, during time Δt:

1. From the state $(\overrightarrow{k} - \widetilde{I}_{2i-1}, \overrightarrow{l}, t + \Delta t)$ with probability $\lambda_{0i}^+ u(k_i^{(p)})\Delta t + o(\Delta)$ positive request enters the control queue of QS S_i, $i = \overline{1,n}$.
2. From the state $(\overrightarrow{k}, \overrightarrow{l} - \widetilde{I}_{2i-1}, t + \Delta t)$ with probability $\lambda_{0i}^- u(l_i^{(n)})\Delta t + o(\Delta)$ negative requests enters the control queue of QS S_i,$i = \overline{1,n}$.
3. From the state $(\overrightarrow{k} + \widetilde{I}_{2i-1} - \widetilde{I}_{2i}, \overrightarrow{l}, t)$ with probability $\mu_i^{(v)} q_i^+ p_i^+ u(k_i^{(s)})\Delta t + o(\Delta t)$ positive request, after successful verification in control queue, moves to service queue of S_i, $i = \overline{1,n}$.
4. From the state $(\overrightarrow{k} + \widetilde{I}_{2i-1}, \overrightarrow{l} - \widetilde{I}_{2i}, t)$ with probability $\mu_i^{(v)} q_i^+ (1 - p_i^+)\times$ $\times u(l_i^{(c)})\Delta t + o(\Delta t)$ positive request is recognized as negative and redirected to quarantine queue of S_i, $i = \overline{1,n}$.
5. From the state $(\overrightarrow{k} + \widetilde{I}_{2i}, \overrightarrow{l} + \widetilde{I}_{2i-1}, t)$ with probability $\mu_i^{(v)}(1 - q_i^+)(1 - p_i^-)\times$ $\times \Delta t + o(\Delta t)$ negative request after verification in control queue is recognized as positive, then it moves in service queue and immediately destroys the positive request, $i = \overline{1,n}$.

6. From the state $(\overrightarrow{k}, \overrightarrow{l} + \widetilde{I}_{2i-1} - \widetilde{I}_{2i}, t)$ with probability $\mu_i^{(v)}(1 - q_i^+)p_i^- \times$ $\times u(l_i^{(c)})\Delta t + o(\Delta t)$ negative request after verification in control queue is recognized as negative and redirected in quarantine queue in S_i, $i = \overline{1,n}$.
7. From the state $(\overrightarrow{k} - \widetilde{I}_{2i}, \overrightarrow{l} + \widetilde{I}_{2i}, t)$ with probability $\mu_i^{(c)} p_i^{(s)} u(k_i^{(s)})\Delta t + o(\Delta t)$ the request, after successful treatment, is sent to the QS S_i for service, $i = \overline{1,n}$.
8. From the state $(\overrightarrow{k}, \overrightarrow{l} + \widetilde{I}_{2i}, t)$ with probability $\mu_i^{(c)}(1 - p_i^{(s)})\Delta t + o(\Delta t)$ request leaves the network if treatment was failed, $i = \overline{1,n}$.
9. From the state $(\overrightarrow{k} + \widetilde{I}_{2i}, \overrightarrow{l}, t)$ with probability $\mu_i p_{i0}\Delta t + o(\Delta t)$, after service in the QS S_i is finished, request leaves networks, $i = \overline{1,n}$.
10. From the state $(\overrightarrow{k} + \widetilde{I}_{2i} - \widetilde{I}_{2j-1}, \overrightarrow{l}, t)$ with probability $\mu_i p_{ij}^+ u(k_j^{(p)})\Delta t + o(\Delta t)$, after service in the QS S_i is finished, request enters the control queue of the QS S_j as positive request, $i = \overline{1,n}$.
11. From the state $(\overrightarrow{k} + \widetilde{I}_{2i}, \overrightarrow{l} - \widetilde{I}_{2j-1}, t)$ with probability $\mu_i p_{ij}^- u(l_j^{(n)})\Delta t + o(\Delta t)$, after service in the QS S_i is finished, request enters the control queue of the QS S_j as negative request, $i = \overline{1,n}$.
12. With probability $1 - \sum_{i=1}^n (\lambda_{0i}^+ + \lambda_{0i}^- + \mu_i + \mu_i^{(v)} + \mu_i^{(c)})\Delta t + o(\Delta t)$ the network state doesn't change.

Using the formula for the total probability of transitions between the states of the described network, it is possible to prove that the system of differential equations for a given network has the form:

$$\frac{dP\left(\overrightarrow{k}, \overrightarrow{l}, t\right)}{dt} = -\sum_{i=1}^{n} \left(\lambda_{0i}^+ + \lambda_{0i}^- + \mu_i^{(v)} + \mu_i^{(c)} + \mu_i\right) P\left(\overrightarrow{k}, \overrightarrow{l}, t\right)$$

$$+ \sum_{i=1}^{n} \Big\{ \lambda_{0i}^+ u\left(k_i^{(p)}\right) P\left(\overrightarrow{k} - \widetilde{I}_{2i-1}, \overrightarrow{l}, t\right) + \lambda_{0i}^- u\left(l_i^{(n)}\right) P\left(\overrightarrow{k}, \overrightarrow{l} - \widetilde{I}_{2i-1}, t\right)$$

$$+ \mu_i^{(v)} q_i^+ p_i^+ u\left(k_i^{(s)}\right) P\left(\overrightarrow{k} + \widetilde{I}_{2i-1} - \widetilde{I}_{2i}, \overrightarrow{l}, t\right)$$

$$+ \mu_i^{(v)} q_i^+ \left(1 - p_i^+\right) u\left(l_i^{(c)}\right) P\left(\overrightarrow{k} + \widetilde{I}_{2i-1}, \overrightarrow{l} - \widetilde{I}_{2i}, t\right)$$

$$+ \mu_i^{(v)} \left(1 - q_i^+\right) \left(1 - p_i^-\right) P\left(\overrightarrow{k} + \widetilde{I}_{2i}, \overrightarrow{l} + \widetilde{I}_{2i-1}, t\right)$$

$$+ \mu_i^{(v)} \left(1 - q_i^+\right) p_i^- u\left(l_i^{(c)}\right) P\left(\overrightarrow{k}, \overrightarrow{l} + \widetilde{I}_{2i-1} - \widetilde{I}_{2i}, t\right)$$

$$+ \mu_i^{(c)} p_i^{(s)} u\left(k_i^{(s)}\right) P\left(\overrightarrow{k} - \widetilde{I}_{2i}, \overrightarrow{l} + \widetilde{I}_{2i}, t\right)$$

$$+ \mu_i^{(c)} \left(1 - p_i^{(s)}\right) P\left(\overrightarrow{k}, \overrightarrow{l} + \widetilde{I}_{2i}, t\right) + \mu_i p_{i0} P\left(\overrightarrow{k} + \widetilde{I}_{2i}, \overrightarrow{l}, t\right)$$

$$+ \sum_{j=1}^{n} \Big[\mu_i p_{ij}^+ u\left(k_j^{(p)}\right) P\left(\overrightarrow{k} + \widetilde{I}_{2i} - \widetilde{I}_{2j-1}, \overrightarrow{l}, t\right)$$

$$+\mu_i p_{ij}^{-} u\left(l_j^{(n)}\right) P\left(\overrightarrow{k}+\widetilde{I}_{2i}, \overrightarrow{l}-\widetilde{I}_{2j-1}, t\right)\Big]\Big\}, \quad (2)$$

where $u(x)=1$ if $x>0$ and 0 otherwise; $\widetilde{I}_r$ – $2n$-dimentional zero vector except component with a number r which equals to 1.

The system (2) is a special case of difference-differention equation. General scheme of its solution by means of successive approximation method is described in [17]:

$$\frac{dP(\overrightarrow{k},\overrightarrow{l},t)}{dt} = -\Lambda(\overrightarrow{k},\overrightarrow{l})P(\overrightarrow{k},\overrightarrow{l},t) + \sum_{\alpha,\beta,\gamma,\theta,\eta=0}^{2n} \Phi_{\alpha\beta,\gamma,\theta,\eta}(\overrightarrow{k},\overrightarrow{l}) \times P(\overrightarrow{k}+\widetilde{I}_\alpha+\widetilde{I}_\beta-\widetilde{I}_\gamma, \overrightarrow{l}+\widetilde{I}_\eta-\widetilde{I}_\theta, t), \quad (3)$$

where functions $\Lambda(\overrightarrow{k},\overrightarrow{l}), \Phi_{\alpha,\beta,b,\gamma,\theta,\eta}(\overrightarrow{k},\overrightarrow{l})$ equal respectively:

$$\Lambda(\overrightarrow{k},\overrightarrow{l}) = \sum_{i=1}^{n}\left(\lambda_{0i}^{+}+\lambda_{0i}^{-}+\mu_i^{(v)}+\mu_i^{(c)}+\mu_i\right),$$

$$\begin{aligned}
\Phi_{\alpha,\beta,b,\gamma,\theta,\eta}(\overrightarrow{k},\overrightarrow{l}) &= \delta_{\alpha 0}\delta_{\beta 0}\delta_{\gamma(2i-1)}\delta_{\theta\eta}\lambda_{0i}^{+}u\left(k_i^{(c)}\right) + \delta_{\alpha 0}\delta_{\beta 0}\delta_{\gamma 0}\delta_{\theta 0}\delta_{\eta(2i-1)}\lambda_{0i}^{-}u\left(l_i^{(c)}\right)\\
&+\delta_{\alpha 0}\delta_{\gamma 0}\delta_{\beta 0}\delta_{\eta 0}\delta_{\theta(2i-1)}\mu_i^{(c)}\left(1-p_i^{(s)}\right)\\
&+\delta_{\eta 0}\delta_{\theta(2i-1)}\delta_{\beta 0}\delta_{\gamma 0}\delta_{\alpha(2i)}\mu_i^{(v)}\left(1-q_i^{+}\right)\left(1-p_i^{-}\right)u\left(k_i^{(p)}\right)\\
&+\delta_{\eta 0}\delta_{\theta(2i-1)}\delta_{\beta 0}\delta_{\gamma 0}\delta_{\alpha 0}\mu_i^{(v)}\left(1-q_i^{+}\right)\left(1-p_i^{-}\right)\left(1-u\left(k_i^{(p)}\right)\right)\\
&+\delta_{\gamma 0}\delta_{\beta 0}\delta_{\eta(2j-1)}\delta_{\theta 0}\delta_{\alpha(2i)}\mu_i p_{ij}^{-}u\left(l_j^{(c)}\right)\\
&+v\delta_{\gamma(2i)}\delta_{\beta 0}\delta_{\theta\eta}\delta_{\alpha(2i-1)}\mu_i^{(v)}q_i^{+}p_i^{+}u\left(k_i^{(p)}\right)+\delta_{\beta 0}\delta_{\gamma 0}\delta_{\theta 0}\delta_{\alpha(2i-1)}\delta_{\eta(2i)}\mu_i^{(v)}q_i^{+}\left(1-p_i^{+}\right)u\left(l_i^{(n)}\right)\\
&+\delta_{\gamma 0}\delta_{\beta 0}\delta_{\theta(2i)}\delta_{\alpha 0}\delta_{\eta(2i-1)}\mu_i^{(v)}\left(1-q_i^{+}\right)p_i^{-}u\left(l_i^{(n)}\right)\\
&+\delta_{\beta 0}\delta_{\gamma(2i)}\delta_{\alpha 0}\delta_{\theta(2i)}\delta_{\eta 0}\mu_i^{(c)}p_i^{(s)}u\left(k_i^{(s)}\right)\\
&+\delta_{\gamma 0}\delta_{\beta 0}\delta_{\theta\eta}\delta_{\alpha(2i)}\mu_i p_{i0}+\delta_{\gamma(2j-1)}\delta_{\beta 0}\delta_{\theta\eta}\delta_{\alpha(2i)}\mu_i p_{ij}^{+}u\left(k_j^{(c)}\right),
\end{aligned} \quad (4)$$

where

$$\delta_{ij}=\begin{cases}1, i=j\\ 0, i\neq j.\end{cases}.$$

Let $P_q(\overrightarrow{k},\overrightarrow{l},t)$ an approximation $P(\overrightarrow{k},\overrightarrow{l},t)$ at the q-th iteration, then $P_{q+1}(\overrightarrow{k},\overrightarrow{l},t)$ - solution of system (2) obtained by the method of successive approximations. Any approximation representable in the form of a convergent power series [17]

$$P_q(\overrightarrow{k},\overrightarrow{l},t)=\sum_{l=0}^{\infty} d_{ql}^{+-}\left(\overrightarrow{k},\overrightarrow{l}\right)t^l, \quad (5)$$

which coefficients satisfy the recurrence relations [17]:

$$d_{q+1z}^{+-}(\overrightarrow{k},\overrightarrow{l}) = \frac{\left[-\Lambda\left(\overrightarrow{k},\overrightarrow{l}\right)\right]^z}{z!}\left\{P\left(\overrightarrow{k},\overrightarrow{l},0\right) + \sum_{u=0}^{z-1}\frac{(-1)^{u+1}\,u!D_{qu}^{+-}\left(\overrightarrow{k},\overrightarrow{l}\right)}{\left[\Lambda\left(\overrightarrow{k},\overrightarrow{l}\right)\right]^{u+1}}\right\},$$

$$d_{q0}^{+-}\left(\overrightarrow{k},\overrightarrow{l}\right) = P\left(\overrightarrow{k},\overrightarrow{l},0\right), d_{0z}^{+-}\left(\overrightarrow{k},\overrightarrow{l}\right) = P\left(\overrightarrow{k},\overrightarrow{l},0\right)\delta_{z0}, z \geq 0,$$

$$D_{qz}^{+-}\left(\overrightarrow{k},\overrightarrow{l}\right) = \sum_{\alpha,\beta,\gamma,\theta,\eta=0}^{2n}\Phi_{\alpha\beta\gamma\theta\eta}\left(\overrightarrow{k},\overrightarrow{l}\right)$$

$$\times\, d_{qz}^{+-}\left(\overrightarrow{k}+\tilde{I}_\alpha+\tilde{I}_\beta-\tilde{I}_\gamma, \overrightarrow{l}+\tilde{I}_\theta-\tilde{I}_\eta\right). \tag{6}$$

Example 1. Let us consider a queuing network, which consists of $n = 2$ systems. Rate of positive and negative arrivals in i-th QS equal: $\lambda_{01}^{+} = 15, \lambda_{01}^{-} = 7, \lambda_{02}^{+} = 13, \lambda_{02}^{-} = 6$. The rates of verification of the requests for standardness, service and treatment in systems equal $\mu_i^{(v)} = 30, \mu_i = 5, \mu_i^{(c)} = 0.1, i = \overline{1,2}$. The probabilistic parameters of the network are as follows: $p_1^{+} = 0.9, p_1^{-} = 0.93, p_2^{+} = 0.95, p_2^{-} = 0.97, p_1^{(c)} = 0.7, p_2^{(c)} = 0.8, p_{ij}^{+} = p_{ij}^{-} = 0.4, p_{i0} = 0.2, i \neq j, i, j = \overline{1,2}$. Probabilities of the network's states $k_1 = (\overrightarrow{k},\overrightarrow{l}) = (1;2;1;1;1;3;1;2)$, $k_2 = (2;2;1;1;1;2;1;3)$, $k_3 = (2;1;1;1;1;2;1;2)$ if state (1;1;1;1;1;1;1;1) is initial, are represented on Fig. 2.

4 Finding the Expected Incomes of the Network's Systems

The queuing network described in second section can be used as a model of a computer network with antivirus software. Antivirus should reduce the risks associated with malicious files. However, in addition to the obvious advantages of antivirus tools, such programs reduce the performance of the entire system at the expense of processor resources. In addition, errors in antivirus software can lead to the loss of valuable information in the case of deleting "good/ benign" files. All this leads to additional costs associated with the use of antivirus programs (drop in user productivity and additional risks associated with security) [18].

A drop in system performance leads to a decrease in the speed of performing routine operations, such as opening files, sending them by e-mail, etc. Antivirus introduces delays in these processes. In addition, completely legitimate system detections can have a negative impact on user productivity (for example, blocking access to an important letter or file), which makes it impossible to quickly get the job done. At the same time, the cost of a particular "unfinished" work can seriously exceed the negative effect of a virus infection on the computer.

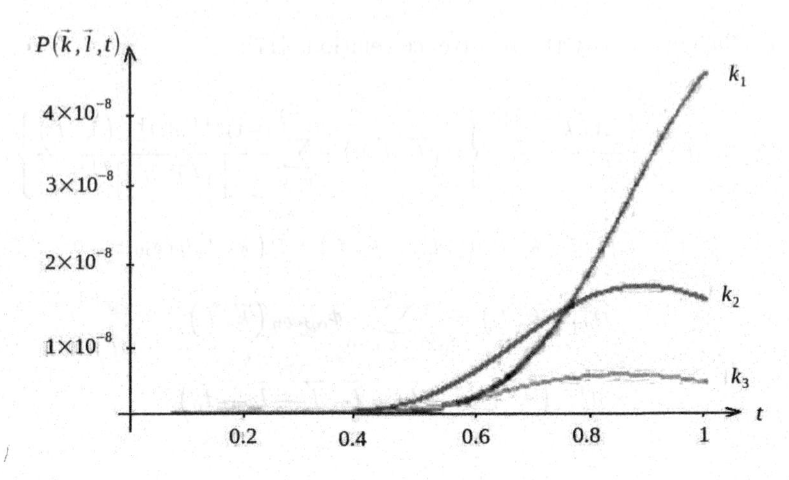

Fig. 2. State probabilities $P(\overrightarrow{k}, \overrightarrow{l}, t)$.

Antivirus programs, like any software, can contain bugs and vulnerabilities. CyberArk Labs, a provider of information security software, is exploring new attack methods and regularly publishes articles on bugs and vulnerabilities identified in antivirus programs. In its latest report, CyberArk reports bugs in products from Kaspersky, McAfee, Symantec, Fortinet and CheckPoint and others, as well as vulnerabilities in products from Microsoft, Avast and Avira, among others [19].

Another negative point is the antivirus false positives. It is not uncommon for an antivirus to block legitimate software or even the entire operating system. The damage from such incidents can often exceed the benefits that antivirus software can bring.

Let $v_i(\overrightarrow{k}, \overrightarrow{l}, t)$ is an expected income of the QS S_i which it recieved during time t if the initial state of the QN was $(\overrightarrow{k}, \overrightarrow{l})$. Considering possible network's transitions between its states, let us introduce expected income of the QS S_i depending on network's state:

1. If positive request enters the control queue of the QS S_i from the outside it generates income for the system $R_{0i}^{+}(\overrightarrow{k} - \tilde{I}_{2i-1}, \overrightarrow{l})$, $i = \overline{1, n}$, which characterizes the amaunt of the information received.
2. If negative request enters the control queue of QS S_i, the income of this system equals $-R_{0i}^{-}(\overrightarrow{k}, \overrightarrow{l} - \tilde{I}_{2i-1})$, $i = \overline{1, n}$ and characterizes the potential danger of infection of a network's node.
3. If positive request, after successful verification in control queue, moves to service queue of the QS S_i, the income of this system equals $-r_i^{+}(\overrightarrow{k} + \tilde{I}_{2i-1} - \tilde{I}_{2i}, \overrightarrow{l})$, which coresponds to drop in the QS S_i performance during request verification, $i = \overline{1, n}$.
4. If negative request, after verification in control queue of the QS S_i, is recognized as positive and redirected in service queue, the income of this system

equals $-R_i^+(\overrightarrow{k}+\widetilde{I}_{2i},\overrightarrow{l}+\widetilde{I}_{2i-1})$, which corresponds to the infection of a file on a network node, due to which it becomes unavailable for processing, $i=\overline{1,n}$. If there are no positive requests in the system at this moment in time, then the negative request leaves the network and the system income equals $r_i^-(\overrightarrow{k},\overrightarrow{l}+\widetilde{I}_{2i-1})$, which corresponds to the elimination of the potential threat of infection of the entire network, $i=\overline{1,n}$.

5. If positive request is recognized as negative and redirected to quarantine node of the S_i, income of the system S_i equals $-R_i^c(\overrightarrow{k}+\widetilde{I}_{2i-1},\overrightarrow{l}-\widetilde{I}_{2i})$, which characterizes the losses associated with the blocking of legitimate software by antivirus, $i=\overline{1,n}$.
6. If negative request, after verification in control queue, is recognized as negative and redirected in quarantine node in S_i, system's income in this case equals $r_i^-(\overrightarrow{k},\overrightarrow{l}+\widetilde{I}_{2i-1}-\widetilde{I}_{2i})$, $i=\overline{1,n}$.
7. If the request, after successful treatment, is sent to the QS S_i for service, the income of the S_i equals $R_i^+(\overrightarrow{k}-\widetilde{I}_{2i},\overrightarrow{l}+\widetilde{I}_{2i})$, which characterizes the amount of restored information, $i=\overline{1,n}$.
8. If treatment was failed, request leaves the network and income of the S_i reduces by $-R_i^+(\overrightarrow{k}-\widetilde{I}_{2i},\overrightarrow{l}+\widetilde{I}_{2i})$ due to the amount of information lost, $i=\overline{1,n}$.
9. If the service in the QS S_j is finished and request enters the control queue of the QS S_i as positive request, the income of the QS S_i increases by $R_j^+(\overrightarrow{k}+\widetilde{I}_{2j}-\widetilde{I}_{2i-1},\overrightarrow{l})$, $i=\overline{1,n}$.
10. If service in the QS S_j is finished and request enters the control queue of the QS S_i as negative request, the income of the QS S_i reduces by $-r_i^-(\overrightarrow{k}+\widetilde{I}_{2j},\overrightarrow{l}-\widetilde{I}_{2i-1})$, which corresponds to the potential threat of infection of the entire network, $i=\overline{1,n}$.
11. If request's service in S_i is finished and request leave the network or enters other system, the income of the system S_i doesn't change, $i=\overline{1,n}$.
12. In addition, for each small time interval, the system S_i incurs losses (drop in performance, decrease in the speed of performing routine operations due to the consumption of processor resources by the antivirus software) in the amount $r_i(\overrightarrow{k},\overrightarrow{l})$, $i=\overline{1,n}$.

Taking into acount probabilities of network's transitions between states, we obtain a system of difference-differential equations for the expected income of the system S_i:

$$\frac{dv_i\left(\overrightarrow{k},\overrightarrow{l},t\right)}{dt}=-\left(\lambda_{0i}^{+}+\lambda_{0i}^{-}+\mu_i^{(v)}+\mu_i^{(c)}+\mu_i\right)v_i\left(\overrightarrow{k},\overrightarrow{l},t\right)$$

$$+\sum_{j=1}^{n}\left\{\lambda_{0j}^{+}u\left(k_j^{(p)}\right)v_i\left(\overrightarrow{k}-\widetilde{I}_{2j-1},\overrightarrow{l},t\right)+\lambda_{0j}^{-}u\left(l_j^{(n)}\right)v_i\left(\overrightarrow{k},\overrightarrow{l}-\widetilde{I}_{2j-1},t\right)\right.$$

$$+\mu_j^{(v)}q_j^{+}p_j^{+}u\left(k_j^{(s)}\right)v_i\left(\overrightarrow{k}+\widetilde{I}_{2j-1}-\widetilde{I}_{2j},\overrightarrow{l},t\right)$$

$$+\mu_j^{(v)} q_j^+ (1-p_j^+)\, u\left(l_j^{(c)}\right) v_i\left(\vec{k}+\widetilde{I}_{2j-1}, \vec{l}-\widetilde{I}_{2j}, t\right)$$
$$+\mu_j^{(v)} (1-q_j^+)(1-p_j^-)\, v_i\left(\vec{k}+\widetilde{I}_{2j}, \vec{l}+\widetilde{I}_{2j-1}, t\right)$$
$$+\mu_j^{(v)} (1-q_j^+)\, p_j^- u\left(l_j^{(c)}\right) v_i\left(\vec{k}, \vec{l}+\widetilde{I}_{2j-1}-\widetilde{I}_{2j}, t\right)$$
$$+\mu_j^{(c)} p_j^{(s)} u\left(k_j^{(s)}\right) v_i\left(\vec{k}-\widetilde{I}_{2j}, \vec{l}+\widetilde{I}_{2j}, t\right)$$
$$+\mu_j^{(c)}\left(1-p_j^{(s)}\right) v_i\left(\vec{k}, \vec{l}+\widetilde{I}_{2j}, t\right)+\mu_j p_{j0} v_i\left(\vec{k}+\widetilde{I}_{2j}, \vec{l}, t\right)$$
$$+\sum_{h=1,h\neq j}^{n}\Big[\mu_h p_{hj}^+ u\left(k_j^{(p)}\right) v_i\left(\vec{k}+\widetilde{I}_{2h}-\widetilde{I}_{2j-1}, \vec{l}, t\right)$$
$$+\mu_h p_{hj}^- u\left(l_j^{(n)}\right) v_i\left(\vec{k}+\widetilde{I}_{2h}, \vec{l}-\widetilde{I}_{2j-1}, t\right)\Big]\Big\}+E_i(\vec{k},\vec{l}), \tag{7}$$

where

$$E_i(\vec{k},\vec{l}) = r_i\left(\vec{k},\vec{l}\right)+\lambda_{0i}^+ u\left(k_i^{(p)}\right) R_{0i}^+\left(\vec{k}-\widetilde{I}_{2i-1}, \vec{l}\right)$$
$$-\lambda_{0i}^- u\left(l_i^{(n)}\right) R_{i0}^-\left(\vec{k}, \vec{l}-\widetilde{I}_{2i-1}\right)$$
$$-\mu_i^{(v)} q_i^+ p_i^+ u\left(k_i^{(s)}\right) r_i^+\left(\vec{k}+\widetilde{I}_{2i-1}-\widetilde{I}_{2i}, \vec{l}\right)$$
$$+\mu_i^{(v)} q_i^+ (1-p_i^+)\, u\left(l_i^{(c)}\right) R_i^c\left(\vec{k}+\widetilde{I}_{2i-1}, \vec{l}-\widetilde{I}_{2i}\right)$$
$$-\mu_i^{(v)} (1-q_i^+)(1-p_i^-)\, R_i^+\left(\vec{k}+\widetilde{I}_{2i}, \vec{l}+\widetilde{I}_{2i-1}\right)$$
$$-\mu_i^{(v)} (1-q_i^+)\, p_i^- u\left(l_i^{(c)}\right) r_i^-\left(\vec{k}, \vec{l}+\widetilde{I}_{2i-1}-\widetilde{I}_{2i}\right)$$
$$+\mu_i^{(c)} p_i^{(s)} u\left(k_i^{(s)}\right) R_i^+\left(\vec{k}-\widetilde{I}_{2i}, \vec{l}+\widetilde{I}_{2i}\right)-\mu_i^{(c)}\left(1-p_i^{(s)}\right) R_i^+\left(\vec{k}, \vec{l}+\widetilde{I}_{2i}\right)$$
$$+\sum_{j=1,j\neq i}^{n}\Big[\mu_j p_{ji}^+ u\left(k_i^{(p)}\right) R_j^+\left(\vec{k}+\widetilde{I}_{2j}-\widetilde{I}_{2i-1}, \vec{l}\right)$$
$$-\mu_j p_{ji}^- u\left(l_i^{(n)}\right) r_i^-\left(\vec{k}+\widetilde{I}_{2j}, \vec{l}-\widetilde{I}_{2i-1}\right)\Big], \tag{8}$$

Let us introduce vector $\overline{V}(\vec{k},\vec{l},t)=\left(v_1(\vec{k},\vec{l},t), v_2(\vec{k},\vec{l},t), ..., v_n(\vec{k},\vec{l},t)\right)$. Then similary to (3) the system (7) can be represent as

$$\frac{d\overline{V}(\overrightarrow{k},\overrightarrow{l},t)}{dt} = -\Lambda(\overrightarrow{k},\overrightarrow{l})V(\overrightarrow{k},\overrightarrow{l},t) + \sum_{\alpha,\beta,\gamma,\theta,\eta=0}^{2n} \Phi_{\alpha\beta,\gamma,\theta,\eta}(\overrightarrow{k},\overrightarrow{l})$$

$$\times \overline{V}(\overrightarrow{k} + \widetilde{I}_\alpha + \widetilde{I}_\beta - \widetilde{I}_\gamma, \overrightarrow{l} + \widetilde{I}_\eta - \widetilde{I}_\theta, t) + \overline{E}(\overrightarrow{k},\overrightarrow{l}), \tag{9}$$

where function $\Phi_{\alpha,\beta,b,\gamma,\theta,\eta}(\overrightarrow{k},\overrightarrow{l})$ can be found from (4) and functions $\Lambda(\overrightarrow{k},\overrightarrow{l})$, $\overline{E}(\overrightarrow{k},\overrightarrow{l})$ equal respectively:

$$\Lambda(\overrightarrow{k},\overrightarrow{l}) = \left(\Lambda_i(\overrightarrow{k},\overrightarrow{l})\right)_{i=\overline{1,n}}, \Lambda_i(\overrightarrow{k},\overrightarrow{l}) = \lambda_{0i}^+ + \lambda_{0i}^- + \mu_i^{(v)} + \mu_i^{(c)} + \mu_i, i = \overline{1,n},$$

$$\overline{E}(\overrightarrow{k},\overrightarrow{l}) = \left(E_1(\overrightarrow{k},\overrightarrow{l}), E_2(\overrightarrow{k},\overrightarrow{l}), ..., E_n(\overrightarrow{k},\overrightarrow{l})\right).$$

In [20,21] the method of successive approximations was studied to find the expected incomes of queueing networks with various peculiarities, such as requests of different types and many classes, unreliable devices, limited residence time of requests in the network, positive and negative requests. Based on the above results, it can be shown that for the studied network with systems with a control queue and quarantine, the expected incomes of the network can be presented in the form of a converging power series:

$$\overline{V}_q(\overrightarrow{k},\overrightarrow{l},t) = \sum_{z=0}^{\infty} \overrightarrow{g}_{qz}^{+-}(\overrightarrow{k},\overrightarrow{l})t^z,$$

where $\overrightarrow{g}_{q0}^{+-}(\overrightarrow{k},\overrightarrow{l}) = \overline{V}(\overrightarrow{k},\overrightarrow{l},0)$, $\overrightarrow{g}_{0z}^{+-}(\overrightarrow{k},\overrightarrow{l}) = \overline{V}(\overrightarrow{k},\overrightarrow{l},0)\delta_{z0}$,

$$\overrightarrow{g}_{qz}^{+-}(\overrightarrow{k},\overrightarrow{l}) = \frac{-\Lambda(\overrightarrow{k},\overrightarrow{l})^z}{z!}$$

$$\times \left\{ \overline{V}(\overrightarrow{k},\overrightarrow{l},0) - \frac{\overline{E}(\overrightarrow{k},\overrightarrow{l})}{\Lambda(\overrightarrow{k},\overrightarrow{l})} + \sum_{u=0}^{z-1} \frac{(-1)^{u+1}u!}{\Lambda(\overrightarrow{k},\overrightarrow{l})^{u+1}} \overrightarrow{G}_{qu}^{+-}(\overrightarrow{k},\overrightarrow{l}) \right\}, z \geq 0,$$

$$\overrightarrow{G}_{qz}^{+-}(\overrightarrow{k},\overrightarrow{l}) = \sum_{\alpha,\beta,\gamma,\theta,\eta=0}^{2n} \Phi_{\alpha\beta,\gamma,\theta,\eta}(\overrightarrow{k},\overrightarrow{l})\overrightarrow{g}_{qz}^{+-}(\overrightarrow{k} + \widetilde{I}_\alpha + \widetilde{I}_\beta - \widetilde{I}_\gamma, \overrightarrow{l} + \widetilde{I}_\eta - \widetilde{I}_\theta).$$

Example 2. Consider the network described in example 1. Let the network incomes from transitions between its states are following: $r_i(\overrightarrow{k},\overrightarrow{l}) = 5$, $R_{0i}^+(\overrightarrow{k},\overrightarrow{l}) = 7$, $R_{i0}^-(\overrightarrow{k},\overrightarrow{l}) = 4$, $r_i^+(\overrightarrow{k},\overrightarrow{l}) = 10$, $R_i^c(\overrightarrow{k},\overrightarrow{l}) = 11$, $R_i^+(\overrightarrow{k},\overrightarrow{l}) = 15$, $r_i^-(\overrightarrow{k},\overrightarrow{l}) = 7$, $i = \overline{1,2}$. Figure 3 represents the change in time of the expected income of the first QS if $v_1(1,1,1,1,1,1,1,1,0) = 0$.

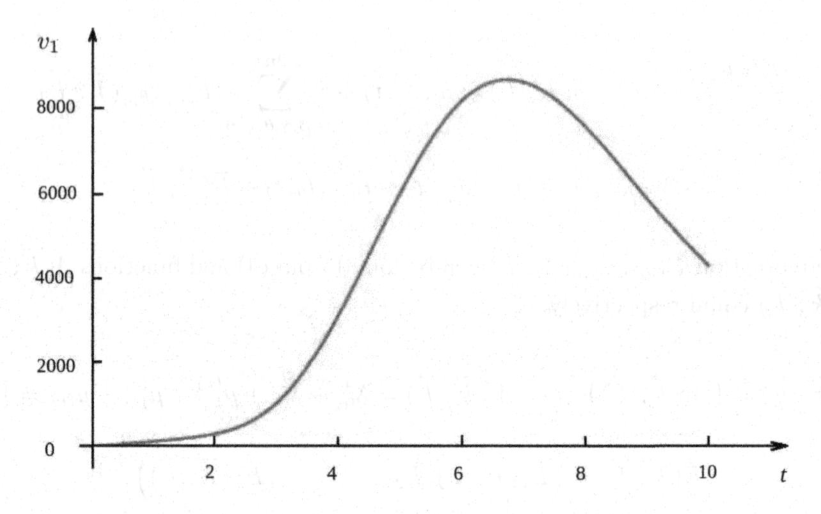

Fig. 3. Expected income $v_1(1, 1, 1, 1, 1, 1, 1, 1, t)$ of the S_1.

5 Conclusion

In this article we described a stochastic model of a computer network with antivirus software. A systems of difference-differential equations for the probabilities of states of such a network and expected incomes of the network's systems were obtained. The technique of solving these systems using the method of successive approximations, combined with the method of series, was described. It was shown that the probabilities of the network's states and the expected incomes of the network's systems can be represented in the form of converging power series. The results obtained can be used to predict the reliability of servicing a computer network and the probabilities of failure of its systems, which will reveal the effectiveness of antivirus software.

References

1. Patil, B.V., Joshi, M.J.: Usages of Selected Antivirus Software in Different Categories of Users in selected Districts. JECET: J. Environ. Sci. Comput. Sci. Eng. Technol. **3**(2), 801–807 (2014)
2. Kabak, I., Sukhanova, N.: Hardware realization of associative memory of arbitrary dimention. Bull. MSTU Stankin **1**, 135–139 (2010)
3. The official website of Kaspersky Lab. https://support.kaspersky.ru/KIS4Mac/16.0/en.lproj/pgs/59231.html. Accessed 24 Jul 2021
4. Letunovich, Yu., Yakubovich, O.: Open Markov queuing networks with control queues and quarantine node. Tomsk State Univ. J. Control Comput. Sci. **41**, 32–38 (2017)
5. Gelenbe, E.: Random neural networks with negative and positive signals and product form solution. Neural Comp. **1**, 502–510 (1989)
6. Fourneau, J.: G-networks of unreliable nodes. Probab. Eng. Inf. Sci. **30**(3), 361–378 (2016)

7. Sohaib, H., et al.: Intrusion detection in IoT using artificial neural networks on UNSW-15 dataset. In: 2019 IEEE 16th International Conference on Smart Cities: Improving Quality of Life Using ICT & IoT and AI (HONET-ICT), pp. 152–156. IEEE, Charlotte, NC, USA (2019). https://doi.org/10.1109/HONET.2019.8908122
8. Günel, G., Loukas, G.: A denial of service detector based on maximum likelihood detection and the random neural network. Comput. J. **50**, 717–727 (2007)
9. Ayyaz-Ul-Haq, Q., et al.: A novel random neural network based approach for intrusion detection systems. In: 2018 10th Computer Science and Electronic Engineering (CEEC), pp. 50–55. IEEE, Colchester, UK (2018). https://doi.org/10.1109/CEEC.2018.8674228
10. Gelenbe, E.: G-Networks with resets. Perform. Eval. **49**, 179–192 (2002)
11. Matalytski, M., Naumenko, V., Kopats, D.: Investigation of G-network with restart at a nonstationary mode and their application. J. Appl. Math. Comput. Mech. **18**(2), 41–51 (2019)
12. Gelenbe, E.: G-networks and the optimization of supply chains. Probab. Eng. Inf. Sci. **35**(1), 62–74 (2019)
13. Yin, Y.: Deep learning with the random neural network and its applications. https://arxiv.org/abs/1810.08653. Accessed 4 Aug 2021
14. Yonghua, Y., Gelenbe, E.: Non-negative autoencoder with simplified random neural network. In: 2019 International Joint Conference on Neural Networks (IJCNN), pp. 1–6. Budapest, Hungary (2019). https://doi.org/10.1109/IJCNN.2019.8851912
15. Koluzaeva, E., Matalytski, M.: Analysis and Optimization of Queueing Networks. LAP Lambert Academic Publishing, Saarbrucken (2011)
16. Rusilko, T., Matalycki, M.: About one method of incomes forecasting in insurance company. Sci. Res. Inst. Math. Comput. Sci. **4**(1), 173–181 (2005)
17. Matalytski, M., Kopats, D.: Finding nonstationary probabilities of open Markov networks with multiple classes of customers and various features. Probab. Eng. Inf. Sci. **34**(1), 158–179 (2021)
18. Drobotun, E., Kozlov, D.: Assessment of influence of anti-virus software on quality of information-computing system functioning (Ocenka stepeni vliyaniya antivirusnyh programmnyh sredstv na kachestvo funkcionirovaniya informacyonno-vychislitel'nyh sistem), Int. Res. Pract. J. Softw. Syst. **4**(116), 129–134 (2016). https://doi.org/10.15827/0236-235X.116.129-134
19. Kaspersky: detecting, understanding and eliminating hidden threats, Oficcial web-site of anti-malware. https://www.anti-malware.ru/news/2020-10-06-111332/33873. Accessed 10 Aug 2021
20. Matalytski, M., Koluzaeva, E.: Investigation of Markov HM-networks with different types of reqests of many classes by the method of successive approximations (Issledovanie markovskih HM-setej s raznotipnymi zayavkami mnogih klassov metodom posledovatel'nyh priblirzenij). In: Proceedings of the National Academy of Sciences of Belarus Physics and Mathematics Series, vol. 4, pp. 113–119 (2008)
21. Kopats, D. Matalytski, M.: Analysis of expected revenues in open Markov networks with various features. Tomsk State Univ. J. Control Comput. Sci. **50**, 31–38 (2020). https://doi.org/10.17223/19988605/50/4

The Automata-Based Model for Control of Large Distributed Systems

Yu. S. Zatuliveter(✉) and E. A. Fishchenko

Institute of Control Sciences of Russian Academy of Sciences,
65 Profsoyuznaya street, Moscow 117997, Russia
zvt@ipu.rssi.ru

Abstract. With the increase in the size of distributed systems implemented in a global computer environment, mathematical methods for formalizing their development and functioning become one of the important problems. The article proposes the automata-based approach to distributed systems representation and control of their operation using balance-based equations. A feature of the automata-based model is that the control complexity in arbitrarily large distributed systems ceases to depend on the size of the distributed network systems and the problems they solve.

Keywords: global computer environment · heterogeneity · large distributed system · automata-based model · balance-based equation

1 Introduction

The Global computer environment (GCE) forms a qualitatively new information infrastructure (in its digital universality) with an extremely large number of network-connected computing nodes in the form of a mobile and stationary computer devices of various classes-from smartphones and PCs to supercomputers and integrated intelligent sensors and actuators. The number of network nodes totals tens of billions and continues to grow rapidly, expanding the scope of the GCE influence.

GCE has already become a factor of total information impact on every person, on societies, and the global sociotechnosphere as a whole. It has brought to the historical processes an unprecedented phenomenon of global information connectivity (everything affects everything at once). The global digital infrastructure de-facto assembles weakly connected socio-systems into a global, strongly connected cybernetic system. Control a sustainable functioning and development of a growing diversity of arbitrarily large, strongly connected distributed systems requires fundamentally new mathematical control models and computer network architectures that ensure their mass implementation and evolution in the GCE environment.

An unprecedented phenomenon of global information strongly- connectivity has formed in the WWW space [1]. We will take into consideration three factors that contributed to the formation of this phenomenon.

V. M. Vishnevskiy et al. (Eds.): DCCN 2021, CCIS 1552, pp. 272–283, 2022.
https://doi.org/10.1007/978-3-030-97110-6_21

The first is associated with the use of mass universal computers with microprocessor architectures and the universal network protocol TCP/IP, which, thanks to cross-platform scalability and high reliability of data transmission, provided a fast growth in the number of nodes and the scale of application of GCE. The second is the appearance of the WWW space with a hypertext interface for accessing distributed information resources of the GCE.

The third is the natural desire of a person and society to maximize the manifestations of information activity.

At the base of every informational processes is three basic types of fundamental actions with information: storage, transmission, and transformation.

The hypertext model became the logical basis of the unified information space WWW in the GCE [1]. The special significance of this simplest model of network globalization of information processes is that it provided users with a single interface for accessing all GCE information and service resources via WWW through hyperlinks.

Providing convenient navigation through network resources, the hypertext model initially assigns semantic processing of information to a person. At the same time, its limited individual and socialized information throughput do not allow for processing exponentially growing flows and volumes of information timely and on the required scale to ensure the sustainable functioning and development of the sociotechnosphere in the conditions of global strongly-connectivity.

In the absence of a general mathematical model of universally programmable distributed computing in arbitrarily large networks, the hypertext model in the existing GCE architecture has implemented network globalization of only two of the three fundamental types of actions with information - storage and transmission. This intra-system disproportion in the development of the GCE has led to a systemically unbalanced formation and expansion of the GCE, which has a growing destabilizing impact on the global social system as a whole.

The fast growth in the intensity of information interactions exceeds the capabilities of existing technologies for system-functional integration of heterogeneous hardware, software, and information resources of the GCE, which are still used to create distributed systems for algorithmic processing of globally distributed information.

Initially, multivariant integration problems of heterogeneous resources have combinatorial complexity. Therefore, increasing the size of such systems at the expense of existing technologies requires an unlimited increase in the cost of their creation. At the same time, existing integration technologies do not meet the growing security requirements of the systems being created. All this fundamentally limits the ability to scale distributed systems.

Global strong-connectedness, which is not controlled within existing computer and network technologies, has caused an exponential growth in the flows/volumes of insufficiently processed information. This becomes the fundamental cause of the global crisis of overproduction of information and, as a result, the instability of social systems. Instability rapidly grows, despite the measures taken to counteract it by known means of financial, and economic regulation.

The absence of systemically holistic and functionally complete globalization of all three types of fundamental actions with information makes it impossible to cumulative universally programmable use of the unlimited growing computational potential of a heterogeneous GCE for timely and sufficiently complete processing of globally distributed information to control sustainable development. The further development of the GCE should be based on the objective laws of joint functioning and development of social and computer environments, as well as on the development of new mathematical methods:

- of reengineering of the computer-network architecture of the GCE, aimed at eliminating system imbalances;
- of building cybernetic models for control of the sustainable functioning and development of the sociotechnosphere in the context of global information strong-connectedness, such models, which focus on implementing in the updated GCE architecture in arbitrarily large networks.

In its development, the GCE becomes fundamentally new, namely, a globally strongly connected, rapidly developing cybernetic object and, at the same time, a universal tool for the global digitalization of the global sociotechnosphere. Digitalization in GCE should be considered an objective cybernetic process of system-wide computer network integration of initially fragmented societies and the global sociotechnosphere (due to their initially weak information connectivity).

From the standpoint of cybernetic sciences, for the sustainable development of the sociotechnosphere, it is necessary to maintain a systematic balance between the growth rates of the flows and volumes of information produced and the available ways and means of its timely and full-scale processing in the processes of managing functioning and development.

In three decades of spontaneous, systemically unbalanced growth of the computer environment, there was a de facto global (in computer-network execution) violation of the Ashby principle [2] - one of the main principles of cybernetics: "For the implementation of control processes, the diversity of the control subsystem must be no less than the diversity of the controlled subsystem."

To a heterogeneous GCE, this means that the diversity of the control part of large distributed socio-technical systems depends on the available functionality and throughput required for timely and sufficiently complete processing of the growing flows/volumes of information coming from the controlled part of such systems.

In the conditions of global information strongly- connectivity, the functioning, and development of the controlled part of large distributed systems are accompanied by the exponential growth of information and, at the same time, a chain reaction of the growth of the number and diversity of uncontrolled degrees of freedom, which re-quire continuous updating of the functionality of the control part of large systems.

Currently, the GCE does not provide such opportunities. The reasons are that the existing GCE, consisting of a large number of locally universal computer devices connected by networks, at the global level of its total resources does not have the system-wide quality of functional completeness (universal programmability). This makes it impossible to time and continuously update the control part of distributed systems according to the exponential growth rate of information coming from the controlled part of the systems. But this is a direct violation of the principles of Ashby's principle of controllability.

One of the main reasons for the inconsistency of modern GCE (in its state of extreme heterogeneity) with the requirements of the Ashby principle is the initial heterogeneity of hardware, software, and information network resources. Creating large distributed systems for processing globally distributed information system-functional integration of heterogeneous resources with increases their scale requires outstripping growth of cost [3]. This is explained by the fact that multivariant problems of integrating heterogeneous resources belong to combinatorially complex problems [4–7].

In conditions of the structural heterogeneity of hardware platforms and forms of data representation, programs, processes, and systems with the increase in the scale of distributed systems the complexity of cybersecurity problems increases with outperforming growth.

GCE as a whole, due to the lack of a system-wide property of functional completeness, cannot be considered as a universally programmable carrier of globally distributed control processes. Functional completeness requires initially seamless expansion of the universal programmability property from intracomputer resources to arbitrarily large networks [4–7].

The global scale of networks and tasks of deep processing of distributed information in GCE require new approaches to solving the problems of building distributed data processing systems in arbitrarily large environments. The property of global strongly- connectedness requires the formation in the GCE of a seamlessly programmable algorithmic space of distributed computing with a single space of globally distributed information [4–7]. It will hide the heterogeneity of machine environments and remove the barriers of combinatorial complexity, which will remove the upper limits on the size of computing environments.

Heterogeneity is the fundamental system-technical reason limiting the further growth of size distributed systems in the GCE.Another fundamental limitation of the size of distributed systems in the GCE is the insufficient development of mathematical methods for formalizing such systems and control processes in them.

Next, using a simple example of managing the parallelism of distributed computing, we present the principles of constructing an automaton-based model that allows us to consider the control of distributed computing in the dynamic parallelization mode based on the principles of Data Flow models [8]. The feature of the automaton-based model is the independence of the computational complexity of the algorithm for controlling the parallelism of distributed computing from the size of the tasks determined by the total number of computational operations.

The model has two levels. The first is the representation of computational problems in the form of information graphs. The second is an automaton-based representation of the control processes of distributed computing with dynamic parallelization.

2 Information Graphs for the Mathematical Presentation of Tasks

We assume that the initial computational tasks view a bipartite information graph, in which directed arcs connect nodes of two different types - operators and objects. Operator nodes represent computational actions, while object nodes represent variables that are arguments or values of operators. Model Data Flow uses this form of task representation, which is non-procedural [8].

The information graph $\mathbf{G} = \langle \mathbf{A} \cup \mathbf{U}, \mathbf{S} \rangle$, where $\mathbf{A} = \{a_1, a_2, \ldots\}$ is the set of operator nodes, $\mathbf{U} = \{u_1, u_2, \ldots\}$ is the set of object nodes, **S=S(A,U)** $\cup$ **S(U,A)** is the set of oriented arcs *S(a,u)* and $S(u,a)$, respectively.

In this case, $\mathbf{S(A,U)} = \{S(a,u) \neq \oslash : a \in \mathbf{A}, u \in \mathbf{U}\}$ are the all arcs from the operators to the objects, $\mathbf{S(A,U)} = \{S(a,u) \neq \oslash : a \in \mathbf{A}, u \in \mathbf{U}\}$ - from the objects to the operators.

All operator nodes $a \in \mathbf{A}$ and object $u \in \mathbf{U}$ have the following nodes types: $b(a) \in \mathbf{B}$ and $d(u) \in \mathbf{D}$, respectively, where $b(a) \in \mathbf{B}$ and $d(u) \in \mathbf{D}$ is a basic set of computational operations/functions $\mathbf{B} = \{b_1, b_2, \ldots, b_m\}$, which we will call operator types, and $\mathbf{D} = \{d_1, d_2, \ldots, d_n\}$ is a basic set of data types. Arcs **S(U,A)** for each operator determine the occurrence of objects (variables) as arguments (input data), arcs **S(U,A)** - the transfer of calculated values (output data) to objects (variables) that take calculated values.

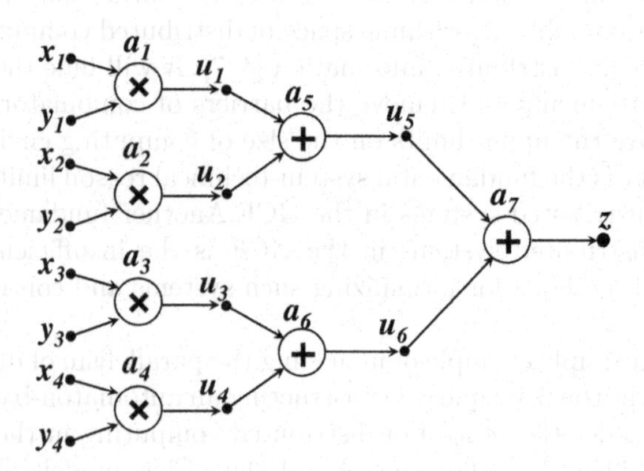

Fig. 1. Information graph of the scalar product of vectors of size 4

Information graphs one-to-one relate with a system of formula expressions in explicit form, in which object nodes have unique variable names, data types, and operator nodes have unique operator identifiers and their types of operations.

Figure 1 shows an example of the information graph for the scalar product of vectors $x = \{x_1, x_2, \ldots\}$ and $y = \{y_1, y_2, \ldots\}$: $z = \sum_i x_i \cdot y_i$. This graph corresponds to the system of formula expressions in the explicit form:

$$\begin{cases} z = u_5 + u_6 \\ u_5 = u_1 + u_2 \\ u_6 = u_3 + u_4 \\ u_1 = x_1 \times y_1 \\ u_2 = x_2 \times y_2 \\ u_3 = x_3 \times y_3 \\ u_4 = x_4 \times y_4 \end{cases}$$

3 Control Automata

The order of actions in the information graph is determined by the rule of automata-based asynchronous parallelization, which is the basis of Data Flow. Namely: each operator can be executed if the calculated values of all its arguments (input variables) are received and entered, and there are free memory cells in which the calculated values of objects store during computing.

The Data Flow model, as is known, provides asynchronous dynamic identification of the maximum possible parallelism of computational operators in information graphs.

Let us build the model of such control in the form of an automata network. We introduce two types of automata for constructing such control networks.

Every $a \in \mathbf{A}$ has two subsets of objects (variables):

$\mathbf{U}^-(a) = \{u \in \mathbf{U} : S(u, a) \neq \oslash\}$ - is the subset of input arguments, and
$\mathbf{U}^+(a) = \{u \in \mathbf{U} : S(a, u) \neq \oslash\}$ - is the output is the subset of values calculated by the operators of $a \in \mathbf{A}$.

Every $u \in \mathbf{U}$ has two subsets of operators:

$\mathbf{A}^-(u) = \{a \in \mathbf{A} : S(a, u) \neq \oslash\}$ - is the input subset of preceding operators, and

$\mathbf{A}^+(u) = \{a \in \mathbf{A} : S(u, a) \neq \oslash\}$ - is the output subset of operators that use the values of $u \in \mathbf{U}$ objects.

For each operator and object, we assign an A-automaton (Fig. 2a) and a U-automaton (Fig. 2b), respectively. During the execution of distributed computing processes, control automata mathematically determine the behavior of operator and object of information graphs and the interactions between neighboring automata.

The automata are connected in a control network. Each arc of the information graph S(a, u) and S(u, a) corresponds to two oppositely directed connections of forward and reverse synchronization, through which interactions between automata of different types are carried out.

Both types of automaton, in the considered simplest case, have three states.

A-automatons implement control by data flows: an operator can be executed if the values of its arguments (input variables) are defined. The alphabet of states

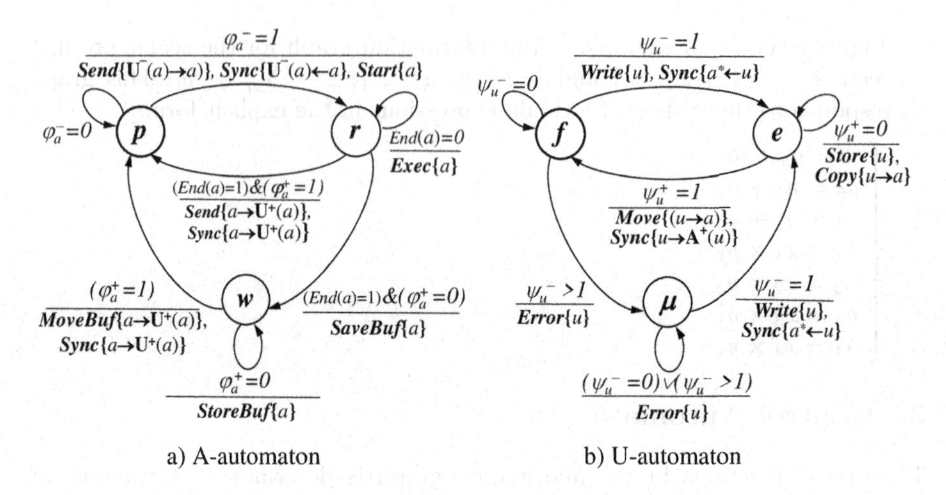

a) A-automaton b) U-automaton

Fig. 2. Control automata

of the **A**-automaton is $\{p, r, w\}$, where the p denotes "operator is passive" (not all input values are calculated still), the r - "operator is executing," and the w - "operator is waiting" for the result to be unloaded.

U-automatons implement the principle of object value protection: new data cannot be written to the object value storage memory until the previous values are fully used. The alphabet of states of the **U**-automaton is $\{f, e, \mu\}$, where the state f corresponds to the "free memory of the object," state e - to the "occupied memory" (storing the value), μ - to the state of uncertainty due to a conflict on the record from two or more operators (the diagnostic result of detecting incorrectness of information graphs).

For direct and reverse synchronization of neighboring automatons, every **A**-automaton attached to node $a \in \mathbf{A}$ and **U**-automaton attached to node $u \in \mathbf{U}$ have, respectively, four numeric variables, which during calculations take the values: $k^+(u), k^-(a), k^+(u), k^-(u) \in N_0 = \{0, 1, 2, \dots\}$.

Before the beginning of the calculations, these variables must have the following initial values:

$k^+(a) = |\mathbf{U}^+(a)|$ - size of subset $\mathbf{U}^+(a)$, $a \in \mathbf{A}$;
$k^-(a) = |\mathbf{U}^-(a)|$ - size of subset $\mathbf{U}^-(a)$, $a \in \mathbf{A}$;
$k^+(u) = |\mathbf{A}^+(u)|$ - size of subset $\mathbf{A}^+(u)$, $u \in \mathbf{U}$;
$k^-(u) = 0$ for all $u \in \mathbf{U}$.

At the moments of transitions (Fig. 2), the automata connected to the network perform internal (inter-automatons) synchronization interactions, as well as "external" synchronization interactions between the automata and their nodes of the information graph.

Inter-automata interactions are carried out by calculating the threshold indicator functions $\varphi_a^-, \varphi_a^+, \psi_u^-, \psi_u^+$ associated with the graph's nodes.

Interactions with the graph nodes are performed using threshold indicator functions that determine the initial and final moments of changing the modes of operation of the graph nodes and other functions of performing information actions related to processing information during the computational process of the information graph.

Threshold indicator functions calculate the transition condition between the states of automata (Fig. 2). Three of the 4 types of these functions are defined as the result of the comparison with 0: $\varphi_a^- = (k^+(a) = 0), \varphi_a^+ = (k^-(a) = 0), \psi_u^+ = (k^+(u) = 0)$. The fourth type of such interaction is defined as the result of a comparison with 1: $\psi_u^- = (k^-(u) = 1)$.

Each **A**-automaton during the automata interactions controls the switching of the operating mode of the operator utilizing the following signal functions (Fig. 2a): ***Start*** $\{a\}$ - beginning execution the operator, *End*(a) - asynchronous signal from this node (*End*$(a) = 1$) about ending execution.

Table 1 shows other actions processing initiated by the A-automata.

Table 1. Actions performed by the controlling A - automaton during state transitions (between Of and In).

Of	In	Condition	Actions on the transitions
p	p	$\varphi_a^- = 0$	No
p	r	$\varphi_a^- = 1$	***Send***$\{\mathbf{U}^-(a) \to a\}$ (to all values of input objects in a)
			Sinc$\{\mathbf{U}^-(a) \leftarrow a\}$ [reverse synchronization across the all input arcs $S(\mathbf{U}^-(a), a) = \{S(u, a) : u \in \mathbf{U}^-(a)\}$), where ***Sinc***$\{\mathbf{U}^-(a) \leftarrow a\} \leftrightarrow$ ***Minus1***$\{K^+(\mathbf{U}^-(a))\}=\{k^+(u){:=}k^+(u) - 1 : \forall u \in \mathbf{U}^-(a)\}$]
			Start $\{a\}$ (initializing the operator)
r	r	$End(a) = 0$	***Exec*** $(\{a\})$ (execution of the operator)
r	p	$End(a) = 1 \wedge \varphi_a^+ = 1$	***Send*** $\{a \to U^+(a)\}$ (sending the results to the output $U^+(a)$)
			Sync $(a \to U^+(a))$ [direct synchronization across all output arcs of $\mathbf{S}(a, \mathbf{U}^+(a)) = \{S(a, u) : u \in \mathbf{U}^+(a)\}$, where ***Sync*** $\{a \to U^+(a)\} \leftrightarrow$ ***Plus1*** $\{K^-(\mathbf{U}^+(a))\} = \{k^-(u) := k^-(u) + 1 : \forall u \in \mathbf{U}^+(a)\}$]
r	w	$End(a) = 1 \wedge \varphi_a^+ = 0$	***SaveB*** $\{a\}$ (sending the result values to the intermediate buffer)
w	w	$\varphi_a^+ = 0$	***StoreB*** $\{a\}$ (storing results in a buffer)
w	p	$\varphi_a^+ = 1$	***MoveB*** $\{a \to \mathbf{U}^+(a)\}$ (moving the result values from the buffer to $\mathbf{U}^+(a)$)
			Sync $\{a \to \mathbf{U}^+(a)\}$ (direct synchronization over all output arcs (see above))

Table 2 shows other actions processing initiated by the U-automata.

Table 2. Actions performed by control U-automaton in the moments of states transition (between Of and In).

Of	In	Condition	Actions on the transitions
f	f	$\psi_u^- = 0$	No
f	e	$\psi_u^- = 1$	***Write*** $\{u\}$ (writing the new value of u)
			Sync $\{a^* \leftarrow u\}$ [reverse synchronization over the engaged input arc $S(a^*, u)$. Here a^*- is the only operator ($\exists_1 a^* \in A^-(u)$, which calculated the new value of u. ***Sync*** $\{a^* \leftarrow u\} \leftrightarrow$ ***Minus1*** $\{\exists_1 a^* \in A^-(a) : k^+(a^*) := k^+(a^*) - 1\}$]
f	μ	$\psi_u^- > 1$	***Error*** $\{u\}$ (detecting a conflict between two or more values received in u)
e	e	$\psi_a^+ = 0$	***Store*** $\{u\}$ (storing the current value of u)
			Copy $\{u \rightarrow a\}$ (copying the current value through the arcs $S(u, a)$ on requests from activated operators $a \in A^+(u)$]
e	f	$\psi_u^+ = 1$	***Move*** $\{u \rightarrow a\}$ (moving the value through the arc S(u,a) with the release of the memory of the object u)
			Sync $\{u \rightarrow A^+(u)\}$ [direct synchronization over all output arcs $\mathbf{S}(u, A^+(u) = (\{S(u,a) : a \in A^+(u)\})$, **Sync** $\{u \rightarrow A^+(u)\} \leftrightarrow$ **Minus1** $\{K^-(A^+(u))\} = \{k^-(a) := k^-(a) - 1 : \forall a \in \mathbf{A}^+(u)\}$]
μ	μ	$(\psi_u^- = 0) \vee (\psi_u^- > 1)$	***Error*** $\{u\}$ (object remains undefined after a conflict of two or more values received).
μ	e	$\psi_u^- = 1$	***Write*** $\{u\}$ (writing the new value u)
			Sync $\{a^* \leftarrow u\}$ [reverse synchronization over the involved input arc $S(a^*, u)$. Here a^* is the only operator ($\exists_1 a^* \in A_u^-$), that calculated new value of u (see above))]

4 Balance-Based Equation of Control Processes in Large Systems

The function of control a distributed process of computing on information graphs implement by the network of **A, U** - automata. Each node of the graph is associated with an automaton of the corresponding type. Each automaton of such a network interacts with the automata of neighboring nodes connected by oriented arcs. Therefore, the size of the automata network is equal to the number of nodes in the graph. In this case, the complexity of the control network significantly depends on the size of the solved problem represented by the graph.

We build a mathematical model using **A, U**-automata that embodies the asynchronous parallelization of calculations that underlie Data Flow. The complexity of control distributed processes in the proposed model does not depend on the size of the information graphs.

This model is the balance-based equation (below referred to as the "balance equation") of control processes, which show the example of information graphs. The model gives an analytical expression of control processes in large distributed systems.

The sets of the **A** and **U** nodes of the graph at each step j, taking into account the current state of the corresponding automata, represent as partitions into disjoint subsets of nodes whose automata are in the identical states:

$$\begin{cases} U \equiv U^f(j) \cup U^e(j) \cup U^\mu(j), \\ A \equiv A^p(j) \cup A^r(j) \cup A^w(j), \\ j = 0, 1, 2, \cdots \end{cases} \quad (1)$$

It should be noted that the set at $\mathbf{A}^r(j)$ each step j defines all operators that can be compute in parallel.

Interaction of the information graph nodes leads to a change in the composition of subsets of nodes (1) in identical states. The following system of recurrence equations defines the rules for updating subsets:

$$\begin{cases} \mathbf{A}^p(j+1) = (\mathbf{A}^p(j) \cup \Delta\mathbf{A}^{wp}(j) \cup \Delta\mathbf{A}^{rp}(j)) \setminus \Delta\mathbf{A}^{pr}(j); \\ \mathbf{A}^r(j+1) = (\mathbf{A}^r(j) \cup \Delta\mathbf{A}^{wp}(j)) \setminus (\Delta\mathbf{A}^{rp}(j)) \cup \Delta\mathbf{A}^{rw}(j)); \\ \mathbf{A}^w(j+1) = (\mathbf{A}^w(j) \cup \Delta\mathbf{A}^{rw}(j)) \cup \Delta\mathbf{A}^{wp}(j)); \\ \mathbf{U}^f(j+1) = (\mathbf{U}^f(j) \cup \Delta\mathbf{U}^{ef}(j)) \setminus (\Delta\mathbf{U}^{fe}(j)) \cup \Delta\mathbf{U}^{f\mu}(j)); \\ \mathbf{U}^e(j+1) = (\mathbf{U}^e(j) \cup \Delta\mathbf{U}^{fe}(j) \cup \Delta\mathbf{U}^{\mu e}(j)) \setminus (\Delta\mathbf{U}^{ef}(j)); \\ \mathbf{U}^\mu(j+1) = (\mathbf{U}^\mu(j) \cup \Delta\mathbf{U}^{f\mu}(j)) \setminus \Delta\mathbf{U}^{\mu e}(j); \\ j = 0, 1, 2, \cdots \end{cases} \quad (2)$$

The composition of the sets in the left part (2) defines the balance of the transition flows of the sets of operators and objects in the right part, which we have designated $\Delta\mathbf{A}^{qs}$ and $\Delta\mathbf{U}^{qs}$, respectively. Their automata transitions from state q to state s on clock cycle j, where either $q, s\epsilon\{p, r, w\}$, or $q, s\epsilon\{f, e, \mu\}$.

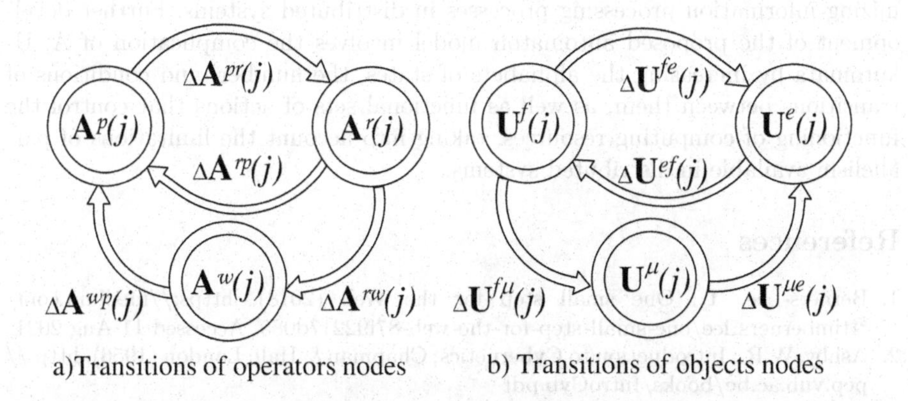

Fig. 3. Diagrams of the balance equations of distributed/parallel computing: (a) Transitions of operators nodes; (b) Transitions of objects nodes.

Balance equation diagrams show that the contents of subsets of nodes determine the state of the process entirely at any moment. The number of such subsets is not related to the size of the graph. Their number depends only on the logical complexity of **A**,**U**-automata and is equal to the size of their alphabets of possible states.

In our case, alphabets of states of each of **A**,**U**-automata (Fig. 3) consist of three symbols: the diagram of the balance equation with six subsets of nodes, which are in identical states (three for each of the two types of automata).

The proposed method can be generalized to cases of more complex behavior of nodes of information graphs, which provides automata with a large number of states.

5 Conclusion

The model of balance equations radically reduces the dimension of the state space of distributed processes compared to the multiplicative number of states of the entire network based on automata. The complexity of control distributed processes ceases to depend significantly on the size of distributed network systems and the tasks they solve. The proposed method of formalizing control processes in distributed systems can solve problems of modeling and controlling distributed processes in arbitrarily large computing environments.

In the considered case, the balance equations describe the distributed processes of computing on information graphs under the assumption that there are no restrictions on the parallelism of access to computational resources for processing operator vertices and access to memory for storing the values of object vertices, as well as on parallelism of data delivery from storage locations to processing locations and vice versa.

There are principle restrictions on the parallelism of access to computing resources in practice. Such limitations should be taken into account when optimizing information processing processes in distributed systems. Further development of the proposed automaton model involves the complication of A, U-automata by increasing the alphabets of states, the number, and conditions of transitions between them, as well as functional sets of actions that control the functioning of computing resources, taking into account the limitations of parallelism available in distributed systems.

References

1. Berners-Lee, T.: One small step for the Web (2018). https://medium.com/@timberners_lee/one-small-step-for-the-web-87f92217d085. Accessed 11 Aug 2021
2. Ashby, W.R.: Introduction to Cybernetics, Chapman & Hall, London (1956). http://pcp.vub.ac.be/books/IntroCyb.pdf
3. Skala, K., Sojat, Z.: The Rainbow: Integrating Computing into the Global Ecosystem. MIPRO, pp. 233–240 (2019). https://doi.org/10.23919/MIPRO.2019.8756947
4. Zatuliveter, Y.S., Fishchenko, E.A.: About the universal algorithmic space of distributed and parallel computing. In: Proceedings of the 11th International Conference "Management of Large-Scale System Development" (MLSD), Moscow, pp. 1–5. IEEE (2018). https://doi.org/10.1109/MLSD.2018.8551799
5. Zatuliveter, Y.S., Fishchenko, E.A.: Evolution of large-scale systems in the universal algorithmic space of digital transformation. In: Proceedings of the 13th International Conference "Management of Large-Scale System Development" (MLSD), Moscow, pp. 1–5. IEEE (2020). https://doi.org/10.1109/MLSD49919.2020.9247834

6. Zatuliveter, Y.S.: Prerequisites for the creation of the universal algorithmic space of digital economy. In: Proceedings of the 5th International Research and Practice Conference-Biennale "System Analysis in Economics" (SAE-2018), Moscow. Prometheus publishing house (2018). https://doi.org/10.33278/SAE-2018.eng.327-332
7. Zatuliveter, Y.S., Fishchenko, E.A.: Cybersecurity in the mathematically uniform algorithmic space of the distributed computing. In: Proceedings of the 10th International Conference "Management of Large-Scale System Development" (MLSD), Moscow, pp. 1–4. IEEE (2017). https://doi.org/10.1109/MLSD.2017.8109713
8. Dennis, J.: Data flow supercomputers. Computer **11**, 48–56 (1980)

Information Spreading with Application to Non-homogeneous Evolving Networks

Natalia M. Markovich(✉) and Maksim S. Ryzhov

V.A. Trapeznikov Institute of Control Sciences Russian Academy of Sciences, Profsoyuznaya Street 65, 117997 Moscow, Russia
markovic@ipu.rssi.ru, maksim.ryzhov@frtk.ru

Abstract. The paper is devoted to finding of leader nodes in evolving directed random networks with regard to the information spreading. We consider non-homogeneous networks consisting of several weakly connected subgraphs having different distributions of node in- and out-degrees. This is a plausible situation for real complex networks. The evolution of the network in time starting from a seed set of nodes is provided by linear preferential attachment schemes. We compare the spreading rate of nodes which share their messages with other nodes when they belong to different subgraphs of the non-homogeneous network. It is found that the nodes of the subgraph with the most heavy tailed out-degree distribution may spread their messages faster. We compared the spreading capacity of the linear preferential attachment used also for the graph evolution with a well-known SPREAD algorithm and found that the latter can disseminate the information faster.

Keywords: non-homogeneous evolving network · information spreading · linear preferential attachment · SPREAD algorithm · leading nodes

1 Introduction

Evolving scale-free network model has been studied in different areas: citation networks [1], the web-page popularity by PageRank during evolutionary changes [2], the evolution of the network [3]. Graphs reflecting a structure of the networks can be directed or undirected.

Information spreading, as a message delivery model in the whole network (the full spreading) [4] or in some community (the partial spreading) [5,6] has an application for the parallel grid calculations in the computation network.

One of our objectives is to study a linear preferential attachment (PA) schemes as a tool to spread the information. The PA has been proposed in [3] for the network evolution. We use the PA also to spread a message among

The reported study was supported by the Russian Science Foundation (grant No. 22-21-00177) (recipient N.M. Markovich, conceptualization, mathematical model development, methodology development; recipient M. S. Ryzhov, numerical analysis, validation.

V. M. Vishnevskiy et al. (Eds.): DCCN 2021, CCIS 1552, pp. 284–292, 2022.
https://doi.org/10.1007/978-3-030-97110-6_22

nodes. We aim to demonstrate that the PA may spread faster than the SPREAD algorithm in [4] for some parameter values of the PA.

Another objective is to find leading nodes which can spread the information fast among nodes of a non-homogeneous graph. On each evolution step, a newly appended directed edge (i, j) may cause the message exchange if a node i with a message communicates to a node j without it. We compare the linear PA schemes with the SPREAD algorithm for directed graphs.

Here, two novelties are implemented. Firstly, the information spreading is studied in a directed evolving graph where the probabilities to choose nodes and create edges for communications are determined by the linear PA schemes α, β and γ. These probabilities depend on the in-/out-degrees and parameters of the PA schemes. Since the in- and out-degrees of nodes may change over time during the evolution, the latter probabilities are evolving, too. Secondly, non-homogeneous graphs are considered. The latter consist from subgraphs whose in- and out-degrees have different power law distributions. We aim to study, does the spreading rate of nodes depend on the heaviness of tail of their node degrees.

The paper is organized as follows. In Sect. 2, related works regarding the spreading information (Sect. 2.1), the PA (Sect. 2.2) are provided. In Sect. 3, our main results concerning for the spreading are presented. The exposition is finalized by Conclusions.

2 Related Works

2.1 Information Spreading

Let us describe the idea of the spreading algorithm SPREAD proposed in [4] for an undirected graph $G = (V, E)$. Here, V and E are sets of graph vertices and edges, respectively. The most realistic situation is that clocks of all nodes are not quite synchronized. Considering an asynchronous time model, a node may initiate a communication by ticks of a global clock which are modelled as a Poisson process of rate $n = |\mathsf{V}|$, [4,5]. Let $k \geq 0$ denote the index of a tick, on which at most one node can receive messages by communicating with another node. To this end, on a clock tick one of n nodes (let say a node i) of the graph is chosen uniformly. Then this node i chooses a node j uniformly among its neighbors with probability $P_{ij} = 1/d_{\max}$, where $d_{\max} = \max_{i \in V} d_i$, d_i is the node degree of node i. In [5] it is proposed to use $P_{ij} = 1/d_i$ to avoid the knowledge of the maximal node degree in the network. As in [4] $S_i(k)$ defines the set of nodes that have the message m_i from node i at the end of the clock tick k. After clock tick $k + 1$, we have either $|\mathsf{S\_i(k+1)}| = |\mathsf{S\_i(k)}|$ or $|\mathsf{S\_i(k+1)}| = |\mathsf{S\_i(k)}| + 1$.

Here, we use the SPREAD for directed graphs considering a partial information spreading, i.e. a node i spreads its message m_i to a part of the rest nodes, only. Then the next node j is proposed to select uniformly in the set of nodes $V \backslash S_i(k)$ without m_i at the clock tick k.

Due to directed graphs we assume that node $i \in S_i(k)$ may share its message with node j, if there is a directed edge $(i \to j)$ from i to j. We consider probabilities $P_{ij} = 1/d_i$, $P_{ij} = 1/I_i$ and $P_{ij} = 1/O_i$, where I_i and O_i are the in- and out-degree of the node i, respectively. Let us explain the difference between the cases. When I_i and O_i are used the node i initiates the communication to one of its nearest neighbors with in-coming and out-going edges of the node i, respectively. The node i may share its message with the node j only in the case of probability $1/O_i$. The probability of $1/I_i$ is excluded since the message cannot be spread.

2.2 Preferential Attachment

The linear PA schemes [3,7] start with an initial directed graph $G(k_0)$ with at least one node and k_0 edges. For the non-negative parameters α, β, γ such as $\alpha + \beta + \gamma = 1$, and $\Delta_{in}, \Delta_{out}$, the model constructs a growing sequence of directed random graphs $G(k) = (V(k), E(k))$. A graph $G(k)$ is produced from $G(k-1)$ by adding a directed edge. Denote the number of nodes at step k as $N(k)$, and in- and out-degree of node w in the graph $G(k)$ with k edges as $I_k(w)$ and $O_k(w)$. Three scenarios of the edge creation are proposed in [3,7], which are activated by flipping a 3-sided coin with probabilities α, β and γ. The i.i.d. trinomial r.v.s with values 1, 2 and 3 and the corresponding probabilities α, β and γ are generated to select schemes.

- According to α-scheme, add a new node w_{new} and an edge $(w_{new} \rightarrow w)$ with probability α. Choose the existing node $w \in V(k-1)$ with probability $P(choose\ w \in V(k-1)) = \frac{I_{k-1}(w)+\Delta_{in}}{k-1+\Delta_{in}N(k-1)}$.
- According to β-scheme, add a new edge $(w_{new} \rightarrow w)$ with probability β, where both existing nodes w_{new} and w are chosen independently and with probability $P(choose\ (w_{new} \rightarrow w)) = \frac{O_{k-1}(w_{new})+\Delta_{out}}{k-1+\Delta_{out}N(k-1)} \cdot \frac{I_{k-1}(w)+\Delta_{in}}{k-1+\Delta_{in}N(k-1)}$.
- According to γ-scheme, add a new node w_{new} and an edge $(w \rightarrow w_{new})$ with probability γ. Choose $w \in V(k-1)$ with probability $P(choose\ w \in V(k-1)) = \frac{O_{k-1}(w)+\Delta_{out}}{k-1+\Delta_{out}N(k-1)}$.

This means that $N(k) = N(k-1)$ for β-schema and $N(k) = N(k-1)+1$ for the others. These scenarios realize a 'rich-get-richer' mechanism, when a node with a large number of in-/out- edges can likely increase them with a high probability. As mentioned in [3], such model can create multiple edges between two nodes and self loops.

3 Main Results

3.1 Comparison of the Linear PA and SPREAD Algorithm

Despite the linear PA is used for the evolution of directed graphs, we will also use it for the information spreading. Let a message exchange between two nodes starts with the initial directed graph G_0 with N_0 nodes and k_0 edges. We assume that the message which is in disposal of one of the nodes is spreading among a fixed number n of nodes of the network. The nodes are assumed to have

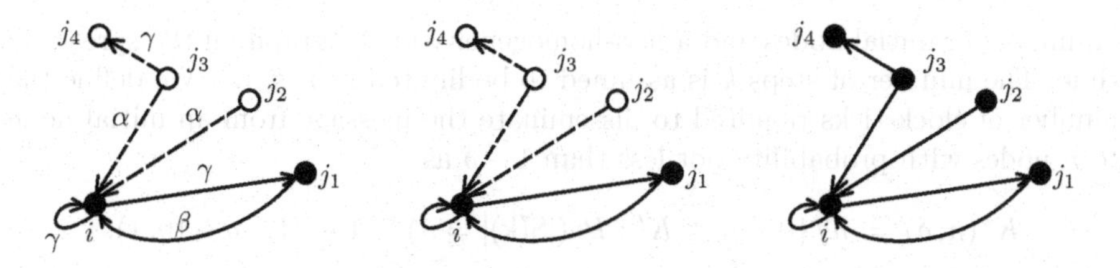

Fig. 1. Example of spreading of the message m_i from the node i by the PA (left), the SPREAD with $P_{ij} = 1/O_i$ (middle) and the SPREAD with $P_{ij} = 1/d_i$ (right). Black filled points mark vertices with the message m_i at step $k = 6$. Doted lines show edges that cannot spread m_i. For the PA, the edges are marked with the names of the schemes which produce them.

asynchronous clocks. We do a step of the linear PA (see, Sect. 2.2) with predefined values of the parameters $\alpha, \beta, \gamma, \Delta_{in}, \Delta_{out}$ by global poissonian clock ticks.

The message m_i of node i can be delivered to a node j without the message only if the directed edge $(i \to j)$ is created. Such edge can be appended to the network by means of $\gamma-$ or $\beta-$ schemes, only. If the node i has no message, then the edge $(i \to j)$ does not spread the message further to the node j. The α-scheme increases the number of appending nodes without the message.

The evolution of the network by the PA schemes in [3] may lead to the appearance of multiple edges and self-loops due to using of the $\beta-$schema. This leads to graphs with cycles. The loops and "bottle-neck" edges may cause a stuck of messages and increase a spreading time.

Example 1. Let us demonstrate the spreading by methods considered above on the small graph. Figure 1 shows that the SPREAD with $P_{ij} = 1/d_i$ can deliver the message m_i from node i to all nodes of the graph since the edge direction is disregarded for the information spreading. The PA and the SPREAD with $P_{ij} = 1/O_i$ spread along directed edges $(i \to j)$, only. The β-scheme leads to multiple edges and self-loops. Thus, the message can circulate between existing nodes having the message beforehand. This leads to the delay. For the SPREAD with $P_{ij} = 1/O_i$ new nodes cannot be selected among nodes with the message, but m_i cannot be spread from nodes without message, e.g., from j_3 to j_4. The PA and the SPREAD with $P_{ij} = 1/O_i$ differ by the number of required rounds and by the order to obtain the message.

At clock tick k we obtain the graph $G(k) = (V(k), E(k))$ with the number of edges $|\mathsf{E(k)}| = k + k_0$ and the number of nodes $|\mathsf{V(k)}| = N(k) + N_0$ nodes.

We compare a spreading ability of the PA schemes and the SPREAD algorithm. To this aim, we first generate a graph by the PA schemes up to step k. Then we apply the SPREAD in the prepared directed graph starting with a node having a message. 100 graphs simulated by the PA are provided for each set of parameters α, β, γ all taken from the interval $[0.04, 0.96]$ with step 0.04 such that $\alpha + \beta + \gamma = 1$, and $\Delta_{in} = \Delta_{out} = 1$.

The graphs evolved start with a triangle of connected nodes and one of these nodes has a message to spread. In this part of the simulation, we do not study an impact of initial nodes and a non-homogeneity of the graph on the spreading time. The number of steps k is assumed to be limited as $k \leq K'$. We define the number of clock ticks required to disseminate the message from an initial node to n nodes with probability not less than $1 - \delta$ as

$$K^*(n, \delta) = inf\{0 < k \leq K' : Pr(|\mathsf{S(k)}| = n) > 1 - \delta\}, \quad \delta \in (0, 1).$$

Let us take K' equal to 3000. If $K^* \leq K'$ holds, then $S(K^*) = n$ is likely held for a sufficiently small δ. If $S(K^*) < n$ holds, then K' steps of the evolution are likely not enough to disseminate the message to n nodes. Results of the comparison of the PA and the SPREAD algorithms for $n = 100$ nodes are presented in Fig. 2.

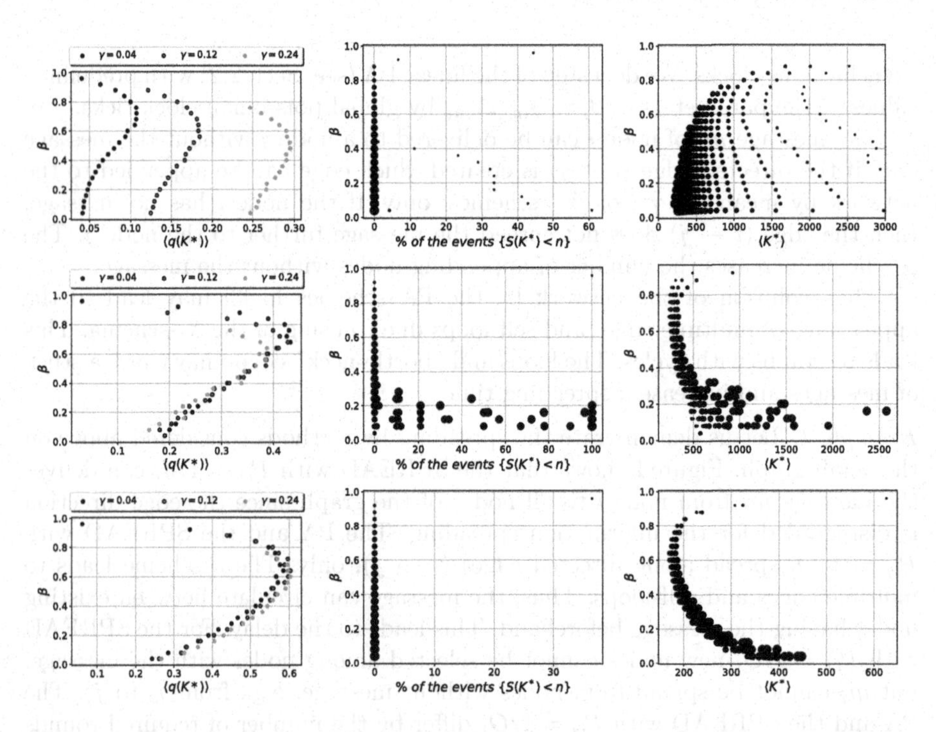

Fig. 2. The β against the $\langle q(K^*)\rangle$ (left column); the parameter β against the proportion of the events $\{S(K^*) < n\}$ (middle column); the β against $\langle K^*\rangle$ (right column) for the PA schemes (top line) and the SPREAD algorithm corresponding to $P_{i,j} = 1/O_i$ (middle line) and $P_{i,j} = 1/d_i$ (bottom line). On the left, the point sizes are increasing with increasing of β values within the interval $[0.04, 0.96]$; in the middle and right figures - with increasing of γ values in the interval $[0.04, 0.96]$.

Let $q(k) = \frac{|\mathsf{S(k)}|}{k}$ be an average number of nodes received the message per tick. The dependence between the averaging clock ticks $\langle K^*\rangle$ and the averaging proportions $\langle q(K^*)\rangle$ over 100 simulations for different γ values as well as the dependence between $\langle K^*\rangle$ and β are shown in Fig. 2 in left and right columns, respectively. For the PA schemes the small or the large β values imply that the information is spreading to newly appended nodes mostly by the γ–schema or the β–schema, respectively. The SPREAD algorithm operates in the graphs created by the PA schemes starting with an initial node having a message and a uniform selection of newly appearing nodes among its nearest neighbors.

The PA schemes spread the information faster for the larger γ, see Fig. 2 (top left and right). The number of ticks is similar for boundary values of β irrespective of γ as shown in Fig. 2 (top left). The events $\{S(K^*) < n\}$ are likely rare except the cases when γ is small, i.e. $\gamma \leq 0.04$, see Fig. 2 (middle top). This implies that $K' = 3000$ ticks are enough to spread the message to $n = 100$ nodes.

In contrast to the PA, the SPREAD can share the message to larger number of nodes irrespective to γ value, namely, the SPREAD strategy with the probability $P_{ij} = 1/d_i$ spreads the message among about 60% of nodes as far as the one with $P_{ij} = 1/O_i$ to 40%.

Comparing both SPREAD strategies, one can see that the choice of $P_{ij} = 1/d_i$ provides the message delivery to all n nodes for any β apart of the single point corresponding to γ not larger than 0.04, see Fig. 2 (middle bottom). For $P_{ij} = 1/O_i$, the number of events $\{S(K^*) < n\}$ grows up for approximately constant $\beta < 0.2$ and $\alpha + \beta > 0.8$, Fig. 2 (middle, the second line). For the same β the number of ticks K^* required to spread the information among n nodes is larger if one uses $1/O_i$ rather than $1/d_i$, see Fig. 2 (right middle and bottom). The minimum $\langle K^* \rangle$ is shown in Table 1. The PA demonstrates the best results for given parameters (α, β, γ).

Table 1. The minimum $\langle K^*(n, \delta) \rangle$ values with the set of parameters (α, β, γ), on which the minimums are received, for the PA and the SPREAD algorithms.

	(α, β, γ)	Minimum $\langle K^*(n, \delta) \rangle$
SPREAD, $1/O_i$	$(0.08, 0.84, 0.08)$	231
SPREAD, $1/d_i$	$(0.2, 0.6, 0.2)$	125
PA	$(0.04, 0.92, 0.04)$	115

3.2 Spreading in Non-homogeneous Graphs

Note that K^* may strongly depend on the choice of the initial node that begins to spread its message. To investigate this problem, we consider a non-homogeneous graph whose nodes belong to communities with different distributions of in- and out-degrees.

To this aim, we simulate three Thorny Branching Tree (TBT) graphs by methodology proposed in [8] with predefined tail indexes (TIs) $(\alpha_{in}, \alpha_{out})$ of in- and out-degree power law distributions with tail distribution function $F(x, \alpha) = P\{X > x\} \sim cx^{-\alpha}$, $c > 0$. The symbol $\sim$ means an asymptotically equal. Simulated graphs are connected by few links to be not completely isolated.

In practice, one can partition a graph into communities and test the distributions of their in- and out-degrees regarding the stationarity.

The TBT is a specific graph with possible cycles such that the sums of in- and out-degrees are the same. Each TBT contains 500 nodes in our experiment. The TBTs are simulated with the following pairs $(\alpha_{in}, \alpha_{out})$: the TBT_1 has $(3.8, 2)$, the TBT_2 - $(2.5, 2.5)$ and the TBT_3 - $(3, 4.5)$. After adding 60 inter-links between the TBTs in such a way to change values of the TIs not much, we estimate the TIs by the Hill's estimator. We obtain that the TBT_1 has $(3.79, 2.06)$, the TBT_2 - $(2.41, 2.61)$ and the TBT_3 - $(3.75, 4.48)$. Then the TBT_1 and TBT_2 have the smallest TIs of the out- and in-degree distributions, respectively. The smallest TI implies the most heavy-tailed distribution. This follows from the properties of regularly varying distributions [9]. Obviously, the distribution of the out-degree is more significant for the spreading than the distribution of the in-degree.

The TI of the TBT node degree, i.e. sum of its mutual independent in- and out-degrees coincides with the TI of the most heavy-tailed term according to properties of regularly varying distributions, [9]. We obtain the following TI $\alpha_{deg} = (2, 2.5, 3)$ and their estimates $\widehat{\alpha}_{deg} = (2.86, 1.99, 3.98)$ for TBT_1, TBT_2 and TBT_3, respectively.

We simulate 50 graphs evolved by the linear PA schemes with the same parameters. Afterwards, the PA and the SPREAD algorithm are used to spread one message from each node belonging to one of the TBT graphs to $n = 100$ other nodes. We examine K^* when the spreading starts from every node of each TBT graph. The following parameters $(\alpha, \beta, \gamma) = (0.4, 0.2, 0.4)$ and $(\alpha, \beta, \gamma) = (0.2, 0.1, 0.7)$, $\Delta_{in} = \Delta_{out} = 1$, $K' = 3000$ are taken. The resulted triples $(\min\{K^*\}, \langle K^* \rangle, \max\{K^*\})$ are presented in Table 2.

Table 2. The triple $(\min\{K^*\}, \langle K^* \rangle, \max\{K^*\})$ shows the minimum, the average and the maximum of K^* values over 50 simulations applying to the SPREAD with $P_{ij} = 1/O_i$ and $P_{ij} = 1/d_i$, and the PA schemes. The row i, $i \in \{1, 2, 3\}$, corresponds to the spreading of one message initiated by each node from TBT_i to other $n = 100$ nodes. Each value of the triple is obtained by the values over all nodes and each simulation.

TBT $(\alpha_{in}, \alpha_{out}, \alpha_{deg})$	SPREAD, $1/O_i$	SPREAD, $1/d_i$	PA
$(\alpha, \beta, \gamma) = (0.4, 0.2, 0.4)$			
$TBT_1(3.8, 2.0, 2.0)$	(1500, 2000, 2500)	(171, 241.1, 330)	(1800, 2325, 2850)
$TBT_2(2.5, 2.5, 2.5)$	(2600, 2700, 2850)	(187, 267.1, 401)	(2150, 2225, 2300)
$TBT_3(3.0, 4.5, 3.0)$	(700, 2012.5, 2850)	(218, 349.7, 669)	(1150, 1883.3, 2350)
$(\alpha, \beta, \gamma) = (0.2, 0.1, 0.7)$			
$TBT_1(3.8, 2.0, 2.0)$	(100, 1566.6, 2800)	(169, 215.1, 289)	(100, 1356.5, 2800)
$TBT_2(2.5, 2.5, 2.5)$	(100, 1498.1, 2950)	(162, 226.3, 333)	(100, 1691.5, 2900)
$TBT_3(3.0, 4.5, 3.0)$	(100, 1531.3, 2900)	(183, 263.4, 421)	(100, 1717.3, 2900)

Comparing the methods, the SPREAD with $P_{ij} = 1/d_i$ demonstrates the triples with the smallest tick numbers which have small variations for all TBTs. The SPREAD with $P_{ij} = 1/O_i$ and the PA work similar, but the PA is faster on the TBT_2 in case $(\alpha, \beta, \gamma) = (0.4, 0.2, 0.4)$. Both SPREAD strategies and the PA show that the best spreading nodes are in the TBT_1 subgraph with the smallest TI value of the out-degree. The PA and the SPREAD with $P_{ij} = 1/O_i$ are more sensitive to the parameter γ than the SPREAD with $P_{ij} = 1/d_i$. The larger γ the smaller the minimum and average of their K^*.

4 Conclusions

We study the linear PA schemes to share one message from one node to a fixed number n of nodes of the network. The information spreading is investigated both for homogeneous (Sect. 3.1) and non-homogeneous (Sect. 3.2) graphs. We compare the PA and the well-known SPREAD algorithm on directed graphs with possible cycles and multiple edges generated by the PA with different sets of parameters.

Considering the homogeneous graphs studied in Sect. 3.1 one may conclude that the SPREAD algorithm that ignores directions of edges spreads the message faster than both the PA and the SPREAD algorithm that takes the edge directions into account. However, the PA may be the best spreader for some sets of its parameters (α, β, γ). Other parameters $(\Delta_{in}, \Delta_{out})$ were taken constant in our simulation study.

Regarding the non-homogeneous graphs (see Sect. 3.2) consisting of the TBT subgraphs with different tail indexes of the in- and out-degrees, we found that the node from the TBT with the smallest tail index of the out-degrees, i.e. having the heaviest tail distribution of the out-degrees, spreads its message faster than nodes from the TBTs with the larger tail index of the out-degrees, i.e. with the lighter tail distribution of the out-degrees. The increasing of the parameter γ leads to the decreasing of the number of clock ticks required to share the message to arbitrary n nodes and hence, to the increasing of the spreading rate.

References

1. Newman, M.E.J.: Networks: An Introduction. 2nd edition, Oxford University Press, Oxford (2018)
2. Avrachenkov, K., Lebedev, D.: PageRank of scale-free growing networks. Internet Math. **3**(2), 207–231 (2006)
3. Wan, P., Wang, T., Davis, R.A., Resnick, S.I.: Are extreme value estimation methods useful for network data? Extremes **23**, 171–195 (2020)
4. Mosk-Aoyama, D., Shah, D.: Computing separable functions via gossip. In: Proceedings of the 25th ACM Symposium on Principles of Distributed Computing (PODC 2006), pp. 113–122. ACM, New York, USA (2006)
5. Censor-Hillel, K., Shachnai, H.: Partial information spreading with application to distributed maximum coverage. In: Proceedings of the 29th ACM Symposium on Principles of Distributed Computing (PODC 2010), pp. 161–170. ACM, New York, USA (2010)

6. Censor-Hillel, K., Shachnai, H.: Fast information spreading in graphs with large weak conductance. In: Proceedings of the 2011 Annual ACM-SIAM Symposium on Discrete Algorithms (SODA), pp. 440–448 (2011). https://doi.org/10.1137/1.9781611973082.35
7. Bollobás B., Borgs C., Chayes J., Riordan O., Directed scale-free graphs. In: Proceedings of the fourteenth annual ACM-SIAM symposium on Discrete algorithms (SODA 2003), Society for Industrial and Applied Mathematics, USA, pp. 132–139 (2003)
8. Chen, N., Litvak, N., Olvera-Cravioto, M.: PageRank in scale-free random graphs. In: Bonato, A., Graham, F.C., Prałat, P. (eds.) WAW 2014, LNCS, vol. 8882, pp. 120–131. Springer, Cham (2014). https://doi.org/10.1007/978-3-319-13123-8_10
9. de Haan, L., Ferreira, A.: Extreme Value Theory: An Introduction. Springer, New York (2006). https://doi.org/10.1007/0-387-34471-3

Machine Learning for Recognition Learning of Control Systems for Autonomous Unmanned Underwater Vehicles of Events in Hostile Environments

Vyacheslav Abrosimov[1] and Ekaterina Panteley[2](✉)

[1] Ministry of Defense of the Russian Federation, Moscow, Russian Federation
[2] Samara State Technical University, Samara, Russian Federation
panteley@kg.ru

Abstract. The paper considers the problem of machine learning for recognizing events that can occur with a control object operating in its environment. Information about the state of the environment is formed by means of technical vision systems as a result of its targeted online monitoring. Such events, triggered by a set of recorded facts and influencing environmental factors, become a training sample. The sample is used for machine learning with deep neural networks. The neural model obtained during the learning process is used in the object control system to recognize and predict potential events and their consequences.

Keywords: Control object · Machine learning · Event · Autonomous unmanned underwater vehicle · Deep neural networks

1 Introduction

Recently, the concept of situational awareness has been actively used in various fields of knowledge. According to the model of M.R. Endsley [1], the concept of situational awareness has three main conceptforming elements: a) information about the surrounding situation in time and space; b) awareness or understanding of the situation and c) forecasting the scenario of situation development in the form of expected events, the user's actions and actions of other participants. In practice, situational awareness systems are implemented in specific models of accumulating, accounting, storing and retrieving information, as well as its analysis and forecasting the consequences of decisions.

Despite all the attention paid to the issues of situational awareness abroad [2–4], this concept is usually considered for people who manage complex systems. Whereas, cases of using these approaches for technical devices are quite rare. In Russian science, the issues of building models of situational awareness are also actively considered [5,6]. However, terminology has not yet fully settled down.

V. M. Vishnevskiy et al. (Eds.): DCCN 2021, CCIS 1552, pp. 293–306, 2022.
https://doi.org/10.1007/978-3-030-97110-6_23

The main efforts in building situational awareness databases are spent on collecting and processing big data in relevant databases (DB). The authors' research area is group management in multiaccent systems.

To ensure situational awareness of the group in the context of occurrence of potential events, the database should be constantly filled with current information about the environment.

Among the most important tasks to be solved by the AUV control system during autonomous navigation and movement under water are obstacle detection, recognition and identification of underwater objects, and obstacle avoidance. For this purpose, AUVs, in addition to scientific and research equipment, have hydroacoustic navigation and communication devices, echo sounders, side-view locators, lighting devices, video cameras and other technical vision systems (TVS) on board. The underwater environment in which AUV operates is considered one of the most dangerous and problematic; it is characterized by nontransparency, turbidity, underwater currents and obstacles, especially in the coastal regions and other areas of active human activity. While AUV is moving along the route, its TVS scans the environment, both in passive and active modes. When an obstacle is detected, it determines its characteristics (bearing, distance, speed, depth, etc.) and analyzes the degree of danger of this object for AUV. If the obstacle is recognized as dangerous, the AUV makes a decision on further actions either independently or together with the person controlling its actions. The obstacle is bypassed by changing the route and/or speed and/or depth. Other situations can also occur with the AUV under water, such as hitting an underwater current, engine failure.

2 Problems of Functioning of an Autonomous Control Object Under Water

Within this study, the authors focused on the well-known and widely used class of underwater monitoring problems. The paper considers insignificant areas of underwater space - up to 2–3 km^2, with a depth of no more than hundreds of meters. Here, to solve the problems of underwater monitoring, the topic related to remotecontrolled and autonomous unmanned underwater vehicles is actively developed [7,8]. Analysis and experience of certain underwater work [9,10] suggests that group work of several devices of this type, especially heterogeneous, in difficult conditions of the underwater environment can become more effective than individual work. Indeed, with the use of technical vision systems, it is possible to increase the degree of situational awareness, provide assistance to each other, both informational and functional, jointly solve the tasks of monitoring and reconnaissance of underwater areas, inspection of various objects, and, if necessary, intervention in underwater activities, etc.

Formation of strategies for AUV underwater behavior has not been widely studied. This is due to several tasks that developers have to solve to create truly autonomous vehicles: solving navigation issues under water in the absence of radio waves [11,12], equipping the AUV with TVS sensors and computing power, accordingly [13,14], and, finally, taking into account local underwater currents and a very aggressive biological environment.

In order to ensure the necessary autonomy, there is currently an urgent task of teaching AUVs to recognize objects, predict their route and choose a strategy of behavior in case of an obstacle.

The paper [15] solves the navigation task using the Underwater docking systems. AUV charging is contactless, therefore, a sufficiently accurate positioning of the device in the charging area is highly important. For this purpose, a sensor system and SLAM technology are used, which are well suited for working at close range. Results of dynamic positioning and visual docking of servo systems are verified by simulation.

Operation in shallow water, where conditions for positioning and navigation are especially difficult due to the large number of obstacles and a small space for maneuvering, is considered in [16]. The paper proposes an intelligent module with the special OP-ELM algorithm - Optimally Pruned Extreme Learning Machine. Credibility of its implementation is verified through modeling of the Flacon AUV using the Matlab software. The proposed algorithm successfully controls the object along the desired route, thanks to the optimal choice of parameters. However, parametric error is estimated at 20.

The problem of planning the AUV route using deep neural networks based on a 3D map is solved in [17]. In the course of the study, a model of the underwater environment has been created to reduce the training time of an AUV as an agent. However, the use of the underwater space model does not take into account many of the obstacles that AUVs encounter in real missions. Detection of static obstacles and learning to bypass them is considered in [18]. Reinforcement learning algorithms are used for training. Such an algorithm does not require an initial training sample, but provides the possibility of autonomous training. The authors have not only simulated the underwater environment and obstacles, but also created an AUV model with a certain set of motor functions in three degrees of freedom.

Learning to make decisions when meeting obstacles is discussed in [19]. AUV receives information about the environment from sensors and forms an obstacle avoidance algorithm. The paper considers three main algorithms used for bypassing tasks: neural networks, fuzzy logic, and reinforcement learning algorithms. The most promising one is an obstacle avoidance algorithm using a 3D representation of the environment.

The problem of group work [20] is more difficult. Cooperation of AUVs and joint search for objects are based on the natural GBNN algorithm: Glasius bio-inspired neural network. The work uses predictive control of a nonlinear model. The efficiency of the proposed algorithm has been compared with the particle swarm optimization (PSO) algorithm. Based on the results of a 50-fold simulation experiment, it has been demonstrated that the GBNN algorithm performs better according to the following criteria: path length, task execution time, energy consumption. All of this confirms the relevance of development of specialized models that would make it possible to train AUV control systems for the tasks of detection, recognition, identification and decisionmaking in an underdetermined and often objectively hostile underwater environment.

3 Environment of the Situational Awareness Databaser

An analysis of practical tasks has shown that during development of a group management strategy, all objects in the group must possess generalized information about the environment and the state of other objects.

The environment can be friendly, neutral, and hostile. It acquires certain properties due to various factors. The first source of such factors is uncontrollable objective environmental conditions that make it difficult to perform a group task (examples: rain, night, fog). The second source is associated with specially organized countermeasures (for example, the defense system of an object being protected from attacks). Due to the object's ability to influence the environment, its negative effect (activity) can be reduced.

The paper [21] considers a general approach to describing the environment. It consists in dividing the environment into fragments that form a kind of spatial "mesh" network. Each area can be described by coordinates of its location, the size of the fragment and associated data tuples or information matrices containing data about the situation in this fragment in three slices: history, current state and prospects. All of them are significant for execution of the collective mission by an object or a group of objects.

The environment is replenished with information coming from the technical vision systems. All the data on fragments of environment are combined into the "Environment" section of the situational awareness database.

Description of objects in the situational awareness database together with pro-posals of including the "Objects" section into the database are presented in [22]. In general, description of objects and environment is sufficient for situational awareness of the group. However, such information requires significant human participation in solving the problems of the second stage - awareness and understanding of the situation. With the general trend towards creation of autonomous control objects, the degree of human participation in the tasks of analysis should be steadily decreasing.

4 Event as a Result of Learning

Control objects can face different events in the environment, and can encounter various situations.

Events are formed by the influencing factors at a given time in the selected fragment of the environment, recorded by the technical vision systems of objects in the form of facts. AUVs solve a variety of practical tasks under water. However, all of them, to one degree or another, require a) detection of objects under water b) recognition of objects c) their identification and d) making decisions and actions.

The process of detecting objects under water consists in organizing a general panoramic view of a certain spatial area. For this, a variety of technical vision means can be used. The most common ones are sensors of acoustic range (sonars) and visible range (cameras).

Sonars are of greater interest for underwater navigation and obstacle avoidance tasks due to their considerable visibility range, which is independent of lighting conditions and water quality. However, experts say that sonars cannot provide high-quality images at close distances; in such cases, cameras are more efficient. At the same time, good operating conditions for video cameras, even with significant resolution under water, are very limited (Fig. 1).

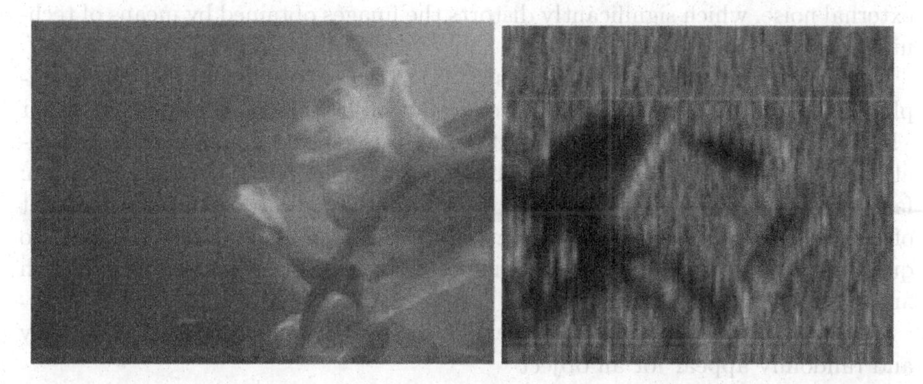

Fig. 1. An example of a low-quality image. View of the underwater environment and objects on the river bed near the railway bridge across the Volga River (Samara) from a video camera and from a side-view sonar (at the distance of 2.5 m).

The tasks of recognizing objects under water are considered classification tasks and consist in assigning the detected object to a certain class. Identification of an unknown object with the known ones from a certain class is carried out based on coincidence of its features. To do this, for each object, group of objects, event, situation and similar phenomena, classification signs are identified and appropriate comparison is carried out. Many methods have been developed for solving discriminant analysis problems, using regressions, decision trees, etc. All of them, to one degree or another, are based on statistical data and selection of informative features of classes.

The task of identifying objects under water is to clearly determine which particular object has fallen into the view of AUV technical means. In particular, for military applications, this is both the well-known aspect of "friend or foe", and the degree of danger of the object for performing the assigned task by AUV. Both on the ground and under water, successful identification requires consideration of the distance to the object, viewing angles and external conditions, which are especially difficult under water. For example, when immersed in a river at just a couple of meters, visibility under water deteriorates considerably. Oddly enough, bright sunlight can also be a negative factor due to the large amount of reflected noise generated by dirt particles. In coastal seas and oceans, the situation is even more difficult.

To solve the above problems of technical vision, machine learning and, in particular, training on neural models, is currently actively used, which is a widespread technology of artificial intelligence. In the available literature on AUV, the number of papers devoted to this topic is quite limited [23–25]. Despite the fact that capabilities of machine learning are quite high, especially when solving classification tasks, the use of neural models is really limited by extremely low quality of information used for training and its incompleteness, while in a number of important applications it is practically completely absence. This is largely due to negative environmental factors and a high level of the so-called external noise, which significantly distorts the images obtained by means of technical vision.

Events can be triggered by factors $F^A(t)$ of the environment itself (for example: factor - unstable weather; fact - reduced visibility of the video camera; event - hitting an obstacle). Moreover, events can be triggered by factors $F^B(t)$ associated with the control objects themselves (for example: factor - engine depletion; fact - engine failure; event - motion stop). In addition, events can occur as a result of active actions of the control object $F^C(t)$ (for example: factor - the need to quickly solve the problem; fact - high speed of movement; event - collision with an obstacle).

An event arises as a result of implementation of a set of factors that externally and randomly appear for an object

$$\sum_{A} F^a \cap \sum_{B} F^b \cap \sum_{C} F^c \rightarrow e \tag{1}$$

The e event, which occurs at the time t^*, can be described by a set of parameters

$$e(t^*) := e_1, e_2, ..., e_n \tag{2}$$

Information about event parameters can be described by numerical, probabilistic, interval characteristics, fuzzy or linguistic variables. The event $e \in E$ is of no interest on its own; it becomes important only as a result of integral influence at a given time t of the factors $F^A(t), F^B(t), F^C(t)$ on the status of the control object $e \Longrightarrow SR(t)$. In practice, the $e \in E$ event is estimated by its attributes: the area sp^e, in which it occurs, the probability of its occurrence p^e, and the degree of danger β_k^e for the r-th control object to perform the set task.

$$e(t^*) := \{sp^e, p^e, \beta_k^e\} \tag{3}$$

Events $e \in E$ can be systematized into classes. The evidence from practive shows that the number of R classes of events for a given environment and typical collective missions can be considered limited. If the event is known, standard and does not affect the operation of control objects, it is simply ignored.

However, often the factors as event sources remain unknown. In opposing active environments, an event can influence the result of a solution to a collective task in different ways.

Therefore, it is advisable to include potential events into the situational awareness database. Then, in addition to the "Objects" and "Environment" sections, the sets of data about events, grouped by classes with estimation of their consequences for control objects, will be combined into the "Events" section of the situational awareness database.

5 Formation of a Training Sample

The training sample for machine learning has two components: real events and simulated events.

Events cannot happen on their own - they are initiated by the state and actions of the control object and manifestation of environment factors. Metainformation about the actions of control objects is systematized in the "Objects" section and certain metainformation of the form

$$A_{object} := \{time, functionused, actionforce, duration\} \quad (4)$$

is associated with objects. Manifestations of environment factors are recorded in the form of facts by means of technical vision systems (video cameras, lidars, etc.), which are also associated with meta-information

$$F_{environment} := \{time, factor, parameters\} \quad (5)$$

Then the event is described by a tuple

$$Event := \{F1_{environment}, .., FN_{environment}, A1_{obj}, .., AM_{obj}\} \quad (6)$$

Occurrence of events does not always obey the laws of probability. It is problematic to obtain statistics of their occurrence in the same conditions, as required by the principles of probabilistic event processing. They are best described in fuzzy categories. With that in mind, it is natural to consider the possibility of event occurrence as a combination of possibilities of manifestation of factors of groups F and A [26].

All events occurring with the object are systematized in the situational awareness database in the "Events" section.

The variety of actions of control objects and environment factors, taking into account conditions for performing missions, is so great that the required statistics amounts to thousands of required images. The solution is on the way of synthesizing model (the so-called "synthesized") events. Their number is practically unlimited, apart from the complexity of creating the corresponding models.

The set of initial data in the form (6), supplemented by synthesized events, constitutes a training sample.

6 Preparation of Initial Data for Machine Learning

Reference database. It should contain reference images of objects that AUVs can encounter under water. At present, the authors do not know not only about the existence of such bases for AUVs, but even information about the very formulation of such a problem. However, it is already clear that the number of such objects is significant, but countable, and the task of creating such a base can indeed be solved.

It is assumed that information contained in the database is labeled by types of its sources (acoustic information, photo and video data, sonar data, etc.) and by object classes. Moreover, it should have a standard format in classes. It is essential, however, that these formats depend on the technical means for which they have been formed; this poses a certain problem for their complete unification.

Sampling of data. To train a neural network, it is necessary to create training and testing data samples. Due to the limited real data for AUVs, it is assumed to use virtual space models and model (or synthetic) data. Modern software tools and well-known physical models make it possible to form reliable virtual spaces and take into account a large number of events that are potentially possible both for an AUV immersed into water and occurring as a result of environmental factors.

Within this study, the training sample is formed from the reference database by means of special spatiotemporal requests related mainly to functionality of AUV. Such a request, according to the authors, should include a) depth of immersion, b) dimensions of objects c) time of obtaining the data d) conditions of the underwater environment, etc.

At the same time, the task of collecting verification data for analyzing learning results, with known responses of AUV and the environment, is still relevant and requires a series of experiments.

Neural model. At the moment, there is a significant number of different architectures of deep neural networks for learning. The choice of a network for the described tasks is currently not nonessential. The only important thing is that there is a possibility of choosing different architectures of neural networks to solve the corresponding given class of AUV tasks, in particular, a classifier (assigning data to given classes), a clusterizer (grouping a set of objects into clusters according to their similarity), etc. In the last 5–7 years, a significant number of frameworks have been created that facilitate development and creation of neural models of various types, in particular, deep neural networks of various architectures.

The neural model is a program that must be "sewn" into the AUV control system. This question refers to the technical issues of data preparation and functioning of the trained neural model on board the AUV, and therefore it goes beyond the scope of this paper. However, there are no fundamental difficulties, limiting the use of machine learning.

AUV model. It is worth emphasizing that to solve the problem of training the AUV control system, its detailed model is not required. Indeed, the key in learning tasks is the equipment that is installed on the object, as well as the standards, type and protocols of information exchange with which such equipment works. Events and situations that AUV may encounter during its mission are actually determined by the environment itself and do not depend on the object moving under water.

AUV missions. Three types of AUV missions are implemented in the virtual environment: monitoring, inspection and intervention. Within such missions, it is possible to perform a large class of scientific and technical tasks, in particular, work on survey sonar detection, mapping and profiling of the seabed relief, conducting search operations with inspection of detected objects, examining bottom structures, trunk pipelines, underwater cable lines, port waters, monitoring sea environments, including burial sites of chemical and explosive substances, inspection and detection of foreign objects, underwater potentially dangerous objects, installation of hydroacoustic markers, searching, capturing and rising to the surface objects of special interest, etc.

Environments. It is necessary to emphasize the variety of environments in which missions can be carried out. These can be large rivers, seas (coastal and remote regions) with a different structure of aquatic environment, actions under ice, in areas of very high or, on the contrary, low human activity, at different depths, etc. All these features lead to the need for synthesis of various synthetic data suited for corresponding environments, and strengthen requirements for the quality of the neural network and its structure in deep machine learning.

Currently, there is very few real information on AUV operation in various underwater environments. In addition, it is not readily available and quite specific. Therefore, acquisition of the so-called synthetic data is required. Synthetic data is obtained as a result of simulation on special models [27]. They make it possible to supplement the training sample with missing data up to the composition required for high-quality training of the neural model.

In [18], methods for generating a training data set are described in sufficient detail. Let us consider their applicability for solving problems of generating synthetic data for training vehicles operating in underwater space.

In software implementation, synthetic data is generated according to a certain algorithm with varying object parameters. However, our experience has shown that this method is applicable only for conditions when the old and new parameters are well distinguishable. In other words, in nontransparent environments, small changes of parameters lead to such blurred perception that the recognition result may classify the object into a completely different class. This method for underwater objects in AUV operation environments seems to be effective only for very simple objects such as underwater buoys at small depths of the southern seas (in Russia, such cases are very limited). However, most underwater objects have a complex configuration (infrastructure of subsea production systems, sunken objects of human activity, etc.) which implies a considerable number of poorly informative options.

In this context, for underwater activities, even sampling (that is, including only those objects that have meaningful elements) does not greatly expand the sample due to fundamental limitation of the initial data set based on results of studying the underwater environment.

To solve this problem of obtaining synthetic data for AUV machine learning, the authors propose to use a hybrid approach. For its application, two main components are required - a virtual 3D environment of the underwater world and typical objects that can be observed by the AUV's technical vision system. The approach itself consists in modifying parameters of a certain set of real typical objects while simultaneously placing them in a virtual environment, modeling its negative factors and showing special data labeling.

The virtual 3D environment developed and presented in this paper simulates a realistic complex environment for various processes of search, survey and functional operations under water, in coastal inland, sea and ocean waters. In this environment, it is possible to simulate the special operating conditions of AUV technical vision systems, in particular, variable visibility depending on illumination, transparency, water impurity, shading of objects in accordance with illumination, splashes of bottom silt sediments, etc.

The list of typical objects that can be observed by AUV technical vision systems is naturally compiled by professionals with experience in various underwater operations. The authors, being specialists of a different profile, have only received limited knowledge from numerous interviews. Nevertheless, at this stage, the following set of objects has been formulated: trunk pipelines and underwater cable lines; farms that have collapsed into water; bridge supports; sunken ships of various types and sizes; unmanned underwater vehicles (of small sizes); sonar buoys; large stones and rocks; port quay walls, etc.

The catalog of underwater objects, developed by the authors, currently includes 14 items. For each object, users can additionally enter up to 3–5 of its types, differing in parameters, mainly in size, shape and color.

The specified objects are introduced into the specified environment using a special workstation in 3D form. By rotating the virtual image of objects in different projections, changing their parameters, introducing various environmental conditions (transparency, shading, immersion into silt, etc.), it is possible to receive a significant number of images of virtual objects that AUV can encounter during the mission.

The sample for training is formed as a mapping of the indicated virtual objects by formats of various AUV technical vision means.

There are, of course, certain difficulties in working with sometimes extremely fuzzy images obtained by technical vision systems when working in an underwater environment, which significantly complicates data labeling. However, these disadvantages become advantages when working with synthetic data due to the fact that, when forming them, researchers clearly understand what object they are currently working with. This makes it possible to develop methods for automated (and in the future, automatic) labelling of synthetic data, which significantly reduces the share of human participation in preparation of data for machine learning.

7 Preparation of Initial Data for Machine Learning

Currently, there is a significant number of ready-made technical solutions and architectures of deep neural networks (see, for example, [28]). Using such solutions, presenting a training sample at the input of a deep neural network, an event classification model will be obtained at the output of the network Fig. 2.

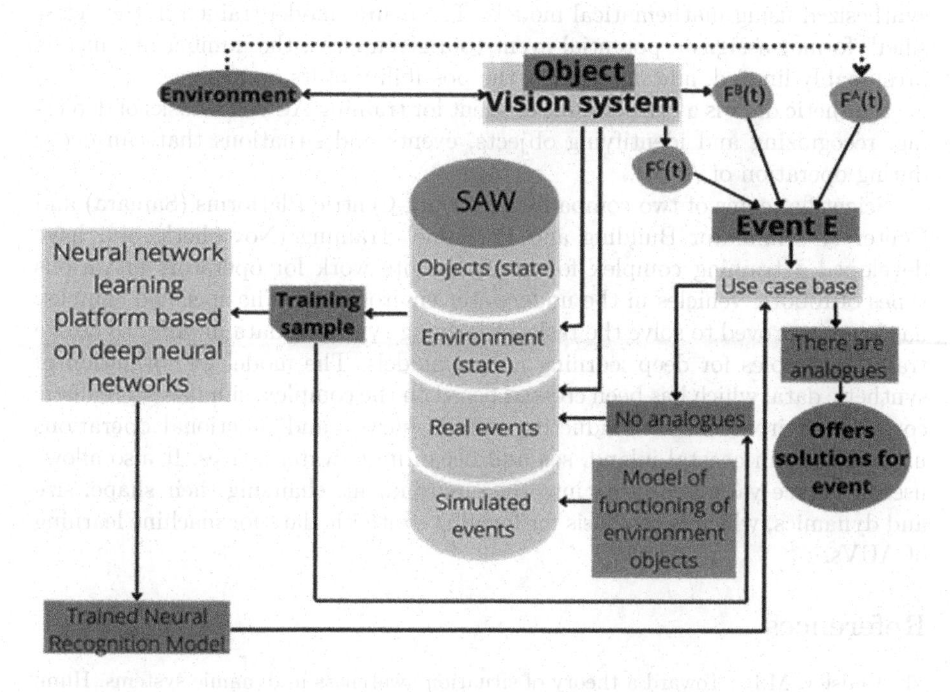

Fig. 2. Machine learning for event recognition

The specified neural model for classification of events is embedded in the object's control system.

When solving problems during fulfilment of the collective mission, the technical vision system of the control object records the facts. Special software compares the observed facts with the situational awareness stored in the database in the online format. When the observed A fact is close, according to a certain set of criteria, it records the possibility of occurrence of events in fact A. Since the A fact during training was among the parameters that determine various events $S_a \ldots S_g$, the possibility of their occurrence is fixed. With further successive coincidences with the observed facts B, C...H, the possibility of occurrence of an event, the A, B, C ... H parameters of which are contained in its meta description, increases until the recognition system issues the decision "high possibility of occurrence of an event S_l from the R-class.

8 Conclusion

The paper, in contrast to the existing traditional solutions for pattern recognition, describes a scientific and methodological approach to machine learning for recognizing events that can occur with a control object in the environment. The paper demonstrates the possibility of forming a training sample based on facts and events observed by the technical vision system, including the ones synthesized using mathematical models. The neural model trained in the "classifier" format assigns a potential event to a given class, the number of which is presumably limited, and determines the possibility of its occurrence.

Synthetic data is a considerable element for training AUVs for tasks of detecting, recognizing and identifying objects, events and situations that can occur during operation of AUVs.

Scientific teams of two companies: Network-Centric Platforms (Samara) and Center for Simulator Building and Personnel Training (Novocherkassk), have developed a training complex for group remote work for operators of various types of robotic vehicles in the underwater environment. The specified complex has been improved to solve the tasks of forming synthetic data in order to create training samples for deep learning neural models. The model for formation of synthetic data, which has been created based on the complex, simulates a realistic complex environment of conducting search, survey and functional operations under water, in coastal inland, sea and ocean underwater spaces. It also allows users to place various objects into the environment, changing their shape, size and dynamics, which is the basis for forming synthetic data for machine learning of AUVs.

References

1. Endsley, M.R.: Toward a theory of situation awareness in dynamic systems. Hum. Factors **37**(1), 32–64 (1995)
2. Salmon, P.M., Stanton, N.A., Walker, G.H., Jenkins, D., Baber, C., McMaster, R.: Representing situation awareness in collaborative systems: a case study in the energy distribution domain. Ergonomics **51**(3), 367–384 (2008)
3. Erlandsson, T., Helldin, T., Falkman, G., Niklasson, L.: Information fusion supporting team situation awareness for future fighting aircraft. Inf. Fusion (FUSION) **13**, 1–8 (2010)
4. Schuster, D., Jentsch, F.: Measurement of situation awareness in human robot teams. http://works.bepress.com/. Accessed 1 Oct 2021
5. Rudianov, N.A., Khrushchev, V.S., Ryabov, A.V., Lapshov, V.S., Noskov, V.P., Rubtsov, I.V.: Increasing situational awareness of units equipped with combat and support robots. In: Materials of the 4th All-Russian Multiconference on Control Problems. Publishing house of TTI SFU 2, pp. 222–225 (2011)
6. Ronzhin, A.L., Basov, O.O., Sokolov, B.V., Yusupov, R.M.: Conceptual and formal models for the synthesis of cyber-physical systems and intelligent spaces. Izvestiya vysshikh uchebnykh zavod **59**(11), 897–905
7. Matvienko, Y.V., Inzartsev, A.V., Kiselev, L.V., Shcherbatyuk, A.F.: Prospects for increasing the efficiency of autonomous underwater robots. Izvestia YuFU. Technical science. Section II. Marine robotics, pp. 123–141 (2016)

8. Martynova, L.A., Mashoshin, A.I.: Construction of a control system for autonomous un-manned underwater vehicles based on multi-agent technology. Izvestiya YuFU. Tech. Sci. **2**, 38–48 (2016)
9. Zanin, V.Y., Kozhemyakin, I.V., Potekhin, Y.P., Putintsev, I.A., Ryzhov, V.A., Semenov, N.N., Chemodanov, M.N.: Development of microclass autonomous unmanned underwater vehicles with group control function. In: Izvestia, Y.FU (ed.) Technical science. Section II. Marine robotics, pp. 55–74 (2017)
10. Abrosimov, V.K., Mochalkin, A.N., Trusilov, V.T., Panteley, E.: Hydroacoustic studies of the underwater part of bridge supports. Railtrack and track facilities, In print (2021)
11. Horgan, J., Toal, D., Ridao, P., Garcia, R.: Real-time vision based AUV navigation system using a complementary sensor suite. IFAC Proc. **40**(17), 373–378 (2007)
12. Xiaokai, M., He, B., Zhang, X., Song, Y., Shen, Y., Feng, C.: End-to-end navigation for autonomous underwater vehicle with hybrid recurrent neural networks. Ocean Eng. **194**, 06602 (2019)
13. Bayat, M., Aguiar, A.P.: SLAM for an AUV using vision and an acoustic beacon. IFAC Proc. Vol. **43**(16), 503–508 (2010)
14. Peng-Fei, L., Bo, H., Jia, G., Yue, S., Tian-Hong, Y., Qi-Xin, S.: Underwater navigation methodology based on intelligent velocity model for standard AUV. Ocean Eng. **202**, 107073 (2020)
15. Wang, T., Zhao, Q., Yang, C.: Visual navigation and docking for a planar type AUV docking and charging system. Ocean Eng. **224**, 108744 (2021)
16. Shen, C., Yang, S.: Distributed implementation of nonlinear model predictive control for AUV trajectory tracking. Automatica **115**, 108863 (2020)
17. Sun, Y., Ran, X., Zhang, G., Xu, H., Wang, X.: AUV 3D path planning based on the improved hierarchical deep Q network. J. Mar. Sci. Eng. **145**(145) (2020)
18. Li, W., Yang, X., Yan, J., Luo, X.: An obstacle avoiding method of autonomous underwater vehicle based on the reinforcement learning. In: 2020 39th Chinese Control Conference (CCC), Shenyang, China, pp. 4538–4543 (2020)
19. Hui, L., Guo, Y.: Intelligent obstacle avoidance algorithms for autonomous underwater vehicle. J. Phys: Conf. Ser. **1693**, 012223 (2020)
20. Cao, X., Sun, H.: Gene Eu Jan: multi-AUV cooperative target search and tracking in unknown underwater environment. Ocean Eng. **150**, 1–11 (2018)
21. Abrosimov, V., Mochalkin, A.: Collective behavior strategy development based on friendship of robots. In: Proceedings of 4th International Conference on Mechatronics and Robotics Engineering, Valenciennes, France, pp. 38–42 (2018)
22. Gorbachenko, V.I.: Intelligent systems: fuzzy systems and networks: a textbook for universities, 2nd edn. Yurayt Publishing House, Rev. and add Moscow (2019)
23. Guo, Y., Qin, H., Bin, X., Han, Y., Fan, Q.-Y., Zhang, P.: Composite learning adaptive sliding mode control for AUV target tracking. Neurocomputing **351**, 180–186 (2019)
24. Himri, K., Ridao, P., Gracias, N., Palomer, A., Palomeras, N., Pi, R.: Semantic SLAM for an AUV using object recognition from point clouds. IFAC-PapersOnLine. **51**, 360–365 (2018)
25. Woolfrey, J., Wenjie, L., Vidal-Calleja, T., Liu, D.: Clarifying clairvoyance: analysis of forecasting models for nearsinusoidal periodic motion as applied to AUVs in shallow bathymetry. Ocean Eng. **190**, 106385 (2019)
26. Kaftannikov, I.L., Parasich, A.V.: Problems of forming a training sample for machine learning. Bulletin of YuUrGU, Series "Computer technologies, control, radio electronics" **16**(3), 15–24 (2016)

27. Abrosimov, V.: Situational awareness formation for large network-systems. Adv. Syst. Sci. Appl. **18**(3), 29–34 (2018)
28. Fedulin, A.M., Gorbatsevich, V.S., Osadchuk, A.V.: Conceptual approach to the creation of ground infrastructure for machine learning of technical vision systems for unmanned aircraft. All-Russ. Sci. Tech. J. "Polet" **11**, 32–38 (2020)

Statistical Model of Graph Structure Based on "VKontakte" Social Network

A. A. Kislitsyn[1] and Yu. N. Orlov[2(✉)]

[1] Keldysh Institute of Applied Mathematics, Miusskaya sq., 4, Moscow 125047, Russia

[2] The Peoples' Friendship University of Russia, Miklukho-Maklaya str. 6, Moscow 117198, Russia

Abstract. The paper presents a study of the graph network structure formed by the friendly connections of the "VKontakte" social network between the cities of Russia. It turned out that the graph node degree distribution is close to uniform. The consequence of this is the existence of a high-dimensional fully connected region and thin periphery. Also, the probability of fully connected structures was estimated for dense and sparse areas of the graph.

Keywords: explanatory dictionary · strongly connected component · clusterization · semantic analysis

1 Introduction

Nowadays, analysis of social networks graph is a very popular mathematical problem, which also has an important theoretical aspect associated with the development of algorithms for fast search of certain configurations. The theoretical side of the problem is interesting primarily because at a present time there are no proofed statements regarding the statistical properties of large connected graphs that allow to build efficient search algorithms. Most of these algorithms are associated with model graphs, or based on heuristics that are checked only after the fact. As examples of theoretical works that discuss the main characteristics of social network graphs and algorithms that develop them, is pointed out in [1], work [2] is about algorithms for visualization and analysis of large graph structures, as well as work [3] with a detailed analysis of methods for studying network structures.

In this paper the structure of connections in undirected graph representing a social network of "friendship" between residents of different territorial units (conditionally "cities") of Russian Federation will be analyzed. The data for the analysis was taken from the VKontakte resource in 2015 year, was processed in [4] and described in [5,6]. Before identifying and analyzing the features of such communication, depending on the individual parameters of individual "subscribers" (gender, age, education, etc.), it is necessary to determine whether there are

V. M. Vishnevskiy et al. (Eds.): DCCN 2021, CCIS 1552, pp. 307–317, 2022.
https://doi.org/10.1007/978-3-030-97110-6_24

purely territorial features of the graph that allow us to highlight some priori structural blocks in it.

The structure of the graph is characterized by a structural matrix with elements A_{ij} that are equal to one if there is a connection between the nodes Γ_i and Γ_j, and zero if there is no connection. Since the graph is undirected, and the intracity network is not considered, the matrix A_{ij} is symmetrical, and there are zeros on its main diagonal. The number of edges Γ_i, leaving a given node, is called the degree n_i of that node. The degree of the node is calculated using the formula $n_i = \sum_{j=1}^{N} A_{ij}$.

The analysis revealed the following features of the considered graph:

1. The graph is connected.
2. The number of nodes (i.e. cities) is equal $N = 2441$.
3. The number of edges $K = 1876564$, average density $\rho = \frac{2K}{N(N-1)} \approx 0,63$. This density is equal to the probability that two random nodes of the graph are connected.
4. The depth of the graph is equal to $H = 1$: there are several nodes (7, exactly), which are directly – in one connection – connected to all other nodes.

Fig. 1. Schematic representation of graph connections

Figure 1 shows clusters of points in the central region with a large number of connections, as well as the existents of a sparse periphery with a relatively small number of connections. Nevertheless, the peripheral points have a rather large number of connections with the central nodes, although they are almost not connected with each other, and in the center of the graph there is a typical picture for the full connectivity.

Since $\sqrt{2K}/N \approx 0,8$, the graph is not complete, but it is very dense, which makes it difficult to search for clusters inside it, i.e. such subgraphs, the density of which would be higher than the average density of the graph, and the density of the remaining part would be lower than the average density. If the graph has a complete subgraph, then any node subsets of that subgraph form a community. In this case, the interest to select the most complete independent subgraphs in the original graph. For this purpose, we use the statistics of the nodes degree.

2 Statistical Properties of the "VKontakte" Network Graph

Let us examine the statistics of the graph nodes degree. The following statistics are of interest: the dependence of the node degree and its rank when ordering nodes in decreasing degrees (Fig. 2) and the distribution of nodes by ranks itself (Fig. 3).

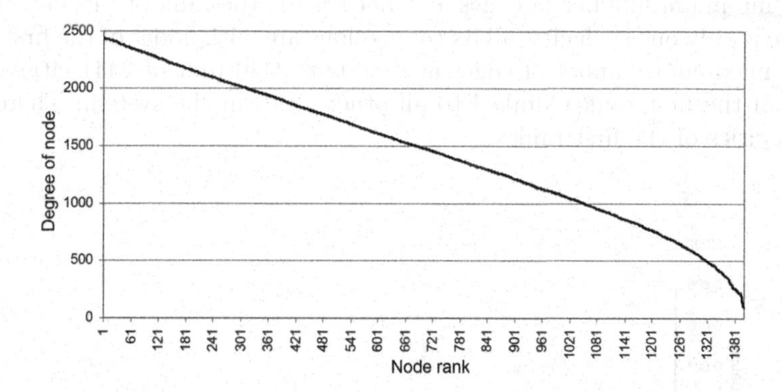

Fig. 2. Dependence of the degree of the graph node from its rank

The graphic in Fig. 2 is weakly nonlinear, the difference from the straight line is noticeable only in the first seven and about one and a half hundred points of last ranks. The total number of cities in these fragments is 55 and 180, respectively; the number of connections (including possible repeats due to double counting of peaks) is 130 thousand in the cities of the first ranks and 65 thousand in the cities of the last ranks.

The dependence of the degree n of a node on its rank r is approximately expressed by the formula

$$n = (N-1)\left(1 - \frac{r-1}{\alpha^2 (N-1)}\right)^{\alpha} \qquad (1)$$

where N is the number of nodes of the graph, and the numerical parameter α is close to 3/4 (its estimation in this case is $\alpha \approx 0,7628$). The root-mean-square deviation of approximation (3) from the quasilinear part of the curve in Fig. 2 is 0,08.

The statistical characteristics of this distribution are as follows:

1. The most probable value of the connections number for a city is 2439, there are 15 of such cities in the system, the corresponding rank of these cities is 2.
2. The median of the distribution is 1550 links; rank 645 corresponds to it.
3. The average node degree is 1537, which corresponds to the rank of 655.
4. The standard deviation of the node degree is 732.
5. The ordinal number of the city in ascending rank, which coincides with the number of connections of this city, is equal to 1423, the corresponding rank is equal to 745. This city number can be called an analogue of the "Hirsch index" of the cities system.
6. The stationary point of the distribution of the degree of the nodes is equal to 1518, which is close to the median and is typical for a uniform distribution. This point corresponds to the rank of the city, equal to 670. Note that the distribution function of nodes by rank is indeed close to linear.
7. The minimum number of edges at a node is 97, the rank of this city is 1400, there is only one such city, all its connections are with nodes of the first ranks.
8. The maximum number of edges at a node is 2440 (out of 2441 cities), i.e. a city of the first rank is linked to all other cities in the system. There are 7 such cities of the first rank.

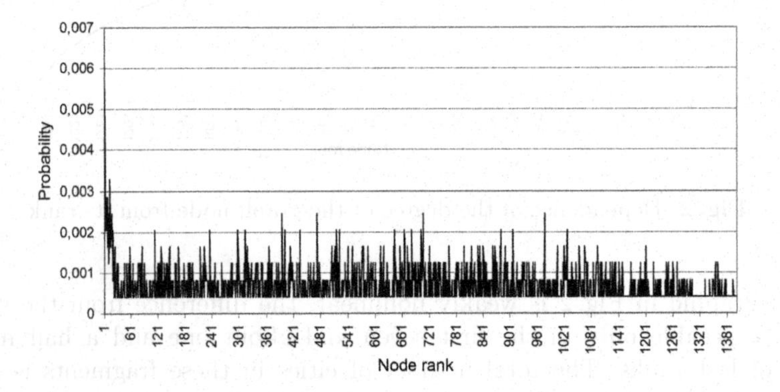

Fig. 3. Distribution of graph nodes by rank

Described statistical properties of this graph differ significantly from the properties of a sparse strongly connected component of a directed graph, which corresponds, for example, to the structure of links within an information hierarchical resource [7]. This difference is important because it allows to discover that different "self-organizing" structures, depending on the density, can have completely dissimilar properties, which requires the development of specific search algorithms in each specific case.

3 The Existence of Fully Connected Subgraphs

From the analysis of the structure of the graph G it follows that the graph is very dense related. It is important to highlight that there is no narrow part along which it could be cut into two disconnected subgraphs of approximately the same size. To separate the periphery (and this is only about 10% of the nodes of the graph) from the core, it is necessary to break about 75 thousand edges.

In addition, since there is a complete subgraph consisting of approximately half of the nodes of the original graph, it is impossible to break this dense part: removing any vertex from this subgraph leaves it complete. Moreover, if we delete this entire complete subgraph, then the remaining half of the cities will still remain connected, although not completely.

Note also that the cities of the first ranks are interconnected, i.e., they form a complete subgraph. The number of cities with which the cities of the first ranks have no connection increases linearly with increasing rank, which makes it possible to estimate the maximum complete subgraph by the maximum number of connections of the last city in terms of rank. This number is approximately estimated from the linear relationship of the degree of the nodes with the ordinal number of the city. As the rank increases, the degree of the nodes, as shown above, decreases almost linearly. In this case, it is convenient to consider the ideal model of the represented system of friendly ties in the form of a linear dependence of the number of ties on the rank (number) of the node.

Consider a model of uniformly distributed degrees of nodes. Differing at the ends from the rank distribution (1), in the main part of the graph it approximately adequately describes its statistical properties.

So, let us assume that the nodes are numbered in descending order of the number of available links, linearly depending on the number. The first node has the maximum number of links $N-1$, and the last (numbered N) has the minimum, equal to N_{min}. Then the number of connections at the node with number k is equal to

$$n\left(k\right)=N-1-\left[\beta\left(k-1\right)\right],\ \beta=1-\frac{N_{min}}{N-1} \tag{2}$$

Here square brackets mean the integer part of the number, and the coefficient β in our case is approximately equal $\beta = 0,96$.

Formula (2) does not completely determine the structure of the graph: it is also necessary to indicate the typical rules for connecting nodes. We distinguish three areas of the graph in accordance with the structure of the incidence matrix. The first area corresponds to nodes with a large number of links; the decrease in the number of connections occurs due to the exclusion of cities of the third region with a small number of connections from the friendship orbit. This area conventionally extends to the border of the most complete subgraph. The third region of cities with a small number of friendly ties is characterized by a zero-diagonal submatrix, indicating that cities with a small number of ties are friends not with each other, but with the cities of the first ranks. Between these two areas is the middle – the second zone, where there is a transition from the center to the periphery. Ideally, this layer does not exist.

Let us estimate the dimension of the maximum complete subgraph, assuming that the second zone is absent. Let the required dimension be k_t. Then the number of edges of such a simplified graph is equal, as it is easy to calculate, to the value

$$K = \frac{k_t\left(k_t-1\right)}{2} + \left(N-k_t\right)\left(2k_t-N\right) + \frac{\left(N-k_t\right)\left(N-k_t+1\right)}{2} \tag{3}$$

For simplicity, ignoring 1 in comparison with much larger N and k_t and introducing a value $x = k_t/N$, we obtain from this the equation for x:

$$x^2 - 2x + \frac{1+\rho}{2} = 0 \tag{4}$$

where $\rho = 2K/N^2$ (graph density). Therefore, the maximum dimension of a complete subgraph is

$$k_t = N\left(1-\sqrt{\frac{1-\rho}{2}}\right) \tag{5}$$

For the system of cities under consideration, we obtain $x = 0,57$ and $k_t = 1390$. This subgraph takes into accounts for about 970 thousand edges, or 52% of all connections in the graph. In the ideal model, there are no other complete subgraphs disjoint with this subgraph.

In a real system, there is a non-empty interlayer in the form of a second zone, where are sporadically scattered complete subgraphs of small dimensions (3, 4, 5, ...), which formed by nodes with an intermediate number of links. These intermediate nodes are connected not only with the center of the graph, but also with each other. This reduces the actual dimension of the zero submatrix in the last ranks. To estimate it, we can calculate the complete low-dimensional subgraphs for the nodes of the last ranks.

Consider the probability of connectivity for three nodes. Random pair is connected with each other with probability $p_2 = \rho$. As for a particular node with the number of connections $n\left(k\right)$, according to (2), if all its connections are in the first zone, the nodes of which are obviously connected with each other, then the probability of forming a triangle with nodes with numbers greater than k_t is equal to zero, and on average probability over the graph for a given node is equal $n\left(n-1\right)/\left(N\left(N-1\right)\right)$, i.e. approximately equal to the square of the pair's probability. In Fig. 4 shows the dependence of the probability of forming a triangle from the nodes, numbers of which do not exceed the indicated on the abscissa axis.

From Fig. 4 it can be seen that the last two hundred peaks (in fact, the periphery, highlighted in paragraph 3) are not connected with each other by friendly ties, and noticeable differences begin after the fourth hundred. The probability grows exponentially, proportionally $\exp^{0,01(2040-n)}$, reaching at $n = k_t$ magnitude $2 \cdot 10^{-3}$. On average, for a node from the peripheral region, the probability

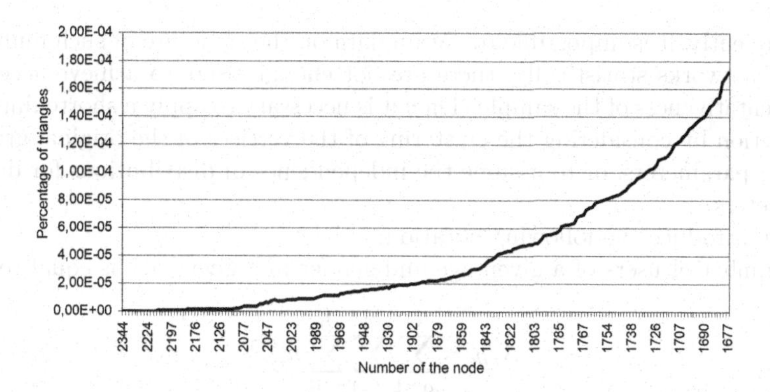

Fig. 4. The likelihood of triple friendly ties, depending on the number of the node, opening the "transition" second zone

of being connected with any node of the graph is 0,45. If all connections of the periphery are in a complete subgraph, then the probability of a triangle, one of the nodes of which lies in the periphery, and two in a dense region, is equal to the square of the indicated probability, i.e., 0,222. As shown by the calculation in Fig. 4, the difference from this value due to triangles in the periphery is 1%, i.e., negligible. Thus, the applied model of the graph G in the form of two zones – a complete subgraph and the periphery associated only with it – is quite adequate.

Note here that although friendships between cities are being studied, this graph is very different from the so-called "friendly relations graph" [8], in which any two nodes have exactly one common neighbor. This shows how far apart can be theoretical and practical objects that have the same semantics.

4 The Analysis of Connections in Multidimensional Vertices

Above, we considered only territorial connections when the "city" parameter is mapped to the vertex. In real social networks, vertices also depend on other attributes, i.e. they are multidimensional. Consider the following relatively simple model. Let each vertex at the level of an individual user be defined by three parameters denoting, r, a, g, respectively, the region of residence, age and gender. The totality of these characteristics of all network users at a given time forms a three-dimensional discrete parameter space, the sampling step is determined by the scale of the measurement tool. The dimension of the totality of user characteristics for each of the three parameters is denoted accordingly $N_1 = N = 2551, N_2 = 100, N_3 = 2$

Let's enter the number of connections between users combined into the specified classes: $n_{ijk}^{\alpha\beta\gamma}$ is a number of users connections between gender i age j and region k and users with gender α age β from region γ. The potential number of adjacency matrix cells requiring filling is estimated by $(N_1N_2N_3)^2 \approx 10^{11}$.

Consequently, it is impossible to obtain data on the structure of such multiparametric networks statistically: there are not enough users to achieve acceptable representativeness of the sample. Then it is necessary to apply a shortening of the description by considering the clustering of the vertices of the original graph by a set of parameters, or to assume the independence of distributions for different parameters.

We introduce the following notation.

The number of users of a given age and gender in a given city is equal to

$$N_{ijk} = \sum_{\alpha=1}^{G} \sum_{\beta=1}^{A} \sum_{\gamma=1}^{R} n_{ijk}^{\alpha\beta\gamma} \tag{6}$$

The number of users of a given gender and age, regardless of the city of residence, is

$$K_{ij} = \sum_{k=1}^{R} N_{ijk} \tag{7}$$

Similarly, the sums

$$M_{ik} = \sum_{j=1}^{A} N_{ijk}, \qquad L_{jk} = \sum_{i=1}^{G} N_{ijk} \tag{8}$$

Are determined the number of users with given gender i or given age j in definite region k. Summarizing by all indexes we achieve the whole number N of users. Normalized by N entities (6–8) represent empirical estimations of the probability density of the distribution of connections between subscribers in a given range of vertex parameters.

If we summarize number of connections $n_{ijk}^{\alpha\beta\gamma}$ by gender and age, we get the number of connections between cities k and γ, which we denote by ω_k^γ

$$\omega_k^\gamma = \sum_{i,j,\alpha,\beta} n_{ijk}^{\alpha\beta\gamma} \tag{9}$$

Similarly, if we summarize amount of links $n_{ijk}^{\alpha\beta\gamma}$ by gender and city, we get the number of connections between users of certain ages:

$$f_j^\beta = \sum_{i,k,\alpha,\gamma} n_{ijk}^{\alpha\beta\gamma} \tag{10}$$

We also introduce a matrix of relationships by gender:

$$g_i^\alpha = \sum_{i,k,\beta,\gamma} n_{ijk}^{\alpha\beta\gamma} \tag{11}$$

To study the relationships between network parameters, it is necessary to study the two-parameter distribution of connections - by gender and age, by gender and city, by age and city. We can determine them, respectively, by the formulas:

$$\begin{aligned} b_{ij}^{\alpha\beta} &= \sum_{k,\gamma} n_{ijk}^{\alpha\beta\gamma}, \\ c_{ik}^{\alpha\gamma} &= \sum_{j,\beta} n_{ijk}^{\alpha\beta\gamma}, \\ d_{jk}^{\beta\gamma} &= \sum_{i,\alpha} n_{ijk}^{\alpha\beta\gamma} \end{aligned} \tag{12}$$

The simplest way to model such a structure is to assume that the distributions are independent, then we can assume that

$$n_{ijk}^{\alpha\beta\gamma} = \frac{1}{N^2} g_i^\alpha f_j^\beta \omega_k^\gamma \tag{13}$$

Factor $\frac{1}{N^2}$ in (13) introduced to preserve the normalization for the total number of network subscribers. Let us investigate to what extent the assumption can be admitted (13). To do this, we construct a model for the evolution of the distribution of the graph nodes by the degree of their multidimensional parameters, in particular, by age.

In order to correctly analyze the age structure of the network, it is necessary to consider the move according to the age characteristic, in other words, to correct the change in the observed connections considering the demographic equation. In a short-term analysis, the physical mortality rates per time unit (year) can be considered constant and equal to the average for the population in a given region, depending on gender and age. Denote $q_{ik}(j)$ the likelihood of death for a person of age j with gender i in region k. In differential form, considering age as a continuous parameter, the demographic equation of population changes $R_{ik}(x,t)$ without taking into account the birth rate has the form

$$\frac{\partial R_{ik}(x,t)}{\partial t} + \frac{\partial R_{ik}(x,t)}{\partial x} = -q_{ik} R_{ik}(x,t) \tag{14}$$

Then the change in population $R_{ik}(x,t)$ by age j in a given region with a one-step time shift equal to a year is described by the equation:

$$R_{ik}(j, t+1) = (1 - q_{ik}(j-1)) R_{ik}(j-1, t) \tag{15}$$

If we now assume that the demographic characteristics are the same both for subscribers of social networks and for the rest of the population of the region, and the subscribers themselves do not change their profile over time, then, due only to the natural course of events, the change in the distribution of paired connections will be described by the equation:

$$
\begin{aligned}
n_{ijk}^{\alpha\beta\gamma}(t+1) = n_{i(j-1)k}^{\alpha(\beta-1)\gamma}(t) - q_{ik}(j-1)\sum_{\lambda\mu\eta} n_{i(j-1)k}^{\lambda\mu\eta}(t) \\
- q_{\alpha\gamma}(\beta-1)\sum_{rsm} n_{rsm}^{\alpha(\beta-1)\gamma}(t) \\
+ q_{ik}(j-1)q_{\alpha\gamma}(\beta-1)\sum_{\lambda\mu\eta} n_{i(j-1)k}^{\lambda\mu\eta}(t)\sum_{rsm} n_{rsm}^{\alpha(\beta-1)\gamma}(t)
\end{aligned}
\tag{16}
$$

Summing expression (16) over one or another pair of indices will give an equation for the evolution of a distribution that depends on a smaller number of parameters, which will make it possible to determine the correctness of representation (13) on factorization as applied to an evolutionary problem. Let us consider, in particular, the evolution of the distribution of the nodes degree for age connections, i.e. value $f_j^\beta(t) = \sum_i ik\alpha\gamma n_{ijk}^{\alpha\beta\gamma}(t)$. From (16) receive:

$$
\begin{aligned}
f_j^\beta(t+1) = f_{j-1}^{\beta-1}(t) - \sum_{ik} q_{ik}(j-1)N_{i(j-1)k}(t) \\
- \sum_{\alpha\gamma} q_{\alpha\gamma}(\beta-1)N_{\alpha(\beta-1)\gamma}(t) \\
+ \sum_{ik\alpha\gamma} q_{ik}(j-1)q_{\alpha\gamma}(\beta-1)N_{i(j-1)k}(t)N_{\alpha(\beta-1)\gamma}(t)
\end{aligned}
\tag{17}
$$

It follows from (17) that summation over the indices on the right-hand side of this equation is possible only if the mortality rates are constants that do not depend on the indices. Then we get that the evolution of age relationships is split off from the rest of the parameters of users and can be analyzed independently:

$$
\begin{aligned}
f_j^\beta(t+1) = f_{j-1}^{\beta-1}(t) - q\sum_{\lambda} f_{j-1}^{\lambda}(t) \\
- q\sum_{m} f_m^{\beta-1}(t) + q^2\sum_{\lambda m} f_m^{\beta-1}(t)f_{j-1}^{\lambda}(t)
\end{aligned}
\tag{18}
$$

Otherwise, the distribution of age-connections of subscribers turns out to be dependent on gender and region. Therefore, assumption (13) is generally incorrect and it is necessary to analyze the structures of graphs in order to cluster users with similar types of connections.

In particular, the distribution of the nodes of the "age" graph by degrees is not uniform, in contrast to the distribution of connections by regions. Most likely the number of connections between fifty ages. Relations with less than 40 connections and more than 60 connections make up about 15% of the distribution. Interestingly, even with 100 million users, none of the peaks is associated with all age peaks. The greatest coverage of friendship by age is roughly 0.6 in the age range, while urban ties, as we have seen, have examples of full coverage. The density of this graph is $\rho_A = 0,487$.

The adjacency matrix of an "age" graph has a significantly different structure than the adjacency matrix of an "urban" graph. If a city graph has a large fully connected core, then in an age graph the maximum dimension of a complete subgraph is 7, there are quite a few such subgraphs and they are weakly connected through one or two vertices.

The actual number of connections by age and city depends on the ranks of cities k and γ. If we fix the rank of the city and sum up all the connections for the rest of the cities, we get the age distribution of connections for the city k. For a city with a large number of connections (first ranks in the city system), the corresponding adjacency matrix forms a connected graph

Also, note that despite a fairly large number of users even in the city of the last rank, only the first 55 cities of a fully connected urban core have connections between all ages. Cities of subsequent ranks have fewer links between fewer ages. This decrease is linear with rank.

5 Conclusion

As a result of the analysis of the "VKontakte" network of friendly ties between the cities of the Russian Federation, the following conclusions can be drawn.

1. The graph of connections is dense and does not contain bottlenecks. It consists of a fully connected core, consisting of about half of the cities, and associated cities in the periphery, which are almost not connected with each other.
2. The distribution of nodes by the number of links is almost uniform, which allows an analytically estimate the dimension of the complete subgraph and of the periphery.
3. Cumulatively, the periphery is connected with all the nodes of the graph, so that the option with cutting the graph with the minimal, in terms of the number of deleted nodes, is absolutely ineffective, since after removing about 200 cities from the periphery, a transition zone and a complete subgraph remain.

References

1. Batura, T.V.: Methods for analyzing computer social networks. Vestnik NSU **10**(4), 13–28 (2012)
2. Kolomeichenko, M.I., Chepovsky, A.N.: Visualization and analysis of large graphs. Bus. Inform. 4, 7–16 (2014)
3. Gusarova, N.F.: Social media analysis, p. 67. ITMO, Basic concepts and metrics // STP (2016)
4. The virtual population of Russia. http://webcensus.ru
5. Chekmyshev, O.A., Yashunsky, A.D.: Extraction and usage of online social network data. Keldysh Institute Preprints, No. 62, 16 (2014). http://library.keldysh.ru/preprint.asp?id=2014-62
6. Zamyatina, N.Y., Yashunsky, A.D.: Virtual geography of the virtual population. Opin. Monitoring: Econ. Soc. Change **143**(1), 117–137 (2018)
7. Kislitsyn, A.A., Orlov, Y.N.: Structure of semantic connections of words of the Russian explanatory dictionary. Keldysh Institute Preprints (252), 28 (2018). http://library.keldysh.ru/preprint.asp?id=2018-252
8. Erdosh, P., Renyi, A., Sos, V.T.: On a problem of graph theory. Studia Sci. Mat. Hungar **1**, 215–235 (1966)

Response Time Estimate for a Fork-Join System with Pareto Distributed Service Time as a Model of a Cloud Computing System Using Neural Networks

A. V. Gorbunova[1(✉)] and A. V. Lebedev[2]

[1] V. A. Trapeznikov Institute of Control Sciences of Russian Academy of Sciences, 65, Profsoyuznaya Street, Moscow 117997, Russian Federation

[2] Lomonosov Moscow State University, GSP-1, Leninskie Gory, Moscow 119991, Russian Federation

Abstract. A cloud computing system that receives complex user tasks involving several subtasks is studied from the point of view of the response time. In order to reduce the service time, the tasks are divided into smaller components and processed in parallel. As a cloud center model we use a fork-join queuing system with Pareto distribution of the service time on the servers. To analyze the mean response time and its standard deviation, a new approach is used combining simulation modeling with one of the machine learning methods. The estimates obtained are much more accurate than the earlier analytical results on fork-join systems.

Keywords: Cloud computing · Parallel computing · Queuing system · Parallel processing · Mean response time · Artificial neural networks · Machine learning

1 Introduction

The present study is a continuation of the research started in [1]. The main goal, as before, is to evaluate the key performance characteristics of a cloud computing system, but in the context of building a more realistic analytical model by expanding the class of service time distributions within the previously selected queuing model, which will be discussed in more detail below.

A cloud center is a collection of several physical servers that can be used together to process complex user tasks through the use of virtualization technology. Here, due to the division of a complex task into smaller components and their further parallel processing, there is an increase in the processing rate speed, and accordingly, an improvement in the quality of service for users of the cloud system.

A fork-join queuing model is used for estimation of the mean response time of a cloud center with parallel data processing. Thus, it should be noted that the

V. M. Vishnevskiy et al. (Eds.): DCCN 2021, CCIS 1552, pp. 318–332, 2022.
https://doi.org/10.1007/978-3-030-97110-6_25

fork-join systems are capable of modeling many real physical systems with parallelization of tasks, starting with the banal warehouse task assembly and ending with simulation of the operation of high-performance applications in industrial, medical, financial, scientific, and in other areas, which are based on the concept of distributed or parallel computing [2–8].

Characteristics of cloud computing systems have been extensively studied (for example, see [9–14]). However, the majority of studies devoted to fork-join systems, in particular, serving as models of cloud computing centers, have been related to the case of exponential service time, which would seem surprising, since a rather critical attitude to this distribution has been adopted for considering problems of description of real physical processes. The explanation of this fact lies in the complexity of the analysis of such systems in the situation of general service times and in the case of correlation of the sojourn time in subsystems. Therefore, it is not surprising, despite the relevance of the problem, that at present there has been only a small number of studies in which such QSs were considered with $M|G|1$-type subsystems.

In the present paper, we will study a fork-join system with branches of $M|G|1$ type — this seems more realistic from the point of view of modeling cloud systems (and not only because of this). As a research method, we choose a fairly new approach, which has recently gained popularity in solving complex problems of queuing theory in which the solution cannot be obtained in an analytical form via classical methods.

The approach consisting in a combination of simulation modeling with machine learning methods was successfully applied in [1], which has motivated us to consider a more complicate analytical model and conduct a new study. A more detailed discussion of this approach and links to recent publications concerned with application of this approach will be given below.

The paper is organized as follows: in Sect. 2 we describe the mathematical model of a cloud center; in Sect. 3 we present main analytical results for fork-join systems with joint distribution of the service time. In Sect. 4, we discuss in more detail the specifics of application of the new approach to the analysis of system characteristics; in Sect. 5, a numerical example is given; the final section summarizes the results and outlines prospects for further research.

2 Mathematical Model of a Cloud Computing Systems

So, as a model of a cloud system, we consider a fork-join type queuing system. Let us recall the basic principles of its operation (Fig. 1) [1]:

1) at the moment of task arrival in the system, the task is instantly split into K ($K \geq 2$) subtasks, each of which joins the tail of the task-specific queue to the server (if it is busy) or is instantly processed (if the corresponding server is empty); in addition, we assume that each subtask has its own type, which corresponds to the number of the server on which it will be processed;

2) once the processing is complete, the subtask enters the synchronization buffer and remains there until all the related subtasks (i.e., the subtasks that originally belonged to the same task) are completely processed, after which the entire task is instantly assembled and only after that the task is considered processed and can leave the system.

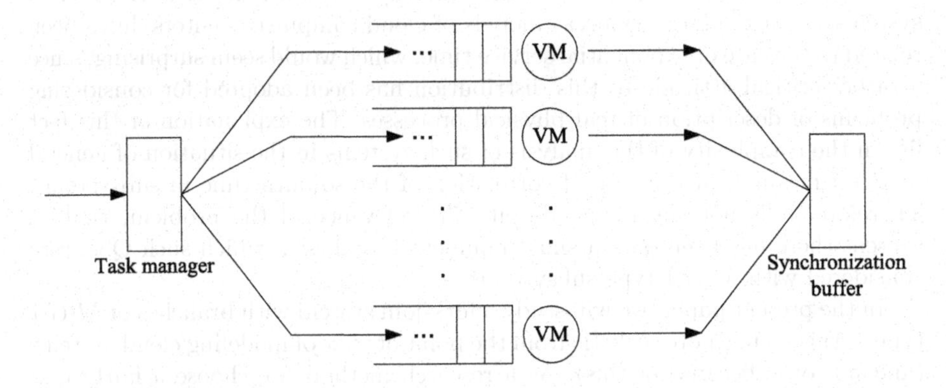

Fig. 1. Model of a cloud fork-join type center.

One of the principal performance indicators of any queuing system is its response time. In this regard a fork-join system is no exception, and so we will specify this concept in the context of task splitting. A task is considered processed only at the end of the service of its last subtask, and so to evaluate the sojourn time of the task in the system, with due account of the fact that the moments of appearance of all its subtasks in the system coincide, it suffices to find the maximum of all the sojourn times of its subtasks [1]. In what follows, this value will be understood as the response time and denoted it by R_K.

3 Analytical Estimates of the Response Time in a Fork-Join System

In the situation involving the Poisson incoming flow and exponential distributions of service times on servers of each of the QS branches, the exact expression for the mean response time was obtained only for $K = 2$, and for $K > 2$, only approximations of various degrees of accuracy are known. For specific expressions, see, for example, [1]; [15] gives a detailed survey of the most significant results that have been obtained in recent years in the study of fork-join systems.

Among the main approaches to derivation of approximate estimates of the mean response time in the conditions of exponential service, we can mention the following: methods of the theory of ordinal statistics, the use of simulation results for preparation of analytical formulas, interpolation under high and weak input loads for QSs, and matrix-geometric methods.

In the case of an arbitrary distribution of service times, i.e., in the case of K branches of type $M|G|1$ in a system with task splitting, not many formulas are known for approximating the mean response time. Moreover, most of these approximations are based on the theory of order statistics, since the random response time is in essence the Kth order statistic,

$$R_K = \xi_{(K)} = \max(\xi_1, ..., \xi_K),$$

where ξ_k is the sojourn time of the subtask in the kth subsystem, $k = \overline{1, K}$. Further, as a rule, to simplify the analysis of extreme values, it is assumed that these random variables are independent (although in fact they are positively dependent).

So, for example, for identically distributed and independent sojourn times in subsystems with expectation $E[\xi_k] = \mu$ and the second moment $E[\xi_k^2] = \mu^{(2)}$, the formula

$$E[R_K] \approx \mu + \frac{\mu^{(2)}}{2\mu}(H_K - 1), \quad H_K = \sum_{i=1}^{K} \frac{1}{i} \tag{1}$$

is given in [16]. An expression for $E[\xi_{(K)}]$ under conditions of various distributions, and, accordingly, different first and second moments of the random variables ξ_k, can be found in [15]. In addition, for some specific distributions of the service times (for example, such as the Erlang distribution), explicit expressions were obtained for the expectation of the extreme value of $\xi_{(K)}$ (see [15,17]).

Moreover, the theory of order statistics is capable of delivering the upper bound for the expectation of the value $\xi_{(K)}$, which in turn can be used to approximate the mean response time [18]:

$$E[R_K] \approx \mu + \sigma \frac{K-1}{\sqrt{2K-1}}; \tag{2}$$

here σ is the standard deviation.

For subsystems of type $M_\lambda|G|1$, it is not always possible to obtain an explicit form for the specific residence time distribution. However, in this situation, in order to use Formula (1) or (2), it suffices to find its first two moments. The formula for the expectation of the sojourn time in $M_\lambda|G|1$ is well known and has the form

$$\mu = b + \frac{\lambda b^{(2)}}{2(1-\rho)}, \tag{3}$$

where b is the mean service time on the server, $b^{(2)}$ is the second moment of this time, and $\rho = \lambda b$ is the load factor of the system. In order to obtain the second moment of the sojourn time in the QS of type $M_\lambda|G|1$, it suffices to twice differentiate the Laplace–Stieltjes transform (LST) of the corresponding distribution function $\psi(s)$ and calculate its limit value at zero, i.e.,

$$\mu^{(2)} = \lim_{s \to 0} \frac{d^2\psi(s)}{ds^2}, \quad \psi(s) = \frac{s(1-\rho)\beta(s)}{s - \lambda + \lambda\beta(s)},$$

where $\beta(s)$ is the LST of the distribution function of the service time. As a result, we get

$$\mu^{(2)} = b^{(2)} + \frac{\lambda b^{(3)} + 3\rho b^{(2)}}{3(1-\rho)} + \frac{\lambda^2 (b^{(2)})^2}{2(1-\rho)^2}, \tag{4}$$

where $b^{(3)}$ is the third moment of the service time.

Note that we only consider analytical estimates that do not use simulation results and are based on moments of the service time. On the one hand, this makes it possible to apply them to arbitrary distributions whose moments exist and are known. On the other hand, estimates do not use more detailed information about distributions, and for different distributions with the same moment values, they may work differently (better or worse).

As an example, which we will used for a comparative analysis of the known analytical results and the new approach, we will consider the Pareto distribution for the service time on the server in an $M_\lambda|G|1$-system in which the distribution function and the first three moments are as follows:

$$F(x) = 1 - \left(\frac{\alpha-1}{\alpha}\cdot\frac{1}{x}\right)^{\alpha}, \quad x \geq \frac{\alpha-1}{\alpha}, \quad \alpha > 3, \tag{5}$$

$$b_{Pa} = 1, \quad b^{(2)}_{Pa} = \frac{(\alpha-1)^2}{\alpha(\alpha-2)}, \quad b^{(3)}_{Pa} = \frac{(\alpha-1)^3}{\alpha^2(\alpha-3)}.$$

Hence the mean service time on the server will be equal to the one conditional time unit (this is generally convenient in the context of numerical analysis) and the first and second moments of the sojourn time of the subtask (with due account of the fact that $\rho = \lambda$) are given by the expressions

$$\mu_{Pa} = 1 + \frac{\rho(\alpha-1)^2}{2\alpha(\alpha-2)(1-\rho)}, \tag{6}$$

$$\mu^{(2)}_{Pa} = \frac{(\alpha-1)^2}{\alpha(\alpha-2)} + \frac{\rho(\alpha-1)^3}{3\alpha^2(\alpha-3)(1-\rho)} + \frac{\rho(\alpha-1)^2}{\alpha(\alpha-2)(1-\rho)} + \frac{\rho^2(\alpha-1)^4}{2\alpha^2(\alpha-2)^2(1-\rho)^2} \tag{7}$$

in accordance with (3) and (4). Besides, the standard deviation of the subtask sojourn time is given by

$$\sigma_{Pa} = \sqrt{\mu^{(2)}_{Pa} - \mu^2_{Pa}}. \tag{8}$$

Note that the characteristics of the sojourn time of the subtask, as described by (6)–(8), can themselves serve as estimates of the corresponding characteristics of the response time (if, under some system parameters, the subtasks are processed fairly synchronously).

The choice of the Pareto distribution for description of the service time on the servers is generally due to the recent interest in it, since this distribution is relevant for modeling of self-similar traffic in information/computing systems

[19–21]. The Pareto distribution is a typical example of heavy-tailed distributions. If the service time distribution has a heavy tail then the response time distribution also has a heavy tail. Heavy-tailed queuing systems have their own specifics which can affect the success of evaluating their characteristics.

On the other hand, to demonstrate the effectiveness of the new approach, it is necessary to conduct a comparative analysis with the previously known results. And since analytical formulas contain the values of the first and second moments of the distributions of the service time, it follows that we, of course, are forced to limit ourselves to the case of finite expectation and variance when choosing specific parameters of the Pareto distribution. In addition, there are no analytical formulas for fork-join systems with subsystems of $G|G|1$ type, which explains the preservation of the Poisson incoming flow in the analytical model. Note that such restrictions can be avoided if the machine learning approach is used.

4 Application of Neural Networks to the Analysis of a Fork-Join System

Application of artificial neural networks (ANN) or any other methods from the rather extensive realm of data mining to the analysis of QSs or queuing networks (QNs) is a combination of simulation on a limited set of input parameters on pre-defined numerical intervals, followed by training on the obtained data of a neural networks, which in the future will allow one to get estimates of the system's characteristics of interest for other (for example, intermediate) values of input parameters.

Simulation modeling is sometimes the only possible method for fairly accurate analysis of complex QSs or QNs: either because for them it is impossible to obtain a closed solution or because the developed approximate algorithms produce an unacceptable approximation error under certain conditions – this applies, for example, to open multiphase networks with nodes of type $\cdot|G|1$. However, as is known, the time spent on simulation can be quite large, and so implementation of simulation not on all the required data, but on a significantly truncated dataset, can significantly save computational and time resources. At the same time, within the specified numerical intervals for input parameters, the number of estimates issued almost instantly by the neural network is not limited by anything.

In [1], it was already mentioned that the new research method was first applied to the analysis of complex queuing models in [22,23] and also in [24] (in Russian), in which the so-called QSs with "warm-up" were investigated [25]. Neural networks were also used to estimate the characteristics of real queues in banks/hospitals [26,27], and in earlier papers [28,29], for modeling a self-similar traffic when analyzing the performance of telecommunications networks.

Some recent publications in this direction are also worth mentioning. So, in [30], estimates of the expectation and variance of the response time of a fork-join system with exponential maintenance were obtained, which turned out to be more advantageous compared to estimates by known approximate formulas, in which, for example, the permissible accuracy was limited by the number K of subsystems. In [30], a comparative analysis of estimates of the mean end-to-end delay of a queuing network was performed with single-server nodes with an arbitrary distribution function. These estimates were also obtained in two ways — either by using a new approach and by using the decomposition method in parallel with diffusion approximation. Since the second approach, as mentioned above, sometimes produces a generally unacceptable approximation error, the advantage of the first approach is obvious.

5 Numerical Experiment

So, to conduct a comparative analysis of the results in a numerical example, it is necessary to obtain estimates in accordance with the described approach. Recall that we consider a fork-join system with K branches of type $M|G|1$ with Pareto distribution of the service time defined in (5). The next step is to develop a simulation model of the specified system in order to obtain a dataset both for training the neural network and for testing it later. To this end one needs to determine the list of input parameters and the interval of values they accept, and specify the output characteristics that will be calculated at the model simulation stage.

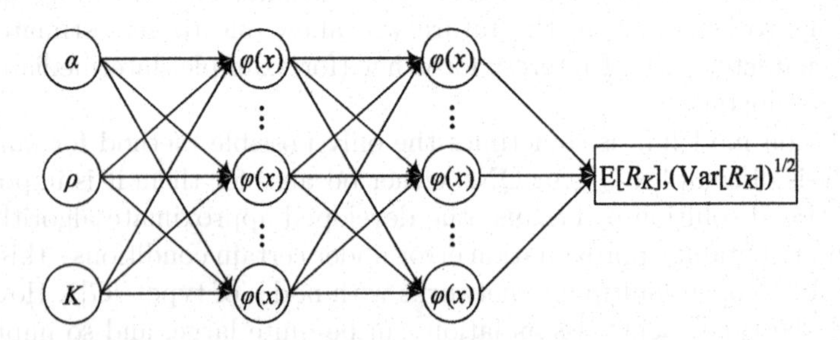

Fig. 2. Perceptron setup for estimation of performance characteristics of a fork-join system.

Assume that the parameter α assumes only integer values from the interval $[4.0, 10.0]$, the number of branches K in the network changes from 2 to 20 inclusively, and the load factor ρ varies from 0.1 to 0.9 with step 0.1 (Table 1). As the performance characteristics of a system, we will choose, which is natural, the mean sojourn time in the QS $E[R_K]$ and the standard deviation of this random variable $\sqrt{\mathrm{Var}[R_K]}$. So, it is required to obtain 1197 datasets.

A simulation model of the above fork-join system was developed in the Python software environment; the artificial neural network was learned in this environment too. Of course, for QS modeling one can use ready-made commercial software products such as GPSS World, AnyLogic, and Arena. To achieve

Table 1. Input data for artificial neural network learning.

no.	1	2	...	19	20	21	...	1197
α	4.0	4.0	...	4.0	4.0	4.0	...	10.0
ρ	0.1	0.1	...	0.1	0.2	0.2	...	0.9
K	2	3	...	20	2	3	...	20

a high accuracy in the estimation of characteristics, we simulated 10 millions of response times for each combination of parameter values. As the structure of the neural network, we chose the three-layer perceptron with 10 neurons in each of the two hidden layers with the logistics activation function $\varphi(x) = 1/(1+e^{-x})$ (Fig. 2). The output layer consists of only one neuron responsible for some one of the estimated characteristics, i.e., two neural networks were actually built, due to which the prediction accuracy was expected to increase. In addition, for the same purpose the input data have been pre-standardized and normalized.

Next, the sample from Table 1 was randomly split into the training and test datasets in the ratio of 80% and 20%, respectively. As a result, the data were trained on 958 conditional units from the training dataset by the Adam method, which is essentially an extension of the classical gradient descent algorithm [31]. During the training process, a validation dataset was selected from the training dataset, which was used to evaluate the quality of training after passing each training epoch. As the criteria for estimating the forecast error, we chose the mean square error (MSE), the mean absolute error (MAE), and the mean absolute percentage error (MAPE)

$$MSE = \frac{1}{n}\sum_{j=1}^{N}(x_j - \widehat{x}_j)^2, \quad MAE = \frac{1}{n}\sum_{j=1}^{N}|x_j - \widehat{x}_j|, \tag{9}$$

$$MAPE = \frac{1}{n}\sum_{j=1}^{N}\left|\frac{(x_j - \widehat{x}_j)}{x_j}\right| \cdot 100\%, \tag{10}$$

where $\widehat{x}_j$ is the estimate of the characteristic under study (the expectation or the standard deviation of the response time), which is obtained either using the neural network or by analytical formulas; besides, x_j is the real value of one of the estimated characteristics, which is obtained as a result of simulation of the system with task splitting, $j = \overline{1, N}$, and N is the number of datasets in the sample aimed for estimating the approximation accuracy.

Table 2 summarizes the values of errors (9)–(10), which were obtained on a test dataset not involved in neural network learning. This already indicates the satisfactory quality of the forecast produced by a neural network on completely new, unfamiliar input data. However, to finally make sure of a kind of stability of the qualitative forecast, we calculate similar errors, but now for an intermediate input data, which are presented in Table 3.

Table 2. Errors of approximations of estimates of the performance characteristics of a fork-join system obtained via ANN on the test dataset.

Estimated characteristic	Error types		
	MSE	MAE	MAPE, %
$E[R_K]$	0.000668	0.013629	0.490643
$\sqrt{\text{Var}[R_K]}$	0.000308	0.012031	0.884236

Table 3. Input data for performance characteristics of a fork-join queuing model.

no.	1	2	...	19	20	21	...	915
α	4.50	4.50	...	4.50	4.50	4.50	...	9.50
ρ	0.15	0.15	...	0.15	0.25	0.25	...	0.85
K	2	3	...	20	2	3	...	20

To perform a comparative analysis, we calculate, on the same dataset, the approximation errors for Formulas (1)–(2), taking into account expressions (6)–(7), and also Formula (6) itself.

Similarly, we compare the approximation error by the analytical Formula (8) with the forecast by a neural network. The results of calculations are presented in Table 4. It is obvious that the analytical formulas produce an unacceptable relative error.

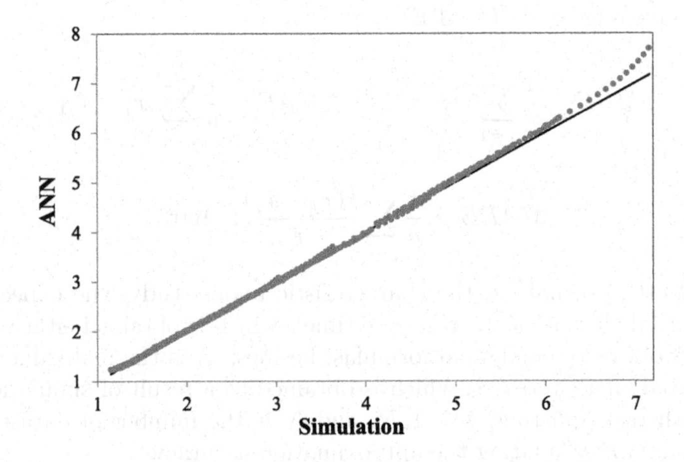

Fig. 3. The mean response time of a fork-join QS obtained by simulation and by application of an ANN.

In order to fully understand the structure of the obtained estimates of the mean values and the standard deviation of the response time, we will build graphs that clearly reflect the deviation of these estimates from the true values of indicators. The values of the mean response time obtained on the dataset from Table 3 via simulation model are plotted on the abscissa axis, and the values obtained either by using a neural network or via Formulas (1), (2), (6), as shown in Figs. 3 and 4, 5, 6, respectively, and plotted on the ordinate axis. Moreover, for completeness, in the case of analytical formulas, the plotting area is expanded by using the input data from the Table 1. The graphs for the standard deviation are constructed similarly (Fig. 7, 8).

Table 4. Errors of approximations of estimates of the performance characteristics of a fork-join system obtained via ANN and analytical formulas on the dataset from Table 3.

Estimated characteristic	Error types		
	MSE	MAE	MAPE, %
$E[R_K]$, ANN	0.001939	0.021409	0.707818
$E[R_K]$, Formula (1)	5.892661	1.878592	67.785755
$E[R_K]$, Formula (2)	6.722487	1.817356	59.983739
$E[R_K]$, Formula (6)	0.621682	0.642618	24.425016
$\sqrt{\text{Var}[R_K]}$, ANN	0.002679	0.038941	3.355246
$\sqrt{\text{Var}[R_K]}$, Formula (8)	0.107970	0.254335	18.701897

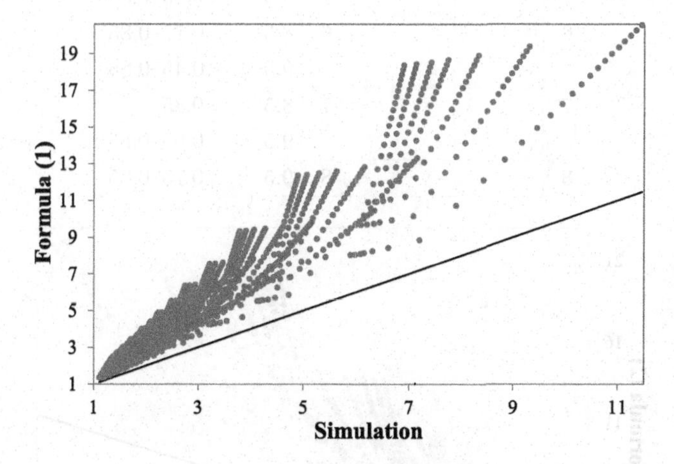

Fig. 4. The mean response time of a fork-join QS obtained by simulation and via formula (1).

As can be seen from the graphs (Figs. 4, 5), for the vast majority of values of input parameters, the analytical Formulas (1)–(2) give an unnecessarily overestimated result, and in the case of expression (1) all the values of relative approximation error exceed 14%, and in the case of (2) the number of point estimates whose relative error would not exceed the threshold of 10% is 6. Note that the estimate obtained by Formula (6) for the mean response time in an $M_\lambda|G|1$ QS gives a more realistic estimate, in the sense that the relative approximation

Table 5. Input data for which the relative approximation error of analytical formulas is less than 10%.

Formula (6)			Formula (8)		
K	α	ρ	K	α	ρ
2	4.5	–	2	4.5	0.65–0.85
	5.5	0.15–0.45		5.5	0.35–0.85
	6.5–9.5	0.15–0.85		6.5–9.5	0.15–0.85
3	6.5	–	3	6.5	0.55–0.85
	7.5	–		7.5	0.35–0.85
	8.5	0.15–0.55		8.5	0.25–0.85
	9.5	0.15–0.85		9.5	0.15–0.85
4	–	–	4	7.5	0.65–0.85
	–	–		8.5	0.45–0.85
	–	–		9.5	0.25–0.85
5	–	–	5	8.5	0.55–0.85
	–	–		9.5	0.35–0.85
6	–	–	6	8.5	0.75–0.85
	–	–		9.5	0.45–0.85
7	–	–	7	8.5	0.85
	–	–		9.5	0.65–0.85
8	–	–	8	9.5	0.75–0.85

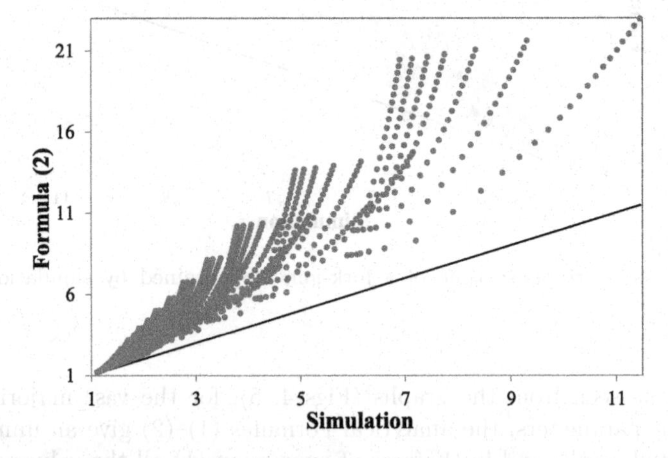

Fig. 5. The mean response time of a fork-join QS obtained by simulation and via formula (2).

error of less than 10% is already valid for 49 values out of 912 (Table 5). The situation in the case of standard deviation (Fig. 8) is better; however here the estimate of this exponent via (8) is in fact a lower estimate, and the threshold of 10% is valid for approximately 11.4% values of the total number of estimates and extends to the area of large values of the parameter α and large load factor ρ (Table 5).

The estimates obtained using trained networks (Figs. 3, 7) demonstrate acceptable approximation quality. So, in the case of mean response time, the relative error of approximation does not exceed 5% for 99.23% of the total number of the obtained estimates, and for the remaining 7 estimates, whose values

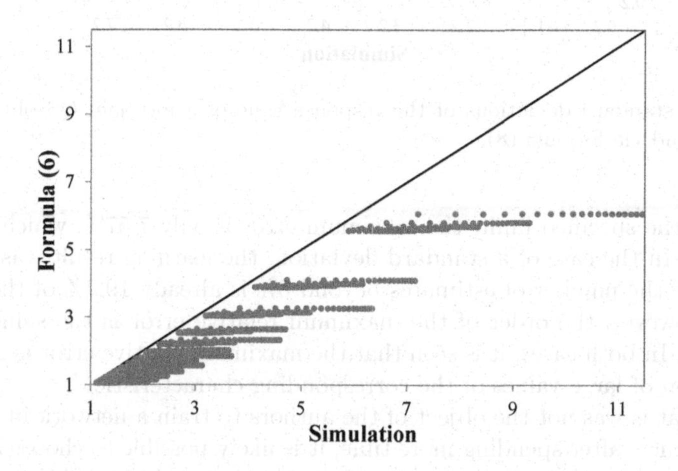

Fig. 6. The mean response time of a fork-join QS obtained by simulation and via formula (6).

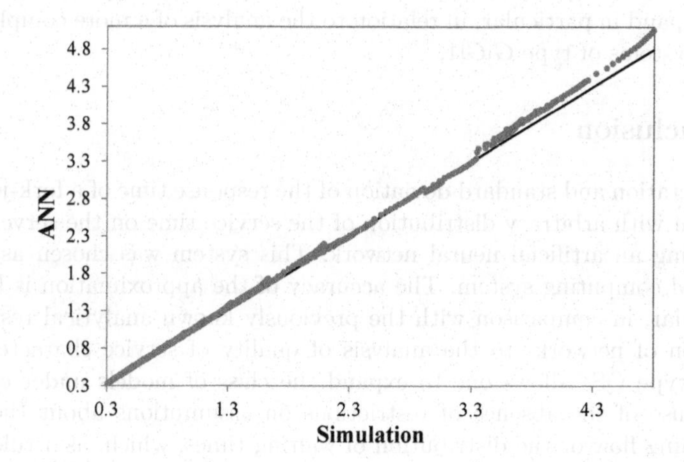

Fig. 7. The standard deviations of the response time of fork-join QS obtained by simulation and by applying an ANN.

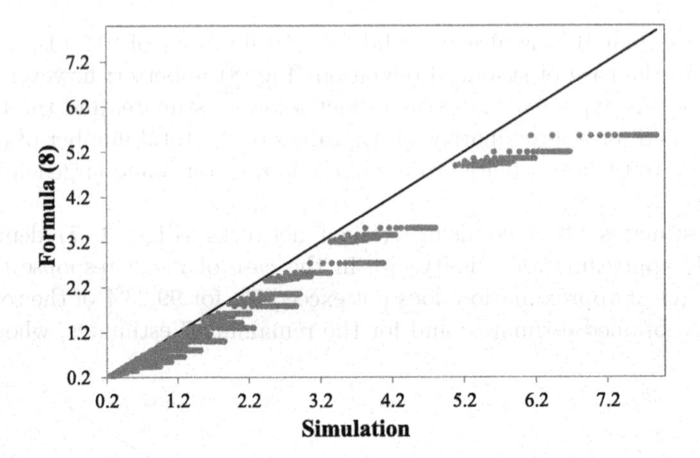

Fig. 8. The standard deviations of the response time of a fork-join QS obtained by simulation and via formula (8).

go beyond the specified limit, the maximum error is only 7.57%, which is quite acceptable. In the case of a standard deviation, the learning result was slightly worse, since the number of estimates beyond 5% is already 19.7% of their total number, however, the order of the maximum relative error is the same and is only 7.61%. In both cases, it is seen that the maximum relative error is observed in the region of large values of the corresponding characteristic.

Note that is was not the object of the authors to train a network in the best way, and hence, after spending more time, it is likely possible to choose an ANN with a different architecture, activation functions, etc., which could give a more accurate prediction. However, it is already clear that due to the lack of analytical results that would give an acceptable error, the proposed approach seems more promising, and in particular, in relation to the analysis of a more complex model with subsystems of type $G|G|1$.

6 Conclusion

The expectation and standard deviation of the response time of a fork-join queuing system with arbitrary distribution of the service time on the servers is estimated using an artificial neural network. This system was chosen as a model for a cloud computing system. The accuracy of the approximation is high, and in particular, in comparison with the previously known analytical results. The application of networks to the analysis of quality of service characteristics in fork-join type QSs allows one to expand the class of models under consideration because of the absence of restrictions on assumptions about the type of the incoming flow or the distribution of waiting times, which, as a rule, greatly complicate the analytical or numerical process of solving the problem, however, under our approach there is no loss in the quality of evaluation of the sought-for parameters.

References

1. Gorbunova, A.V., Vishnevsky, V.M.: Estimating the response time of a cloud computing system with the help of neural networks. Adv. Syst. Sci. Appl. **20**(3), 105–112 (2020)
2. Chen, R.J.: An upper bound solution for homogeneous fork/join queuing systems. IEEE Trans. Parallel Distrib. Syst. **22**(5), 874–878 (2011)
3. Kemper, B., Mandjes, M.: Mean sojourn time in two-queue fork-join systems: bounds and approximations. OR Spectrum **34**, 723–742 (2012)
4. Axer, P., Quinton, S., Neukirchner, M., Ernst, R., Dobel, B., Hartig, H.: Response-time analysis of parallel fork-join workloads with real-time constraints. In: Proceedings of the ECRTS, pp. 215–224. IEEE, Los Alamitos, CA, USA (2013)
5. Fiorini, P.M.: Analytic approximations of fork-join queues. In: Proceedings of the IDAACS, pp. 966–971. IEEE, Warsaw, Poland (2015)
6. Jacob, A., et al.: Efficient fork-join on GPUs through warp specialization. In: Proceedings HiPC, pp. 358–367. IEEE, Jaipur, India (2017)
7. Rashid, Z.N., Zebari, S.R.M., Sharif, K.H., Jacksi, K.: Distributed cloud computing and distributed parallel computing: a review. In: Proceedings of the ICOASE, pp. 167–172. IEEE, Duhok, Iraq (2018)
8. Gorbunova, A.V., Lebedev, A.V.: Bivariate distributions of maximum remaining service times in fork-join infinite-server queues. Probl. Inf. Transm. **56**(1), 73–90 (2020)
9. Gorbunova, A.V., Zaryadov, I.S., Matushenko, S.I., Sopin, E.S.: The estimation of probability characteristics of cloud computing systems with splitting of requests. In: Vishnevskiy V., Samouylov K., Kozyrev D. (eds.) DCCN 2016. CCIS, vol. 678, pp. 467–472. Springer, Cham (2016). https://doi.org/10.1007/978-3-319-51917-3_37
10. Keller, M., Karl, H.: Response-time-optimized service deployment: MILP formulations of piece-wise linear functions approximating bivariate mixed-integer functions. IEEE Trans. Netw. Serv. Manage. **14**(1), 279–294 (2017)
11. Alhamad, M., Dillon, T., Wu, Ch., Chang, E.: Response time for cloud computing providers. In: Proceedings of the IIWAS, pp. 603–606. Association for Computing Machinery, Paris, France (2010)
12. Xiong, K., Perros, H.: Service performance and analysis in cloud computing. In: Proceedings of the Congress on Services - I, pp. 693–700. IEEE, Los Angeles, CA, USA (2009)
13. Gorbunova, A.V., Zaryadov, I.S., Matushenko, S.I., Samuylov, K.E.: The approximation of response time of a cloud computing system. Inform. Appl. **9**(3), 32–38 (2015)
14. Jitendra, S.: Study of response time in cloud computing. Int. J. Inf. Eng. Electron. Bus. **6**, 36–43 (2014)
15. Thomasian, A.: Analysis of fork/join and related queueing systems. ACM Comput. Surveys **47**(2), 17:1–17:71 (2014)
16. Harrison, P., Zertal, S.: Queueing models with maxima of service times. In: Kemper, P., Sanders, W.H. (eds.) TOOLS 2003. LNCS, vol. 2794, pp. 152–168. Springer, Berlin, Heidelberg (2003). https://doi.org/10.1007/978-3-540-45232-4_10
17. Thomasian, A., Menon, J.: RAID5 performance with distributed sparing. IEEE Trans. Parallel Distrib. Syst. **8**(6), 640–657 (1997)
18. David, H.A., Nagaraja, H.N.: Order Statistics, 3rd edn. John Wiley & Sons, New York (2003)

19. Ageyev, D., Qasim, N.: LTE EPS network with self-similar traffic modeling for performance analysis. In: Proceedings of the PIC S&T, pp. 275–277. IEEE, Kharkov, Ukraine (2015)
20. Lozhkovskyi, A., Levenberg Y.: Investigation of simulating methods for self-similar traffic flows: the QoS-characteristics depend on the type of distribution in self-similar traffic. In: Proceedings of the PIC S&T, pp. 410–413. IEEE, Kharkov, Ukraine (2017)
21. Savu-Jivanov, A., Isar, A., Stolojescu-Crisan, C., Gal J.: Network self-similar traffic generator with variable Hurst parameter. In: Proceedings of the ISETC, pp. 1–4. IEEE, Timisoara, Romania (2020)
22. Sivakami Sundaria, M., Palaniammalb, S.: Simulation of $M|M|1$ queuing system using ANN. Malaya J. Matematik: Special Issue **1**, 279–294 (2015)
23. Sivakami Sundaria, M., Palaniammalb, S.: An ANN simulation of single server with infinite capacity queuing system. Int. J. Innov. Technol. Explor. Eng. **8**(12), 4067–4071 (2019)
24. Khomonenko, A.D., Yakovlev, E.L.: Nejrosetevaya approksimaciya harakteristik mnogokanal'nyh nemarkovskih sistem massovogo obsluzhivaniya [Neural network approximation of characteristics of multi-channel non-Markovian queuing systems]. Trudy SPIIRAN [SPIIRAS Proceedings] **4**(4), 81–93 (2015). (in Russian)
25. Khomonenko, A.D., Adadurov, S.E., Gindin, S.I.: CHislennyj raschet mnogokanal'noj sistemy massovogo obsluzhivaniya s rekurrentnym vhodyashchim potokom i "razogrevom" [Numerical calculations of multichannel queuing system with recurrent input and "warm up"] // Izvestiya Peterburgskogo universiteta putey soobscheniya [Proceedings of Petersburg Transport University] **37**(4), 92–101 (2013). (in Russian)
26. Hermanto, R.P.S., Suharjito, S., Nugroho, A.: Waiting-time estimation in bank customer queues using RPROP neural networks. Procedia Comput. Sci. **135**, 35–42 (2018)
27. Curtis, C., Liu, Ch., Bollerman, T.J., Pianykh, O.S.: Machine learning for predicting patient wait times and appointment delays. J. Am. Coll. Radiol. **15**(9), 1310–1316 (2018)
28. Habib, I.W.: Applications of neurocomputing in traffic management of ATM networks. Proc. IEEE **84**(10), 1430–1441 (1996)
29. Aussem, A., Murtagh, F.: A neuro-wavelet strategy for Web traffic forecasting. Res. Official Stat. **1**(1), 65–87 (1998)
30. Gorbunova, A.V., Vishnevsky, V.M., Larionov, A.A.: Evaluation of the end-to-end delay of a multiphase queuing system using artificial neural networks. In: Vishnevskiy, V.M., Samouylov, K.E., Kozyrev, D.V. (eds.) DCCN 2020. LNCS, vol. 12563, pp. 631–642. Springer, Cham (2020). https://doi.org/10.1007/978-3-030-66471-8_48
31. Kingma, D.P., Ba, J.: Adam: a method for stochastic gradient descent. In: Proceedings of the ICLR, pp. 1–15, San Diego, CA, USA (2015)

Approximation of the Two-Dimensional Output Process of a Retrial Queue with MMPP Input

Alexey Blaginin and Ivan Lapatin(✉)

Tomsk State University, Lenina pr. 36, 634050 Tomsk, Russia
rector@tsu.ru
http://www.tsu.ru

Abstract. In this paper, we review a retrial queue with MMPP input and two-way communication. Incoming requests arriving at the server and finding it busy join the source of retrial calls and try to enter the server again after some exponentially distributed time. While idle, the server makes outgoing calls and serves them with another delay parameter. MMPP (Markov Modulated Poisson Process) is an input process in which control is driven by a continuous Markov chain. Changing its state entails a change in the intensity of the input process. For this model, we present an asymptotic approximation of the two-dimensional characteristic function under the condition of a long delay of requests in the source of retrial calls. For this approximation, we carried out a numerical experiment, where asymptotic results were compared to computations obtained via simulation.

Keywords: Output process · Retrial queue · Two-way communication · Asymptotic analysis method · Simulation · Markov modulated poisson process

1 Introduction

The specific property of RQ systems [10,16] with two-way communication [16] is the presence of different request types, which gives rise to many new service disciplines. For this reason, RQ systems with two-way communication are a powerful tool in design and optimization of real-life systems with multiple random access to a resource. Despite that these systems are well studied, their output process is still a complex and insufficiently explored area to research.

In modern telecommunication networks, there are also point processes with a varying rate of calls incoming. To simulate such jobs within the framework of queuing theory, the Markov Modulated Poisson Process (MMPP) [2,10] is used. It has a mechanism for taking into account the temporal inhomogeneity of the

V. M. Vishnevskiy et al. (Eds.): DCCN 2021, CCIS 1552, pp. 333–345, 2022.
https://doi.org/10.1007/978-3-030-97110-6_26

arrival rate of requests and also gives analytically processable queuing results [11]. For this reason, MMPP is widely used in Internet research, in particular, using MMPP in [13], a traffic model that accurately approximates the LRD (Long Range Dependence) characteristics of Internet traffic traces, was built. Using the concepts of sessions and streams, the proposed MMPP model simulates the actual hierarchical behaviour of Internet users generating packets. It allows traffic simulation with the desired characteristics, that have a clear physical meaning. The results prove that the queuing traffic behaviour generated by the MMPP model is consistent with the model created by the actual traces of packets collected at the edge router under various scenarios and loads.

Earlier, we presented a similar work, where a retrial queue with Poisson input process is described [3]. In this paper, we take into consideration an improved model with MMPP, which is more suitable for modelling real optimization problems. We find the approximation of the characteristic function of the number of served requests in the considered system using the method of asymptotic analysis. Subsequently, we determine the applicability of the asymptotic results by comparing them to calculations provided with simulation software, which was designed especially for this research.

2 Mathematical Model

MMPP is qualified with two matrices. Matrix of infinitesimal characteristics $\boldsymbol{Q}$ defines the state. Value q_{ij} determines the intensity of the transition of the process from the state i to the state j, and the value $-q_{ii}$ is the intensity of leaving the state i. The matrix $\boldsymbol{Q}$ has property $\sum_j q_{ij} = 0$. The diagonal matrix $\boldsymbol{\Lambda}$ specifies the rate of calls for each of the states of the process.

Let us consider the RQ system with MMPP input. An incoming request takes the server if it is idle. The server, in turn, starts serving it for some exponentially distributed time with parameter μ_1. When an incoming request cannot access the server, it travels to the source of retrial calls, where waits for exponentially distributed time with parameter σ. While free from serving incoming requests, the server produces requests itself with the intensity α and serves them with parameter μ_2.

We denote the following notations: *i(t)* is the number of requests in the orbit at the moment t, *k(t)* is the state of the server: *0*—idle, *1*—busy serving an incoming request, *2*—busy serving an outgoing request; $m_1(t)$ is the number of served input process requests at the moment t, $m_2(t)$ is the number of served outgoing requests at the moment t, $n(t)$ is the state of the input process at the moment t.

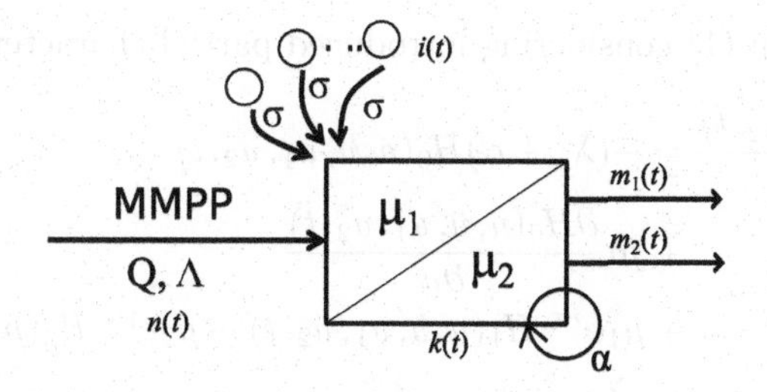

Fig. 1. RQ system with two-way communication

3 Kolmogorov Differential Equations System

We consider the five-dimensional Markov process

$$\{k(t), n(t), i(t), m_1(t), m_2(t)\}$$

Based on the formulated Markov process, we introduce probabilities

$$P\{k(t) = k, n(t) = n, i(t) = i, m_1(t) = m_1, m_2(t) = m_2\}$$

and write down for them the Kolmogorov differential equations system

$$\begin{aligned}
\frac{\partial P_0(n,i,m_1,m_2,t)}{\partial t} &= -(\lambda_n + i\sigma + \alpha)P_0(n,i,m_1,m_2,t) \\
&+ P_1(n,i,m_1-1,m_2,t)\mu_1 + P_2(n,i,m_1,m_2-1,t)\mu_2 \\
&+ \sum_{v=1}^{N} P_0(v,i,m_1,m_2,t)q_{vn}, \\
\frac{\partial P_1(n,i,m_1,m_2,t)}{\partial t} &= -(\lambda_n + \mu_1)P_1(n,i,m_1,m_2,t) \\
&+ (i+1)\sigma P_0(n,i+1,m_1,m_2,t) + \lambda_n P_0(i,m_1,m_2,t) \\
&+ \sum_{v=1}^{N} P_1(v,i,m_1,m_2,t)q_{vn}, \\
\frac{\partial P_2(n,i,m_1,m_2,t)}{\partial t} &= -(\lambda_n + \mu_2)P_2(n,i,m_1,m_2,t) \\
&+ \lambda_n P_2(n,i-1,m_1,m_2,t) + \alpha P_0(n,i,m_1,m_2,t) \\
&+ \sum_{v=1}^{N} P_2(v,i,m_1,m_2,t)q_{vn}.
\end{aligned} \tag{1}$$

Since the obtained system is infinite, we introduce the partial characteristic functions, denoting $j^2 = -1$. Such wise we passed to the system, having only three equations.

$$H_k(n,u,u_1,u_2,t) = \sum_{i=0}^{\infty} \sum_{m_1=0}^{\infty} \sum_{m_2=0}^{\infty} e^{jui} e^{ju_1 m_1} e^{ju_2 m_2} P_k(n,i,m_1,m_2,t).$$

We rewrite system (1) considering introduced partial characteristic functions

$$\begin{aligned}\frac{\partial H_0(n,u,u_1,u_2,t)}{\partial t} &= -(\lambda_n+\alpha)H_0(n,u,u_1,u_2,t)\\&+j\sigma\frac{\partial H_0(n,u,u_1,u_2,t)}{\partial u}\\&+\mu_1 e^{ju_1}H_1(n,u,u_1,u_2,t)+\mu_2 e^{ju_2}H_2(n,u,u_1,u_2,t)\\&+\sum_{v=1}^{N}H_0(v,u,u_1,u_2,t)q_{vn},\\\frac{\partial H_1(n,u,u_1,u_2,t)}{\partial t} &= -(\lambda_n+\mu_1)H_1(n,u,u_1,u_2,t)\\&-j\sigma e^{-ju}\frac{\partial H_0(n,u,u_1,u_2,t)}{\partial u}\\&+\lambda_n H_0(n,u,u_1,u_2,t)+\lambda_n e^{ju}H_1(n,u,u_1,u_2,t)\\&+\sum_{v=1}^{N}H_1(v,u,u_1,u_2,t)q_{vn},\\\frac{\partial H_2(n,u,u_1,u_2,t)}{\partial t} &= -(\lambda_n+\mu_2)H_2(n,u,u_1,u_2,t)\\&+\lambda_n e^{ju}H_2(n,u,u_1,u_2,t)\\&+\alpha H_0(n,u,u_1,u_2,t)+\sum_{v=1}^{N}H_2(v,u,u_1,u_2,t)q_{vn}.\end{aligned} \tag{2}$$

For further analysis let us denote

$$\boldsymbol{H}_k(u,u_1,u_2,t)=\{H_k(1,u,u_1,u_2,t),H_k(2,u,u_1,u_2,t),\ldots,H_k(N,u,u_1,u_2,t)\},$$

diagonal unit matrix $\boldsymbol{I}$ with size N. Then (2) will be rewritten in the following form

$$\begin{aligned}\frac{\partial \boldsymbol{H}_0(u,u_1,u_2,t)}{\partial t} &= (\boldsymbol{Q}-\boldsymbol{\Lambda}-\alpha\boldsymbol{I})\boldsymbol{H}_0(u,u_1,u_2,t)\\&+\mu_1 e^{ju_1}\boldsymbol{H}_1(n,u,u_1,u_2,t)\\&+\mu_2 e^{ju_2}\boldsymbol{H}_2(u,u_1,u_2,t)+j\sigma\frac{\partial \boldsymbol{H}_0(u,u_1,u_2,t)}{\partial u},\\\frac{\partial \boldsymbol{H}_1(u,u_1,u_2,t)}{\partial t} &= \boldsymbol{\Lambda}\boldsymbol{H}_0(u,u_1,u_2,t)\\&+(\boldsymbol{Q}+(e^{ju}-1)\boldsymbol{\Lambda}-\boldsymbol{I}\mu_1)\boldsymbol{H}_1(u,u_1,u_2,t)\\&-j\sigma e^{-ju}\frac{\partial \boldsymbol{H}_0(u,u_1,u_2,t)}{\partial u},\\\frac{\partial \boldsymbol{H}_2(u,u_1,u_2,t)}{\partial t} &= \alpha\boldsymbol{H}_0(u,u_1,u_2,t)\\&+(\boldsymbol{Q}+(e^{ju}-1)\boldsymbol{\Lambda}-\boldsymbol{I}\mu_2)\boldsymbol{H}_2(u,u_1,u_2,t).\end{aligned} \tag{3}$$

4 Asymptotic Analysis Method

We solve the obtained system with the asymptotic analysis method with the limit condition of a long delay of requests in the orbit ($\sigma \to 0$).

Denoting $\epsilon = \sigma, u = \epsilon w, \boldsymbol{F}_k(w, u_1, u_2, t, \epsilon) = \boldsymbol{H}_k(u, u_1, u_2, t)$ we (3) as

$$\begin{aligned}
\frac{\partial \boldsymbol{F}_0(w, u_1, u_2, t, \epsilon)}{\partial t} &= (\boldsymbol{Q} - \boldsymbol{\Lambda} - \alpha \boldsymbol{I})\boldsymbol{F}_0(w, u_1, u_2, t, \epsilon) \\
&+ \mu_1 e^{ju_1}\boldsymbol{F}_1(w, u_1, u_2, t, \epsilon) + \mu_2 e^{ju_2}\boldsymbol{F}_2(w, u_1, u_2, t, \epsilon) \\
&+ j\frac{\partial \boldsymbol{F}_0(w, u_1, u_2, t, \epsilon)}{\partial w}, \\
\frac{\partial \boldsymbol{F}_1(w, u_1, u_2, t, \epsilon)}{\partial t} &= \boldsymbol{\Lambda}\boldsymbol{F}_0(w, u_1, u_2, t, \epsilon) \\
&+ (\boldsymbol{Q} + (e^{j\epsilon w} - 1)\boldsymbol{\Lambda} - \boldsymbol{I}\mu_1)\boldsymbol{F}_1(w, u_1, u_2, t, \epsilon) \\
&- je^{-j\epsilon w}\frac{\partial \boldsymbol{F}_0(w, u_1, u_2, t, \epsilon)}{\partial w}, \\
\frac{\partial \boldsymbol{F}_2(w, u_1, u_2, t, \epsilon)}{\partial t} &= \alpha \boldsymbol{F}_0(w, u_1, u_2, t, \epsilon) \\
&+ (\boldsymbol{Q} + (e^{j\epsilon w} - 1)\boldsymbol{\Lambda} - \boldsymbol{I}\mu_2)\boldsymbol{F}_2(w, u_1, u_2, t, \epsilon).
\end{aligned} \tag{4}$$

The solution for system (4) is formulated in Theorems 1 and 2.

Theorem 1. *Let $i(t)$ is the number of requests in the orbit at the moment t, then in the stationary regime we obtain*

$$\lim_{\epsilon \to 0}\{\sum_{k=0}^{2} \boldsymbol{F}_k(w, 0, 0, t, \epsilon)\} = \lim_{\sigma \to 0} Me^{jw\sigma i(t)} = e^{jw\kappa},$$

where κ is a positive root of the equation

$$\kappa \boldsymbol{R}_0(\kappa)\boldsymbol{e} = [\boldsymbol{R}_1(\kappa) + \boldsymbol{R}_2(\kappa)]\boldsymbol{\Lambda}\boldsymbol{e}.$$

Vectors $\boldsymbol{R}_k$ are defined as

$$\begin{cases}
\boldsymbol{R}_0(\kappa) = \boldsymbol{r}\{\boldsymbol{I} + [\boldsymbol{\Lambda} + \kappa \boldsymbol{I}](\mu_1 \boldsymbol{I} - \boldsymbol{Q})^{-1} + \alpha(\mu_2 \boldsymbol{I} - \boldsymbol{Q})^{-1}\}^{-1}, \\
\boldsymbol{R}_1(\kappa) = \boldsymbol{R}_0(\kappa)[\boldsymbol{\Lambda} + \kappa \boldsymbol{I}](\mu_1 \boldsymbol{I} - \boldsymbol{Q})^{-1}, \\
\boldsymbol{R}_2(\kappa) = \alpha \boldsymbol{R}_0(\kappa)(\mu_2 \boldsymbol{I} - \boldsymbol{Q})^{-1}.
\end{cases}$$

The row vector $\boldsymbol{r}$ is the stationary probability distribution of the background process $n(t)$, which is obtained as the unique solution for the system $\boldsymbol{rQ} = 0, \boldsymbol{re} = 1$.

Proof. In (4), we denoted $u_1 = u_2 = 0$, which allows us to remove processes $m_1(t)$ and $m_2(t)$ from consideration. Thus we get a system of equations yet for the three-dimensional process $\{n(t), k(t), i(t)\}$ and consider it in the stationary regime, which spares us from the time derivative t.

Let us denote

$$\boldsymbol{F}_k(w,\epsilon) = \lim_{t\to\infty} \boldsymbol{F}_k(w,0,0,t,\epsilon).$$

Then we obtain

$$\begin{aligned}
&(\boldsymbol{Q} - \boldsymbol{\Lambda} - \alpha\boldsymbol{I})\boldsymbol{F}_0(w,\epsilon) + \mu_1\boldsymbol{F}_1(w,\epsilon) + \mu_2\boldsymbol{F}_2(w,\epsilon) + j\boldsymbol{F}_0'(w,\epsilon) = 0,\\
&\boldsymbol{\Lambda}\boldsymbol{F}_0(w,\epsilon) + (\boldsymbol{Q} + (e^{j\epsilon w} - 1)\boldsymbol{\Lambda} - \boldsymbol{I}\mu_1)\boldsymbol{F}_1(w,\epsilon) - je^{-j\epsilon w}\boldsymbol{F}_0'(w,\epsilon) = 0,\\
&\alpha\boldsymbol{F}_0(w,\epsilon) + (\boldsymbol{Q} + (e^{j\epsilon w} - 1)\boldsymbol{\Lambda} - \boldsymbol{I}\mu_2)\boldsymbol{F}_2(w,\epsilon) = 0.
\end{aligned} \tag{5}$$

Making the passage to the limit $\epsilon \to 0$ in (5) results in

$$\begin{aligned}
&(\boldsymbol{Q} - \boldsymbol{\Lambda} - \alpha\boldsymbol{I})\boldsymbol{F}_0(w) + \mu_1\boldsymbol{F}_1(w) + \mu_2\boldsymbol{F}_2(w) + j\boldsymbol{F}_0'(w) = 0,\\
&\boldsymbol{\Lambda}\boldsymbol{F}_0(w) + (\boldsymbol{Q} - \boldsymbol{I}\mu_1)\boldsymbol{F}_1(w) - j\boldsymbol{F}_0'(w) = 0,\\
&\alpha\boldsymbol{F}_0(w) + (\boldsymbol{Q} - \boldsymbol{I}\mu_2)\boldsymbol{F}_2(w) = 0.
\end{aligned} \tag{6}$$

Solution for the system will be found as

$$\boldsymbol{F}_k(w) = \Phi(w)\boldsymbol{R}_k, \tag{7}$$

where $\boldsymbol{R}_n$ is the server's state stationary probability distribution, and $\Phi(w)$ is the asymptotic approximation of the characteristic function of the number of requests in the orbit under the condition of their long delay. Substituting (7) in (6) and dividing it by $\Phi(w)$, we get

$$\begin{aligned}
&(\boldsymbol{Q} - \boldsymbol{\Lambda} - \alpha\boldsymbol{I})\boldsymbol{R}_0 + \mu_1\boldsymbol{R}_1 + \mu_2\boldsymbol{R}_2 + j\frac{\Phi'(w)}{\Phi(w)}\boldsymbol{R}_0 = 0,\\
&\boldsymbol{\Lambda}\boldsymbol{R}_0 + (\boldsymbol{Q} - \boldsymbol{I}\mu_1)\boldsymbol{R}_1 - j\frac{\Phi'(w)}{\Phi(w)}\boldsymbol{R}_0 = 0,\\
&\alpha\boldsymbol{R}_0 + (\boldsymbol{Q} - \boldsymbol{I}\mu_2)\boldsymbol{R}_2 = 0.
\end{aligned} \tag{8}$$

Since w only appears in $\frac{\Phi'(w)}{\Phi(w)}$, other equation terms do not depend on w. It means that $\Phi(w)$ is exponential. Taking into account that $\Phi(w)$ has the meaning of an asymptotic approximation of the characteristic function of the number of requests in the source of retrial calls, we can clarify the form of this function

$$\frac{\Phi'(w)}{\Phi(w)} = \frac{e^{j\kappa w}j\kappa}{e^{j\kappa w}}, \tag{9}$$

which follows to $j\frac{\Phi(w)'}{\Phi(w)} = -\kappa$. Let us substitute this expression into (8). Then we obtain

$$\begin{aligned}
&(\boldsymbol{Q} - \boldsymbol{\Lambda} - \alpha\boldsymbol{I})\boldsymbol{R}_0 + \mu_1\boldsymbol{R}_1 + \mu_2\boldsymbol{R}_2 - \kappa\boldsymbol{R}_0 = 0,\\
&\boldsymbol{\Lambda}\boldsymbol{R}_0 + (\boldsymbol{Q} - \boldsymbol{I}\mu_1)\boldsymbol{R}_1 + \kappa\boldsymbol{R}_0 = 0,\\
&\alpha\boldsymbol{R}_0 + (\boldsymbol{Q} - \boldsymbol{I}\mu_2)\boldsymbol{R}_2 = 0.
\end{aligned} \tag{10}$$

Let us write down the normality condition for the stationary distribution of the number of served requests

$$\boldsymbol{R}_0 + \boldsymbol{R}_1 + \boldsymbol{R}_2 = \boldsymbol{r}.$$

Based on this equation, as well as on the last two equations of (10), we write the system as

$$\begin{cases} \boldsymbol{R}_1 = \boldsymbol{R}_0[\boldsymbol{\Lambda} + \kappa \boldsymbol{I}](\mu_1 \boldsymbol{I} - \boldsymbol{Q})^{-1}, \\ \boldsymbol{R}_2 = \alpha \boldsymbol{R}_0(\mu_2 \boldsymbol{I} - \boldsymbol{Q})^{-1}, \\ \boldsymbol{R}_0 + \boldsymbol{R}_1 + \boldsymbol{R}_2 = \boldsymbol{r}. \end{cases} \tag{11}$$

Let us sum up the equations of system (5)

$$[\boldsymbol{F}_0(w,\epsilon) + \boldsymbol{F}_1(w,\epsilon) + \boldsymbol{F}_2(w,\epsilon)]\boldsymbol{Q} + \boldsymbol{F}_1(w,\epsilon)(e^{jw\epsilon} - 1)\boldsymbol{\Lambda} + \boldsymbol{F}_2(w,\epsilon)(e^{jw\epsilon} - 1)\boldsymbol{\Lambda} + je^{-jw\epsilon}(e^{jw\epsilon} - 1)\boldsymbol{F}_0'(w,\epsilon) = 0.$$

Multiplying the resulting equations by the unit column vector $\boldsymbol{e}$, we obtain

$$\{\boldsymbol{F}_1(w,\epsilon) + \boldsymbol{F}_2(w,\epsilon)\}\boldsymbol{\Lambda}\boldsymbol{e} + je^{-jw\epsilon}\boldsymbol{F}_0'(w,\epsilon)\boldsymbol{e} = 0.$$

Then we substitute product (7) into the resulting equation

$$[\boldsymbol{R}_1 + \boldsymbol{R}_2]\boldsymbol{\Lambda}\boldsymbol{e} + j\frac{\Phi'(w)}{\Phi(w)}\boldsymbol{R}_0\boldsymbol{e} = 0$$

and make the replacement

$$[\boldsymbol{R}_1 + \boldsymbol{R}_2]\boldsymbol{\Lambda}\boldsymbol{e} - \kappa\boldsymbol{R}_0\boldsymbol{e} = 0. \tag{12}$$

From (12) we can express κ with $\boldsymbol{R}_0$,$\boldsymbol{R}_1$ and $\boldsymbol{R}_2$. In addition, we can rewrite system (11) as follows

$$\begin{cases} \boldsymbol{R}_0(\kappa) = \boldsymbol{r}\{\boldsymbol{I} + [\boldsymbol{\Lambda} + \kappa\boldsymbol{I}](\mu_1\boldsymbol{I} - \boldsymbol{Q})^{-1} + \alpha(\mu_2\boldsymbol{I} - \boldsymbol{Q})^{-1}\}^{-1}, \\ \boldsymbol{R}_1(\kappa) = \boldsymbol{R}_0(\kappa)[\boldsymbol{\Lambda} + \kappa\boldsymbol{I}](\mu_1\boldsymbol{I} - \boldsymbol{Q})^{-1}, \\ \boldsymbol{R}_2(\kappa) = \alpha\boldsymbol{R}_0(\kappa)(\mu_2\boldsymbol{I} - \boldsymbol{Q})^{-1} \end{cases}$$

Theorem 1 is auxiliary since the general solution for the system is stated in Theorem 2 and needs the results obtained at this stage, namely, the normalized average amount of requests in the source of retrial calls κ and the stationary probability distribution of the server's state $\boldsymbol{R}_k$.

Theorem 2. *The asymptotic approximation of the two-dimensional characteristic function of the number of served requests of the MMPP input process and the number of served outgoing requests for some time t has the form*

$$\lim_{\sigma\to 0} M\{\exp(ju_1 m_1(t))\exp(ju_2 m_2(t))\}$$

$$= \lim_{\epsilon\to 0}\{\sum_{k=0}^{2}\boldsymbol{F}_k(0,u_1,u_2,t,\epsilon)\}\boldsymbol{e} = \boldsymbol{R}\cdot\exp\{\boldsymbol{G}(u_1,u_2)t\}\boldsymbol{e}\boldsymbol{e},$$

where matrix $\boldsymbol{G}(u_1, u_2)$ *can be written as*

$$\boldsymbol{G}(u_1,u_2) = \begin{bmatrix} \boldsymbol{Q}-\boldsymbol{\Lambda}-(\alpha+\kappa)\boldsymbol{I} & \mu_1 e^{ju_1}\boldsymbol{I} & \mu_2 e^{ju_2}\boldsymbol{I} \\ \boldsymbol{\Lambda}+\kappa\boldsymbol{I} & \boldsymbol{Q}-\mu_1\boldsymbol{I} & 0 \\ \alpha\boldsymbol{I} & 0 & \boldsymbol{Q}-\mu_2\boldsymbol{I} \end{bmatrix}^T,$$

row vector $\boldsymbol{R} = \{\boldsymbol{R}_0, \boldsymbol{R}_1, \boldsymbol{R}_2\}$ *is the two-dimensional stationary probability distribution of the process* $\{k(t), n(t)\}$*, where* $\boldsymbol{R}_k$ *has dimension* N*,* κ *is the normalized average number of requests in the orbit, and e and ee are unit vector columns of dimensions* N *and* $N \cdot K$*, where* K *is the number of server's states.*

Proof. After making the passage to the limit $\lim_{\epsilon\to 0} \boldsymbol{F}_k(w, u_1, u_2, t, \epsilon) = \boldsymbol{F}_k(w, u_1, u_2, t)$ in resulting system (4), it will be written as follows

$$\begin{aligned} \frac{\partial \boldsymbol{F}_0(w,u_1,u_2,t)}{\partial t} &= (\boldsymbol{Q}-\boldsymbol{\Lambda}-\alpha\boldsymbol{I})\boldsymbol{F}_0(w,u_1,u_2,t) \\ &+ \mu_1 e^{ju_1}\boldsymbol{F}_1(w,u_1,u_2,t) \\ &+ \mu_2 e^{ju_2}\boldsymbol{F}_2(w,u_1,u_2,t) + j\frac{\partial \boldsymbol{F}_0(w,u_1,u_2,t)}{\partial w}, \\ \frac{\partial \boldsymbol{F}_1(w,u_1,u_2,t)}{\partial t} &= \boldsymbol{\Lambda}\boldsymbol{F}_0(w,u_1,u_2,t) + (\boldsymbol{Q}-\boldsymbol{I}\mu_1)\boldsymbol{F}_1(w,u_1,u_2,t) \\ &- j\frac{\partial \boldsymbol{F}_0(w,u_1,u_2,t)}{\partial w}, \\ \frac{\partial \boldsymbol{F}_2(w,u_1,u_2,t)}{\partial t} &= \alpha\boldsymbol{F}_0(w,u_1,u_2,t) + (\boldsymbol{Q}-\boldsymbol{I}\mu_2)\boldsymbol{F}_2(w,u_1,u_2,t). \end{aligned} \tag{13}$$

Solution for (13) will be found as

$$\boldsymbol{F}_k(w,u_1,u_2,t) = \Phi(w)\boldsymbol{F}_k(u_1,u_2,t). \tag{14}$$

Substituting (14) into (13) and dividing both parts of equations by $\Phi(w)$ we obtain

$$\begin{aligned} \frac{\partial \boldsymbol{F}_0(u_1,u_2,t)}{\partial t} &= (\boldsymbol{Q}-\boldsymbol{\Lambda}-\alpha\boldsymbol{I})\boldsymbol{F}_0(u_1,u_2,t) + \mu_1 e^{ju_1}\boldsymbol{F}_1(u_1,u_2,t) \\ &+ \mu_2 e^{ju_2}\boldsymbol{F}_2(u_1,u_2,t) + j\frac{\Phi'(w)}{\Phi(w)}\boldsymbol{F}_0(u_1,u_2,t), \\ \frac{\partial \boldsymbol{F}_1(u_1,u_2,t)}{\partial t} &= \boldsymbol{\Lambda}\boldsymbol{F}_0(u_1,u_2,t) + (\boldsymbol{Q}-\boldsymbol{I}\mu_1)\boldsymbol{F}_1(u_1,u_2,t) \\ &- j\frac{\Phi'(w)}{\Phi(w)}\boldsymbol{F}_0(u_1,u_2,t), \\ \frac{\partial \boldsymbol{F}_2(u_1,u_2,t)}{\partial t} &= \alpha\boldsymbol{F}_0(u_1,u_2,t) + (\boldsymbol{Q}-\boldsymbol{I}\mu_2)\boldsymbol{F}_2(u_1,u_2,t). \end{aligned} \tag{15}$$

Function $\Phi(w)$ has form (9). After substituting it, (15) will have the form

$$\begin{aligned}\frac{\partial \boldsymbol{F}_0(u_1,u_2,t)}{\partial t} &= (\boldsymbol{Q}-\boldsymbol{\Lambda}-(\alpha+\kappa)\boldsymbol{I})\boldsymbol{F}_0(u_1,u_2,t)\\ &+\mu_1 e^{ju_1}\boldsymbol{F}_1(u_1,u_2,t)\\ &+\mu_2 e^{ju_2}\boldsymbol{F}_2(u_1,u_2,t),\\ \frac{\partial \boldsymbol{F}_1(u_1,u_2,t)}{\partial t} &= (\boldsymbol{\Lambda}+\kappa\boldsymbol{I})\boldsymbol{F}_0(u_1,u_2,t)+(\boldsymbol{Q}-\boldsymbol{I}\mu_1)\boldsymbol{F}_1(u_1,u_2,t)\\ &+0\boldsymbol{F}_2(u_1,u_2,t),\\ \frac{\partial \boldsymbol{F}_2(u_1,u_2,t)}{\partial t} &= \alpha\boldsymbol{F}_0(u_1,u_2,t)+0\boldsymbol{F}_1(u_1,u_2,t)\\ &+(\boldsymbol{Q}-\boldsymbol{I}\mu_2)\boldsymbol{F}_2(u_1,u_2,t).\end{aligned} \tag{16}$$

Let us denote

$$\boldsymbol{FF}(u_1,u_2,t)=\{\boldsymbol{F}_0(u_1,u_2,t),\boldsymbol{F}_1(u_1,u_2,t),\boldsymbol{F}_2(u_1,u_2,t)\},$$

$$\boldsymbol{G}(u_1,u_2)=\begin{bmatrix}\boldsymbol{Q}-\boldsymbol{\Lambda}-(\alpha+\kappa)\boldsymbol{I} & \mu_1 e^{ju_1}\boldsymbol{I} & \mu_2 e^{ju_2}\boldsymbol{I}\\ \boldsymbol{\Lambda}+\kappa\boldsymbol{I} & \boldsymbol{Q}-\mu_1\boldsymbol{I} & 0\\ \alpha\boldsymbol{I} & 0 & \boldsymbol{Q}-\mu_2\boldsymbol{I}\end{bmatrix}^T,$$

$\boldsymbol{G}(u_1,u_2)$ is the transposed matrix of system coefficients (16). Then we obtain the following matrix equation

$$\frac{\partial \boldsymbol{FF}(u_1,u_2,t)}{\partial t}=\boldsymbol{FF}(u_1,u_2,t)\boldsymbol{G}(u_1,u_2),$$

the general solution of which is

$$\boldsymbol{FF}(u_1,u_2,t)=\boldsymbol{C}e^{G(u_1,u_2)t}. \tag{17}$$

Finding a unique solution corresponding to the functioning of the system under consideration requires us to set the initial condition

$$\boldsymbol{FF}(u_1,u_2,0)=\boldsymbol{R}, \tag{18}$$

where row vector $\boldsymbol{R}$ is the two-dimensional stationary probability distribution of server's state $k(t)$, which was found in Theorem 1. With (18) described, we solve can solve the Cauchy problem (17)

$$\boldsymbol{FF}(u_1,u_2,t)=\boldsymbol{R}e^{G(u_1,u_2)t}.$$

Since we are focusing on the probability distribution of requests in output processes the marginal distribution is needed. For this, we multiply row vector $\boldsymbol{FF}(u_1,u_2,t)$ by unit vector-column $\boldsymbol{e}$ of size N and the right part of the equation by unit vector-column $\boldsymbol{ee}$ of size $K \cdot N$. We obtain

$$\boldsymbol{FF}(u_1,u_2,t)\boldsymbol{e}=\boldsymbol{R}e^{G(u_1,u_2)t}\boldsymbol{ee}. \tag{19}$$

(19) is the solution for the considered system.

5 Explicit Probability Distribution

Characteristic function (19) allows us to move to an explicit formula for calculating probabilities of the number of served requests in output processes $m_1(t)$ and $m_2(t)$. (19) contains the matrix exponent, for which we apply the similarity transformation [4]

$$\boldsymbol{G}(u_1, u_2) = \boldsymbol{T}(u_1, u_2)\boldsymbol{GJ}(u_1, u_2)\boldsymbol{T}(u_1, u_2)^{-1},$$

where $\boldsymbol{T}(u_1, u_2)$ is an eigenvector matrix of $\boldsymbol{G}(u_1, u_2)$, and $\boldsymbol{GJ}(u_1, u_2)$ is a diagonal eigenvalue matrix of $\boldsymbol{G}(u_1, u_2)$. This conversion is objective for any power m of some matrix $\boldsymbol{A}^m$, which follows it is also valid for the matrix exponent

$$e^{\boldsymbol{G}(u_1,u_2)t} = \boldsymbol{T}(u_1, u_2) \cdot \begin{bmatrix} e^{t\Lambda_1(u_1,u_2)} & 0 & 0 \\ 0 & e^{t\Lambda_2(u_1,u_2)} & 0 \\ 0 & 0 & e^{t\Lambda_3(u_1,u_2)} \end{bmatrix} \cdot \boldsymbol{T}(u_1, u_2)^{-1},$$

where Λ_n is an eigenvalue of $\boldsymbol{G}(u_1, u_2)$. Then the distribution is written as follows

$$\boldsymbol{F}(u_1, u_2, t)\boldsymbol{E} = \boldsymbol{R} \cdot \boldsymbol{T}(u_1, u_2) \cdot \begin{bmatrix} e^{t\Lambda_1(u_1,u_2)} & 0 & 0 \\ 0 & e^{t\Lambda_2(u_1,u_2)} & 0 \\ 0 & 0 & e^{t\Lambda_3(u_1,u_2)} \end{bmatrix} \cdot \boldsymbol{T}(u_1, u_2)^{-1} \cdot \boldsymbol{E}.$$

To restore the distribution, we use the inverse Fourier transform for discrete values

$$P(m_1, m_2, t) = \frac{1}{(2\pi)^2} \int_{-\pi}^{\pi} \int_{-\pi}^{\pi} e^{-i \cdot u_1 \cdot m_1} e^{-i \cdot u_2 \cdot m_2} \boldsymbol{FF}(u_1, u_2, t)\boldsymbol{e}\, du_1 du_2. \quad (20)$$

The resulting formula characterizes the probability of servicing m_1 input process requests and m_2 outgoing requests at the moment t in the system under consideration.

6 Numerical Examples

Let us compare simulation output with the calculations based on the obtained asymptotic results. σ affects accuracy, since the solution of the system was obtained under the asymptotic condition of a long delay of requests in the orbit.

We measure the accuracy of the results with the Kolmogorov-Smirnov distance, which is calculated as

$$\Delta = \max_{0 \le i \le \infty} \left| \sum_{v=0}^{i} (P_0(v) - P_1(v)) \right|,$$

where $P_0(v)$ and $P_1(v)$ are comparable probability distributions.

Let us set the following parameters

$$\alpha = 0.6, \mu_1 = 2, \mu_2 = 1.5, t = 15,$$

$$\boldsymbol{Q} = \begin{bmatrix} -0.5 & 0.2 & 0.3 \\ 0.15 & -0.2 & 0.05 \\ 0.3 & 0.4 & -0.7 \end{bmatrix}, \boldsymbol{\Lambda} = \begin{bmatrix} 1 & 0 & 0 \\ 0 & 0.6 & 0 \\ 0 & 0 & 0.7 \end{bmatrix}.$$

The input process intensity can be written in form $\boldsymbol{r} \cdot \boldsymbol{\Lambda} \cdot \boldsymbol{e}$, after calculation of which, we get the value 0.72. For the parameters set, we obtained the following results.

Let us denote: Δ_S is the KS distance values for the summary distribution, which implies, that served incoming and outgoing requests are homogeneous, and Δ_{TD} is the KS distance values for the two-dimensional distribution of served requests, which are, in the two-dimensional case, of different types.

Table 1. KS distance values for various σ

σ	10	1	0.6	0.4	0.2	0.1	0.05	0.01
Δ_S	0.053	0.045	0.04	0.036	0.028	0.023	0.018	0.016
Δ_{TD}	0.059	0.049	0.042	0.035	0.024	0.015	0.01	0.003

In Table 1, we can notice that for lower values of σ asymptotic results of the two-dimensional distribution are more accurate. Let us raise system load by setting up the new intensity matrix with greater values of diagonal elements

$$\boldsymbol{\Lambda} = \begin{bmatrix} 1.2 & 0 & 0 \\ 0 & 0.9 & 0 \\ 0 & 0 & 1.5 \end{bmatrix}.$$

For these parameters, the overall input process intensity is 1.07. Calculations for the new set of parameters are

Table 2. KS distance values for various σ with high system load

σ	10	1	0.6	0.4	0.2	0.1	0.05	0.01
Δ_S	0.037	0.029	0.024	0.02	0.015	0.01	0.008	0.008
Δ_{TD}	0.066	0.048	0.039	0.031	0.019	0.01	0.006	0.002

Based on the performed experiments, we can conclude that a tendency towards an accuracy increase of asymptotic results is always observed when decreasing σ. For a value of σ exceeding the intensity of the input process, the accuracy does not exceed 0.066 (the longest KS distance, which is observed in the case of a two-dimensional probability distribution), which indicates a high degree of accuracy of the obtained approximation. Raising system load with input process requests, as can be seen in Table 2, has a positive effect on the accuracy of the asymptotic results. It is because more events occur within a fixed time interval during simulation.

7 Conclusion

In this paper, we have described the process of finding the asymptotic approximation of the two-dimensional characteristic function of the number of incoming and outgoing requests that have finished serving in retrial queue with two-way communication under the condition of a long delay in the source of retrial calls. This allows retrieving different performance characteristics, including the correlation of the processes in the system output. Moreover, we used it to calculate probability values for further experiments.

Carried out numerical experiments show that obtained approximation gives high accuracy results, and for this reason, it can be used for further research of this type of system.

References

1. Artalejo, J.R., Gómez-Corral, A.: Retrial Queueing Systems: A Computational Approach. Springer-Verlag, Heidelberg (2008). https://doi.org/10.1007/978-3-540-78725-9
2. Baiocchi, A., Blefari-Melazzi, N.: Steady-state analysis of the MMPP/G/1/K queue. IEEE Trans. Commun. **41**(4), 531–534 (1993)
3. Blaginin, A., Lapatin, I.: The two-dimensional output process of retrial queue with two-way communication. In: Dudin, A., Nazarov, A., Moiseev, A. (eds.) ITMM 2020. CCIS, vol. 1391, pp. 279–290. Springer, Cham (2021). https://doi.org/10.1007/978-3-030-72247-0_21
4. Bronson, R.: Matrix Methods: An Introduction. Gulf Professional Publishing, London (1991)
5. Burke, P.: The output process of a stationary m/m/s queueing system. Ann. Math. Stat. **39**(4), 1144–1152 (1968)
6. Daley, D.: Queueing output processes. Adv. Appl. Probab. **8**(2), 395–415 (1976)
7. Fischer, W., Meier-Hellstern, K.: The Markov-modulated poisson process (MMPP) cookbook. Perform. Eval. **18**(2), 149–171 (1993)
8. Gharbi, N., Dutheillet, C.: An algorithmic approach for analysis of finite-source retrial systems with unreliable servers. Comput. Math. Appl. **62**(6), 2535–2546 (2011)
9. Kulkarni, V.G.: On queueing systems with retrials. J. Appl. Probab. **20**, 380–389 (1983)
10. Lapatin, I., Nazarov, A.: Asymptotic analysis of the output process in retrial queue with Markov-modulated Poisson input under low rate of retrials condition. In: Vishnevskiy, V.M., Samouylov, K.E., Kozyrev, D.V. (eds.) DCCN 2019. CCIS, vol. 1141, pp. 315–324. Springer, Cham (2019). https://doi.org/10.1007/978-3-030-36625-4_25
11. Meier-Hellstern, K.S.: A fitting algorithm for Markov-modulated Poisson processes having two arrival rates. Eur. J. Oper. Res. **29**(3), 370–377 (1987)
12. Mirasol, N.M.: The output of an m/g/∞ queuing system is poisson. Oper. Res. **11**(2), 282–284 (1963)
13. Muscariello, L., Meillia, M., Meo, M., Marsan, M.A., Cigno, R.L.: An MMPP-based hierarchical model of internet traffic. In: 2004 IEEE International Conference on Communications (IEEE Cat. No. 04CH37577), vol. 4, pp. 2143–2147. IEEE (2004)

14. Nazarov, A.A., Paul, S., Gudkova, I., et al.: Asymptotic analysis of Markovian retrial queue with two-way communication under low rate of retrials condition (2017)
15. Paul, S., Phung-Duc, T.: Retrial queueing model with two-way communication, unreliable server and resume of interrupted call for cognitive radio networks. In: Dudin, A., Nazarov, A., Moiseev, A. (eds.) ITMM/WRQ -2018. CCIS, vol. 912, pp. 213–224. Springer, Cham (2018). https://doi.org/10.1007/978-3-319-97595-5_17
16. Phung-Duc, T.: Retrial queueing models: a survey on theory and applications (2019)

Method of Analyzing the Availability Factor in a Mesh Network

Alexander Dagaev[1(✉)], Van Dai Pham[1], Ruslan Kirichek[1], Olga Afanaseva[2], and Ekaterina Yakovleva[3]

[1] Bonch-Bruevich Saint-Petersburg State University of Telecommunications, St. Petersburg, Russian Federation
fam.vd@spbgut.ru, kirichek@sut.ru

[2] Staint-Petersburg Mining University (SPMU), St. Petersburg, Russian Federation

[3] Ivangorod Humanitarian and Technical Institute (Branch) of Federal State Autonomous Educational Institution of Higher Education, St. Petersburg State University of Aerospace Instrumentation, St. Petersburg, Russian Federation

Abstract. The development of network technologies leads to the introduction of data transmission systems in all areas of human activity. Problems cover large areas with reliable networks, intellectualizing devices for receiving and transmitting data, increasing the speed, quality and spectrum of transmitted information. Requirements for the given characteristics of the reliability of networks are among the most promising and essential tasks since the cost of equipment, service life, and network maintenance strategy depend on this. Analytical and simulation models play an essential role in determining the reliability characteristics; mathematical apparatus and programming become an inevitable factor in a successful project. This paper presents a methodology and an example of calculating the reliability of networks in a mesh topology.

Keywords: simulation · system · mesh network · mean availability factor · probability

1 Introduction

Mesh network topology is designed to solve a wide range of tasks, from monitoring the battlefield and analyzing weather conditions to redistribute smart sustainable city technologies' load. The mesh network structure can be represented in a graph or an $m \times n$ matrix. Determination of reliability characteristics depends on the initial structure of the network and, depending on the reliability of its elements, and their maintenance strategies can take different values. Today, many sources are known on the topic of determining the characteristics of the system reliability [1,9].

Nowadays, the applications of communication networks has been rapidly increasing. The reliability and cost of these systems are important and are largely

V. M. Vishnevskiy et al. (Eds.): DCCN 2021, CCIS 1552, pp. 346–358, 2022.
https://doi.org/10.1007/978-3-030-97110-6_27

determined by network topology [4]. Also in [12] authors studied the global reliability of communication networks.

Authors in [6] studied and compared the reliability of networks using Wiener index. The paper presented the topology invariant which calculates the reliability of the newly constructed network using graph operations tensor product and Cartesian product in Topology theory.

In [10] authors assumed that the failure probability of a failure set, a set of network elements that simultaneously fails at a single disaster, can be decreased according to cost for protection. Hence, the network failure probability defined as the probability that the entire information network is not connected, is decreased. They defined a network design problem that determines cost assigned to each failure set so that the sum of the cost assigned to each failure set is minimized under the constraint that the network failure probability must be within a threshold.

The paper [11] discussed new trends in the design of reliable communication systems with special focus on software failure mitigation, reliability of wireless communications, robust optimization and network design, multilevel and multirealm network resilience, multiple criteria routing approaches in multilayer networks, resilience options of the fixed IP backbone network in the interplay with the optical layer survivability, reliability of cloud computing networks, as well as resiliency of software-defined networks.

The paper [8] presented an effective optimization solution by designing the dual redundancy warm-standby module of the mission computer and I/O port, the algorithm of selecting output path of the mission computer in network nodes, the decision-making algorithm upon the on-duty host and output, and the video output decision-making algorithm upon the upper host to optimize the network node architecture of amphibious combat simulation system.

In this book [5] authors presented the core concepts and methods of network reliability analysis. The book explains the modeling and critical analysis of systems and probabilistic networks, and requires only a minimal background in probability theory and computer programming.

The paper [3] discussed conventional consecutive k systems and considered failure criteria (single failure criterion and multiple failure criteria), geometric structure of the system, states of components and the system, weight of each component, dependency of components.

The paper [14] presented a method for obtaining lower and upper bounds on the required value, which produces a functional (symbolic) form for the answer, especially useful for subsequent sensitivity analyses.

Authors in [7] provided the development and implementation of a new methodology for expanding existing computer networks. A genetic algorithm was proposed to optimize a specified objective function (reliability measure) under a given set of network constraints. The proposed algorithm can be applied to solve various network expansion problems (optimize diameter, average distance and computer network reliability for network expansion).

Papers [2,13,15,16] are devoted to studying the reliability characteristics of network structures and computer systems, such as calculations for the exponential distribution of reliability characteristics, packet coding in communication channels, and reliability calculation for Cisco equipment. In these articles, asymptotic estimates of reliability characteristics were used, however, evaluation of complex structures and non-asymptotic models were not given attention. Therefore, let us describe the functioning strategy more detail in this paper.

2 Description of the Functioning Model

The model assumes the presence of built-in control with the detection of failures in the system and its complete recovery. In the event of a failure, the system is repaired and is idle until it is restored.

At the initial moment $t_0 = 0$, the system starts to work, and the availability has a maximum value. After that, the system works to failure – ξ_i then recovery is performed, which lasts a period – $n_{i\ fr}$.

After recovery – τ_{ir} the system continues its work until the next moment of failure, then there is a recovery and transition to an operational state. This cycle is repeated until the selected time t. The presented strategy is shown in Fig. 1.

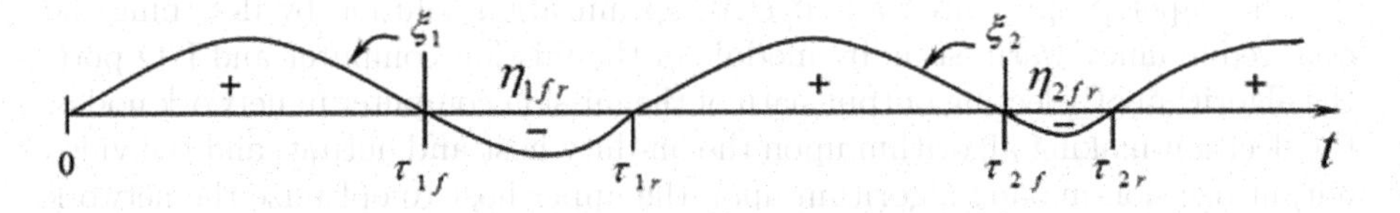

Fig. 1. The strategy takes into account the built-in control

The pros and cons of the above designations are the periods of operation and repair of the system; ξ_i – i-th time to failure; $n_{i\ fr}$ – the duration of the i-th disaster recovery; τ_{if} and τ_{ir} – time intervals from the start of work to the i-th failure and the i-th recovery. These values can be written in terms of several other random variables:

$$\tau_{0r} = 0; \begin{cases} \tau_{1f} = \xi_1 \\ \tau_{1r} = \xi_1 + \eta_{fr} \end{cases}; \begin{cases} \tau_{2f} = \tau_{1r} + \xi_2 \\ \tau_{2r} = \tau_{1r} + \xi_2 + \eta_{fr} \end{cases}; \begin{cases} \tau_{if} = \tau_{i-1,r} + \xi_i \\ \tau_{2r} = \tau_{i-1,r} + \xi_i + \eta_{fr} \end{cases}$$

The mean availability is the sum of the probabilities of the system being in a working state:

$$\begin{aligned} K(t) &= \sum_{i=1}^{\infty} P(\tau_{i-1,r} < t < \tau_{i,f}) = P_1(t < \xi_1) + \sum_{i=1}^{\infty} P(\tau_{i,r} < t < \tau_{i+1,f}) \\ &= (1 - F_{\xi}(t)) + \sum_{i=1}^{\infty} P(\tau_{i,r} < t < \tau_{i+1,f}) \end{aligned} \quad (1)$$

After performing some transformations, we obtain the convolution by doing the inverse Laplace transform. Then, we obtain the equation of the non-asymptotic system availability factor. Thus, the availability can be written as follows:

$$K(t) = [1 - F_{\xi}(t)] + \int_0^t f_{\eta_{fr}} \int_0^{t-\chi} f_{\xi}(y) K(t - x - y) dy dx \tag{2}$$

The asymptotic availability equation is found under the condition $t \to \infty$ and can be derived from this equation. The asymptotic availability is the ratio of the mathematical expectation of the failure time to the sum of the mathematical expectations of the failure and recovery times.

$$K_a = \frac{M(\xi)}{M(\xi) + M(\eta_{fr})} \tag{3}$$

2.1 Identical Elements in the System

If under the condition of the same elements in the graph and a parallel data transfer condition, we can assume that the reliability of elements in each chain can be calculated as a sequential structure (Fig. 2).

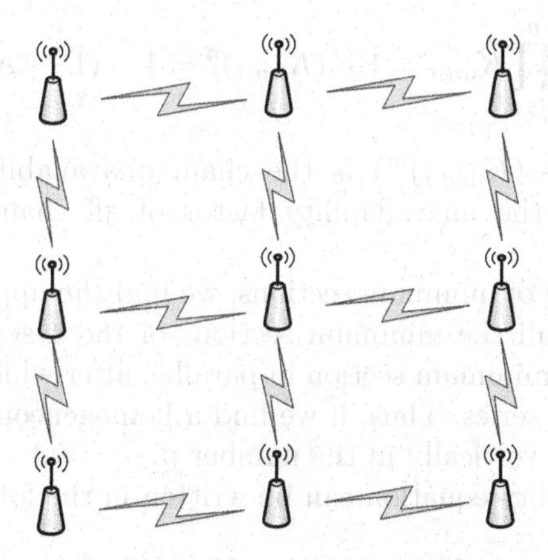

Fig. 2. The mesh network structure

Therefore, the reliability calculation (the probability of no-failure operation and the availability) can be calculated using a similar formula for non-recoverable and recoverable systems. Using the method of dimension reduction of the original problem, first, we calculate the reliability of independent paths of the system (chains) in the graph. Next, we determine the reliability of parallel chains. Finally, we use the method of minimal paths and minimal sections. A path is any set of elements for which the system is operable. Thus, the availability for an individual chain will have the following product form:

$$K_i = \prod_{i=1}^{m} K_{nod\,i} = K_{nod\,i}{}^{m} \tag{4}$$

where $K_{nod\,i}$ is the availability of the i-th element of the chain, in this case it should be taken into account that $K_{nod\,i}$ are the same for $\forall i \in (1 \ldots m)$

As can be seen from the equation, chain availability is the availability of an individual element to the power m. Therefore, to receive the upper-reliability estimate based on the minimum path method, we first need to find all the minimum paths of the system. Then, we connect the elements of each minimum path in series, and all the resulting chains with a series connection of the elements are connected in parallel.

Then, we will find the system availability. For this, we write the availability (and the probability of no-failure operation) for a parallel connection. It is known that the element unavailability ratio is determined first for parallel connection of recoverable elements, or the element failure probability is determined for non-recoverable systems.

Thus, the system availability factor K_s is calculated as the deviation from the unit of the system unavailability factor $\prod_{i=1}^{n} K_{un.i}$, i.e.:

$$K_s = 1 - \prod_{i=1}^{n} K_{un.i} = 1 - (K_{un.i})^n = 1 - (1 - (K_{nod\,j})^m)^n \tag{5}$$

where $K_{un.i} = (1 - (K_{nod\,j})^m)$ is the chain unavailability factor; $K_{un.s} = (1 - (K_{nod\,j})^m)^n$ is the unavailability factor of all chains or system, where $j \in (1 \ldots n)$.

Using the method of minimum sections, we find the upper availability value. To do this, we find all the minimum sections of the system, then we connect the elements of each minimum section in parallel, after which we connect all the minimum sections in series. Thus, if we find a homogeneous network section, it contains all elements vertically in the number n.

Then the availability equation can be written in the following form:

$$K_s = (1 - (1 - K_{nod\,j})^n)^m \tag{6}$$

where $(1 - K_{nod\,j})$ – the unavailability factor of the j-th element.

Next, we give an example of calculating the system under the conditions of the asymptotic formulation of the problem. In this case, the asymptotic availability value is calculated using the formula (3). The initial data are the mathematical expectations of the time of failure and recovery. Based on the real data of the equipment passports, it is known that the mathematical expectation of failure is 90000 h, and the recovery time is two days. In this case, the value of the asymptotic availability factor of the element will be equal to $K_{nod\,j} = 0.999467$.

It should be noted that this value is the worst estimate of the availability coefficient of a system element. As a mesh topology, we take a network with dimension $n = 8$ and $m = 7$ cells. Substituting this value into the Eq. (5), we receive:

$$K_s = (1 - (1 - 0.999467^7))^8 = 1 - (3.71E^{-20}) \cong 1 \tag{7}$$

where the unavailability coefficient of a single chain is 0.003725. In this case, the system availability will tend to unity since the chain availability slightly decreases with an increase in its length (with an increase in the degree m of Eq. (5)).

We consider the lower value of the system availability coefficient in this case. We calculate the unavailability of an element, it will be as follows: $(1 - K_{nod\,j} = 5.33e^{-4}$. Then the lower value of the availability coefficient will be as follows:

$$K_s = (1 - (1 - 0.999467)^8)^7 \cong 1$$

Due to the high availability coefficient of one element, the upper and lower availability coefficient will be very high and tend to unity. Let us show using a test example that the availability can take values different from unity. Assume that we have a system with low reliability $Mean.failure = 1000$, $Mean.recovery = 300$. Then the asymptotic availability ratio will be 0.769. Based on this, substituting the value in (5) and (6), we get the availability value equal to 0.751 and 0.99994.

2.2 Failure of Some Elements

In this section, we present formulas that can be used to failure some elements, where the rest of the elements send data, and the system performs its function. For example, suppose that half of the elements of the system fail and evenly. After a failure, a different number of elements may remain in the system horizontally and vertically. Depending on how many elements - even or odd remain, the original formula can be converted to one of the followings presented below:

$$\textbf{even n, m} \rightarrow K_s = 1 - (1 - K_{nod\,j}^{m/2})^{n/2}$$

$$\textbf{odd n, m} \rightarrow K_s = 1 - (1 - K_{nod\,j}^{(m+1)/2})^{(n+1)/2}$$

$$\textbf{even n, odd m} \rightarrow K_s = 1 - (1 - K_{nod\,j}^{(m+1)/2})^{n/2}$$

$$\textbf{odd n, even m} \rightarrow K_s = 1 - (1 - K_{nod\,j}^{m/2})^{(n+1)/2}$$

where, n and m represent the initial number of nodes horizontally and vertically in the graph representing the mesh topology.

2.3 Different Elements of the System

In this case, the availability coefficient cannot be written in a concise form. However, it is possible to write a general formula for the entire system in the following form:

$$K_s = 1 - \prod_{i=1}^{n} \left(1 - \prod_{j=1}^{m} K_{nod\,j_i} \right) \tag{8}$$

3 Example of Availability Analysis

We describe the results of calculating the availability factor of the mesh network using the Esari-Proshaan method. Figure 3 shows three simple mesh structures with a minimum number of elements. The formulas for calculating the upper (6) and lower (5) boundaries of the system availability factor were used in the calculations.

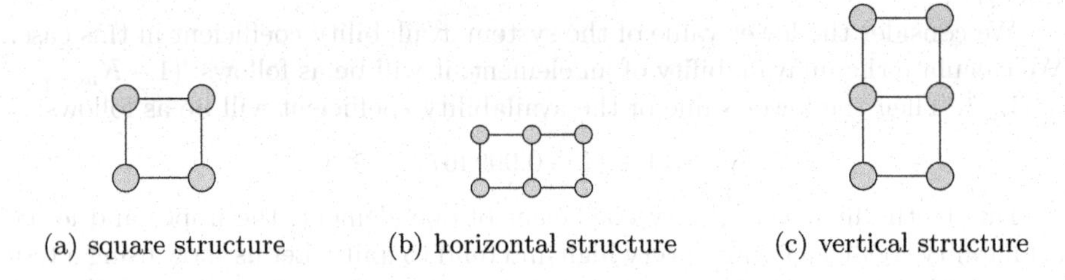

(a) square structure (b) horizontal structure (c) vertical structure

Fig. 3. Simple mesh structures

The following analytical formulas were obtained for calculating the boundaries of the availability factor of the presented structures. For the first structure Fig. 3a), simple equations for the lower and upper bounds of the availability factor are derived, respectively:

$$\begin{aligned} K_{s.up.b} &= (K_{nod})^2 \times (2 - (K_{nod})^2) \\ K_{s.low.b} &= (K_{nod})^2 \times (2 - K_{nod})^2 \end{aligned} \tag{9}$$

The figure below shows the behaviour of the boundaries of the availability factor for structure Fig. 3a) in identical system elements. The dotted line shows the average value of the availability factor, which can be considered close to the actual value. As can be seen from the graph, the upper limit of the availability factor is due to the initial parallel connection of single vertical elements, the resulting section of the circuit and their further serial connection. Moreover, the lower boundary is obtained by serial connection of horizontal elements and their further parallel connection.

For the second structure (Fig. 3b, using formulas (5) and (6), the following equations for the boundaries of the availability factor were obtained:

$$\begin{aligned} K_{s.up.b} &= (K_{nod})^3 \times (2 - (K_{nod})^3) \\ K_{s.low.b} &= (K_{nod})^3 \times (2 - K_{nod})^3 \end{aligned} \tag{10}$$

Figure 5 below shows the boundaries of the availability factor with orange and dark blue lines and its average estimate with a dash-dotted line.

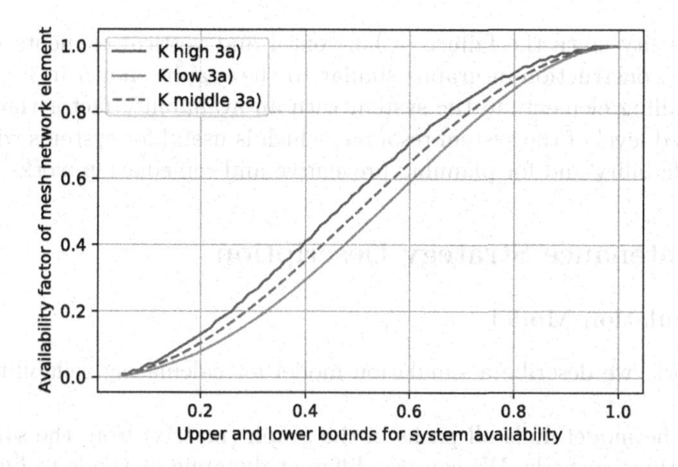

Fig. 4. Dependence of the boundaries and the average estimate of the system (Fig. 3a) availability factor on the network element availability factor

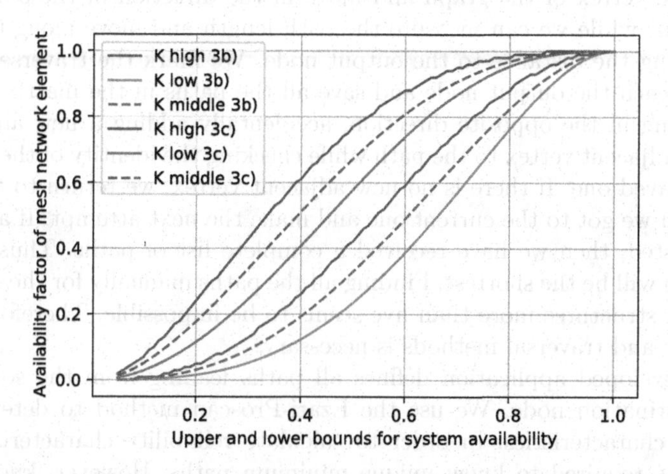

Fig. 5. Dependence of the boundaries and the average estimate of the (Fig. 3b and 3c) systems availability on the element availability factor (Color figure online)

For the third structure Fig. 3c), the following equations for the bounds of the availability factor were obtained:

$$\begin{aligned} K_{s.up.b} &= (K_{nod})^2 \times (3 - 3 \times (K_{nod})^2 + (K_{nod})^4) \\ K_{s.low.b} &= (K_{nod})^2 \times (3 - 3 \times K_{nod} + (K_{nod})^2)^2 \end{aligned} \tag{11}$$

As can be seen from the graph above, the reliability of the scheme for the third structure (Fig. 3c) is on average higher than for the second. This is because, during the section in the third structure, there will be three parallel elements, while there are only two of them in the first cases. It should be noted that the average value of the availability factor should be used since it will give values close to reality since the failure of horizontal and vertical elements is equally likely. The construction of graphs similar to the graphs shown in Fig. 3, 4 will allow installing elements in the system with an availability factor that satisfies the required level of the system resource, which is useful for systems with a high level of reliability and for planning preventive and remediation works.

4 Maintenance Strategy Description

4.1 Simulation Model

In this work, we describe a simulation model for calculating reliability characteristics.

First, the model finds all paths in the graph (matrix) from the source node to the destination node. We can use different dynamic methods to find a path, such as breadth-first and depth-first traversal. In the latter case, we take the initial node vertex of the graph and move in the direction of the output node to the right, while we can measure the path length and move along the vector of decreasing the distance to the output node. We mark the traversed vertices when we reach the output node and save all the paths in the matrix. Then we start moving in the opposite direction, accidentally adding a new and not yet traversed adjacent vertex to the path while checking the identity of the new path with the saved one. If there is no new adjacent vertex, we return to the vertex from which we got to the current one and make the next attempt. If all vertices are exhausted, then we have received a complete list of paths. Thus, the first path found will be the shortest. Finding all the paths manually for the dimension of network structures more than five seems to be impossible. Therefore the use of software and traversal methods is necessary.

The developed application defines all paths leading from the source node to the destination node. We use the Ezari-Proscan method to determine the reliability characteristics. In order to calculate reliability characteristics from above, it is required to know unique minimum paths. However, knowledge of all paths is necessary to determine both the least reliable nodes and paths and determine the most reliable ones.

After the paths have been determined, the minimum paths required to determine the upper-reliability estimate are determined. Then, the operation of the elements of the system is simulated, with a given functioning strategy. In the implemented version, the presence of built-in control is taken into account, as described above. The simulation of a random value of failure and recovery was carried out according to the normal distribution with the parameters: $M.f = 100000$, $Var.f = 25000$, $M.rec = 100$, $Var.rec = 25$. The size of the mesh structure was taken equal to 7×7. The total simulation time is taken equal to ten times the recovery period $(Mo.f + Mo.rec) * 10$.

We used the Box-Muller method for generating random variables. The method is convenient because it allows one to obtain two independently distributed random variables with zero mean and unit variance. Below are the formulas for generating random variables using this method.

$$X_1 = R_1\sqrt{\frac{-2lnD}{D}}; X_2 = R_2\sqrt{\frac{-2lnD}{D}} \quad (12)$$

where X_1, X_2 – the desired values, $D = R_1 + R_2$, where R_1, R_2 – uniformly distributed random variables on $(-1, 1)$. It should be noted that the presented random variables must satisfy the condition $D \in [0...1]$,

4.2 Methodology for Reliability Evaluation

1. A structural diagram of the system is constructed in the form of a graph or matrix.
2. Find all the paths leading from the source node to the destination node.
3. The functioning strategies of the elements are set.
4. For modelling, periods of preventive maintenance, distribution characteristics of the time of failure, recovery, determination of the failure place, etc. are set.
5. Simulation of work with statistics is performed for each element at least ten thousand times. The average value of the simulated random variable is taken as the calculated estimate. Simulation is performed over the entire specified time interval. The random values of failure and recovery are stored in the element state matrix, which contains the times of the element state change. The number of elements in this matrix is defined as the ratio of the simulation time to the estimate of the mathematical expectation of the failure and recovery cycle, multiplied by two.
6. Using the element state matrix, a binary matrix (or vector for each) state is created for the entire simulation time. Binarization is performed to simplify calculations of reliability characteristics using functions of logic algebra and logical operations. The sampling step was taken equal to one minute, the size of the binary matrix is calculated as the size of the matrix of states multiplied by 60. When passing from the matrix of states of elements to the matrix of binarization, the state of the elements is checked. The working sections of the first matrix go to the unit elements of the second, non-working ones to zero.

4.3 Model Calculation Example

We present the results of simulation for the element and the cellular system. Below is a graph of the availability factor of one element with a value of a mathematical expectation of a failure time of 100000 h, a recovery time of 50 h, a standard deviation of a failure of 25000 h and a recovery of 10 h. The mean availability of the element according to the presented data was 0.9995.

When modelling the mesh network, the parameters of the distribution laws presented above were taken, except the mathematical expectation of the recovery time, which was taken equal to 100 h. As a result, the mean non-asymptotic availability was obtained approximately equal to one as:

$$K_c = 1 - (3.18e^{-25}) \cong 1 \quad (13)$$

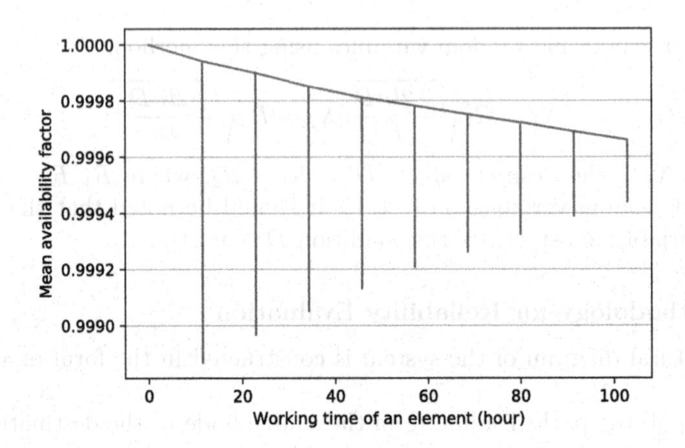

Fig. 6. Element availability

As can be seen from Eqs. (5)–(13), the calculated values of the asymptotic availability and the mean availability obtained by simulation converge to one. The rapid recovery of the system determines high values of the asymptotic and non-asymptotic availability coefficient of the element and the system as a whole in the event of the failure and the high value of the system operation time to failure.

Calculations of the probability of no-failure operation were performed in the case of non-recoverability of the elements in the mesh network with the parameters presented above. The graph of the probability of failure-free operation is shown in Fig. 7.

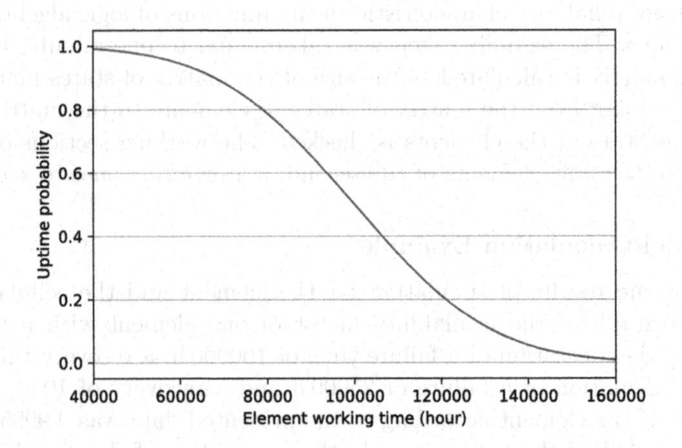

Fig. 7. Probability behavior of the uptime element

For non-recoverable elements, the probability of failure-free operation of the system can be calculated by analogous to formula (5) for the uptime probability.

At the moment, one hundred thousand hours, the uptime probability will be 0.939, which is a sufficiently high value for 11 years of system operation.

5 Conclusion

In this research, the following results were obtained:

1. The analysis of the operation and functioning of the elements has been carried out in the network with mesh topology.

2. The technique of analytical determination of the availability is described for a homogeneous network. The calculation model of minimum paths and section was considered.

3. A method of availability determination has been developed using a simulation model.

4. The simulation model of the system has been developed, it is shown that it converges to an analytical model 5.

It should be noted that the simulation model can be used to analyze the reliability characteristics and other indicators of systems of any dimension and complexity. It can be used to model subsystems of hazardous and expensive objects, such as aerospace, telecommunications, energy, transport and other areas of human activity.

Acknowledgment. The publication has been prepared with the support of the grant from the President of the Russian Federation for state support of leading scientific schools of the Russian Federation according to the research project SS-2604.2020.9. The reported study was funded by RFBR, project number 20-37-70059.

References

1. Beichl, I., Moseman, E., Sullivan, F., Bowie, M.: Computing network reliability coefficients. In: Proceedings of the 42nd Southeastern International Conference on Combinatorics, Graph Theory and Computing, vol. 207, pp. 111–127 (2011)
2. Cisco: Network availability: how much do you need? How do you get it? pp. 1–17. https://www.cisco.com/web/IT/unified_channels/area_partner/cisco_powered_network/net_availability.pdf
3. Cui, L., Dong, Q.: Consecutive k and related models—a survey. In: Li, Q.-L., Wang, J., Yu, H.-B. (eds.) JHC80 2019. CCIS, vol. 1102, pp. 3–18. Springer, Singapore (2019). https://doi.org/10.1007/978-981-15-0864-6_1
4. Dengiz, B., Altiparmak, F., Smith, A.: Efficient optimization of all-terminal reliable networks, using an evolutionary approach. IEEE Trans. Reliab. **46**(1), 18–26 (1997). https://doi.org/10.1109/24.589921
5. Gertsbakh, I., Shpungin, Y.: Network Reliability. SECE, Springer, Singapore (2020). https://doi.org/10.1007/978-981-15-1458-6
6. Joshi, A., Subedha, V.: Systematization of reliable network topologies using graph operators. In: Satapathy, S.C., Bhatt, Y.C., Joshi, A., Mishra, D.K. (eds.) Proceedings of the International Congress on Information and Communication Technology. AISC, vol. 439, pp. 263–270. Springer, Singapore (2016). https://doi.org/10.1007/978-981-10-0755-2_28

7. Kumar, A., Pathak, R., Gupta, Y.: Genetic-algorithm-based reliability optimization for computer network expansion. IEEE Trans. Reliab. **44**(1), 63–72 (1995). https://doi.org/10.1109/24.376523
8. Liu, Yu., Zhang, L., Luo, R.: Research on architecture design with high reliability of network nodes in the amphibious combating simulation system. Wuhan Univ. J. Nat. Sci. **24**(6), 537–548 (2019). https://doi.org/10.1007/s11859-019-1433-3
9. Maalel, N., Natalizio, E., Bouabdallah, A., Roux, P., Kellil, M.: Reliability for emergency applications in internet of things. In: 2013 IEEE International Conference on Distributed Computing in Sensor Systems. IEEE (May 2013). https://doi.org/10.1109/dcoss.2013.40
10. Morino, Y., Miwa, H.: Reliable network design considering cost to decrease failure probability of simultaneous failure. In: Barolli, L., Li, K.F., Miwa, H. (eds.) INCoS 2020. AISC, vol. 1263, pp. 493–502. Springer, Cham (2021). https://doi.org/10.1007/978-3-030-57796-4_47
11. Rak, J., Pickavet, M., Yoshino, H.: Reliable networks design and modeling (foreword). Telecommun. Syst. **60**(4), 419–421 (2015). https://doi.org/10.1007/s11235-015-9986-8
12. Rodríguez-Velázquez, J.A., Kamišalić, A., Domingo-Ferrer, J.: On reliability indices of communication networks. Comput. Math. Appl. **58**(7), 1433–1440 (2009). https://doi.org/10.1016/j.camwa.2009.07.019
13. Sattiraju, R., Schotten, H.D.: Reliability modeling, analysis and prediction of wireless mobile communications. In: 2014 IEEE 79th Vehicular Technology Conference (VTC Spring). IEEE (May 2014). https://doi.org/10.1109/vtcspring.2014.7023170
14. Shier, D.R., Liu, N.: Bounding the reliability of networks. J. Oper. Res. Soc. **43**(5), 539–548 (1992). https://doi.org/10.1057/jors.1992.79
15. Xin, J., Guo, L., Huang, N., Li, R.: Network service reliability analysis model. Chem. Eng. Trans. **33**, 511–516 (2013). https://doi.org/10.3303/CET1333086
16. Zhu, X., Lu, Y., Han, J., Shi, L.: Transmission reliability evaluation for wireless sensor networks. Int. J. Distrib. Sens. Netw. **12**(2), 1346079 (2016). https://doi.org/10.1155/2016/1346079

The Importance of Conference Proceedings in Research Evaluation: A Methodology for Assessing Conference Impact

Dmitry Kochetkov[1,2(✉)], Aliaksandr Birukou[3,4], and Anna Ermolayeva[3]

[1] Centre for Science and Technology Studies, Leiden University, Leiden, The Netherlands
d.kochetkov@cwts.leidenuniv.nl
[2] Ural Federal University, Ekaterinburg, Russia
kochetkovdm@hotmail.com
[3] Peoples' Friendship University of Russia (RUDN University), 6 Miklukho-Maklaya Street, Moscow 117198, Russian Federation
ermolaevaanna@bk.ru
[4] Springer Nature, Heidelberg, Germany
aliaksandr.birukou@springer.com

Abstract. Conferences are an essential tool for scientific communication. In disciplines such as Computer Science the majority of original research results are published in conference proceedings. In this study, we have analyzed the role of conference proceedings in various disciplines and propose an alternative approach to research evaluation based on conference proceedings sources indexed in Scopus and Scimago Journal Rank (SJR). This allows one to categorize conference proceedings in quartiles $Q1$–$Q4$ by analogy with SJR journal quartiles. Out of 171 conference proceedings sources analyzed, 38 conference proceedings in Engineering (45% of the list) and 23 in Computer Science (32% of the list) have an SJR level corresponding to the first quartile journals in these areas, which emphasizes the exceptional importance of conferences in these disciplines The comparison of this bibliometric-driven ranking with the expert-driven CORE ranking in Computer Science showed a 62% overlap, as well as a significant average rank correlation of the category distribution.

Keywords: conference proceedings · research evaluation · research impact · SJR · CORE · conference rankings · bibliometrics

This paper has been supported by the Ural Federal University Strategic Academic Leadership Program.

V. M. Vishnevskiy et al. (Eds.): DCCN 2021, CCIS 1552, pp. 359–370, 2022.
https://doi.org/10.1007/978-3-030-97110-6_28

1 Introduction

In many countries (for instance, China [1], India [2], Russia, Turkey[1], UK [3]) research evaluation is based on the indicators of the sources, i.e., journals, conference proceedings, book series, in which the results are published. This often leads to labeling publication sources with several predefined classes and judging the importance of a publication based on the class of the source. As in many research areas original results are published in journals [4], research evaluation policies are often biased towards journals.

As an example of such a policy, we can mention the ongoing discussion on the Comprehensive Methodology for Evaluating Publication Performance (CMEPP) [5] in Russia. The methodology is based on the evaluation of publications depending on the Impact Factor (IF) quartile of journals in Web of Science: a journal article published in the first quartile journal gets 20 points, the second quartile - 10 points, the third quartile - 5 points, and the fourth quartile - 2.5 points. All other publications, including conference papers, book chapters, journal articles, indexed only in Scopus or Russian Science Citation Index, receive 1 point. This scale applies to the natural sciences, engineering, and life sciences. There is a flat scale for social sciences and humanities: all publications in Scopus or Web of Science get 3 points regardless of the quartile. The advantages of this technique include its simplicity and the ability to evaluate the re-search manuscript directly at the time of publication, as the process of gathering citations takes time. The disadvantages of the method are discussed in the Declaration on Research Assessment, DORA [6] and include the possibility of manipulation with quantitative journal metrics [7], although this is true of all purely quantitative methods of evaluation. However, the main drawbacks of the methodology are the possible discrepancy between the citation count of a specific article and the value of the journal where it is published and high variance in the citations to the articles in the same journal [8]. More recent research [9] shows that it is possible that the Impact Factor is accurate predictor of the significance of an article than the number of citations it has received. However, the design of the study does not allow the authors to make conclusions about whether in practice the Impact Factor is indeed more accurate than the number of citations. Without going into further details of article-level vs journal-level metrics arguments, we will try to reconcile the debates for the purpose of this paper. IF and other journal-level metrics are widely used in research evaluation to judge the impact of research contributions or researchers. If an article was published in a Q1 journal, it most probably has a higher probability of being superb than an article in a Q4 journal in the same discipline. However, counter-examples could be found, i.e., a weak article in the Q1 journal or a superb article in the Q4 journal. Therefore, article-level metrics, or even better, qualitative research methods should be used next to the quantitative journal-level indicators. The statement that it is necessary to use not only quantitative, but also qualitative methods for quality assessment is supported by the REF Guidance on submissions [10], which recommends relying

[1] https://www.urapcenter.org/Methodology, last accessed 03.07.2020.

more on qualitative research assessments than on citations. Now, let us turn back to the purpose of the paper and stress that the CMEPP approach, as well as most of journal-level approaches neglects the conferences: even the most prestigious conferences receive a lower rating than articles in the journals of the fourth quartile.

Let us recall the advantages of conferences. First, the publishing process is usually faster, which can be especially important in rapidly changing fields such as computer science. Secondly, conference reports can be a good way to get early feedback on current work or promote its results. Thirdly, attending conferences is a great way to build connections in your field, learn about the latest work of other groups, and meet people who may be interested in your results or working with you in the future. Problems that may arise when participating in conferences include paying a fee, preparing a presentation and speaking, which takes quite a long time, as well as the fact that publications in conference proceedings often have less weight for the researcher in his resume than journal articles. The problem of underestimating conferences in the researcher's resume should be solved, since in areas where most scientific results are published in the proceedings of the conference, they have a great weight for scientific communities. The above advantages and disadvantages are listed in [11]. Also, one of the disadvantages may be the rapid review process, which may lead to a less thorough selection of works [12].

Unwillingness to participate in conferences, which is provoked by the above factors, can lead to more serious consequences. So [13] calls the lack of scientific communication skills as one of the factors that negatively affects the effectiveness of re-searchers' publishing activities. The lack of scientific communication in some countries leads to the creation of platforms for the exchange of scientific information [14,15], which do not always work effectively. The Research Collaboration System and the SECI model among scientists in Thailand were studied in [14]. This was the first step to improve knowledge sharing. In turn, in the study [15], the authors examined whether Turkish scientists use an academic social media service, such as ResearchGate, Academia.edu, Google Scholar Profile, LinkedIn, ResearcherID and Mendeley. The study found that these services are widely used, but most scientists do not use them for knowledge sharing and collaboration.

The authors [16] emphasize the importance of conference participation for re-search scientists in computer science and engineering and the speed of information transfer outweighs the many disadvantages of conferences. And the fact that scientists or teachers are evaluated in the traditional way (through publications in journals) hinders their career [16], and despite the fact that this work was published more than 20 years ago, the trend continues to this day.

In this paper we 1) review the current practices of using conferences in the re-search evaluation; 2) identify scientific disciplines, where conference proceedings play a significant role in the communication of primary research results; 3) propose a new methodology for the assessment of conference proceedings based on Scopus and Scimago Journal Rank (SJR) data; 4) show that such bibliometric-driven methodology produces classification of conferences similar to the classification designed by domain experts, such as CORE. This article is structured as follows: Sect. 2 presents the role of conferences in the scientific community. Section 3 describes the methodology of the study, Sect. 4 presents the results of the study and its limitations, and Sect. 5 describes the conclusions and prospects for the development of this study.

2 The Role of Conferences

Conferences play an important role in some areas of science - for example, in Computer Science, more than 60% of the research results are published in conference proceedings [17]. The authors of this study also identified several problems related to evaluating impact of conference proceedings: first, rankings such as Google Scholar and Microsoft Academic miss or underestimate some conferences due to the young age of the conference or the unreliable size, which forces researchers to look for new methods of ranking conferences. For conferences that are at least four years old[2], CiteScore provides equal opportunities for all conferences, symposia, and seminars, regardless of their size or age. Secondly, the authors show that Scopus' CiteScore is an effective method of evaluating computer science conferences. Thirdly, the CiteScore method shows that conference papers also have a high impact, as do articles in leading computer science journals. Research [4] show that a small number of elite conferences have higher average citations rate than elite journals. The authors of [18] looked at the bibliometrics of conferences and concluded: conferences are more popular than journals; about 78% of publications reflecting the most relevant research are published in conference proceedings; publications in journals are cited more than those in conferences (57% vs. 43%, respectively); the distribution of publications in conferences vs journals is country-dependent. While there is a range of metrics for evaluating journals, e.g., Impact Factor, SCImago Journal Rank (SJR), there is no common metric for evaluating conferences [19].

The first attempts to create conference rankings appeared in Computer Science. The most frequently used rankings, based on the experience of the authors, are the Computing Research and Education Association of Australasia, CORE and ERA Ranking [20], Qualis (2012), China Computer Federation (CCF) ranking (2012), MSAR (2014), and GII-GRIN-SCIE (2014). The CORE Conference Ranking[3] provides evaluation of major computer science conferences. Decisions are made by academic committees based on data requested as part of the submission process. ERA Ranking (2010) was created in the framework of Excellence in Research in Australia (ERA)[4]. It incorporates data from a previous ranking

[2] More specifically, "conferences, whose proceedings have been indexed in Scopus for at least 4 years", but as we mostly interested in conference that have proceedings we will use "conferences" and "conference proceedings" interchangeably through the paper.

[3] Available at http://www.core.edu.au/conference-portal (date accessed 01.03.2020).

[4] Available at https://www.arc.gov.au/excellence-research-australia (date accessed 01.03.2020).

attempt done by CORE. Qualis [21] has been published by the Brazilian ministry of education and uses the H-index (strictly speaking, H-index percentiles) as a performance measure for conferences. The latest edition was released in 2016. The fifth edition of the list of journals and conferences recommended by the CCF[5] was released in April 2019 and assigns A, B, C ranks to conferences in each CS subarea. Interestingly, the release mentions that "the influence of journals and conferences is not directly related to the influence of a single paper published there. Therefore, it is not recommended to use this list as a basis for academic evaluation" - see our earlier discussion on journal- vs. article-level metrics. MSAR[6] is the Microsoft Academic's field rankings for conferences. It is similar to the h-index and calculates the number of publications by an author and the distribution of citations to the publications. Field ranking only calculates publications and citations within a specific field and shows the impact of the scholar or journal within that specific field. GII-GRIN-SCIE conference rating[7] is an attempt to develop a unified rating of computer science conferences led by GII (Group of Italian Professors of Computer Engineering), GRIN (Group of Italian Professors of Computer Science), and SCIE (Spanish Computer-Science Society), see [22]. The latest version ranks all conferences in four tiers (top notch, very high quality, good, and work in progress) based on the CORE, Qualis and MSAR rankings and it was released in 2018. Last but not the least, there is also Google Scholar Top publications [23] tool which lists both conferences and journals in the same ranking.

Now that we reviewed why conferences are important in Computer Science and how this is tackled with different rankings, let us check if conferences only matter in Computer Science. As in previous research on conference [17,24] we used Scopus to compute the share of conference proceedings in the total number of publications for particular subject categories in 2015–2019. Figure 1 shows the subject categories where the share of conference proceedings is more than 10% of all publication sources. The highest percentage of conferences is observed in Computer Science, as well as Mathematics and Decision Sciences. Note that given that publications in Scopus can belong to several categories, there might be overlap. There is a substantial share of publications in conference proceedings for Engineering and Energy.

The conference rankings mentioned above provide an initial basis for the evaluation of research published in conference proceedings. While most of the current rankings are based on citations, alternative metrics can be used to identify influential articles and authors, e.g., PageRank-like algorithms [26]. A promising research direction is whether such rankings could be used to address the phenomenon of predatory or fake conferences [27]. However, most of the existing rankings have been created for specific local communities and were never meant to be used globally, like journal IFs. In the following, we will propose a metric

[5] Available at https://www.ccf.org.cn/c/2019-04-25/663625.shtml (date accessed 28.04.2020).

[6] Available at https://academic.microsoft.com/home (date accessed 04.03.2020).

[7] Available at http://gii-grin-scie-rating.scie.es/ (date accessed 28.04.2020).

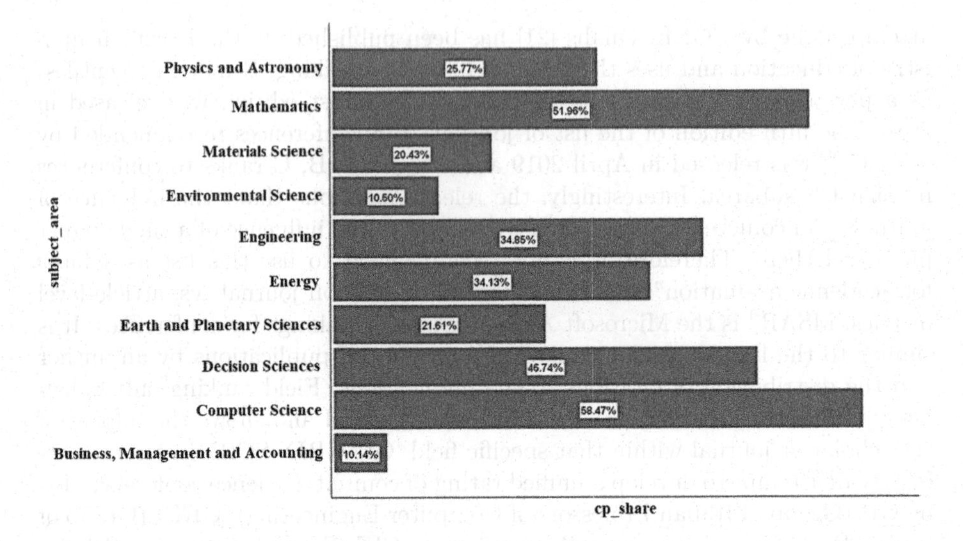

Fig. 1. The share of conference proceedings in the total number of publications for particular subject categories 2015–2019. Source: authors' own calculations based on Scopus [25] data (data retrieved on 19.03.2020)

for the conferences indexed in Scopus and compare it with the most frequently used CS conference ranking, CORE. Thus, the overall goal of this study is to use the bibliometrics methods to estimate the role of conference proceedings in different disciplines.

3 Materials and Methods

At the first stage, in the Scopus sources list[8] we selected those which are as conference proceedings (Conference Proceedings post-1995). In terms of data collection time and completeness, Scopus is the optimal tool for conducting research [28]. The authors [29] who conducted a similar study, but much later than the first one, made the same conclusions. We then focused on those which are currently indexed (i.e., have the ongoing status), and for which an SJR[9] score is available. This selection resulted in 171 sources with conference proceedings. Note that the way Scopus indexes conferences depends on the conference and the publication outlet (journal, book series, conference proceedings). So the 171 sources we selected contained a much bigger number of conferences, as sources like ACM International Conference Proceeding Series or CEUR Workshop Proceedings publish several hundreds conference proceedings per year.

The SCImago Journal Rank (SJR) is not just a citation indicator such as Impact Factor or CiteScore; it is based on a PageRank-like algorithm, which

[8] Available at https://www.scopus.com/ (date accessed 04.03.2020).
[9] Available at https://www.scimagojr.com/ (date accessed 04.03.2020).

is an iterative process of prestige transfer among the publication sources. The calculation is an iterative process in which the prestige of each source depends on the prestige of the sources which cite it. The final SJR value is normalized over the number of documents published in the citation window [30].

Given that the SJR is computed based on Scopus data, we also used this database in our analysis. Out of the 171 conference proceedings sources, 153 were assigned one or more subject categories (third level ASJC[10]) in Scopus. For the 18 conferences proceedings sources that were not assigned any subject category; we deduced the categories based on publications in Scopus.

Next, for each of the subject categories, we computed the threshold SJR values for the quartiles, in the same way SCImago calculates them for journals. This was necessary because SCImago does not assign quartiles to the conferences proceedings sources, only to journals and book series. This allowed us to assign each conference source to the corresponding quartile ($Q1, Q2, Q3, Q4$) in each subject category. For example, the minimum SJR for journals and book series of the first quartile is 0.261, the second is 0.139, the third is 0.104, and the fourth is 0.1. The IOP Conference Series: Materials Science and Engineering has an SJR of 0.195, so we can classify the source as $Q2$. We emphasize that this is not a quartile itself; it is a conditional assignment of conference proceedings source to a quartile based on the SJR value. For journals covering several subjects, CMEPP research evaluation guidelines suggest using the maximum of the quartiles in those subjects. However, the importance of the same conference in different communities varies, as also mentioned in the CCF release notes. We therefore would like to stress the importance of using subject-specific quartiles for conferences, i.e., a conference can belong to several subject categories and can have different quartiles there.

4 Results

The distribution of conferences proceedings sources across subject categories is shown in Fig. 2. Note that one conference can belong to several subject categories. Out of 171 sources, one was assigned to five subject categories, one to four, 13 to three, 66 to two, and 90 conferences had only one subject category.

Figure 3 shows the distribution of conference proceedings across quartiles in the context of subject categories. If one considers all sources (journals, book series, conference proceedings), the share of each quartile is obviously 25%. However, as our selection is limited to the conference proceedings, the distribution between the categories Q1–Q4, is very different for each subject category.

From the graph, it is evident that Engineering and Computer Science have not only the highest share of conference proceedings but also the largest number of high-impact conference proceedings. This once again confirms the thesis that conference proceedings must be considered when evaluating research in these

[10] All Science Journal Classification Codes. Available at [31] (date accessed 04.03.2020).

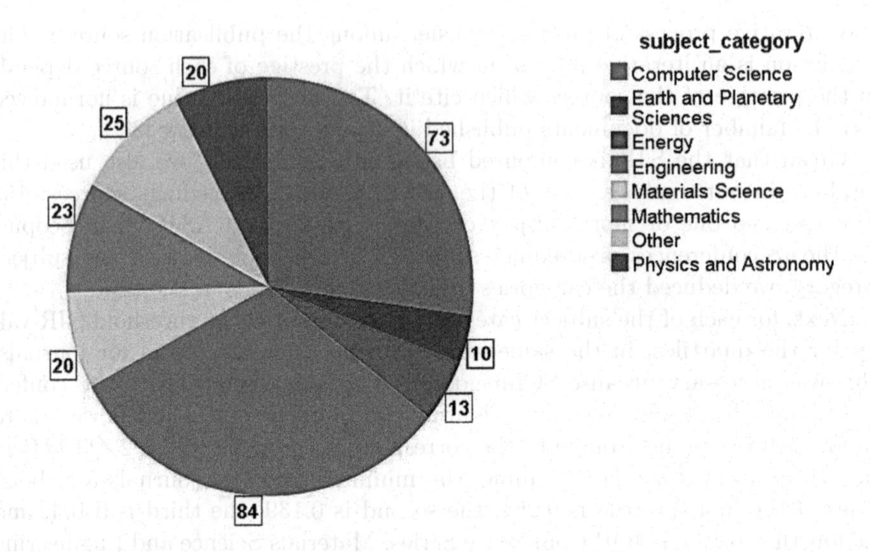

Fig. 2. The distribution of conferences proceedings sources across subject categories. Source: authors' own calculations

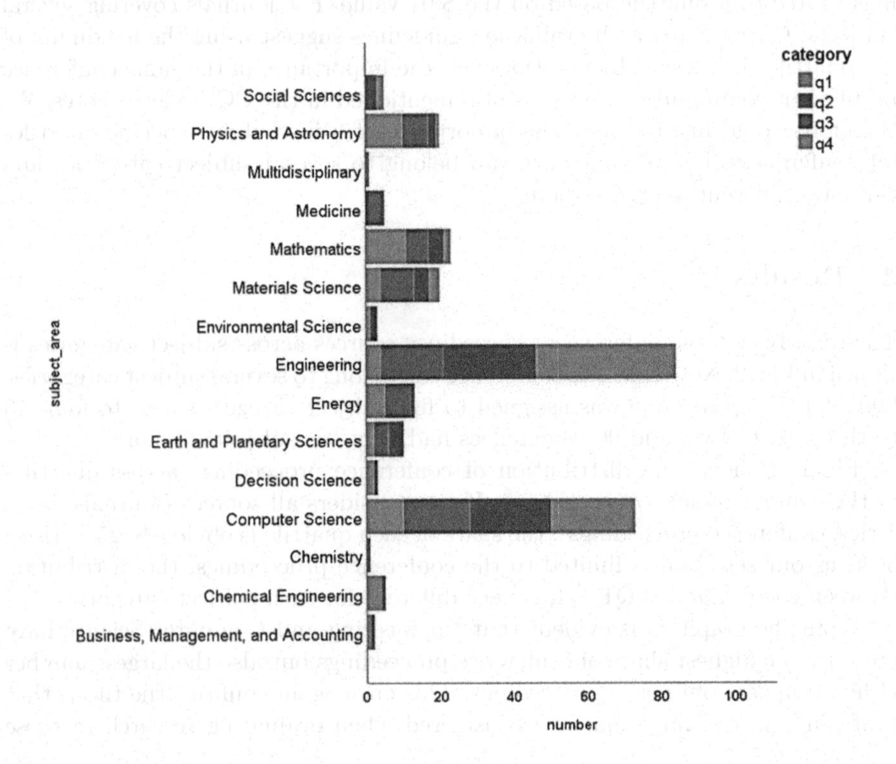

Fig. 3. Distribution of conference proceedings sources into categories. Source: authors' own calculations

areas. Our results are in line with the results of earlier studies [17]. The difference in the number and quality of conference proceedings between these subject categories and the rest is substantial. The source data is available in [32].

Out of the 73 proceedings of Computer Science conferences, 45 conferences (62%) are in the CORE ranking; 10 sources are aggregators that publish the proceedings of many conferences (e.g., Procedia Computer Science, ACM International Conference Proceeding Series, etc.); and 18 conference proceedings are not core CS conferences, but are from related fields (for example, IEEE MTT-S International Microwave Symposium Digest). The latter appear in our dataset because according to the ASJC classification conferences can fall simultaneously into several subject areas/categories. Such conferences, however, are out of scope for CORE, which focuses exclusively on Computer Science conferences. For the 45 conferences from our list, which are also present in CORE, we compared the distribution by category (Fig. 4, Q1 for SJR corresponds to A * for CORE, Q2 - A, Q3 - B, Q4 - C). Spearman's rank correlation coefficient was 0.452, which suggests an average correlation dependence. This is an interesting fact, given the fundamentally different approaches to the formation of lists, bibliometric and expert. The full table is also presented in the dataset available online (see Footnote 8).

SJR 2018/CORE 2018	A*	A	B	C	NA
Q1	11	4	4	1	3
Q2	5	7	2	1	7
Q3	-	2	6	1	9
Q4	-	-	-	1	9
NA	51	407	402	793	

*Source: author's calculations.

Fig. 4. A comparative analysis of distribution of conferences into categories

4.1 Limitations

The study has several limitations:

1. We have evaluated conference proceedings sources, not the conferences themselves. If one would like to evaluate conferences, they should take into account not only the bibliometric data, but also various other parameters: topical scope, program committee, authors, the peer review process, proceedings publication culture, etc. However, a quantitative assessment presented here may be a convenient auxiliary tool, even though it does not eliminate the need for expert evaluation.

2. The list reflects only non-journal and non-book sources. Conference proceedings published in journals and book series (e.g., Journal of Physics Conference Series, Lecture Notes in Computer Science) can use the SJR quartile of the corresponding journal or the book series.
3. The list only reflects conference proceedings with the serial ISSN; some conferences do not receive it due to the oversight of the organizing committee. Such conferences are not included in the list of serials in Scopus and could not be included in the analysis.
4. The list includes not only the proceedings of individual conferences but also aggregators such as CEUR Workshop Proceedings, Leibniz International Proceedings in Informatics (LIPIcs). The level of conferences within such publications may vary significantly. Unfortunately, the data granularity in Scopus does not allow for the conference-level analysis within these sources.

Even though the CCF recommends not using conference rankings for academic evaluation, [33] shows how such rankings influence publishing behavior of scientists. Therefore, it is important to provide more transparency in how rankings are created, what is included, which metrics are used, etc. The methodology proposed in this paper represents a step in this direction, as it combines transparent bibliometric indicators and correlates with expert opinions.

The authors will continue research on the evaluation of conferences and conference papers. We would like to move towards paper-level metrics, as different papers in the same conference proceedings have different quality, citations, importance. In this regard we would like to mention several projects that aim at providing open identification of conferences, which is the first step before doing any bibliometrics. ConfIDent aims at developing a crodwsourcing platform for providing semantically structured metadata of scientific events [34]. The ConfRef.org project, which was created to provide information on scientific conferences and provide standard identifiers for conferences. The current prototype provides data on 40,000 conferences, mainly from computer science, provided by Springer Nature and DBLP. The primary purpose of ConfRef is to provide trusted information about the history, dates, venues, places of publication/past issues of a series of conferences (and related conferences) in various disciplines (Computer Science, Electrical Engineering, Mathematics), as well as information about upcoming conferences and invitations, dates and information about program committee. On top of this, ConfRef will deal with identifying predatory or fake conferences.

5 Conclusions

In this paper we made an attempt to review the role of conferences in the research evaluation and to identify scientific disciplines, where conference proceedings are an important outlet for publishing original research results. Next to the "usual suspects", i.e. Computer Science, conference proceedings often used for publishing results in Engineering, Mathematics, Energy, Decision Sciences. We also presented a new methodology for applying Scopus and Scimago Journal Rank (SJR) data for the assessment of conference proceedings and showed that it provides similar results to expert-designed ranking, such as CORE. The methodology shows that some conference proceedings in Computer Science, Engineering, Material Science, Physics and Astronomy, and Mathematics are comparable with Q1–Q2 journals.

Future work includes development of tools which would implement the proposed methodology and work on removing the limitations such as the different granularity of conference proceedings sources.

References

1. Editorial in Nature: China's research-evaluation revamp should not mean fewer international collaborations. Nature **579**, 8 (2020)
2. Madhan, M., Gunasekaran, S., Arunachalam, S.: Evaluation of research in India-are we doing it right. Indian J. Med. Ethics **3**(3), 221–229 (2018)
3. Koya, K., Chowdhury, G.: Metric-based vs peer-reviewed evaluation of a research output: lesson learnt from UK's national research assessment exercise. PLoS ONE **12**(7), e0179722 (2017)
4. Vrettas, G., Sanderson, M.: Conferences versus journals in computer science. J. Am. Soc. Inf. Sci. **66**(12), 2674–2684 (2015)
5. Methodology for calculating the qualitative indicator of the state task comprehensive methodology for evaluating publication performance for scientific organizations subordinate to the Ministry of Science and Higher Education of Russia for 2020 (2020). https://docs.cntd.ru/document/564894817?marker=64U0IK
6. Cagan, R.: San Francisco declaration on research assessment (2013)
7. The PLoS Medicine Editors: The impact factor game. PLoS Med. **3**(6), e291 (2006)
8. Seglen, P.O.: Why the impact factor of journals should not be used for evaluating research. BMJ **314**(7079), 497 (1997)
9. Waltman, L., Traag, V.A.: Use of the journal impact factor for assessing individual articles need not be statistically wrong. F1000Research **9**, 366 (2020)
10. Index of revisions to the 'guidance on submissions' (May 2019). https://www.ref.ac.uk/media/1447/ref-2019_01-guidance-on-submissions.pdf
11. Ernst, M.: Choosing a venue: conference or journal (2006). https://homes.cs.washington.edu/~mernst/advice/conferences-vs-journals.html
12. Bowyer, K.W.: Mentoring advice on "conferences versus journals" for CSE faculty. University of Notre Dame, Notre Dame (2012)
13. Kwanya, T.: Publishing and perishing? Publishing patterns of information science academics in Kenya. Inf. Dev. **36**(1), 5–15 (2020)
14. Julpisit, A., Esichaikul, V.: A collaborative system to improve knowledge sharing in scientific research projects. Inf. Dev. **35**(4), 624–638 (2019)
15. Bardakcı, S., Arslan, Ö., Ünver, T.K.: How scholars use academic social networking services. Inf. Develop. **34**(4), 334–345 (2018)
16. Patterson, D., Snyder, L., Ullman, J.: Evaluating computer scientists and engineers for promotion and tenure. Comput. Res. News (1999)
17. Meho, L.I.: Using Scopus's CiteScore for assessing the quality of computer science conferences. J. Inf. **13**(1), 419–433 (2019)
18. Franceschet, M.: The role of conference publications in CS. Commun. ACM **53**(12), 129–132 (2010)

19. Almendra, V.S., Enăchescu, D., Enăchescu, C.: Ranking computer science conferences using self-organizing maps with dynamic node splitting. Scientometrics **102**(1), 267–283 (2014). https://doi.org/10.1007/s11192-014-1436-y
20. Butler, L.: ICT assessment: moving beyond journal outputs. Scientometrics **74**(1), 39–55 (2008)
21. QUALIS CONFERNCIAS (2016). https://qualis.ic.ufmt.br/
22. Cabitza, F., Locoro, A.: Exploiting the collective knowledge of communities of experts. In: KMIS, pp. 159–167 (2015)
23. Google Scholar conference ranking. https://scholar.google.com/citations?view_op=top_venues&hl=en
24. Jia, Y., et al.: Trends and characteristics of global medical informatics conferences from 2007 to 2017: a bibliometric comparison of conference publications from Chinese, American, European and the Global Conferences. Comput. Meth. Programs Biomed. **166**, 19–32 (2018)
25. Scopus website. https://www.scopus.com/. Accessed 19 Mar 2020
26. Dunaiski, M., Visser, W., Geldenhuys, J.: Evaluating paper and author ranking algorithms using impact and contribution awards. J. Informet. **10**(2), 392–407 (2016)
27. Taking the con out of conferences (February 2017). https://www.crossref.org/blog/taking-the-con-out-of-conferences/
28. Meho, L.I., Yang, K.: Impact of data sources on citation counts and rankings of LIS faculty: Web of Science versus Scopus and Google Scholar. J. Am. Soc. Inform. Sci. Technol. **58**(13), 2105–2125 (2007)
29. Guerrero-Bote, V,P., Chinchilla-Rodríguez, Z., Mendoza, A., de Moya-Anegón, F.: Comparative analysis of the bibliographic data sources dimensions and scopus: an approach at the country and institutional levels. Front. Res. Metrics Anal. **5**, 19 (2021)
30. Description of Scimago J. Rank Indic. (2020). https://www.scimagojr.com/SCImagoJournalRank.pdf
31. What is the complete list of Scopus Subject Areas and All Science Journal Classification Codes (ASJC)? (2020). https://service.elsevier.com/app/answers/detail/a_id/15181/supporthub/scopus/
32. Kochetkov, D., Birukou, A., Ermolayeva, A.: Methodology for conference proceedings assessment: a conference proceedings dataset (2020). Mendeley data, v5
33. Li, X., Rong, W., Shi, H., Tang, J., Xiong, Z.: The impact of conference ranking systems in computer science: a comparative regression analysis. Scientometrics **116**(2), 879–907 (2018). https://doi.org/10.1007/s11192-018-2763-1
34. Hagemann-Wilholt, S., Plank, M., Hauschke, C.: ConfIDent-an open platform for FAIR conference metadata. In: GL Conference Series, vol. 21. TextRelease, Amsterdam (2020)

Cardiac Arrhythmia Disorders Detection with Deep Learning Models

Eugene Yu. Shchetinin[1], Leonid A. Sevastianov[2,3](✉), Anastasia V. Demidova[3], and Anastasia G. Glushkova[4]

[1] Financial University, Government of the Russian Federation 49, Leningradsky Prospect, 125993 Moscow, Russian Federation
[2] Peoples' Friendship University of Russia (RUDN University), 6 Miklukho-Maklaya Street, 117198 Moscow, Russian Federation
sevastianov-la@rudn.ru
[3] Joint Institute for Nuclear Research 6, Joliot-Curie Street, Moscow region, 141980 Dubna, Russian Federation
demidova-av@rudn.ru
[4] Oxford University, Oxford, UK

Abstract. In this paper, the research of computer algorithms for automatic detection of heart rhythm disorders based on the analysis of electrocardiograms have been conducted. A new model of the electrocardiogram classifier is proposed, as an ensemble of a two-dimensional convolutional neural network and a long short-term memory model, which includes an attention layer. Computer experiments conducted on the MIT-BIH Physionet collection of ECG showed its high performance compared to other models of machine and deep learning. Proposed model successfully detected cardiac arrhythmia classes with an accuracy of 99.34%, AUC = 99% and recall = 99%.

Keywords: cardiac arryhythmia · cardiovascular diseases · convolutional neural network · long short-term memory · attention module · MIT-BIH · ECG

1 Introduction

Cardiovascular disease (CVD), according to the World Health Organization, is one of the most common causes of death in the world [1]. Problems of the cardiac system can lead to aberrations in electrical pulses that disrupt the normal heart rate and rhythm. This abnormality is widely known as arrhythmia, a dangerous disease that threatens human life, therefore, timely diagnosis of arrhythmia is of great importance in the prevention of cardiovascular diseases. The most effective clinical method for visualizing the electrical activity of the heart is electrocardiography (ECG). In addition to being non-invasive, it is also fast and easy to use,

This paper has been supported by the RUDN University Strategic Academic Leadership Program.

V. M. Vishnevskiy et al. (Eds.): DCCN 2021, CCIS 1552, pp. 371–384, 2022.
https://doi.org/10.1007/978-3-030-97110-6_29

providing enough information to diagnose and treat heart disease [2,3]. Manual analysis of the ECG signal is a complex task, which justifies the need to develop methods for the automated detection of cardiac arrhythmias. Automated ECG computer analysis has been a subject of great interest in the field of biomedical technology for many years, and it is still a challenging theoretical and practical task.

The volume of studies of electrocardiograms and the demand for computer-aided diagnostics of cardiovascular diseases is constantly growing, so the computer methods of processing and analyzing electrocardiograms, as well as their use in real time, as well as in the form of remote applications, are becoming popular. The formulation of the problem of automated detection of cardiac arrhythmias can be reduced to the problem of classifying ECG signals. When developing approaches to automating the ECG classification process, such methods as hidden Markov model (HMM) [4], discrete wavelet transform (DWT), machine learning algorithms, as well as artificial neural networks were used [5,6]. With the introduction of deep learning models, computer diagnostics of ECG signals has reached a new level. In particular, the authors of the paper [7] used a deep convolutional neural network to identify arrhythmias according to ECG data of 5 different heartbeat classes, having obtained an accuracy of 92%. In paper [8], a model of a one-dimensional convolutional neural network 1D_CNN was developed for the classification of 5 types of cardiac arrhythmias by electrocardiograms and an accuracy of 92.70% was obtained. In addition, in paper [9], two-dimensional model of convolutional neural network was implemented to detect heart arrhythmias.

In this paper, a number of deep neural network models, including convolutional, recurrent networks and their ensembles, are investigated and implemented in software for the classification of ECG signals. The ensemble of deep neural networks is proposed based on a one-dimensional convolutional network 1D_CNN and a recurrent network with a long short-term memory (LSTM) unit. A comparative analysis of these deep models with popular machine learning algorithms is carried out and the effectiveness of using deep neural networks for the automated detection of cardiac arrhythmias is approved. The accuracy of the best deep model is achieved 99.34%.

2 Development of Deep Learning Models for ECG Signals Classification

2.1 Deep Convolutional Neural Networks

Convolutional networks were originally designed for image recognition, however, due to their advanced ability to extract the features of classes from the objects under study, they are also used for processing time sequences and digital signals [5–7]. The traditional architecture of such a neural network consists of two parts: in the first step, it performs feature extraction from raw data, including a convolution layer and pooling layers. Then, in a second step, the dense layer performs classification based on the extracted features. A convolutional layer convolves into localized regions by transforming an incoming layer into a subsequent one. This approach is used specifically to extract features from raw data. Union layers are used after convolutional layers, which minimizes the number of convolution parameters and computational complexity. In addition, a batch normalization layer is used.

2.2 Recurrent Neural Networks

Another class of deep neural networks, often used in the analysis of one-dimensional digital signals and time sequences, are recurrent neural networks. They were originally created to solve the problem of text processing and natural language. In this work, for the classification of ECG signals, we used a fairly well-known model with a long short-term memory (LSTM) unit.

3 Description of the MIT-BIH Database of ECG Signal Samples

There are fifteen diagnosed types of arrhythmias, which are divided into five large classes [3]. Based on this, one can judge about pathological changes in various parts of the heart in a wide variety of cardiovascular diseases, and this can help doctors in making specific clinical decisions. To validate and test the proposed methods, we used the MIT-BIH database [10,11], a freely available dataset that is widely used to assess the effectiveness of ECG signal classification algorithms. In accordance with the Association for the Advancement of Medical Instrumentation (AAMI) standard EC57, each ECG signal can be divided into 5 types of heartbeats [11]:

- **N** – normal rhythm;
- **S** – Supraventricular ectopic beats (atrial premature): atrial (supraventricular) extrasystole, a violation of the heart rhythm, characterized by the occurrence of single or paired premature heart contractions (extrasystoles) caused by excitation of the myocardium. Frequent atrial extrasystoles can be harbingers of atrial fibrillation or atrial paroxysmal tachycardia, accompanying overload or changes in the atrial myocardium;
- **V** – Ventricular ectopic beats: ventricular extrasystole, premature ventricular contraction. Ventricular arrhythmia can be a manifestation of coronary heart disease;
- **F** – Fusion beats: fusion of ventricular and normal rhythm;
- **Q** – undefined rhythms.

4 Analysis of the Effectiveness of the Deep Learning Models Application to the Classification of Cardiac Arrhythmias

In this section, we proceeded to the development and research of deep learning models for multi-class classification of ECG signals from a data set MIT-BIH [10].

The analyzed dataset consists of 109446 samples of ECG signals. Let us for our convenience to enumerate the records by classes as 'N': 0, 'S': 1, 'V': 2, 'F': 3, 'Q': 4. Then the number of ECG records in each class are as follows: 0-72471, 1-2223, 2-5788, 3-641, 4-6431. The distribution of records over classes is shown in Fig. 1. It could be seen from this, that the classes of ECG signals are extremely misbalanced and it is a real problem for us, because most of the machine learning classifiers ignores minor classes and this can lead to their poor performance. But, as a rule, the accuracy for minor classes is most important for us, so how exactly do they describe heart rhythm disorders.

One approach to solve this problem is to duplicate examples from the minority class in the training dataset before fitting the model. This can balance the class distribution, but does not provide any additional information for the model. The improvement with regard to duplication of examples from the minority class is to synthesize new examples from the minority class. This is a type of data augmentation for a minority class and is called the Synthetic Minority Technique (SMOTE) method [19]. So, in this paper we applied SMOTE algorithm to balance the distribution of ECG-signal classes realized in [15]. As a result of its application, the following distribution of classes ECG-signals was obtained: 0-26.3%, 1-18.4%, 2-18.4%, 3-18.4%, 4-18.4%. The following data are used as input data for the computer algorithms studied in the work: training data $X\_train = (87554, 187)$, $y\_train = (87554, 187)$; validation data $X\_val = (21892, 187)$, $y\_val = (21892, 1)$.

In this paper we proposed one-dimensional 1D_CNN model of the deep convolutional network for ECG-signal classification as follows: a vector of input data of dimension (187, 1) is fed to the input of the neural network. The first layer of the network is a one-dimensional convolutional layer with 32 filters of size (6×1), a convolution step of 1 and a non-linear activation function ReLU. This is followed by the Max_Pooling layer with a kernel size (3×1) and a step equal 1. Its application halves the number of parameters, choosing only neurons with the maximum activation value within the region (3×1). This is followed by the second block, consisting of a one-dimensional convolutional layer with 64 filters, a kernel (3×1), a convolution step equal 1 and an ReLU as activation function, as well as a Max_Pooling layer with a kernel size (3×1) and step 1. The third block also consists of a one-dimensional convolutional layer with 128 filters, a kernel (6×1), a convolution step of 1, and a ReLU as activation function. Next, a Dropout layer is added with a coefficient of 0.25. Dropout is one of the most effective and common neural network regularization techniques. Dropout applied to a layer consists in removing (assigning zero) to randomly selected features. The dropout coefficient is the proportion of the nulled features, usually it is selected in the range from 0.2 to 0.5 [12].

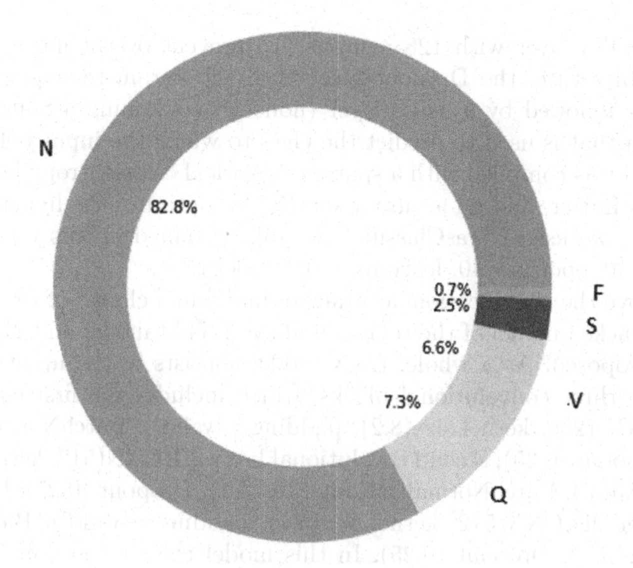

Fig. 1. Distribution classes of records of ECG signals in MIT-BIH base

Then comes the Flatten layer, which converts the multidimensional feature vector to the one-dimensional vector, preparing the output for a Fully Connected (FC) layer. The output of the Flatten layer is then passed to a dense layer of 100 neurons and a ReLU activation function. This is followed by another dense layer with 50 neurons and the ReLU activation function. Finally, another dense layer is included in the network with the Softmax activation function, which is used to predict the class to which the input belongs. The output size of this layer is 5 because there are 5 classes of ECG signal patterns.

As the hyperparameters of the neural network, we have chosen the learning rate, the number of training epochs, the optimizer, and a number of others. As a result of their optimization, the optimal learning rate was 0.001, the number of training epochs was 150, the value of the dropout layer parameter was obtained equal to 0.2, batch_size = 32. To train the network, the Adam optimizer was used, and a sparse categorical cross-entropy as a loss function. Neural network models are implemented in the Keras and Tensorflow environment, and the basic libraries of the Python programming language was also used in the implementation of machine learning algorithms [12,13]. To optimize the hyperparameteres of deep learning models we used KerasTuner library [20].

Next, we proposed the recurrent neural network (RNN) model based on long short-term memory (LSTM) as follows. The introduction of recurrent neural networks (LSTM) has become a popular and effective approach in modeling not only in natural language processing systems, but also digital signals in general [23]. So, the input layer is proposed to convert ECG signals into a structure of dimension (187, 1) corresponding to the length of the ECG signal. Next comes the LSTM layer, which contains an LSTM cell with 187 neurons. Next we add one more LSTM layer with 128 neurons. To prevent overfitting in the neural network architecture, the Dropout layer with the parameter equal 0.2 is also used. This is followed by a dense layer (none, 5) containing a Softmax activation function that is used to predict the class to which the input belongs. Then LSTM model was compiled with a sparse categorical cross-entropy loss function, an Adam optimizer, and an accuracy metric. To optimize the hyperparameters of this model we used KerasClassifier [12,13]. Obtained in this way values are batch_size = 40, epochs = 40, learning rate = 0.0001.

To improve the classification accuracy of individual classes of the data under study, the stacked model of the reviewed above LSTM model and 2D_CNN network was proposed. As a whole, CNN model consists of the input layer, next it follows by three convolutional blocks, which includes the first convolutional layer 2D_CNN (256, kernel_size(8,2), padding = 'valid'), BatchNormalization(), ReLU(), Dropout (0.25); second covolutional layer 2D_CNN(512, kerne_size(5,1), padding = 'valid'), BatchNormalization(), ReLU(), Dropout (0.25); third convolutional layer 2D_CNN(512, kernel_size(5,1), padding = 'valid'), BatchNormalization(), ReLU(), Dropout (0.25). In this model the convolutional block first analyzes the ECG signals, selects the key features of the classes from them and transmits them to the subsequent part of the model of. Further, the received features enter the LSTM unit, where they are analyzed and classification is done.

To this convolutional unit we attached LSTM (128) and Dense (5, activation = 'softmax') layers. In classical deep models, all neurons of the input layer have the same weight. However, different areas of the heartbeat rhythm have different interpretations and affect the results of detecting rhythm disorders. To take into account this feature of the studied data, we proposed to include an attention layer in order to adjust the model to a more attentive perception of the most important parts of the ECG signal [24]. The model constructed in this way is compiled with optimizer = 'Adam', loss = 'categorical_crossentropy' and trained on epochs number = 100.

In addition, the following machine learning algorithms were used: support vector machine (SVM), decision trees (DT), random forest (RF) and extreme gradient boosting classifier (XGB). We used Grid_Search method to optimize the hyperparameters of these algorithms: Best Parameter of SVM: (C = 1, gamma = 0.8, kernel = rbf'), Best Parameter of DT: ('criterion': 'gini', 'max_depth' = 12), Best Parameter of RF: ('criterion': 'gini', 'max_depth': 8, 'max_features': 'auto', 'n_estimators' = 500), Best Parameter of XGB: ('colsample_bytree' = 0.8, 'gamma' = 1, 'max_depth' = 4, 'min_child_weight' = 1, 'subsample' = 0.6).

The main results of the classification of ECG signals using machine learning algorithms and deep neural networks are shown in the Table 1. To analyze the performance of the algorithms, we used next metrics:

$$Accuracy = (TP + TN)/(TP + TN + FP + FN),$$

$$Precision = TP/(TP + FP),$$

$$Recall = TP/(TP + FN),$$

$$F1\_score = 2 * (Precision * Recall)/(Precision + Recall),$$

where TP-true positives (data examples labeled as positive that are actually positive), FP- false positives (data examples labeled as positive that are actually negative), TN-true negatives (data examples labeled as negative that are actually negative), FN-false negatives (data examples labeled as negative that are actually positive) [12].

As we can see from the Table 1, almost all of the studied algorithms have demonstrated high accuracy in the classification of ECG signal samples. On the other hand, since that our main task is the detection of cardiac arrhythmia, which is mainly characterized by classes S, V and F, we are primarily interested in the accuracy of classification of these classes. As can be seen from Fig. 1, they contain an extremely small number of samples of ECG signals compared to other classes.

In this situation, an effective indicator of the performance of the machine classifier is the confusion matrix (CM) [15]. The accuracy of the classification of each class can be visually found by its main diagonal in the corresponding row. After computing of the CM for corresponding algorithms, it can be argued that despite the overall high accuracy of classification, machine learning algorithms do not allow to detect heart rate disorders with high quality. For example, Fig. 2 shows a graph of the confusion matrix for the DT algorithm. It follows that the classes of heart rhythm disorders S, F are classified with low accuracy (58% and 57% correspondently). Figure 3 shows the CM for the RF algorithm, which implies that, despite the high accuracy of the classification of classes N and Q (98% and 94%, respectively), but classes S and F are still characterized by low accuracy (65% and 59%, respectively).

Table 1. The results of classification of ECG classes

Classifier model	Accuracy _macro,%	Precision _macro,%	Recall _macro,%	F1-score _macro%	AUC _macro,%
SVM	90.76	79.22	77.3	78.12	81.18
DT	95	81.22	80.13	81.4	83.16
RF	97	96	79.2	85.4	88.55
XGB	96.69	95.22	74.81	81.934	89.6
1D_CNN	97.36	97.66	97.54	97.68	97.8
LSTM	97.49	97.3	96.8	97.35	97.87
2D_CNN_LSTM	99.34	99.25	99.247	99.43	99.7

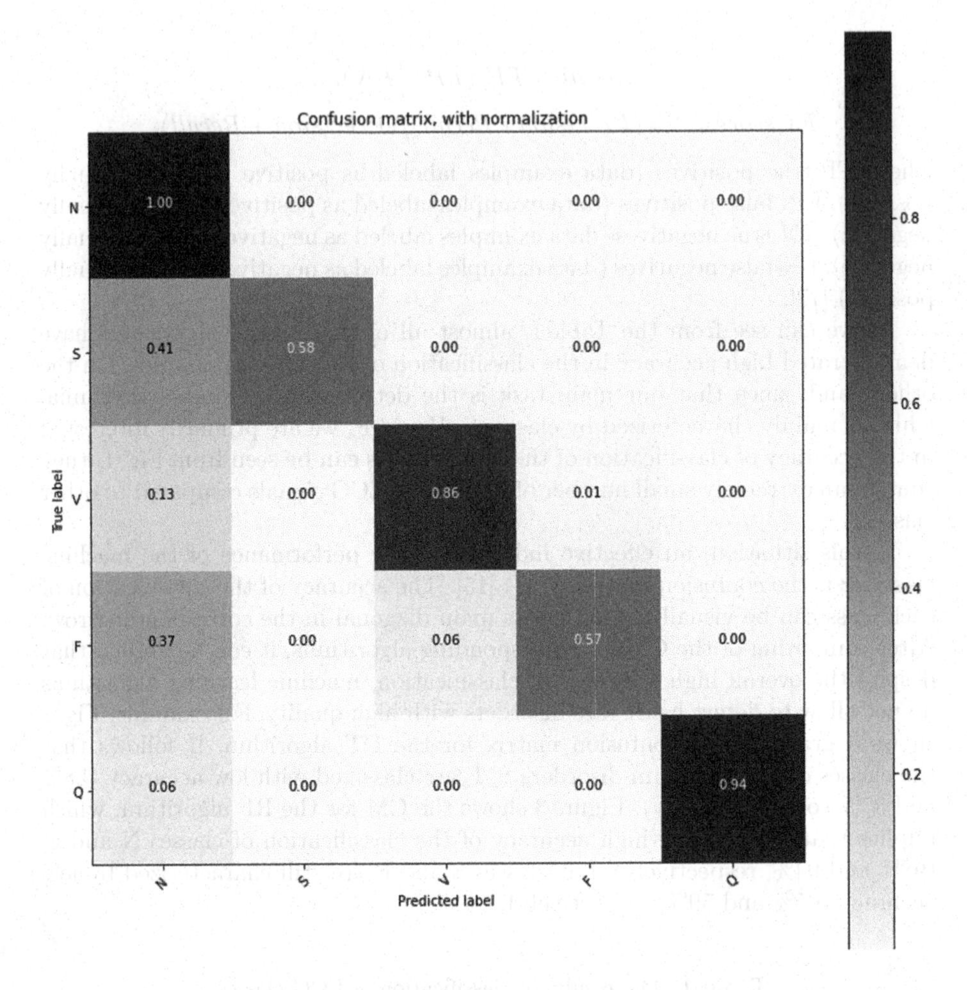

Fig. 2. Confusion matrix of the DT model

For deep learning models, confusion matrices were also constructed and analyzed. Figure 4 shows the confusion matrix for the 1D_CNN deep learning model. It follows that with a sufficiently high accuracy of classification of all classes of samples, the accuracy estimates for classes S, F were less accurate (81% and 84%), compared to the other classes N, Q, V, but significantly higher, compared to the results obtained by the SVM, DT, RF models. Analyzing the confusion matrix of the 2D_CNN_LSTM ensemble model shown in Fig. 5, it can be argued that it performs very well the classification of ECG signal samples, the overall classification accuracy was 99.34% (Table 1), the classification accuracy of classes S, F is turned to be 100%.

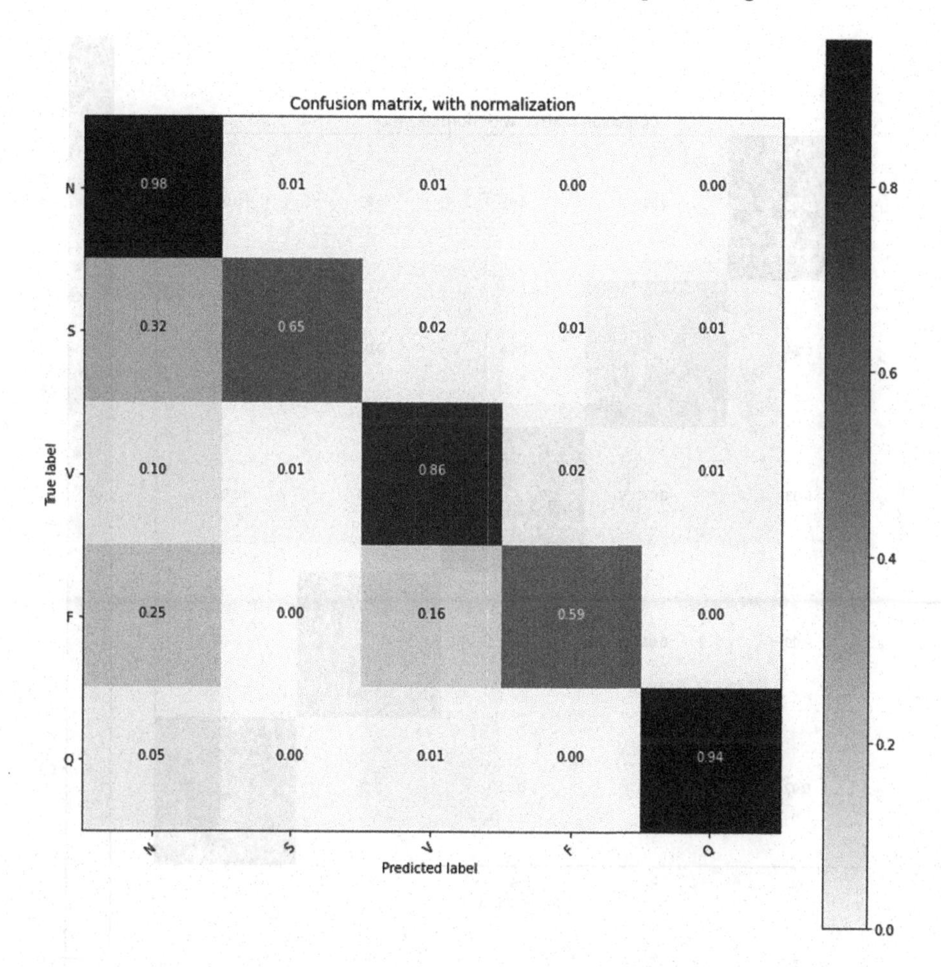

Fig. 3. Confusion matrix of the RF model

When analyzing the classification performance of the machine learning algorithms on imbalanced classes, the ROC_AUC curve graph and the area under curve (AUC) are also effective indicators [15]. Figure 6 shows the ROC curves for each class, and the AUC values for each class of the 2D_CNN_LSTM ensembled model. The analysis of these values shows that the classification quality was very high for all classes of cardiac arrhythmias, compared to the other proposed machine models.

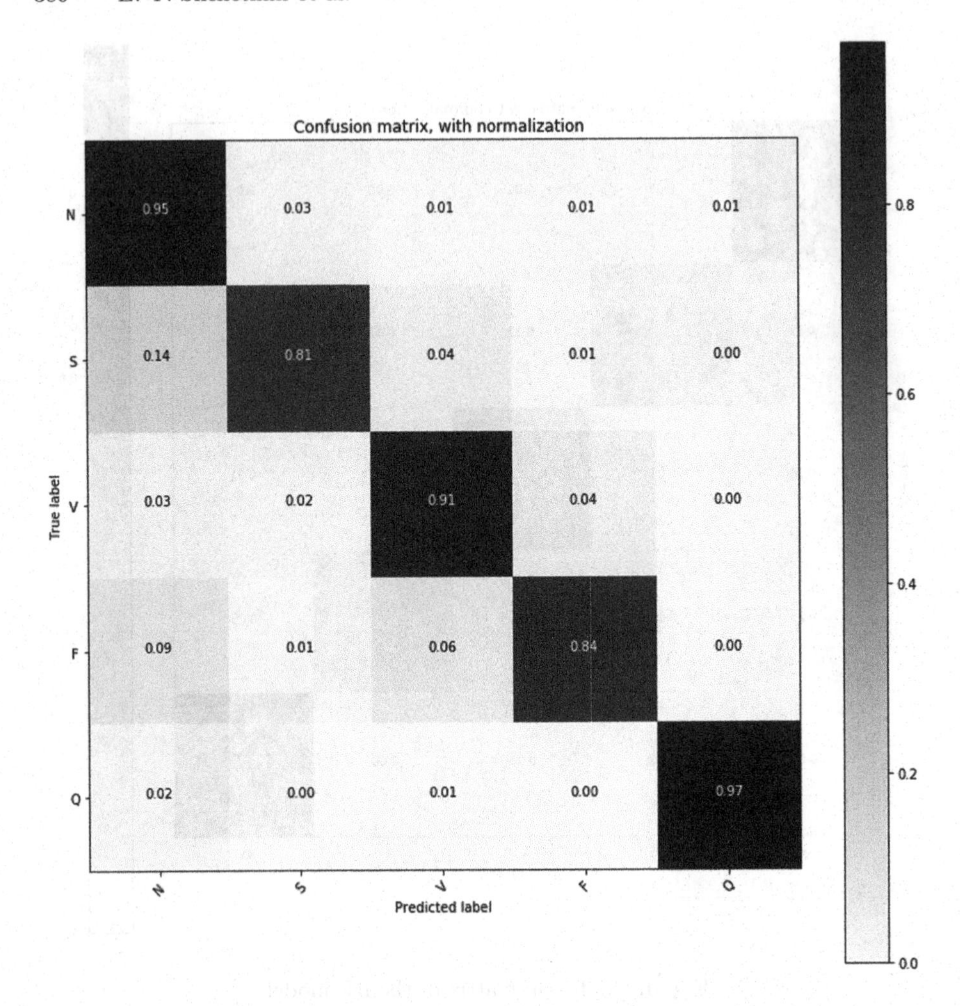

Fig. 4. Confusion matrix of the 1D CNN model

Thus, summarizing up our studies of classification quality analysis by various computer algorithms for individual classes of heart rythm disorders, we can say that almost all machine learning algorithms classify well samples from classes with a large volume of samples and demonstrate low classification accuracy for classes containing a small number of samples. At the same time, analyzing and comparing the performance of various neural network models based on the obtained classification accuracy estimates, it can be argued that the

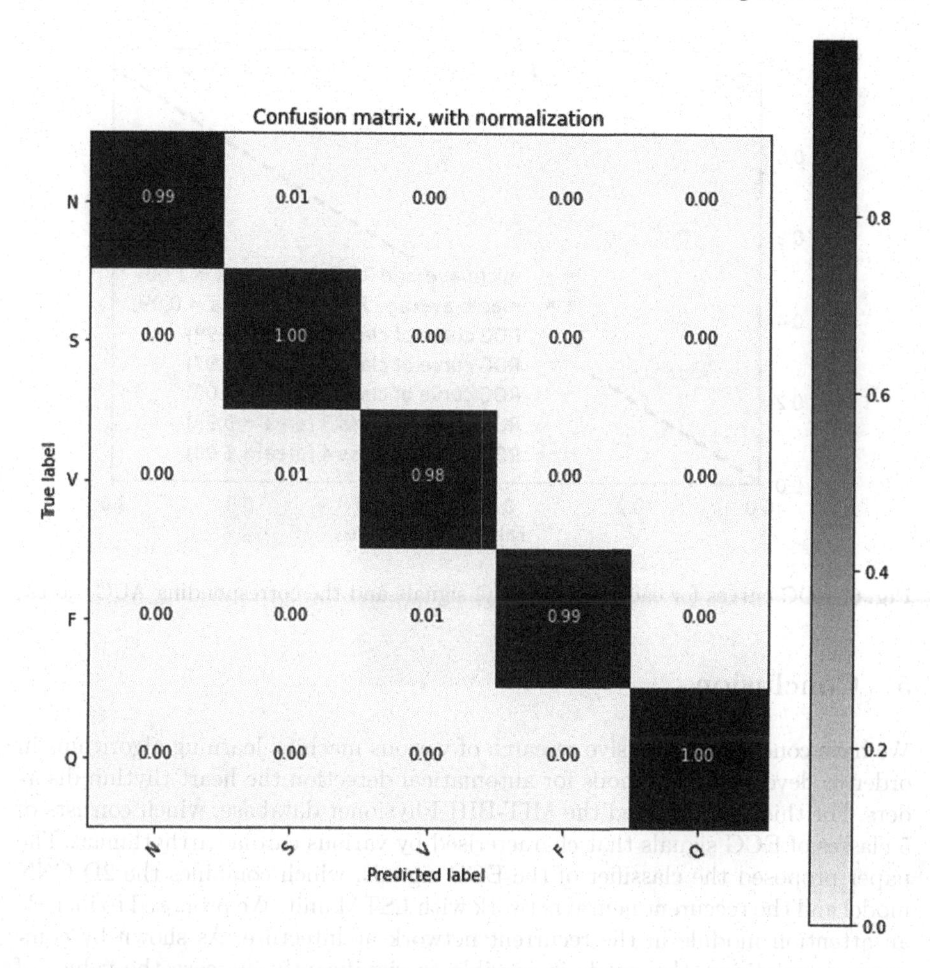

Fig. 5. Confusion matrix of the ensemble of 2D_CNN_LSTM model

2D_CNN_LSTM stacked model allows not only to obtain high classification accuracy, accuracy = 99.34%, but also high values of other classification metrics F1-metric, precision, recall and ROC_AUC curve for the minor classes. This could be explained by the special properties of the attention module, which is used by us in the architecture of the proposed stacked deep model.

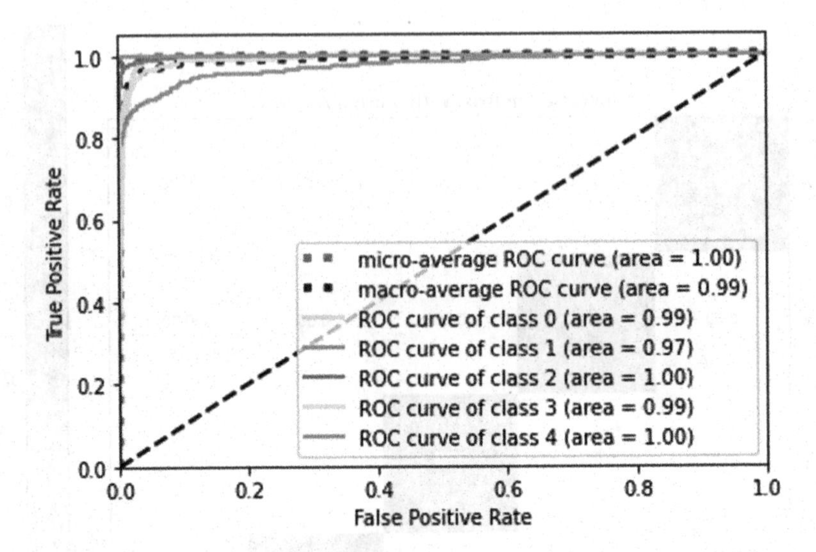

Fig. 6. ROC curves for each class of ECG signals and the corresponding AUC scores

5 Conclusion

We have conducted extensive research of various machine learning algorithms in order to develop the methods for automatical detection the heart rhythm disorders. For this aim, we used the MIT-BIH Physionet database, which consists of 5 classes of ECG signals that characterised by various cardiac arrhythmias. The paper proposed the classifier of the ECG signals, which combines the 2D_CNN model and the reccurent neural network with LSTM unit. We proposed to include an attention module in the recurrent network architecture. As shown by computer calculations, this made it possible to significantly increase the values of the accuracy scores of the classification of minor classes that characterize heart rhythm disorders compared to other computer models. Computer experiments showed that the proposed model successfully classifies cardiac arrhythmias with an overall accuracy of 99.34%.

The most important practical application of the results of the work is the promotion of the developed algorithms for the study of other bases of electrocardiograms in order to create algorithms for the transfer of learning to recognize arrhythmia [22]. The most important line of development of our research is the elaboration of a mobile application that allows solving the problem of remote detection of arrhythmias using an electrocardiogram sample uploaded by a doctor in the form of two-dimensional images. Another significant problem is the imbalance of the classes of ECG signal databases. Standard approaches to its solution in this case can hardly be effective, since artificial samples of digital signals created, for example, by the SMOTE algorithm [14], may simply not be a human electrocardiogram. We will also put our further intellectual efforts in this direction.

References

1. WHO (2018). http://www.who.int/mediacentre/factsheets/fs317/en/
2. Society, H., Heart diseases and disorders (2018). https://www.hrsonline.org/Patient-Resources/Heart-Diseases-Disorders
3. Benjamin, E.J., Virani, S.S., Callaway, C.W., et al.: Heart disease and stroke statistics, 2018 update: a report from the American Heart Association. Curculation **137**(12), 67–492 (2018)
4. Coast, D.A., Stern, R.M., Cano, G.G., Briller, S.A.: An approach to cardiac arrhythmia analysis using hidden Markov models. IEEE Trans. Biomed. Eng. **37**(9), 826–836 (1990)
5. Isin, A., Ozdalili, S.: Cardiac arrhythmia detection using deep learning. Procedia Comput. Sci. **120**, 268–275 (2017)
6. Zhai, X., Tin, C.: Automated ECG classification using dual heartbeat coupling based on convolutional neural network. IEEE Access **6**, 27465–27472 (2018)
7. Acharya, U.R., et al.: A deep convolutional neural network model to classify heartbeats. Comput. Biol. Med. **89**, 389–396 (2017)
8. Kiranyaz, S., Ince, T., Gabbouj, M.: Real-time patient-specific ECG classification by 1-D convolutional neural networks. IEEE Trans. Biomed. Eng. **63**(3), 664–675 (2015)
9. Jun, T.J., Nguyen, H.M., Kang, D., Kim, D., Kim, D., Kim, Y.-H.: ECG arrhythmia classification using a 2-D convolutional neural network. arXiv Preprint arXiv:1804.06812 (2018)
10. Mark, R.G., Schluter, P.S., Moody, G.B., Devlin, P.H., Chernoff, D.: An annotated ECG database for evaluating arrhythmia detectors. IEEE Trans. Biomed. Eng. **29**(3), 600 (1982)
11. Association for the Advancement of Medical Instrumentation. https://www.aami.org/
12. Chollet, F.: Deep Learning with Python, 384 p. Manning Publications (2017)
13. Muller, A.C., Guido, S.: Introduction to Machine Learning with Python. O'Reilly Media, Inc. (2018). ISBN 13:9781449369415
14. Raschka, S., Mirjalili, V.: Python Machine Learning: Machine Learning and Deep Learning with Python, Scikit-learn, and TensorFlow, 3rd edn. Packt Publishing (2019)
15. Sevastyanov, L.A., Shchetinin, E.Y.: On methods of improving the accuracy of multiclass classification on unbalanced data. Informatika i yeye primeneniya **14**(1), 63–70 (2020). (in Russian)
16. Shchetinin, E.Y., Sevastyanov, L.A., Demidova, A.V., Kulyabov, D.S.: Classification of skin lesions according to dermoscopy using deep learning methods. Matematicheskaya biologiya i bioinformatika **2**(15), 180–194 (2020). (in Russian). https://doi.org/10.17537/2020.15.180
17. Rajpurkar, P., Hannun, A.Y., Haghpanahi, M., Bourn, C., Ng, A.Y.: Cardiologist-level arrhythmia detection with convolutional neural networks. arXiv preprint arXiv:1707.01836 (2017)
18. Kachuee, M., Fazeli, S., Sarrafzadeh, M.: ECG heartbeat classification: a deep transferable representation. arXiv arXiv:1805.00794v2 (2018)
19. Chawla, N.V., Bowyer, K.W., Hall, L.O., Kegelmeyer, W.P.: SMOTE: synthetic minority over-sampling technique. J. Artif. Intell. Res. **16**, 321–357 (2002)
20. Keras tuner documentation. https://keras-team.github.io/keras-tuner/

21. Chen, D., Li, D., Xu, X., Yang, R., Ng, S.-K.: Electrocardiogram classification and visual diagnosis of atrial fibrillation with DenseECG, arXiv:2101.07535v1 [eess.SP], 19 January 2021
22. Tajbakhsh, N., et al.: Convolutional neural networks for medical image analysis: full training or fine tuning? IEEE Trans. Med. Imaging **35**(5), 1299–1312 (2016)
23. Shchetinin, E.Y., Sevastianov, L.A., Kulyabov, D.S., Ayrjan, E.A., Demidova, A.V.: Deep neural networks for emotion recognition. In: Vishnevskiy, V.M., Samouylov, K.E., Kozyrev, D.V. (eds.) DCCN 2020. LNCS, vol. 12563, pp. 365–379. Springer, Cham (2020). https://doi.org/10.1007/978-3-030-66471-8_28
24. Vaswani, A., et al.: Attention is all you need. In: Advances in Neural Information Processing Systems, pp. 5998–6008 (2017)

Distributed Systems Applications

Autonomous Infrared Guided Landing System for Unmanned Aerial Vehicles

Mainak Mondal(✉), S. V. Shidlovskiy, D. V. Shashev, and Mikhail Okunsky

National Research Tomsk State University, Tomsk 634050, Russia
mainakme2140@gmail.com, dekanatfit@tic.tsu.ru
http://fit.tsu.ru

Abstract. This article highlights the requirements for precision landing systems in multi rotors and proposes a 3 point IR-guided Landings system for a standard portable landing pad. The proposed system uses a monocular camera with an IR Filter and Open CV. The geometrical Centroid formula is used to anchor itself over the landing pad.

Keywords: Multi-rotor · Guided Landing · Autonomous Navigation

1 Introduction

Today most off-the-shelf consumer multi-rotors are equipped with features like autonomous flight, GPS way-point-mission, optical-flow stabilization and a lot more. An important aspect of autonomous flight is GPS or Global Positioning System. It uses GNSS to triangulate a position approximate to 5 m. Using this, UAVs have achieved great feats in the past decade by using this to navigate far-away territories without human intervention. These features enable multi-rotors to navigate through the skies with ease, and in recent years the industry of multi-rotors has grown due to the consumer interest in these devices. Manual control of these multi-rotors are as safe as the pilots but autonomous fights depend on the flight controller as well as the on-board computer on the UAV itself. Since it is already established that GPS is approximately accurate to 5 m [1] and which is in the best of conditions, landing autonomously in tight or dangerously small spots is fairly risky. Camera Assisted landing makes autonomous flights in these UAVs more reliable. The main objective of this research is to use a monocular camera (with IR Lens) in a UAV to detect visual cues (IR Beacons), and use it as an anchor to align itself while landing at the spot, to avoid unnecessary hitches or movement due to wind or other susceptible causes like COG.

1.1 Design of a Quad-copter

A Multi rotor is made up of multiple thrust-generating engines. A quad-copter, as the name suggests has four rotors or 4 thrust-generating engines. These 4

V. M. Vishnevskiy et al. (Eds.): DCCN 2021, CCIS 1552, pp. 387–395, 2022.
https://doi.org/10.1007/978-3-030-97110-6_30

rotors have to be places such that 2 rotors spin clockwise and 2 rotors spin anti-clockwise. There is flight controller in the quad-copter, usually near the center of gravity, which has an IMU or Inertial Measurement Unit, which measures linear acceleration and the angular velocity. This data can be processed by the controller to produce, the pitch and roll rates/angles and this data can be used by the controller to command the 4 rotors accordingly to stabilize the UAV [2]. The yaw rate is also measured by it, but its not very accurate, and thus a magnetometer is used and to physically yaw the reactive torque generated by the motors is used. A general 'x' quad-copter as seen in Fig. 1 usually has the following equations for distributing thrust to the motors.

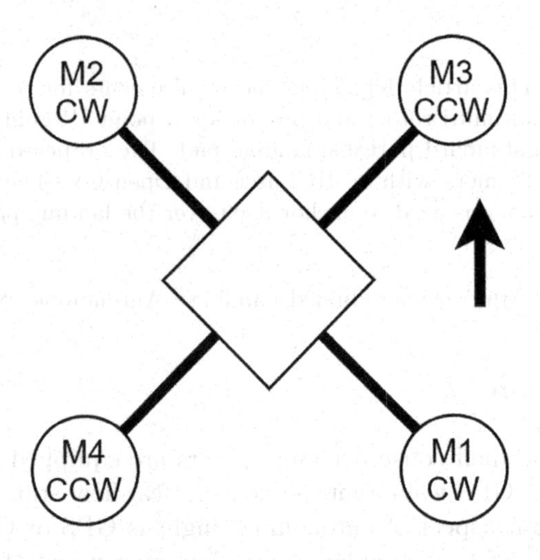

Fig. 1. X Quad-copter

$$M1 = Thrust - Roll + Pitch - Yaw \quad (1)$$

$$M2 = Thrust + Roll - Pitch - Yaw \quad (2)$$

$$M3 = Thrust - Roll - Pitch + Yaw \quad (3)$$

$$M4 = Thrust + Roll + Pitch + Yaw \quad (4)$$

1.2 Control of a Quad-copter

The control system of the Quad-copter can be seen in Fig. 2 [3] where various sensors are used to estimate the local position of the drone in a local/global space. The Eqs. (1), (2), (3) and (4) make up the MMA or Motor Mixer Algorithm. MMA or Motor Mixer Algorithm, which translates the inputs received (pitch, roll, yaw, thrust) to the values understandable by the Electronic Speed Controllers concerning the frame of reference of the quadcopter.

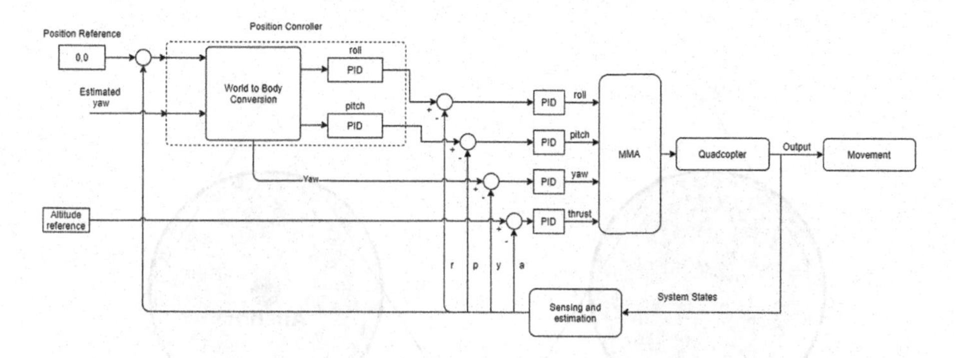

Fig. 2. Control System of a UAV

2 Concept

Assisted landing has existed for a while in open-source flight controllers like the pixhawk, using an infrared emitter beacon and receiver. The controller is connected to an IR-LOCK sensor which is a slightly modified camera to detect IR light. The sensor spits out the position of the detected IR light and the controller uses this data to align itself and land at approximately 1 m/s. This mode of landing can be difficult to use on sunny days as the IR Sensor might recognize the sunlight as a landing beacon. In the tests conducted, on a fairly sunny day, it was observed that 3 out of 4 times the sensor mistook the spot formed by crepuscular rays under a tree as a landing beacon. One can suggest the use of specialized colors and landing markers to use for this process but it might also be mistaken by the camera in many situations.

In usual cases, a brightly colored landing pad can be used by the UAV as the landing area. Recognizing a bright color like orange is simple using Open CV and it works well because there is not a lot of computation involved. This method however fails to work in a dimly lit day and completely fails during the night when it's dark. Keeping this scenario in mind, High Intensity Infrared Beacons were chosen because they are visible from far away distances and are visible in the daylight as well as in the dark. (Note: Visible to a camera with the IR Lens). This method improves on the "Assisted landing" and uses 3 Infrared Beacons around the landing area and a special IR Lens/Filter to let only the IR Light pass.

It is theorized that using a monocular camera (with a IR filter) and three infrared beacons as a marker will provide ample guidance for a multirotor to land at a designated location.

3 Placement of the High Intensity IR Beacons

The 3 High Intensity IR beacons should be placed as marked in Fig. 3. The IR Beacons should be placed such that the center of the triangle (formed by them) and the center of the landing pad fall over each other.

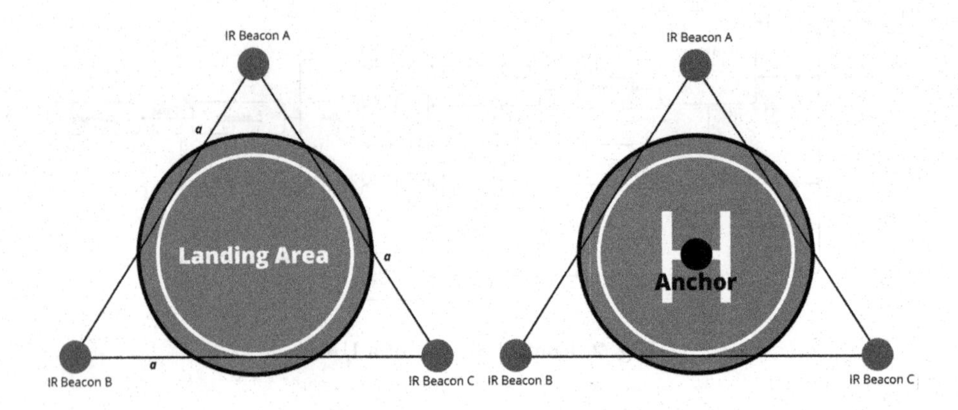

Fig. 3. Landing Pad with beacons

3.1 Calculation: To Find the Distance Between the IR Beacons

Let's assume the Landing Pad to be 1.5 m in diameter, and the landing area (white circle) has a diameter of 1 m. The triangle formed around the landing area is an Equilateral Triangle, i.e. the length of sides (a) are equal to each other and the angle formed in each vertex is 60°.

$$r = (\sqrt{3}a)/6 \tag{5}$$

or,

$$a = 2\sqrt{3}r \tag{6}$$

In (6), r is the radius of the inscribed circle or 0.5 m. Thus, the length of each side is **1.73 m**. Thus, the IR Beacons should be at least 1.73 m apart from each other.

4 Anchor Point Calculation

The Anchor Point is shown in Fig. 3. It is the center of the Landing Pad and it marks the in-center of the triangle formed by the High Intensity IR Beacons.

The anchor is the centroid of the triangle. Given the coordinates of the three vertices of the triangle ABC, the coordinates of the Anchor(O), are given by the Eqs. (7) and (8).

$$O_x = (A_x + B_x + C_x)/3 \tag{7}$$

$$O_y = (A_y + B_y + C_y)/3 \tag{8}$$

Where, Ax and Ay are the x and y coordinates of the point A (IR Beacon). Bx and By are the x and y coordinates of the point B (IR Beacon). Cx and Cy are the x and y coordinates of the point C (IR Beacon).

5 Algorithm

Figure 4 outlines the process of assisted autonomous landing. This landing sequence can be initiated in the air, which will wait for the UAV to navigate to the final waypoint in its mission parameters. Once the final waypoint is reached, it will look for the IR beacons in the ground using the monocular camera (with IR Lens) pointed downwards.

Usually, drones are controlled manually, using radio equipment. In this case, the operator has full control over the flight of the aircraft from takeoff to landing. In the case of automated missions, the operator is less involved in the flight process. Here GPS coordinates are used in order to follow the set trajectory, and takeoff and landing are performed automatically. In the flowchart shown in Fig. 4, the sequence of actions can be described as follows -

1. Start the algorithm.
2. Takeoff in automatic mode (or guided mode, depending on the firmware) and traverse the set points, using the mission-provided GPS coordinates.
3. As soon as the flight mission is completed, the multi-rotor stops and hovers in the mission-set point.
4. The Infrared assisted Landing Sequence module starts searches for beacons.
5. If the aircraft then obtains the coordinates of the beacons, it calculates the coordinates of the central point relative to the triangle, on the top of which the beacons are located. (If the beacons aren't visible in the image received, the last mission-set GPS coordinates are sent back to the controller to verify if the mission is finished and if the aircraft is in the right place).
6. The coordinates calculated using the given method, are sent to the flight controller, the distance to the point is calculated (correction values) and passed through the PID controller to smooth en the transition to the desired point.
7. The transition movements begins using the correction values until the correction values are negligible (nearly zero).
8. If the correction values reach zero (0), then the desired point is reached, and henceforth the position is maintained in the x and y axis, while the thrust(z) is slowly reduced.
9. When the thrust is reduced, the aircraft looses altitude and eventually lands at the desired spot.
10. Stop.

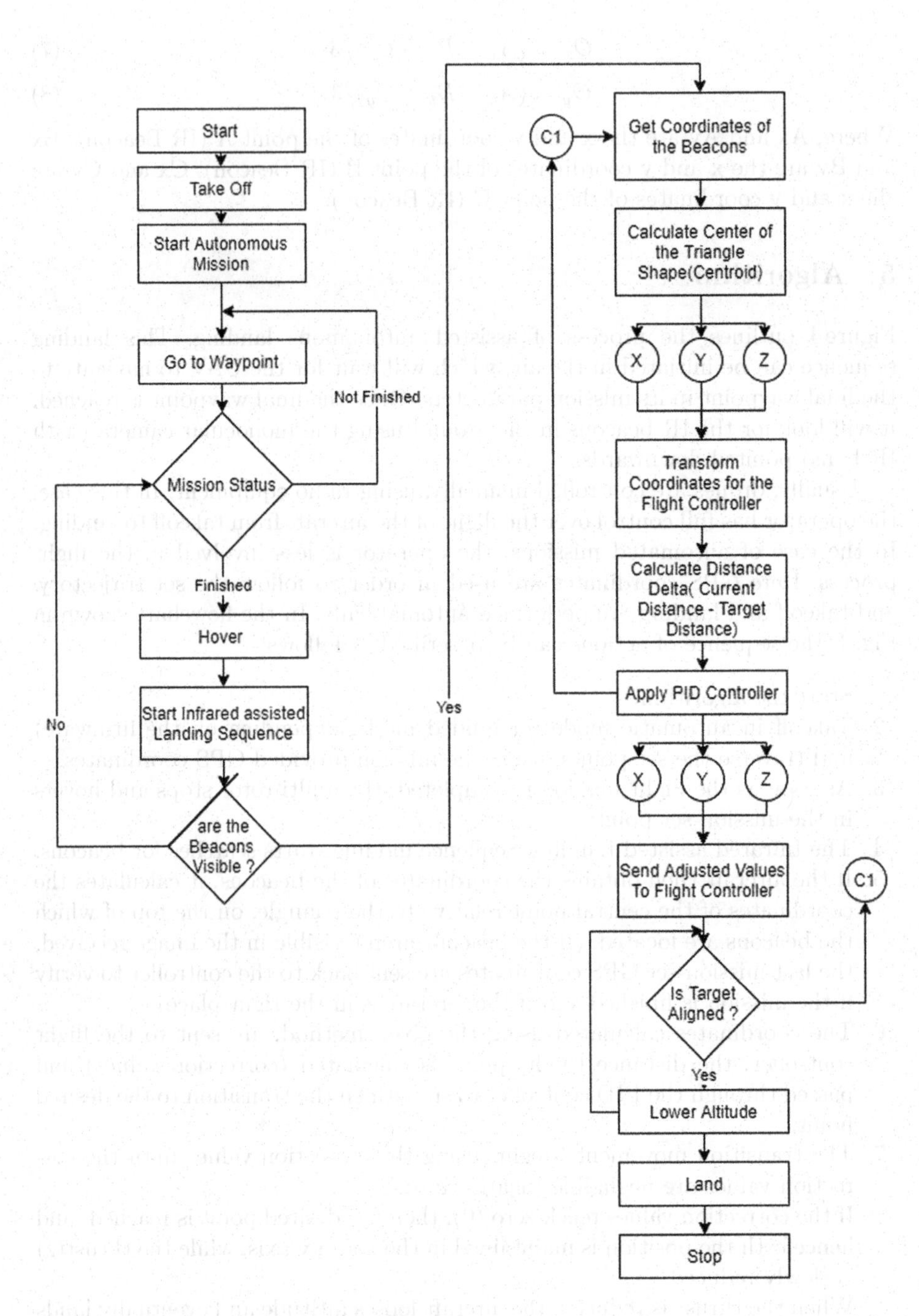

Fig. 4. Flowchart

6 Results

The output of the algorithm should look like Fig. 5 from about 10 m altitude. Once the Beacons are located, the OpenCV will broadcast (publish) the detected location in the ROS network. This data is read by another subscriber which does the necessary Centroid Calculations and transformations required and publishes delta values to the flight controller.

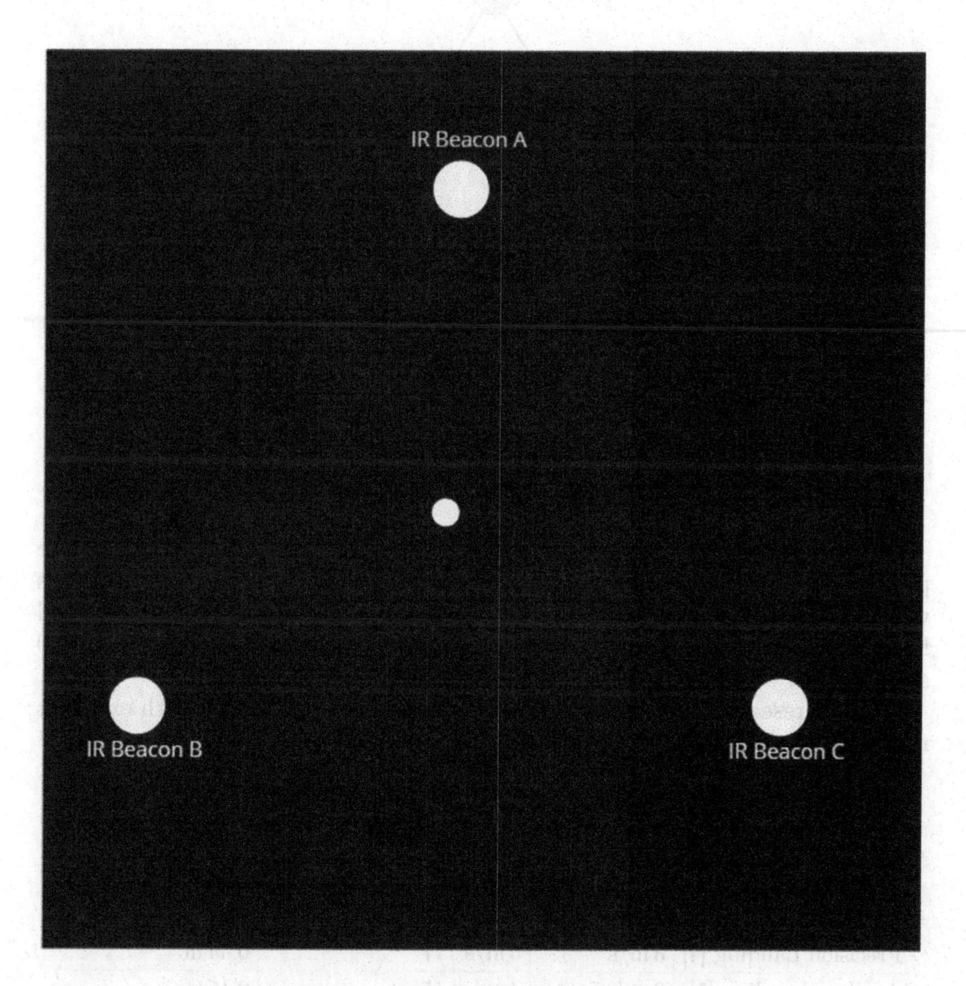

Fig. 5. Image from the Camera with IR Filter

The *landing area* shown in Fig. 6 is the ideal place where the multi-rotor should land after the program finishes. The beauty of this algorithm lies within its simplicity as its not very mathematically or computationally intensive. The size of the multi-rotor also fairly irrelevant as the size of the landing pad as well as the space between the IR Beacons can be scaled up and down easily.

In our tests a standard landing pad was used which is about 1.5 m, like this instance and the largest multi-rotor we landed was a DJI Matrice 600 Pro which has a Diagonal Wheelbase of 1133 mm (1.13 m) but it can easily fit a larger aircraft.

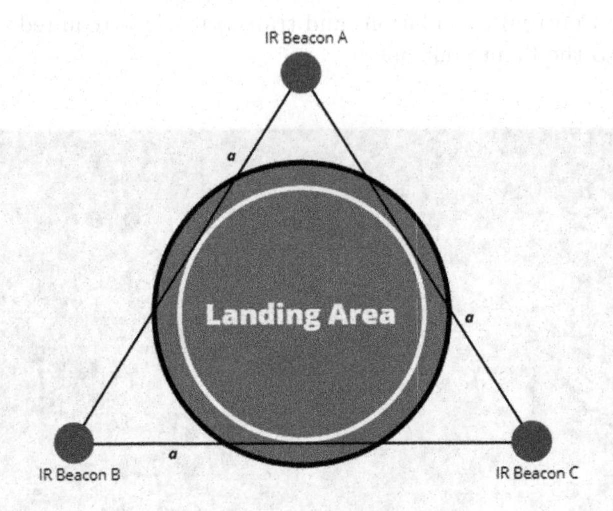

Fig. 6. Ideal Landing Area

A PID controller is also applied for accurate reaction to the changes in the position of the UAV and to account for external factors. The delta distance values will be sent to the flight controller, so the UAV aligns itself over the marker and lowers altitude, while locked on the marker. Eventually, the altitude reaches 0 and the UAV will land.

Similar research presented in [4] and [5] can be seen in Table 1 with excellent results.

Table 1. Comparison of the Landing Methods

Method	Wind Speed	Gusts	Number of Trials	Average Accuracy
GPS Mission	4 m/s	8 m/s	15	3.9 m
Precision Landing [4]	3 m/s	5 m/s	17	0.44 m
Apriltag Landing [5]	3 m/s	4 m/s	15	0.15 m
IR Landing	4 m/s	5 m/s	15	0.19 m

Remark 1. The wind speed and wind gusts have been recorded from windy [6]

The IR Landing method presented in this article is very robust and reliable as seen in the Table 1. It is worth noting that the precision landing presented by the authors of [4] is at a disadvantage as it was performed in an uneven surface and it is safe to assume that the method presented might also be inconsistent when the conditions are less than perfect.

In comparison to the Apriltag Landing, this method was less accurate by 0.04 m but that may be due to the margin of error in experiments conducted.

7 Conclusion

Today, multirotors are not only expected to perform military tasks like payload delivery and remote recon but are also expected to fulfill general civilian needs like simple point to point delivery. Tech giants like DHL and Amazon have invested a huge chunk of money into unmanned delivery systems. Many cities have already allowed these companies to test their platform and adding a precision landing system like the one mentioned in this article, will significantly make these multirotors safer to use in a populated area. A precision landing system will also build confidence with the local aviation authorities as the Landing Areas (landing pad + ir sensors) can be marked as the only designated places where a multirotor could land/take off.

References

1. Van, D., Frank, E.: The world's first GPS MOOC and Worldwide Laboratory using Smartphones. In: Proceedings of the 28th International Technical Meeting of the Satellite Division of the Institute of Navigation, Tampa, Florida, pp. 361–369 (2015)
2. Siciliano, B., Sciavicco, L., Villani, L., Oriolo, G.: Robotics. McGraw-Hill (2009)
3. Mondal, M., Poslavskiy, S.: Offline navigation (homing) of aerial vehicles (quadcopters) in GPS denied environments. Unmanned Syst. **9**, 119–127 (2020)
4. Pluckter, K., Scherer, S.: Precision UAV landing in unstructured environments. In: Xiao, J., Kröger, T., Khatib, O. (eds.) ISER 2018. SPAR, vol. 11, pp. 177–187. Springer, Cham (2020). https://doi.org/10.1007/978-3-030-33950-0_16
5. Mondal, M., Shidlovskiy, S., Shashev, D.: Camera assisted autonomous UAV Landing. In: CEUR Workshop Proceedings, vol. 2744. ITMO University, Russia (2020)
6. Windy Homepage. http://windy.com

Distributed Computing of R Applications Using RBOINC Package with Applications to Parallel Discrete Event Simulation

S. N. Astafiev[1,2] and A. S. Rumyantsev[1,2](✉)

[1] Petrozavodsk State University, 33 Lenina Street, Petrozavodsk, Russia
ar0@krc.karelia.ru
[2] Institute of Applied Mathematical Research, Karelian Research Centre of Russian Academy of Sciences, 11 Pushkinskaya Street, Petrozavodsk, Russia

Abstract. R programming language is commonly used for statistical computing, data science and stochastic simulation. Existing packages for R allow to run parallel code on various parallel architectures, however, the support for distributed (volunteer) computing is rather weak. This article describes a new R package `RBOINC` that allows to run parallel code on desktop grid systems via the BOINC open source system for grid computing, which is a promising approach for parallel stochastic simulation. Among the possible ways to utilize the new package, an approach to parallel regenerative stochastic simulation within the generalized semi-Markov processes framework is suggested.

Keywords: Distributed Computing · Volunteer Computing · BOINC · R Software · Generalized Semi-Markov Processes

1 Introduction

Parallel and distributed computing are widely used technologies of computationally consuming software application speedup used both for the science and for production. These technologies within the field of scientific research are actively used for modeling and simulation, estimation and data analysis.

In terms of the hardware architecture used to implement the aforementioned technologies, the corresponding software applications can be classified by the degree of interdependence of the so-called parallel threads, and the amount of synchronizations (e.g. for data exchange) needed within the application. As such, the parallel hardware architectures correspond to various classes of software applications based on the demand of synchronization. These are: Uniform Memory Access (UMA), Non-Uniform Memory Access (NUMA), and Distributed Memory, ordered by the possible synchronization speed decreasingly.

The latter class of hardware includes various realizations of grid computing systems, including the Desktop Grids and Volunteer Computing. The software

V. M. Vishnevskiy et al. (Eds.): DCCN 2021, CCIS 1552, pp. 396–407, 2022.
https://doi.org/10.1007/978-3-030-97110-6_31

applications most suitable for this class belong to the class of so-called *embarrassingly parallel*. This means that a large computing task can be decomposed into a large number of small independent subtasks that allow to aggregate the results of their computation into the solution of the original task. In the field of stochastic modeling and simulation, a few examples are the perfect simulation technique [18], time-parallel simulation [4,5], discrete-event simulation [7] and heuristic optimization [6], to name a few.

One of the popular environments both for data analysis [19] and for stochastic modeling in queueing [3] is the R language environment [13]. The language is extensible by over 18000 packages in various application fields available at CRAN repository. Existing packages for R support all the aforementioned architectures of parallel machines. There are two main types of parallel processing for R:

- creation a subprocess via fork system call (at unix-like machines, UMA architectures).
- running an additional R interpreter and establishing a connection with it at network level.

In particular, the following packages are widely used: `doMC` (UMA, unix-like systems only), `rslurm` (supercomputing backend), `snow` (simple network of workstations, connected by a conventional network), `parallel` (the one included into the basic distribution by default). At the same time, there are, to the best of our knowledge, almost no packages specific to the Volunteer Computing systems (which is a good choice for establishing a sustainable computing environment at low cost).

BOINC is a Volunteer Computing system [2] used for utilization of the idle CPU time and reducing the energy waste. All the resources are donated by the volunteers at no cost. Any BOINC project has a central server controlled by the project maintainers who create and upload the necessary applications, tasks, and summarize the results of computations. Usually the number of tasks in a project far exceeds the number of BOINC users. The tasks are distributed to the computing resources donated by the users with a certain level of redundancy. Upon the completion of computations, the results are uploaded to the server and summarized afterwards. As such, organizing a BOINC project may allow one to obtain significant computing resources at almost no cost.

In summary, using BOINC for certain tasks in simulation modeling and optimization is promising. Thus, it is important to develop an R software package to equip the researchers with a software solution for parallel simulation over the volunteer computing resources. In this paper we introduce such a software package named `RBOINC`, following the concept introduced in [15].

The structure of the paper is as follows. We introduce the architecture of our framework and highlight the functions implemented in the package in Sect. 2. The general scheme of the working process is briefly outlined in Sect. 3. A possible application of the approach to an embarrassingly parallel implementation of the discrete-event simulation approach is presented in Sect. 4. Finally we draw some conclusions in Sect. 5.

2 Parallel Backend for R Using BOINC High-level Architecture

The presented software package allows to organize a seamless connection from the R environment to the BOINC computing platform. To do so, a backend is used, which contains the following three parts:

- **Client part** - an R package that is installed by the package users (researchers) on their computers.
- **Virtual machine** - a specific virtual machine that runs the independent jobs on the donated computing resources organized and managed by the BOINC software.
- **Server part** - a set of shell scripts, programs and configuration files that need to be run on the BOINC server.

The package is distributed within the so-called R-Forge resource hosting the packages in the development phase, and is planned for distribution within the so-called CRAN network (the package distribution service for R software). The project summary page https://r-forge.r-project.org/projects/rboinc/ contains the necessary usage and licensing information. Next, we briefly describe the structure of the aforementioned backend parts.

2.1 Client Part

The client part is implemented by the `RBOINC.cl` package available at the R environment upon completion of the installation procedure run from R console as follows

```
install.packages("RBOINC.cl", repos = "http://R-Forge.R-project.org")
```

The package is used by the researchers aiming to utilize the BOINC network as a computing resource. This package is intended for non-BOINC specialists who need to run some computations in parallel. Long latency while transmitting, validating and assimilating the data cause a huge communication overhead that forces one to split the initial task into small subtasks (the so-called *workunits*) in such a way to guarantee relatively large computational time per workunit.

Below we briefly summarize the meaning and the parameters of the functions provided by `RBOINC.cl` package. Firstly we give the corresponding name in bold, and the parameters are explained afterwards.

- `create_connection` – the function is used to create a connection to the BOINC server with a project hosting the computations. The possible parameters are:
  - *server* an ssh or http server URL of the form "<protocol>://<full domain name/IP address>:<port>"
    The protocol can take one of the following self-explaining values, "ssh", "http" or "https".

- *dir* is an `RBOINC` project directory on the BOINC server. For the ssh connection, this is the directory where the BOINC project is located. For the http/https connection, this is the full path to the project page (omitting the server name).
- *username* is a string containing the username. For the ssh connection, this is the user login. For the http/https connection, this is the user email.
- *password* is a string containing the corresponding user's password. If this parameter is equal to NULL, then a window will be displayed prompting the user to enter the password.
- *keyfile* is a path to a private key file. This parameter is used for ssh connection only.

The return value of this function is a list, which needs to be passed to all other functions interacting with the server.

– `create_jobs` – the function is responsible for creation and transmission of tasks to the BOINC server. The parameters are (only the first three listed below are required, while the subsequent can be NULL):
- *connection* is a connection (list) created by the `create_connection` function call.
- *work_func* is a data processing function. This function runs for each element in the *data* object. This function can be recursive.
- *data* is the data for processing. Must be a numbered list or vector object.
- n is the requested number of jobs. This parameter must be less than or equal to the length of the data. If not specified, then the number of jobs will be equal to the length of the *data*.
- *init_func* is an initialization function. This function runs once at the beginning of the job computation before the job is split into separate threads. It may be used to implement some specific initialization procedure. This function can not be recursive.
- *global_vars* is a list of global variables in the format <variable name> = <value>. These variables will be available to the functions *work_func* and *init_func*.
- *packages* is a string vector with imported packages names. These packages will be included into R environment at the BOINC host machines (at volunteer's computers) at the initialization phase.
- *files* is a string vector with the files and directories names that should be available for jobs. They will be located in the working folder of the job.

This function makes a tar.xz archive with files required for creation of batch of jobs and uploads it to the BOINC server available by the *connection* object passed as the function parameter. Subsequently, the BOINC server unpacks the archive and registers the unpacked files as the data and the scripts that will be distributed as the workunits to the volunteers upon request. Then the function creates a batch of jobs at the BOINC server. The return value is a list containing the current states of jobs.

– `update_jobs_status` – the function is responsible for updating the status information of the jobs submitted earlier by sending the corresponding request to the BOINC server. The parameters are:

- *connection* is a connection created by `create_connection`.
- *jobs_status* a list returned by `create_n_jobs` or `update_jobs_status`.
- *callback_function* is a function that is run for each single result available after downloading it from the BOINC server. This function must take one argument which is the result of the completed workunit. The value returned by this function is placed in the results list.

This function communicates with the BOINC server, updates the status of the uploaded workunits results and then downloads the results available. Upon sequential calls of this function, the workunits with results already downloaded are skipped. After downloading the result, it is loaded into memory (in R environment on the client side) and passed to the *callback_function* function. The results of the *callback_function* are then placed into the corresponding positions of the result array which is considered as the return value.

– `close_connection` - graceful shutdown of the connection to the BOINC server previously opened. The parameters are:
  - *connection* is a connection created by `create_connection`.

  This function is needed for the correct release of the resources when sending jobs or receiving results is no longer required.

– `test_jobs` is a debug function. Its parameters are similar to the functions `create_jobs` and `update_jobs_status`, except for *connection* and *jobs_status* parameters, which are not used in this function. The parameters are:
  - *work_func* is a data processing function. This function runs for each element in *data*. This function can be recursive.
  - *data* is a data for processing. Must be a numerable list or vector.
  - *n* is a number of jobs. This parameter must be less than or equal to the length of the data. If not specified, then the number of jobs will be equal to the length of the *data*.
  - *init_func* is a initialization function. This function runs once at the start of a job before the job is split into separate threads. Necessary for additional initialization. This function can not be recursive.
  - *global_vars* is a list of global variables in the format <variable name> = <value>. These variables will be available to functions *work_func* and *init_func*.
  - *packages* is a string vector with imported packages names.
  - *files* is a string vector with the files and directories names that should be available for jobs. They will be located in the working folder of the job.
  - *callback_function* is a function that is called for each result after loading. This function must take one argument, which is the result of the work performed. The value returned by this function is placed in the result list.

  This function makes a batch of jobs and runs it locally on the client's computer. In the working process, the function allows one to gain information about the actions performed and the events occurred. This function is intended for quick debugging of the code that uses the `RBOINC.cl` package. Note however that if the jobs are performed correctly within this function, this does not guarantee that the same process will be performed correctly using a real BOINC server, `create_jobs` and `update_jobs_status` functions due to a specific environment and possibly sophisticated configuration of the BOINC server.

A necessary condition to use the `RBOINC.cl` package is the pre-installed R environment. It is also necessary to establish an account with rights for job creation at the BOINC server. Connection to the BOINC server may use http, https or ssh protocols.

2.2 Virtual Machines

This part of the backend must be specially prepared before running the compute tasks. We use the VirtualBox as a hypervisor. Below we list specific conditions and requirements that need to be satisfied to prepare the virtual machine.

- Firstly, a virtual machine must be bootable for IA32 and AMD64 processor architectures (one machine for each architecture). We recommend to install a Gentoo Linux on these machines due to space limits induced by necessity of internet transmission, and since many R packages require C++ headers and link libraries for the system libraries at build time. However, other Linux distributions will work as well.
- Secondly, R environment must be installed on these machines. We also recommend to install all the packages needed for computations beforehand, however, this is optional. If the required packages are not available on the virtual machine, the job will try to install them from the resources rforge.net or cran.rstudio.org.
- Thirdly, a regular user must be created. This user must automatically login into the system after boot.
- Fourthly, the so-called VirtualBox Guest Additions must be installed on this VMs. Latest Linux kernels contain `vboxsf` and `vboxguest` drivers and they can replace the guest additions. It is also needed to add the user to the `vboxguest` and `vboxsf` groups.
- Fifthly, VirtualBox shared directory with the name `shared` must be mounted in the user home directory as `shared`. In the user home directory there should exist a directory with the name `workdir` having full access privileges. We recommend to move `workdir` to RAMFS.
- Finally, a shell scripts provided as a part of the package for virtual machines must be copied into the user home directories at VMs and added to the autoload.

When all these conditions are met, the virtual machine is ready to run the jobs. We have also prepared ready-made images of virtual machines that can be used for most types of tasks immediately or after a slight modification.

The virtual machine is downloaded, configured and launched by the BOINC client software installed at the host machine by a volunteer. The algorithm of the virtual machine is as follows:

1. Configure network at boot time with dhcp.
2. Mount `vboxsf` shared folder as ~/shared and RAMFS as ~/workdir. The BOINC client puts the workunit files in this folder.
3. Copy all files from ~/shared to ~/workdir.
4. Unpack the archive common.tar.xz. This archive contains common files for all jobs. If not exists, create directory ~/workdir/files.
5. Run file ~/workdir/code.R in `Rscript` program. This script will load the necessary packages, load the code and data, change the working directory to ~/workdir/files, call the initialization function and start the job. After the completion of the calculations, the result will be saved in ~/shared/result.rda.
6. Shutdown the virtual machine.

After the virtual machine is shut down, the BOINC client uploads the result file (result.rda) to the BOINC server.

2.3 Server Part

It is a set of scripts and programs that provide functions which are not available in BOINC. These functions are:

- File uploading to BOINC server via http and https protocols[1].
- Unique name generation for the uploaded files.
- Getting the state of the job if the connection to the BOINC server uses the ssh protocol.

Besides, the server part provides the templates for BOINC applications and the so-called *validator* that may be used for arbitrary R jobs.

This part of package must be installed on the BOINC server before it can run the jobs from the client part of the backend. Installation requires copying the files to a BOINC project directory and editing the application templates and configuration files. Virtual machines required for the application are not included in the server part and must be built manually.

To use the validator supplied with the server part of the backend, then the R interpreter needs to be installed on the server. This validator performs the minimal necessary checks: it checks that the result file is a valid *.rda file and that it has an object named "result". Alternatively, the sample_trivial_validator from the standard BOINC server distribution can be used.

Figure 1 demonstrates how the R package is integrated into BOINC, using the services provided by BOINC where possible. The job creation starts on the clients (researchers) computers. R package collects all common files, generates the necessary code file and saves the tasks into the files used by the BOINC server to create jobs. All obtained files are packed in a tar archive, compressed

[1] In fact, BOINC server is able to download jobs files through these protocols, but in our case it was more convenient to pack all the necessary files into an archive, transfer it to the server, unpack it, resolve name conflicts, register the files in BOINC, and only after that return the list of files back to the client.

by the lzma algorithm and uploaded to the BOINC server. Server part unpacks the archive, generates unique names for the job files and registers these files in BOINC, returning these names back to the researcher's computer. Subsequently, requests are issued to the BOINC server to create a batch of jobs. The name for the batch is generated based on the current date and time and is always unique. The server returns the names of the jobs in the batch upon a batch successful creation.

3 General Working Scheme

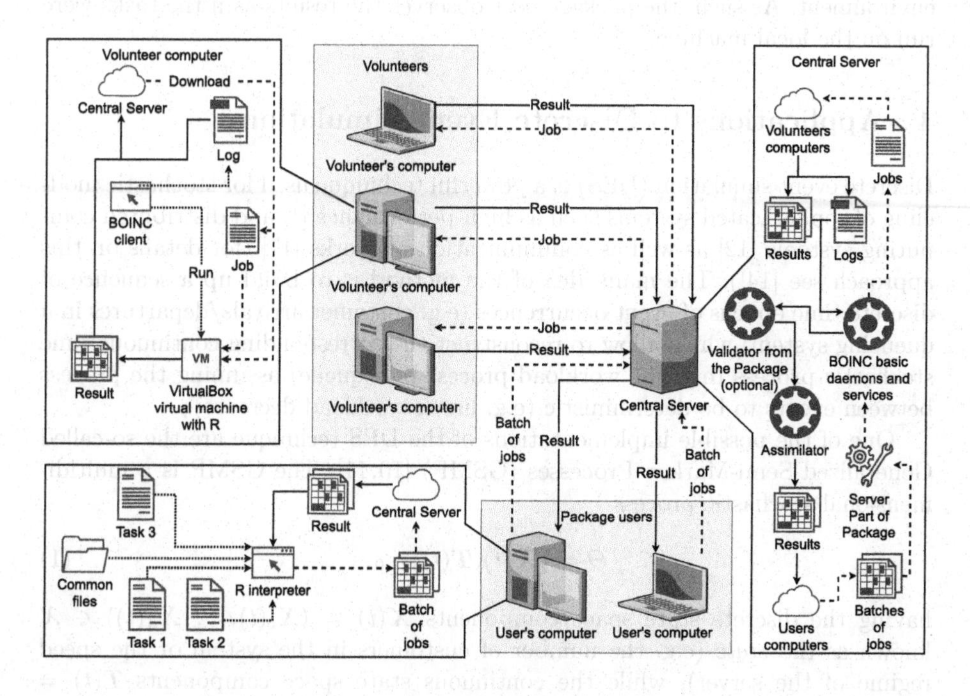

Fig. 1. General scheme of work

Volunteers install the standard BOINC client and configure it. BOINC client connects to the BOINC server and downloads the VirtualBox virtual machine with R and the corresponding assigned job. The virtual machine is not downloaded every time a new job is received. Instead, it is downloaded once at the first connection to the server, if a new version of the application is available or if virtual machine was deleted from the volunteer's computer.

The virtual machine is booted and a job is started. Using the virtual machine it is allowed to provide running code isolation from the host that is good for security reasons. In addition, the job is independent on the software installed on the volunteer computer, such as the R interpreter, and only BOINC client,

VirtualBox and VirtualBox Extension Pack are mandatory. The disadvantage, though, is a slight performance degradation.

Upon computation completion, the virtual machine is shut down. BOINC client uploads the result to the BOINC server where the result is processed and validated by the validator part provided by server part of the backend. Since every task generates only one job, assimilator just copies any valid job result to the results folder.

Users can update local status of the jobs on their computers by the function provided by package. Being called, this function sends messages to the BOINC server to get a list of jobs complete and having results not yet downloaded. After that R downloads the results from the server and loads it into the users environment. As such, the package user observes the results as if the tasks were run on the local machine.

4 Applications to Discrete Event Simulation

Discrete event simulation (DES) is a powerful technique used for stochastic modeling of sophisticated systems such as high-performance [1] and distributed computing systems [12] as well as communication networks [17] (for details on this approach see [14]). The main idea of the method is to build up a sequence of discrete time epochs of event occurrences (e.g. customer arrivals/departures in a queueing system) which allow to reconstruct the corresponding continuous-time stochastic process (e.g. the workload process of a queue) assuming the process between events to be deterministic (e.g. linear workload decrease).

One of the possible implementations of the DES technique are the so-called Generalized Semi-Markov Processes (GSMP) [10,11]. The GSMP is a multidimensional stochastic process

$$\Theta = \{\boldsymbol{X}(t), \boldsymbol{T}(t)\}_{t \geq 0}, \tag{1}$$

having the discrete state space components $\boldsymbol{X}(t) = (X_1(t), \ldots, X_n(t)) \in \mathcal{X}$ known as the *state* (e.g. the number of customers in the system or the speed regime of the server), while the continuous state space components $\boldsymbol{T}(t) = (T_1(t), \ldots, T_m(t)) \geq \boldsymbol{0}$ are known as *clocks* (e.g. residual service and inter-arrival times), each having the corresponding *event*. The clocks decrease linearly with time, each of them having a distinct state-dependent *rate* $\boldsymbol{r}(\boldsymbol{x}) = (r_1(\boldsymbol{x}), \ldots, r_m(\boldsymbol{x})) \geq \boldsymbol{0}$, whereas for each $\boldsymbol{x} \in \mathcal{X}$ the set of *active events* are those having positive clock rates $A(\boldsymbol{x}) = \{i : 1 \leq i \leq m, r_i(\boldsymbol{x}) > 0\}$. At least one event is active for each $\boldsymbol{x} \in \mathcal{X}$, that is, $A(\boldsymbol{x}) \neq \varnothing$. As such, for $h \leqslant \min_{i \in A(\boldsymbol{X}(t))} T_i(t)/r_i(\boldsymbol{X}(t))$ it follows that

$$\boldsymbol{T}(t+h) = \boldsymbol{T}(t) - h\boldsymbol{r}(\boldsymbol{X}(t)).$$

The time epoch $t \geq 0$ for clock i hitting zero, $T_i(t-) = 0$, is known as the event epoch of type $i \in A(\boldsymbol{X}(t))$ (for simplicity, assume that only one such event per unit time is possible). At such an epoch, the state $\boldsymbol{X}$ may make a transition from the state $\boldsymbol{x} \in \mathcal{X}$ to $\boldsymbol{x}' \in \mathcal{X}$ on event $i \in A(\boldsymbol{X}(t))$ according to stochastic matrix $\boldsymbol{P}^{(i)} = ||P^{(i)}_{\boldsymbol{x},\boldsymbol{x}'}||_{\boldsymbol{x},\boldsymbol{x}' \in \mathcal{X}}$, where

$$P^{(i)}_{\boldsymbol{x},\boldsymbol{x}'} = \mathrm{P}\{\boldsymbol{X}(t) = \boldsymbol{x}' | \boldsymbol{X}(t-) = \boldsymbol{x}, T_i(t-) = 0\}.$$

Upon transition caused by event $i \in A(\boldsymbol{x})$, the set of active events may change. At that, the clocks of *new* active events $i' \in A(\boldsymbol{x}') \setminus (A(\boldsymbol{x}) \setminus \{i\})$ are initialized from some given distribution with density

$$f_{i'}(u, \boldsymbol{x}, \boldsymbol{x}', i) = \mathrm{P}\{T_{i'}(t) \in du | \boldsymbol{X}(t-) = \boldsymbol{x}, \boldsymbol{X}(t) = \boldsymbol{x}', T_i(t-) = 0\}, \quad (2)$$

whereas all other clocks remain unaffected. It may be noted that if $i \notin A(\boldsymbol{x}')$, then ith clock remains at zero state. However, since $i \notin A(\boldsymbol{x}')$, event i is not duplicated at time t, while the corresponding clock will be initialized upon appearance of the event i in the set of active events. Moreover, events from the set $\{1, \ldots, m\} \setminus A(\boldsymbol{x}')$ may have positive clock values and zero clock rates. Such clocks may be interpreted as being *suspended* which is a useful property for stochastic modeling of, say, preemptive-resume, vacation and breakdown systems.

It is known that the GSMP model has the so-called one-dependent regenerative structure [9], that is, the corresponding stochastic process may be decomposed into identically distributed one-dependent cycles (the neighboring cycles may be dependent). Moreover, conditions are known for such a process to be positive recurrent, in particular, when the *state* space is finite. At the same time, it is possible to construct a stochastically equivalent process with independent regenerative cycles [8]. This opens the possibility to use the so-called time parallel simulation [4] to construct the independent regenerative cycles in an embarrassingly parallel way, which can be done using the proposed `RBOINC` package. It remains to note that the corresponding confidence estimation using the regenerative simulation [16] approach can be done during the assimilation phase.

5 Conclusion

We introduced a software package for computing of embarrassingly parallel applications over BOINC environment using R language. We find this approach promising in the field of stochastic simulation and optimization, however, the package is rather general and allows various applications including bioinformatics, Monte-Carlo methods etc. As preliminary experiments show, the package allows seamless integration into the R environment. We plan to continue this research by using and enhancing the package for the sake of parallel stochastic simulation. In particular, it is planned to simplify the organization of iterations over the parameter space. It might be also useful to complete the practical implementation of the suggested parallel regenerative GSMP implementation, which we plan to do in future research.

Acknowledgements. The publication has been prepared with the support of Russian Science Foundation according to the research project No. 21-71-10135 https://rscf.ru/en/project/21-71-10135/.

References

1. Rumyantsev, A., Morozova, T., Basmadjian, R.: Discrete-event modeling of a high-performance computing cluster with service rate control. In: 2018 22nd Conference of Open Innovations Association (FRUCT), pp. 224–231 (2018). https://doi.org/10.23919/FRUCT.2018.8468284
2. Anderson, D.P.: BOINC: a system for public-resource computing and storage. In: Proceedings of the 5th IEEE/ACM International Workshop on Grid Computing, GRID 2004, Washington, DC, USA, pp. 4–10. IEEE Computer Society (2004). https://doi.org/10.1109/GRID.2004.14
3. Ebert, A., Wu, P., Mengersen, K., Ruggeri, F.: Computationally efficient simulation of queues: the R package queuecomputer. J. Stat. Softw. **95**(5), 1–29 (2020). https://doi.org/10.18637/jss.v095.i05
4. Fourneau, J.M., Quessette, F.: Tradeoff between accuracy and efficiency in the time-parallel simulation of monotone systems. In: Tribastone, M., Gilmore, S. (eds.) EPEW 2012. LNCS, vol. 7587, pp. 80–95. Springer, Heidelberg (2013). https://doi.org/10.1007/978-3-642-36781-6_6
5. Fourneau, J.-M., Quessette, F.: Monotone queuing networks and time parallel simulation. In: Al-Begain, K., Balsamo, S., Fiems, D., Marin, A. (eds.) ASMTA 2011. LNCS, vol. 6751, pp. 204–218. Springer, Heidelberg (2011). https://doi.org/10.1007/978-3-642-21713-5_15
6. Fox, B.L.: Integrating and accelerating tabu search, simulated annealing, and genetic algorithms. Ann. Oper. Res. **41**(2), 47–67 (1993). https://doi.org/10.1007/BF02022562
7. Fujimoto, R.M.: Parallel discrete event simulation. Commun. ACM **33**(10), 30–53 (1990)
8. Glynn, P.W.: Some topics in regenerative steady-state simulation. Acta Appl. Math. **34**(1–2), 225–236 (1994). https://doi.org/10.1007/BF00994267
9. Glynn, P.W., Haas, P.J.: Laws of large numbers and functional central limit theorems for generalized semi-Markov processes. Stochast. Models **22**(2), 201–231 (2006). https://doi.org/10.1080/15326340600648997
10. Glynn, P.W., Haas, P.J.: On transience and recurrence in irreducible finite-state stochastic systems. ACM Trans. Model. Comput. Simul. **25**(4), 1–19 (2015)
11. Glynn, P.W., Iglehart, D.L.: Simulation methods for queues: an overview. Queueing Syst. **3**(3), 221–255 (1988). https://doi.org/10.1007/BF01161216
12. Ivashko, E., Nikitina, N., Rumyantsev, A.: Discrete event simulation model of a desktop grid system. In: Voevodin, V., Sobolev, S. (eds.) RuSCDays 2020. CCIS, vol. 1331, pp. 574–585. Springer, Cham (2020). https://doi.org/10.1007/978-3-030-64616-5_49
13. R Core Team: R: A Language and Environment for Statistical Computing. R Foundation for Statistical Computing, Vienna, Austria (2021). https://www.R-project.org
14. Ross, S.M.: Simulation, 4th edn. Elsevier Academic Press, Amsterdam, Boston (2006)

15. Rumyantsev, A., Sukhoroslov, O., Eparskaya, A., Blanzieri, E., Cavecchia, V.: Parameter sweep experiments in hybrid computing systems with R language. Int. J. Innov. Technol. Exp. Eng. **8**(7S2), 590–596 (2019)
16. Sharma, V.: Reliable estimation via simulation. Queueing Syst. **19**(1–2), 169–192 (1995). https://doi.org/10.1007/BF01148945
17. Szczerbicka, H., Trivedi, K.S., Choudhary, P.K.: Discrete event simulation with application to computer communication systems performance. In: Reis, R. (ed.) Information Technology. IIFIP, vol. 157, pp. 271–304. Springer, Boston, MA (2004). https://doi.org/10.1007/1-4020-8159-6_10
18. Vincent, J.M., Vienne, J., Id-Imag, L.: Perfect simulation of monotone systems with variance reduction. In: Proceedings of the 6th International Workshop on Rare Event Simulation, RESIM 2006, Bamberg, Germany, pp. 275–285 (2006)
19. Wickham, H., Grolemund, G.: R for Data Science: Import, Tidy, Transform, Visualize, and Model Data, 1st edn. O'Reilly Media Inc. (2017)

Algorithm for Calculating and Using the Characteristics of a Binary Image Intended for Implementation on RCE

A. S. Bondarchuk(✉), D. V. Shashev, and S. V. Shidlovskiy

National Research Tomsk State University, 36 Lenin Avenue, 634050 Tomsk, Russia

Abstract. The article outlines a new approach to constructing a feature vector for implementation on computers with a parallel pipeline architecture. The feature vector consists of the calculated characteristics of the gradient of a binary image (binary gradient) by analogy with the operation of the HOG algorithm. The proposed algorithm detects features of the contour pixels of objects in a binary image, which are further used for pattern recognition. After using the newly generated feature vectors for training a support vector machine (SVM) classifier, the speed of processing and classifying objects of interest on an image with a size of 1280 × 720 pixels increased by 3.5 times, in comparison with using the classical HOG descriptor.

Keywords: binary gradient · descriptor · reconfigurable computing environments

1 Introduction

In the field of creating automatic motion control systems for mobile platforms, the development of computer vision systems (CVS) has a great priority. To solve the problems of interaction with the environment, it is necessary to analyze the external situation in real time. CVS are used in modern underwater, surface, ground, aviation and space mobile robotic objects. For such platforms, the problem of reducing the time of image processing at high speeds of movement, as well as reducing power consumption, is urgent.

The solution to these problems is the implementation of image processing algorithms on parallel-pipelined computing architectures, such as Field-Programmable Gare Arrays (FPGA), Graphic Processing Unit (GPU), and Central Processing Unit (CPU). Using FPGAs will allow you to achieve lower power consumption and higher performance through parallel computing.

At the moment, the most common use of hybrid computing devices based on FPGA and GPU [1–6]. However, used for digital image processing, computing

The reported study was funded by RFBR, project number 19-37-90110, project number 19-29-06078.

V. M. Vishnevskiy et al. (Eds.): DCCN 2021, CCIS 1552, pp. 408–419, 2022.
https://doi.org/10.1007/978-3-030-97110-6_32

architectures have drawbacks that significantly limit their capabilities, especially in the presence of low computing resources. Such disadvantages include: the lack of an automatically reconfigurable architecture to achieve adequacy to the structures and parameters of the tasks being solved, the invariability of communication lines, the difficulty of distributing tasks between parallel processors, etc. For these reasons, intensive research is being carried out, the purpose of which is to find new, effective architectural solutions that provide high quality characteristics of computers for processing, segmentation and recognition of images, as well as the development of new algorithms and software [7–10].

An alternative approach was used in the work to develop a specialized, high-speed algorithm for processing and subsequent classification of binary images. The developed algorithm is intended for hardware implementation on computers with a parallel pipeline architecture, namely, reconfigurable computing environments (RCE), which, due to their unique architectural properties, make it possible to achieve high technical and economic indicators.

2 Calculation of the Parameters of the Image Gradient and Constructing the Feature Vector

A descriptor is an identifier of an image or image area, consisting of a set of features. Features are a descriptive element that characterizes an image. A feature vector is a numerical or binary vector of certain parameters. The type of parameters and the length of the vector depends on the algorithm used. Descriptors are used for pattern recognition and object detection in the image. One of these descriptors is the HOG (Histogram of Oriented Gradients) descriptor, commonly used for image processing and object detection in computer vision systems. The HOG descriptor is constructed by calculating the directions of the gradient in the local areas of the image. The main idea of the algorithm is the assumption that the appearance and shape of an object in the image can be described by the distribution of intensity gradients.

To extract the features of a binary image, it is proposed to use the algorithm for finding the gradient by analogy with the HOG algorithm. In this case, the binary gradient will be the value m and the direction ϕ of the change in the brightness of the neighboring image pixels from 0 to 1 or vice versa. The direction of the gradient can take three values: 180°, 225° and 270°. If there is no change in the brightness of pixels (between the current pixel and pixels of its neighbors), then $m = 0$, which means there is no gradient. In Fig. 1 shows 4 possible variants of the gradient values for the considered pixel I in relation to the neighboring pixels x and y.

Figure 2 shows the result of determining the direction of the binary gradient, where Fig. 2a - initial binary image, Fig. 2b - visualization of a binary gradient in each pixel of the image.

The value of the direction of the gradient was encoded with a two-digit binary number $\phi = \phi_1\phi_2$ according to Table 1.

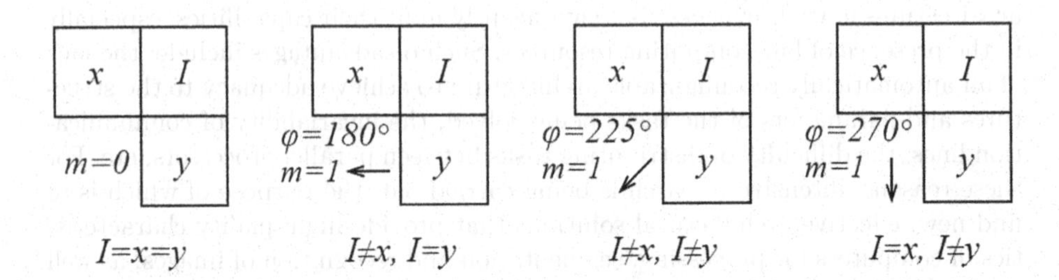

Fig. 1. Variants for the values of the binary gradient.

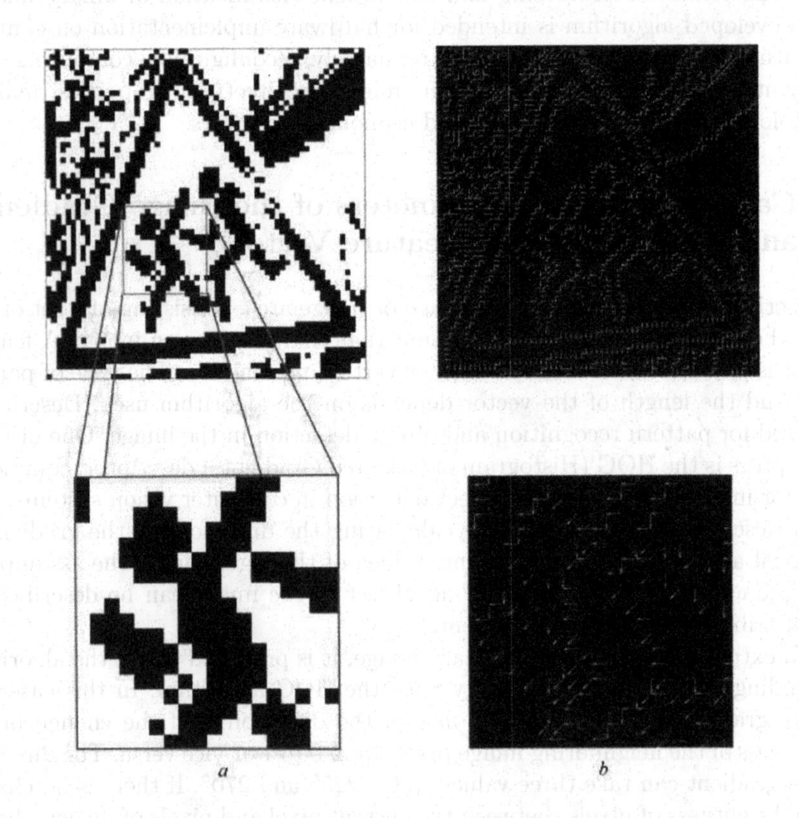

Fig. 2. Visualization of a binary gradient on the image.

Table 1. Encoding the directional values of the binary gradient.

ϕ	"No gradient"	180°	225°	270°
ϕ_1	0	0	1	1
ϕ_2	0	1	0	1

The feature vector, consisting of the values of m, ϕ_1, and ϕ_2 of the binary gradient of the image pixels, is further used to classify objects using the Support vector machine.

3 Support Vector Machine (SVM)

The SVM algorithm solves the classification problem by training on features from the training sample for which class labels are known in advance. Then the already trained algorithm can predict the class label for each vector with N features $X = (x_1, x_2, \ldots, x_n)$ in the space R^n. Class labels can take values less than -1 or greater than 1. During training, the algorithm constructs a function $F(X) = Y$, which takes a feature vector X from the space R^n and produces a label of the class Y. To do this, it is necessary to find the equation of the separating hyperplane $W * X + b = 0$ in the space R^n, which would separate the two classes in some optimal way. The weight vector W is perpendicular to the dividing hyperplane. The parameter $\frac{b}{||W||}$ is equal in absolute value to the distance from the hyperplane to the origin. If b is zero, the hyperplane passes through the origin.

The algorithm maximizes the distance between the optimal and boundary hyperplanes. The feature vectors that are closest to the optimal hyperplane are called support vectors, and the hyperplanes passing through them are called boundaries. These boundaries can be described by the following equations: $W * X + b = -1$ and $W * X + b = 1$. The width of the strip between them is easy to find for reasons of geometry, it is equal $\frac{2}{||W||}$, then in SVM the weights W and b are adjusted in such a way as to minimize $||W||$.

As a result, substituting the vector of weights W and the parameter b obtained during training into the formula $W * X + b = Y$, as well as the vector of features X of the image (or image area) in question, we get a value Y less than -1 or more than 1, which indicates belonging to one of the two trained classes. The construction of a feature vector from the binary gradients of a binary image, and the multiplication of these gradients by the corresponding weight coefficients, can be implemented using the RCE concept.

4 A Reconfigurable Computing Environment

RCE is a discrete mathematical model of a high-performance computing system, consisting of identical and equally connected to each other, the simplest universal elements (elementary calculators, ECs), programmatically tuned to perform any function from a complete set of logical functions, memory and any connection with its neighbours [11–13].

The fundamental principles of creating RCE are: parallelism, reconfigurability, homogeneity and pipelining of information processing. RCE has the form of a geometrically regular lattice, having at least two symmetry axes, with ECs located at the nodes, which contain a certain set of operations performed. A setting code is supplied to the input of each elementary calculator, with the help of which the reconfigurability of the RCE is carried out and it is determined which of the pledged operations will be performed. All elementary calculators are of the same type and are geometrically similarly connected with neighboring ones, and each of the ECs can be conventionally considered the center of symmetry with relation to its connections with the surrounding ECs. The cell of the elementary calculator has functional and connective completeness, i.e., can be configured to perform at a given moment any one function of at least one complete basis and function of the signal transmission channel in a given direction. The operations performed in the EC and the connections between them are intended to ensure the hardware execution of the algorithm being implemented.

The work of RCE can be considered from the point of view of the theory of automata. An automaton is called reconstructible if a set of automaton mappings implemented by it is given and an algorithm for tuning to implement each of these automaton mappings is defined [11]. An automatic mapping is an unambiguous mapping of the dependence of the output vector of the automaton on the vector of inputs, and tuning for the implementation of each of the automatic mappings is carried out by determining their tuning codes.

5 A Model of ECs of the RCE for Calculating Binary Gradients and Their Products by Weight Coefficients

The paper proposes an implementation of the previously described algorithm on the RCE architecture, where each EC is responsible for parallel processing of one of the pixels of a binary image. Thus, the dimension of the RCE coincides with the dimension of the processed image, while the elementary calculators are connected in the same way. Automata mappings that implement the algorithm for calculating the binary gradient are shown in Fig. 3. Each automaton mapping was assigned a tuning code (z_4, z_3, z_2, z_1), at which the automaton is rebuilt to it, as well as systems of output formulas:

1) for $z_4 = 0$, $z_3 = 0$, $z_2 = 0$, $z_1 = 0$ (Fig. 3, *a*)

$$\begin{cases} f_x = I, \\ f_y = 0, \\ W = w_m(I \vee (I \cdot \bar{y} \vee \bar{I} \cdot y)) + w_{\phi 1}(I \cdot \bar{y} \vee \bar{I} \cdot y) \\ + w_{\phi 2}(I \cdot \overline{(I \cdot \bar{y} \vee \bar{I} \cdot y)} \vee \bar{I} \cdot (I \cdot \bar{y} \vee \bar{I} \cdot y)); \end{cases} \quad (1)$$

2) for $z_4 = 1$, $z_3 = 0$, $z_2 = 0$, $z_1 = 0$ (Fig. 3, *b*)

$$\begin{cases} f_x = I, \\ f_y = 0, \\ W = w_m((I \cdot \bar{x} \vee \bar{I} \cdot x) \vee (I \cdot \bar{y} \vee \bar{I} \cdot y)) + w_{\phi 1}(I \cdot \bar{y} \vee \bar{I} \cdot y) \\ + w_{\phi 2}((I \cdot \bar{x} \vee \bar{I} \cdot x) \cdot \overline{(I \cdot \bar{y} \vee \bar{I} \cdot y)} \vee \overline{(I \cdot \bar{x} \vee \bar{I} \cdot x)} \cdot (I \cdot \bar{y} \vee \bar{I} \cdot y)); \end{cases} \quad (2)$$

3) for $z_4 = 0$, $z_3 = 1$, $z_2 = 0$, $z_1 = 0$ (Fig. 3, c)

$$\begin{cases} f_x = 0, \\ f_y = 0, \\ W = w_m((I \cdot \bar{x} \vee \bar{I} \cdot x) \vee (I \cdot \bar{y} \vee \bar{I} \cdot y)) + w_{\phi 1}(I \cdot \bar{y} \vee \bar{I} \cdot y) \\ + w_{\phi 2}((I \cdot \bar{x} \vee \bar{I} \cdot x) \cdot \overline{(I \cdot \bar{y} \vee \bar{I} \cdot y)} \vee \overline{(I \cdot \bar{x} \vee \bar{I} \cdot x)} \cdot (I \cdot \bar{y} \vee \bar{I} \cdot y)); \end{cases} \quad (3)$$

4) for $z_4 = 1$, $z_3 = 1$, $z_2 = 0$, $z_1 = 0$ (Fig. 3, d)

$$\begin{cases} f_x = 0, \\ f_y = I, \\ W = w_m((I \cdot \bar{x} \vee \bar{I} \cdot x) \vee (I \cdot \bar{y} \vee \bar{I} \cdot y)) + w_{\phi 1}(I \cdot \bar{y} \vee \bar{I} \cdot y) \\ + w_{\phi 2}((I \cdot \bar{x} \vee \bar{I} \cdot x) \cdot \overline{(I \cdot \bar{y} \vee \bar{I} \cdot y)} \vee \overline{(I \cdot \bar{x} \vee \bar{I} \cdot x)} \cdot (I \cdot \bar{y} \vee \bar{I} \cdot y)); \end{cases} \quad (4)$$

5) for $z_4 = 0$, $z_3 = 0$, $z_2 = 1$, $z_1 = 0$ (Fig. 3, e)

$$\begin{cases} f_x = 0, \\ f_y = I, \\ W = w_m((I \cdot \bar{x} \vee \bar{I} \cdot x) \vee I) + w_{\phi 1} I \\ + w_{\phi 2}((I \cdot \bar{x} \vee \bar{I} \cdot x) \cdot \bar{I} \vee \overline{(I \cdot \bar{x} \vee \bar{I} \cdot x)} \cdot I); \end{cases} \quad (5)$$

6) for $z_4 = 1$, $z_3 = 0$, $z_2 = 1$, $z_1 = 0$ (Fig. 3, f)

$$\begin{cases} f_x = I, \\ f_y = I, \\ W = w_m((I \cdot \bar{x} \vee \bar{I} \cdot x) \vee I) + w_{\phi 1} I \\ + w_{\phi 2}((I \cdot \bar{x} \vee \bar{I} \cdot x) \cdot \bar{I} \vee \overline{(I \cdot \bar{x} \vee \bar{I} \cdot x)} \cdot I); \end{cases} \quad (6)$$

7) for $z_4 = 0$, $z_3 = 1$, $z_2 = 1$, $z_1 = 0$ (Fig. 3, g)

$$\begin{cases} f_x = I, \\ f_y = I, \\ W = w_m I + w_{\phi 1} I; \end{cases} \quad (7)$$

8) for $z_4 = 1$, $z_3 = 1$, $z_2 = 1$, $z_1 = 0$ (Fig. 3, h)

$$\begin{cases} f_x = I, \\ f_y = I, \\ W = w_m(I \vee (I \cdot \bar{y} \vee \bar{I} \cdot y)) + w_{\phi 1}(I \cdot \bar{y} \vee \bar{I} \cdot y) \\ + w_{\phi 2}(I \cdot \overline{(I \cdot \bar{y} \vee \bar{I} \cdot y)} \vee \bar{I} \cdot (I \cdot \bar{y} \vee \bar{I} \cdot y)); \end{cases} \quad (8)$$

9) for $z_4 = 0$, $z_3 = 0$, $z_2 = 0$, $z_1 = 1$ (Fig. 4, a)

$$\begin{cases} f_x = I, \\ f_y = I, \\ W = w_m((I \cdot \bar{x} \vee \bar{I} \cdot x) \vee (I \cdot \bar{y} \vee \bar{I} \cdot y)) + w_{\phi 1}(I \cdot \bar{y} \vee \bar{I} \cdot y) \\ + w_{\phi 2}((I \cdot \bar{x} \vee \bar{I} \cdot x) \cdot \overline{(I \cdot \bar{y} \vee \bar{I} \cdot y)} \vee \overline{(I \cdot \bar{x} \vee \bar{I} \cdot x)} \cdot (I \cdot \bar{y} \vee \bar{I} \cdot y)); \end{cases} \quad (9)$$

On the basis of the structural automaton method [11], we obtain the following system of equations, which will describe the work of the EC of the RCE:

$$\begin{cases} f_x = I(\bar{z_1}\bar{z_2}\bar{z_3} + z_2 z_4 \bar{z_1} + z_2 z_3 \bar{z_1} + \bar{z_2}\bar{z_3}\bar{z_4}), \\ f_y = I(z_3 z_4 \bar{z_1} + z_2 \bar{z_1}\bar{z_3} + z_2 \bar{z_1}\bar{z_4} + z_1 \bar{z_2}\bar{z_3}\bar{z_4}), \\ W = (w_m(I \vee (I \cdot \bar{y} \vee \bar{I} \cdot y)) + w_{\phi 2}(I \cdot \overline{(I \cdot \bar{y} \vee \bar{I} \cdot y)} \vee \bar{I} \cdot (I \cdot \bar{y} \vee \bar{I} \cdot \\ \cdot y)))(\bar{z_1}\bar{z_2}\bar{z_3}\bar{z_4} + z_2 z_3 z_4 \bar{z_1}) + w_{\phi 1}(I \cdot \bar{y} \vee \bar{I} \cdot y)(\bar{z_1}\bar{z_2}\bar{z_3} + z_3 \bar{z_1}\bar{z_2} + z_3 z_4 \bar{z_1} \\ + \bar{z_2}\bar{z_3}\bar{z_4}) + (w_m((I \cdot \bar{x} \vee \bar{I} \cdot x) \vee (I \cdot \bar{y} \vee \bar{I} \cdot y)) + w_{\phi 2}((I \cdot \bar{x} \vee \bar{I} \cdot x) \cdot \\ \cdot \overline{(I \cdot \bar{y} \vee \bar{I} \cdot y)} \vee \overline{(I \cdot \bar{x} \vee \bar{I} \cdot x)} \cdot (I \cdot \bar{y} \vee \bar{I} \cdot y)))(z_4 \bar{z_1}\bar{z_2} + z_3 \bar{z_1}\bar{z_2} + z_1 \bar{z_2}\bar{z_3}\bar{z_4}) \\ + (w_m((I \cdot \bar{x} \vee \bar{I} \cdot x) \vee I) + w_{\phi 2}((I \cdot \bar{x} \vee \bar{I} \cdot x) \cdot \bar{I} \vee \overline{(I \cdot \bar{x} \vee \bar{I} \cdot x)} \cdot I)) z_2 \bar{z_1}\bar{z_3} \\ + I(w_{\phi 1}(z_2 \bar{z_1}\bar{z_3} + z_2 \bar{z_1}\bar{z_4}) + w_m z_2 z_3 \bar{z_1}\bar{z_4}). \end{cases} \tag{10}$$

In Fig. 3 automaton mappings describe the operations in the EC, necessary to take into account in the processing of a binary image, edge pixels that have no neighbors. In the description of automata mappings basic logical functions "AND, OR, NOT", as well as classical algebraic operations of multiplication and summation are reflected.

The EC of the RCE model performs the above-described system of equations, rebuilding to perform the specified automatic mappings using the tuning code (z_4, z_3, z_2, z_1). In a simplified form, the structure of the RCE model is shown in Fig. 4*b*.

In Fig. 4*b*, only a part of the 3 × 3 EC model is shown, which reflects the processing of the upper left corner of the original 3 × 3 pixel image. Here I is an array of brightness values of pixels of a binary image; Z - an array of tuning codes for ECs, and the figure has already marked the correct codes in accordance with the automatic mappings; w is an array of values of the weighting coefficients obtained after training using the SVM algorithm; W is an array of products of weight coefficients and binary gradient parameters. The communication bus is used to transmit data about the pixel value to another ECs, between the outputs f_x, f_y and the inputs x, y.

When calculating the characteristics of the gradient of each pixel of a binary image, each EC RCE is four-connected and bi-directionally connected with its neighbors. The connections between the ECs RCE are shown in the Fig. 5, where i and j are the coordinates of the image pixel being processed by the EC.

Thus, in the RCE model, simultaneous parallel pixel-by-pixel processing of a binary image is carried out and, for each pixel, the product of the binary gradient parameters by the corresponding weight coefficient is calculated. Subsequently, summing up the values of the array W, we obtain the result of the product of the feature vector by the vector of weights. Adding to the calculated number the parameter b, obtained during training using the SVM algorithm, we determine the class to which the considered image (or image region) belongs.

For training, 5459 images without an object of interest and 4000 images of vehicles with a size of 128 × 128 pixels obtained from the Berkeley Deep Drive 100 K dataset were used. After that, the classifier was tested on 16 test images

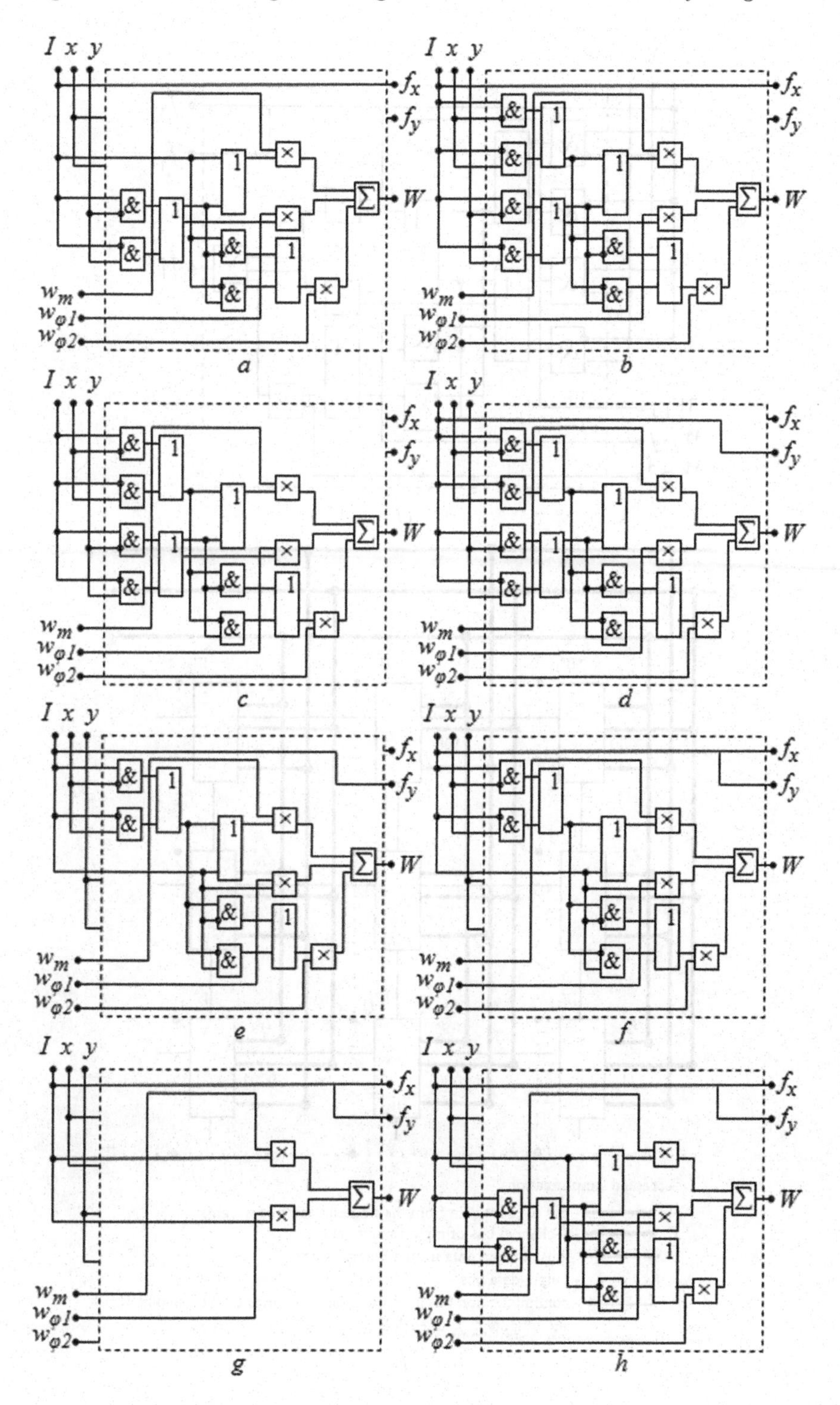

Fig. 3. Automatic mappings: $I, x, y, w_m, w_{\phi 1}, w_{\phi 2}$ - information inputs; W - automaton output; f_x, f_y - outputs of inter-automaton connections.

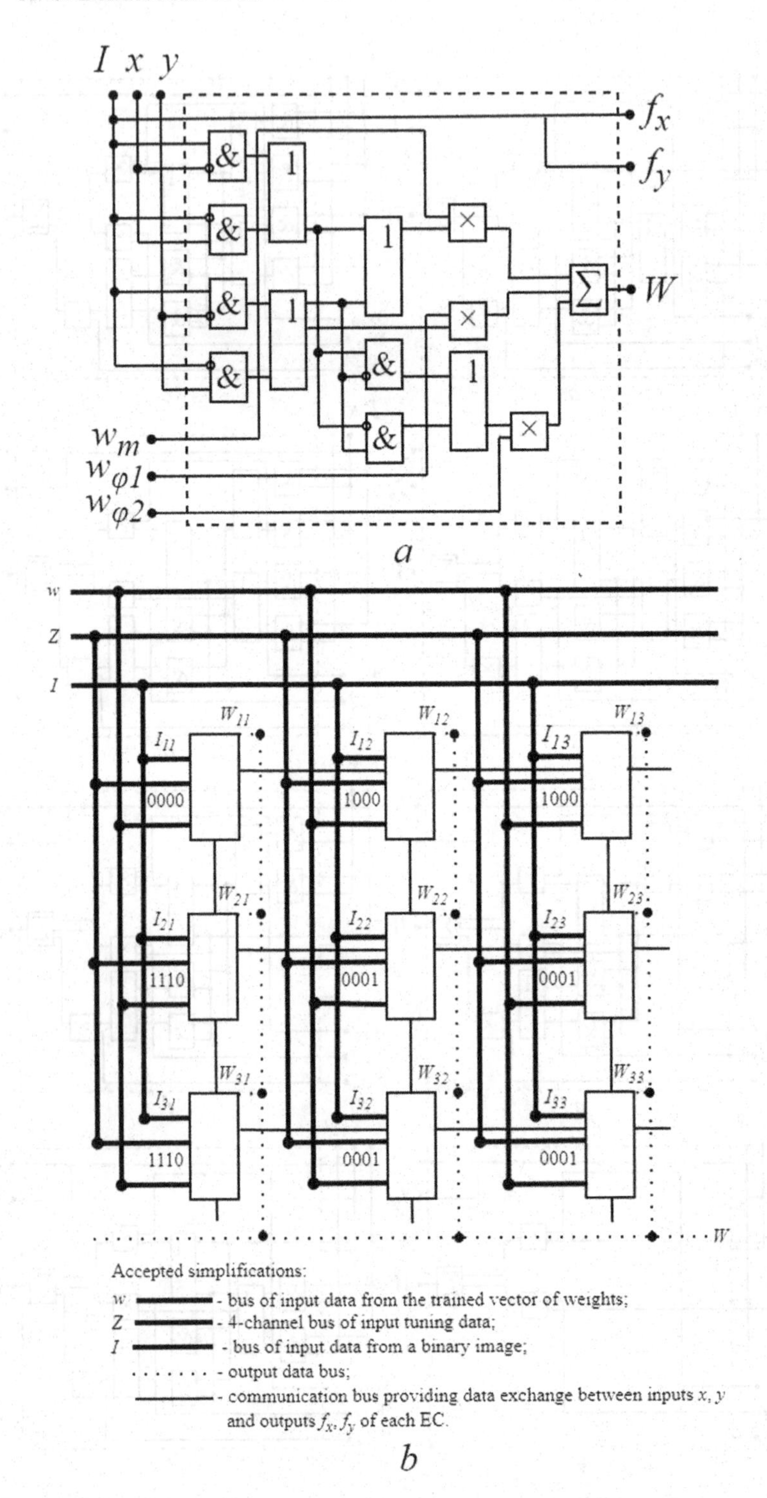

Fig. 4. RCE architecture.

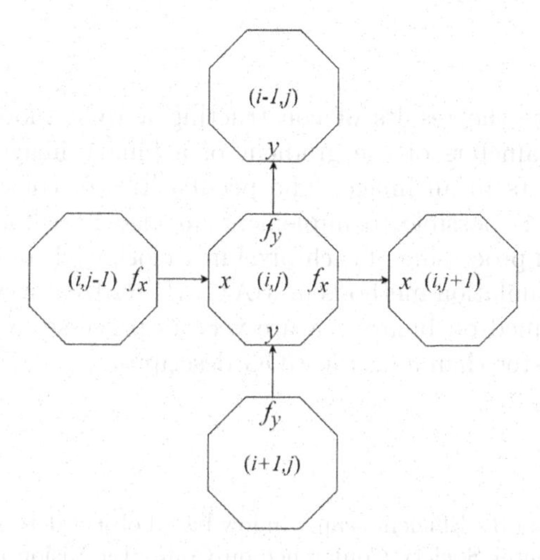

Fig. 5. Connections between elementary calculators in a reconfigurable computing environment.

where the object of interest is present. On all these images, the location of the object was determined correctly, and the number of false positives was equal to 0. To compare the results, two more classifiers were trained. The first classifier uses HOG-descriptor feature vectors, and also correctly determines the object of interest on test images. The second classifier uses vectors consisting of sequential values of the magnitudes and directions of the HOG intensity gradients. As a result, the location of the object is determined correctly on 10 out of 16 test images. In 3 images, there are false positives along with the object, and in the remaining 3, the object was not detected.

The time of processing and recognition of vehicles was measured on an image with a size of 1280 × 720 pixels (Table 2).

Table 2. Comparison of the processing speed of one image.

Feature vectors	Binary	HOG descriptor	Magnitudes and directions of HOG intensity gradients
Processing time, sec	196.5	694	109

The analysis of classification algorithms was carried out in MATLAB R2020a using a computer with the following characteristics:

- Intel(R) Core(TM) i9-9880H CPU @ 2.30 GHz;
- 32 Gb RAM;
- NVIDIA Geforce RTX 2080.

6 Conclusion

The paper presents the results of constructing a RCE model for calculating and using the parameters of the gradient of a binary image in the problems of classifying objects in an image. The peculiarities of constructing the RCE architecture make it possible to implement an algorithm for finding a binary gradient in parallel processing of each pixel in 1 clock cycle of the EC operation. Using computer simulation methods in MATLAB R2020a, it was shown that an SVM classifier trained on binary feature vectors processes a 1280 × 720 pixel image 3.5 times faster than a classic HOG descriptor.

References

1. Zhang L., Nevatia R.: Efficient scan-window based object detection using GPGPU. In: IEEE Computer Society Conference on Computer Vision and Pattern Recognition Workshops, pp. 1–7. Anchorage, USA (2008)
2. Hiromoto M., Miyamoto R.: Hardware architecture for high-accuracy real-time pedestrian detection with CoHOG features. In: IEEE 12th International Conference on Computer Vision Workshops, pp. 894–899. Kyoto, Japan (2009)
3. Shipitko O.S., Grigoryev A.S.: Gaussian filtering for FPGA based image processing with high-level synthesis tools. In: Proceedings of the IV International Conference and Youth School Information Technology and Nanotechnology, pp. 2922–2027. Samara, Russia (2018)
4. Reiche O., Akif Özkan M., Membarth R., Teicha J., Hannig F.: Generating FPGA-based image processing accelerators with Hipacc. In: Proceedings of the 36th International Conference on Computer-Aided Design, pp. 1026–1033. Irvine, California (2017)
5. Qasaimeh M., Denolf K., Lo J., Vissers K., Zambreno J., Jones P.H.: Comparing energy efficiency of CPU, GPU and FPGA implementations for vision kernels. In: Proceedings of 2019 IEEE International Conference on Embedded Software and Systems, vol. 2017, pp. 1–8. Las Vegas, USA (2019)
6. Lu, R., Liu, X., Wang, X., Pan, J., Sun, K., Waynes, H.: The design of FPGA-based digital image processing system and research on algorithms. In: International Journal of Future Generation Communication and Networking, vol. 10, no. 2, pp. 41–54. Australia (2017)
7. Dioşan L., Andreica A., Boros I., Voiculescu I.: Avenues for the use of cellular automata in image segmentation. In: Squillero, G., Sim, K. (eds.) Applications of Evolutionary Computation. EvoApplications 2017. Lecture Notes in Computer Science, vol. 10199. Springer, Cham (2017). https://doi.org/10.1007/978-3-319-55849-319
8. B. Gülmez, F. Demirtas, İ. Yıldırım, U. Leloğlu, M. Yaman, E. Güneyi: a method to enhance homogeneous distribution of matched features for image matching. Int. J. Environ. Geoinf. **7**(1), 102–107 (2020). https://doi.org/10.30897/ijegeo.710634
9. Xiaowen C., Zhonghai L., Axel J., Shuming C., Yang G., Shenggang, C., Hu C.: Performance analysis of homogeneous on-chip large-scale parallel computing architectures for data-parallel applications. J. Electr. Comput. Eng. **2015** (2015). https://doi.org/10.1155/2015/902591

10. Torres-Huitzil, C., Arias-Estrada, M.: FPGA-based configurable systolic architecture for window-based image processing. EURASIP J. Adv. Signal Process. **2005**(7), 1–11 (2005). https://doi.org/10.1155/ASP.2005.1024
11. Shidlovskiy S. V.: Boolean differentiation equations applicable in reconfigurable computational medium. In: MATEC Web of Conference 2016, vol. 79, 01014, 5 p (2016)
12. Shashev, D.V., Shidlovskiy, S.V.: Morphological processing of binary images using reconfigurable computing environments. Optoelectron. Instrum. Data Process. **51**(3), 227–233 (2015). https://doi.org/10.3103/S8756699015030036
13. Khoroshevsky V., G.: Architecture and functioning of large-scale distributed reconfigurable computer systems. In: International Conference on Parallel Computing in Electrical Engineering, pp. 262–267. Dresden, Germany (2004)

Evaluation of Trust in Computer-Computed Results

Alexander Grusho(✉), Nikolai Grusho, Michael Zabezhailo, and Elena Timonina

Federal Research Center "Computer Science and Control" of the Russian Academy of Sciences, Vavilova 44-2, 119333 Moscow, Russia
info@itake.ru

Abstract. The paper addresses the problem of evaluating the trust in the results of complex computer data analysis. The approach of constructing empirical dependencies based on similarity of precedents in the training sample, which has already become classical, is used. Two types of approximation of the causal bases have been constructed. The first type of approximation uses the functions of causal influence and, when there is only one cause, provides a confident interpretation of the results of computer analysis. The second type of approximation is based on simulation of the training data by random sampling from an unknown distribution. Both approaches implement approximate causal analysis and have advantages and disadvantages.

Keywords: Information security · Artificial intelligence · Trust in distributed computing

1 Introduction

System administrators and security officers face the need to make a responsible decision based on empirical and sometimes incomplete data in large distributed information systems (DIS). The question arises whether the result of information technology execution (IT) in DIS is reliable. That is, whether the extraordinary results of monitoring can indicate the presence of an anomaly or are they merely the result of correct calculations of extraordinary input data.

Classical approaches to the evaluation of trust in the decision making process aided by computer modeling may in general appear to be inadequate for Big Data, process-real time, IT openness regarding new input data. Serious problems arise both due to incompleteness of information about the object being studied, which often occurs in information security (IS) applications, and due to the limited expressive capabilities of such models, as well as their effective analytical or numerical solvability.

The efficient alternative approach to this class of problems is to use the so-called interpolation-extrapolation mathematical models. With this approach, the

Partially supported by Russian Foundation for Basic Research (project 18-29-03081).

V. M. Vishnevskiy et al. (Eds.): DCCN 2021, CCIS 1552, pp. 420–432, 2022.
https://doi.org/10.1007/978-3-030-97110-6_33

initial data is presented as precedent descriptions, i.e. the examples and counterexamples of the target phenomenon being studied, combined into a so-called training sample. Further, this sample is interpolated by empirical dependencies (ED) of a specific form. Conclusion about presence or absence of target effect in newly analyzed precedent is formed by means of check of extrapolation on it of ED. To date, a variety of actively used implementations of this approach are known, which are based on both statistical and deterministic mathematical techniques of data analysis and decision support.

In many applications of computer-based data analysis, responsibility for the consequences of decisions made is critical. In particular, such tasks include the information security of the critical information infrastructure. As the result, meaningful informal explainability must be ensured for the conclusions and recommendations being formed. That is, not only the question of "how?", but also to the question "why?" must be both answered. The required properties of understandability, interpretability and stability of the results should be naturally provided. It can be done in those versions of computer data analysis that are based on the causal relationships.

Thus, within these information security (IS) constraints, it is efficient to employ intelligent data analysis (IAD), presented by interpolation-extrapolation analysis of computer data based on the detection and study of causal relationships. This seems to be the only method to build responsible decisions for ensuring the IS in large heterogeneous DIS. It should be added that methods for solving such problems are complicated, and most often use brute force. This means that it is necessary to search for approximations of causes analysis, most often reducing it to special cases of polynomial complexity problems.

Experts in the field of artificial intelligence (AI) over the past 2–3 years increasingly discuss the so-called third "wave" of AI evolution as a field of research and development [1]. The first "wave" is associated with the so-called rule-based AI-systems, the second "wave" - with statistical models and data analysis methods widely ranging from classical Bayesian inductive inference to various types of artificial neural networks. The attention of the third "wave" [1] is focused at the possibility of automated forming of so-called partial theories that are generated based on empirical data being received and updated in "portions". In fact, we are talking about the development of mathematical techniques and their implementation in appropriate software and hardware solutions, which make it possible to represent the knowledge and effectively make use of it in form of partial theories being reconstructed as new data arrives. For example, this includes computer analysis results interpretation based on empirical causal relationships and the sustainability of empirical conclusions.

The answer to the question "How is the transition to the third "wave" motivated?" is associated primarily with the:

- critical role in applications of the problem of responsibility for decisions made during human-machine interaction;
- requirement for explanations and trust in the conclusions and recommendations [2–4];

- limitations of the first and second "waves" of AI development in terms of the above requirements (see, for example, program [3] preceding program [5]).

The ability to generate adequate explanations is a fundamental characteristic of effective IAD solutions in IS support applications. IAD in such situations implies human-machine interaction, in which the responsibility for decisions being made lies on the decision making person (DMP), for example, the system administrator or the security officer. Often the DMP does not understand how to justify the results generated by the computer system. In these cases, he simply has to resort to what is available at that moment, that is, "manual" calculations, intuition, etc. Thus, the requirement of explainability of the results generated by the computer system in the process of data analysis in understandable terms and "formats" is essential.

The creation of an explanation involves at least two components:

1. the generation of a variant of the reasoned, i.e. substantiated, non- contradictory to the factual data, causal "scheme," which provides an appropriate answer to the question "Why should the result be trusted?"
2. the generation of an informal interpretation in a meaningful language, not duplicating the method used to produce the result, for the answer to the question "Why...?" generated within the framework of the causal "scheme."

The problem of explanation is well known in the literature (see, for example, [6]). The most common approaches here are:

- the deductive-nomological model of explanation proposed by K. G. Hempel [7,8];
- the diagram of C.S. Pierce's abductive explanation [9,10].

In methodological terms, both of these schemes can be compared with corresponding clarifications of the concept of truth. Specifically, the first scheme can be compared with the semantic concept of truth by A. Tarsky [11], and the second scheme - with the concept by A. S. Yessenin-Volpin [12] of truth as undeniable by the existing empirical material.

When dealing with open subject areas constantly updated with new data, and working with Big Data, it turns out that the classic explanation scheme by K. G. Hempel is already insufficient when dealing with a number of important practical applications. For example, when applied to IS, medical and technical diagnostics, this technique is simply not applicable due to the absence of a sufficiently "capacious" set of universal statements from which the relevant conclusions could logically follow. In turn, the study and formalization of abductive reasoning (procedures for generating particular dependencies from conclusion to such assumptions, the logical consequence of which is this conclusion) showed a number of interesting possibilities for generation of effective explanations (see, for example, [13,14]), including causal schemes for explaining the results generated during the IAD process (see, for example, [15]).

The analysis of the causes of the studied phenomena and effects plays a critical role in ensuring the explanability and interpretability of the computer-generated results. Accentuation of causality factors from among the available empirical data, which lead to the presence of target effects, allows to form stable empirical dependencies (ED) in relation to the expansion of the analyzed data with new information. The usage of such ED makes it possible to form sustainable measures which do not change upon data replenishment. For example, ED could help form targeted countermeasures against harmful consequences of monitored effects and phenomena [16,17].

The concept of precedent similarity used in IAD can be refined in various ways (see, for example, [18,19]). Two classes of similar formal constructions are given as examples:

1. metrical, i.e. understood as distance and proximity measures;
2. representing similarity as a binary algebraic operation.

Within the framework of the widely known algebraic approach developed by the school of Yu. I. Zhuravlev - K.V. Rudakov [20,21], the task of training by precedents in the IAD process is considered to be a non-classical version of the interpolation-extrapolation scheme for data analysis and prediction. Here, a number of dependencies are built based on the elements of the initial training sample of precedents. The portability (or, conversely, non-portability) of these dependencies is based on new precedents and there is a procedural base for the corresponding prediction. The correctness of such prediction is provided by the feasibility of a number of conditions and constraints, based on formal proofs of the existence of correct solutions for the corresponding mathematical problems (for example, the procedural technique of correct algebras over sets of heuristic algorithms [22,23]). The explanation of the generated computer results, as the answer to the question "Why...?" in this approach is based on the proven existence of correct solutions to the corresponding mathematical problems. For example, in [22], such a proof is based on the existence of the basis in the corresponding metric space. At the same time, the refinement of similarity by metric means used here (i.e., meaningful ideas about similarity as the proximity of the analyzed objects in the corresponding space) allows generation of informal interpretations of computer results that are easily understood by the DMP.

When the similarity is considered as a binary algebraic operation [15,24,25], the non-classical interpolation-extrapolation scheme presented above takes the following formA binary relation of similarity is formed according to the operation of similarity, taking into account the results of its application. Then, standard similarity classes are formed for the obtained similarity relation, and then all equivalence classes covering the similarity classes are restored. outclasses are formed according to all fixed results of existing similarity operation being applied on elements of the initial training sample of precedents. Each equivalence class obtained this way is compared with the corresponding empirical dependency. The problem of extrapolatability of any of the existing dependencies onto the description of a new precedent is solved by checking whether it is "included" into the equivalence class corresponding to this dependence.

This paper examines the approximate approach to causal analysis, which only provides additional arguments to DMP for making a responsible decision.

Therefore, for such problems, similarities can be described in terms of probability statistical models.

For simplicity, the task of finding interpretation and increasing of trust in the extraordinary IT results from the system administrator's point of view is considered. Namely, whether the IT result is generated by extraordinary initial data, or it has appeared as a result of a covert anomaly.

2 The Method for Evaluation of Trust Based on Functions of Causal Influence

Suppose that the system administrator observes and accumulates the results of IT execution. Each j-th copy of IT has its own source data x^j that belongs to the definition range D. The results of the IT^j execution are denoted as y^j and belong to the range of values B. Each range B is divided into extraordinary range of IT outcome data B^+ and range of ordinary IT outcome data B^-. Similarly, the input data is divided into D^+ and D^-. If y^j belongs to B^+, then this can be caused either by extraordinary source data, or by an implicit anomaly of the computational process. In this case, additional information is required to make a decision.

Here are some examples of obtaining such information.

1) If an extraordinary result of IT^j has been obtained, one can take the known precedents IT^{j_1} and IT^{j_2}, and repeat the calculations in the following sequence IT^{j_1}, IT^j, IT^{j_2}.
   Obtaining the previous extraordinary result, which matches with y^j, means that computer IT support works, but there may be an anomaly in the IT^j procedure.
2) The k-programs method uses similar logic, but also depends on implementation of IT^j.
3) Let there be an independent copy of IT. Then you can run this copy with the same input data what IT^j was using. In this case, obtaining the same result is a sufficient reason that the result is not an anomaly. However, another implementation of the hardware-software platform may round fractional numbers differently and uses different elements of the input data. Therefore, the result may differ from the previous one, which under certain conditions can alter the trust in the first result. In addition, such method is expensive and time consuming.

Next, we will consider the method of evaluation of the trust with the help of causality influence functions. Suppose that training was performed on precedents of extraordinary data being obtained in IT realizations, denoted as $x_1^+, ..., x_n^+$. Let all the other precedents be marked with (-). Suppose that there is a single cause for including the correct result into the set of extraordinary data. Let the input data be described by subsets of characteristics from the finite set U. Then the simplest measure of the similarity of the input data is defined as the intersection of characteristics of this input data. Among all the inputs collected for training, we highlight those that have the following property.

For each pair $x^+_{r_1}$, $x^+_{r_2}$, if $x^+_{r_1} \cap x^+_{r_2} = V$, then no pair x^-_l, x^-_t exists such that $x^-_l \cap x^-_t = V$. This property is called the ban on counterexamples (BC) [26].

Let's denote as $T(V)$ the set of pairs of input data from $x^+_1,..., x^+_n$ which have the measure of proximity V. The sets $T(V)$ will be called causal influences on the extraordinary values of IT.

Statement 1. If there is a single cause for the occurrence of an extraordinary IT result, then there exists V such that $T(V)$ contains the entire set $x^+_1,...,$ x^+_n, $n > 1$.

Proof. Assuming that the cause is expressed through characteristics describing precedents, the occurrence of an extraordinary value is associated with the occurrence in the source data of the cause of this value. If V cause is the only one, then this is the set of characteristics that is present in all $x^+_1,..., x^+_n$. The statement is proven.

Corollary. If only one cause V exists for the appearance of an extraordinary IT result and x is the new input data, then under BC conditions $x \in D^+$, if and only if x contains V.

The considered case of the only one cause fully determines the trust in this scheme.

If there is more than one cause, then the new data x under BC conditions should belong to one of the sets $T(V)$. In this case, we are talking only about increasing trust. Consider some of the important properties of this causal analysis method.

1. Let the cause be the only one, but $n = 1$. Then the new data x, which generated an extraordinary result in B^+, have a common set V at the intersection. However, such V can appear randomly, and the extraordinary result y is the result of an anomaly [27].
2. Let there be several causes for the creation of an extraordinary value of y and $x^+_1,..., x^+_n$, $n > 1$, however, the empirically estimated set D^- is small. Then, in addition to the sets of causal influence, there may exist V, which belongs to the new data, has no common intersections with $x^+_1,..., x^+_n$, $n > 1$, and satisfies the BC condition. Again, an uncertainty appears in the estimation of the causal base, since V may be an element of some cause y, but also a random fragment of data not yet obtained from D^-.
3. Sometimes it is difficult to substantiate the BC condition.
4. The considered approach is applicable to the class of finite sets of characteristics U and provided that the cause is a subset of U. This may not hold true when the characteristics are aggregated (unions of other characteristics) and then the BC condition may not be satisfied on them. In these cases, it is necessary to change parametrization of the observed data [28].

Example 1. In this example, we will show how the described method works and what issues rise. Let $y = x^2$, $D = [0, 2\}$, $B^+ = \{y \geq 1\}$, $B^- = \{y < 0\}$. Consider the set of characteristics (parametrization) of the input data, considering that the values of x are obtained from real numbers, leaving k decimal places. Obviously, the BC requirement only takes place only when $k = 0$. At the same time, it is easy to see the ED: if the integer part of x equals to 1, then the result $y \in B^+$. Then if the new data x has an integer part that equals to 1, then this is the causal rationale that the extraordinary result of the calculations is not an anomaly. In this example, the value of 1 of the integer part of the original data is the cause why the result belongs to B^+, and the value 0 of the integer part of the original data is the cause why the result belongs to B^-. Note that when $k \gg 0$ the proving of BC becomes a labor-intensive task.

3 Evaluation of Trust Using Probabilistic Statistical Methods

Section 2 discusses the simplest measure of similarity (measure of proximity). More generally, choosing the proximity measure is a complex problem. Later in this paper, the approximate method of estimating the causal bases of the obtained IT result based on probabilistic-statistical analysis of precedents is considered. The main idea of the method is as follows.

In [29], the association is constructed between the cluster structure of empirical data and the measure of proximity associated with the probabilistic distribution, according to which the data is selected. Suppose that the initial IT data is derived from probabilistic distributions on D^+ and D^-. Let these probabilistic distributions have single-humped density of probabilities in space with a measure of proximity ρ. Then (see [29]) the closer data corresponds to inclusion into the high probability area and therefore gets included into a single cluster. This is due to the fact that significantly more data is included into the high-probability area than into the low-probability area. The data entering the low-probability area is at a greater distance from each other.

Let the ordinary data possess its own distribution that generates its own cluster. Therefore, including data into different clusters corresponds to different distributions (different statistical hypotheses). Causal bases for estimation of

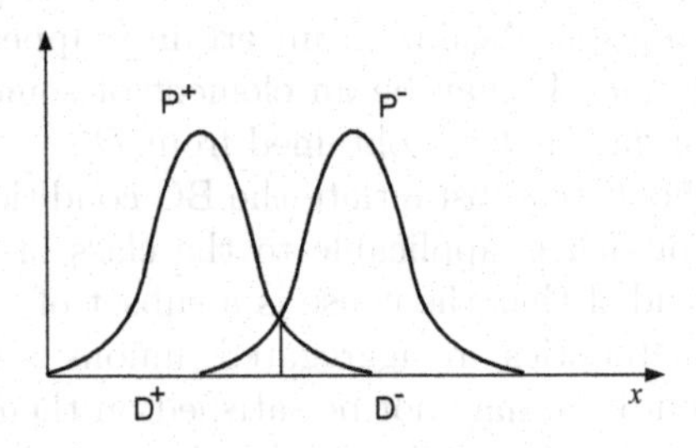

Fig. 1. The example of a contradiction in case of single-humped distributions.

trust in computer results can be substantiated by comparing the densities of single-humped distributions. Consider the example of such substantiation.

In Fig. 1 P^+ denotes the probability distribution density of the occurrence of extraordinary data, P^- - the density of the probability distribution of the occurrence of ordinary data. If computer calculation result $y \in B^+$, and x is located near the maximum of P^+ density, then it follows from the difference in densities P^+ and P^- that the difference ordinate $P^+(x) - P^-(x) > 0$. This condition does not contradict to $y \in B^+$, and therefore increases the trust in the result of computer calculation. If $y \in B^+$ and x is located near the maximum of density P^-, i.e. far from $\max P^+$, then the difference ordinate $P^+(x) - P^-(x) < 0$. Hence with a high probability $x \in D^-$, which contradicts the condition $y \in B^+$.

Since we cannot point out exact values of functions in the tasks under consideration, it is convenient to perform comparative analysis of causal bases in terms of the concept of "contradiction," as shown above.

In case of multidimensional data, cluster methods based on this proximity ideology are easier computed and require less input data then in the case of estimation of distribution density. The main problem of using this method is the choice of an appropriate measure of proximity, consistent with the "humps" of distributions.

To evaluate the causal bases, we construct the method that at least depends on the conjugation of "humps" of distribution density with measures of proximity. Let $x_1^+, ..., x_n^+$, $n > 1$, be the data of the training sample from D^+. Suppose that space D^+ is arranged so that for D^+ is divided into local areas $D_{(1)}, ..., D_{(s)}$, $D_{(1)} \cup ... \cup D_{(s)} = D^+$, $D_{(i)} \cap D_{(j)} = \emptyset$, $i \neq j$, which allows to efficiently distribute which data belongs to which of these areas. These arias have approximately the same size according to the proximity ρ. Then, according to the data $x_1^+, ..., x_n^+$, $n > 1$, it is possible to calculate the frequencies of appearance of data in these areas. Let these frequencies be ordered in descending order, i.e. the set of order statistics is constructed as follows:

$$\nu_{[1]} \geq \cdots \geq \nu_{[s]}. \tag{1}$$

Let's order the local areas $D_{(1)}, ..., D_{(s)}$ in accordance to the set of order statistics (1):

$$D_{[1]} \succ \cdots \succ D_{[s]}. \tag{2}$$

where $D_{[i]} \succ D_{[j]}$ in only case when $\nu_{[i]} \geq \nu_{[j]}$.

Then, according to the relationship between the "hump" of distribution density and the concentration of data around this "hump", we obtain the empirical order of local areas with descending probabilities. From here we get the approximate order of data $x_1^+, ..., x_n^+$, $n > 1$, and the method of determining the data belonging to the "hump" of probability density for the case of extraordinary data.

Let x be new data that generates the extraordinary result. If the data x belongs to the local area having a low number in order (2), this is the causal base that the IT result is derived from the original data and not from the anomalies.

If data x belongs to the area with a high number in order (2), then it can refer to both D^+ and D^-. In this approach, the BC condition is not required, but it is required that the local areas with small numbers in (2) are strictly ordered.

Example 2. Let data space D^+ be a family of balls $D_1, ..., D_m$ in the space of a finite dimension, and the radii of all balls are the same. The IT result is one of two values y_1, y_2, where y_1 is the extraordinary value. If the radii of balls are not large, then there exists an effective algorithm for calculating whether data x belongs to a particular ball.

The one-humped distribution in this scheme can be constructed, for example, using probabilities $p_{i_1} > ... > p_{i_m}$ that describe the independent sampling of balls $D_1, ..., D_m$ in the case of D^+. If all radii are equal to each other, the distributions inside the balls $D_1, ..., D_m$ are uniform, then obviously we get a one-humped distribution on D^+.

In this example, it is assumed that the distributions on D^+ and D^- are not known, but are both one-humped. Then the initial training data $x_1^+, ..., x_n^+$, $n > 1$, effectively determines the frequency of balls occurrence and the order (2). Therefore, the appearance of new data x can be considered as the causal base of an extraordinary result if the ball to which x belongs has a low number in the order (2).

Example 2 shows that analysis of causal bases can be performed simultaneously using different information spaces from which IT input data can be taken. The BC region corresponds to the intersection of distributions on sets D^+ and D^-. At the same time, the probabilistic evaluation of the causal bases for the appearance of extraordinary data allows errors of the 1st and 2nd kind, the estimation of which requires Big Data for training.

Consider some positive and negative aspects of the probability-statistical approach to the analysis of causal bases. This approach can be applied when it is possible to construct (at least indirectly) the sequence (2). In turn, this is possible when the source data is considered to be a part of large areas. However, increasing the size of the area leads to complication of the algorithm that calculates whether the data belongs to the corresponding areas.

It should be noted that to confirm the causal bases of the extraordinary data, it is not necessary to require that the areas do not intersect. In this case, it is possible to simplify the aforementioned algorithm. But the data that appears at the intersections of areas must either be discarded or taken into account in (2) only once.

Since the estimates of causal bases are approximate, the assumption that areas may intersect should not affect the changes in order (2). It should be noted that the one-humped probabilistic distributions are required to substantiate the monotonicity of probabilistic estimates.

4 Extension of the Statistical Approach to the Evaluation of Causal Bases

The concept of "contradiction" can be used in several other cases of the causal bases of extraordinary results of computer calculations.

Let the original data be $x = (x(1), ..., x(r))$, $x(i) \in S(i)$, $i = 1, ..., r$, where $S(i)$ is some isolated information space. This model corresponds to the fact that several partial input data portions can be used to "enhance" trust.

Let the input data has already appeared in $S(1)$ earlier and $x_1^+(1), ..., x_n^+(1)$ correspond to extraordinary results from B^+. Suppose that the space $S(1)$ is arranged so that for $D^+(1)$ is partitioned into local areas $D_{(1)}(1), ..., D_{(s_1)}(1)$, $D_{(1)}(1) \cup ... \cup D_{(s_1)}(1) = D^+(1)$, $D_{(i)}(1) \cap D_{(j)}(1) = \emptyset$, $i \neq j$, that allow efficient calculation of the data belonging to each of these areas. Then, according to the data $x_1^+(1), ..., x_n^+(1)$, $n > 1$, it is possible to calculate frequencies of appearance of areas $\nu_{(1)}, ..., \nu_{(s_1)}$ containing this data, and the set of order statistics

$$\nu_{[1]} \geq \cdots \geq \nu_{[s_1]}. \tag{3}$$

Let's order the local areas $D_{(1)}(1), ..., D_{(s_1)}(1)$ in according to the set of order statistics (3):

$$D_{[1]}(1) \succ \cdots \succ D_{[s_1]}(1). \tag{4}$$

$D_{[i]}(1) \succ D_{[j]}(1)$ in only case when $\nu_{[i]} \geq \nu_{[j]}$.

If the coordinate of the new data $x(1)$ does not contradict the extraordinary IT result, then this is the rationale for the correctness of computer calculations.

The same reasoning can be made for the remaining coordinates of the input data x. In each case, the contradiction "enhances" the trust in the computing results. If only one coordinate does not conflict with an extraordinary result, then additional research is needed on the possibility of a strong effect of the value of this coordinate on IT results.

If all coordinates show a contradiction with the extraordinary result y in the considered IT, then computer calculations cannot be trusted.

Let $x \in D^-$, but $y \in B^+$. Then, in search of the root cause of the anomaly, the IT model can be divided into parts, that is, the IT model is considered to be a form of a DAG (Direct Acyclic Graph). For example, the model of the following form:

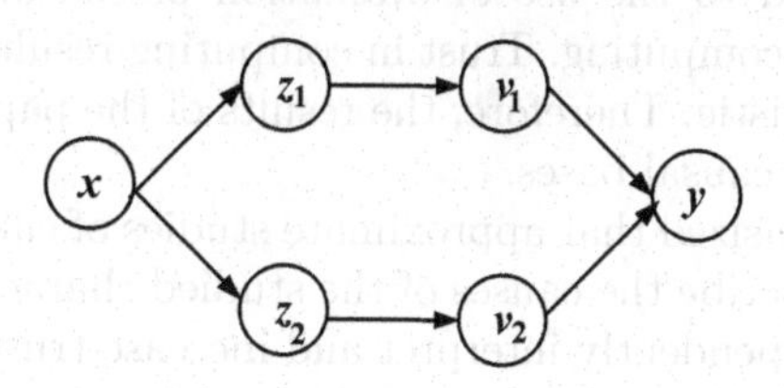

Fig. 2. IT model in the form of DAG.

If $x \in D^-$ and $z_1 \in B^-(z_1)$, then the transformation $x \to z_1$ (Fig. 2) has completed without contradictions. If $x \in D^-$ and $z_2 \in B^+(z_2)$, an anomaly has occurred during transformation $x \to z_2$. If in this example, $y \in B^-$, but $x \in D^+$, then this case can also be considered an anomaly. However, in this case, when looking for the root cause of the anomaly, it is necessary to move from y to v_1 and v_2, and further to z_1 and z_2. Thus, with the known mathematical model of IT and the contradiction between input and output, this method of the statistical approach to causal bases estimation can be used to approximate the localization of the anomaly.

The concept of contradiction of the causal bases with results of computer calculations can also be used to determine a preference when computer calculation present various options and a solution has to be chosen by the DMP. In fact, let solutions y_1 and y_2 be the results of data processing $x \in D$ offered for the DMP to choose from. Consider the solution for which we build on known precedents $x_1^+, ..., x_n^+$, $n > 1$, separately from others. This data can be used for the causal bases of the solution y_1. Assume that the data generates the certain position $[i]$ in the order (2). Similar reasoning can be made to solve y_2 and get the position $[j]$ in (2). Then, if $i < j$, then by its causal base the solution y_1 is preferable.

This statistical approach allows you to build causal bases without reference to the analysis of examples and counterexamples.

If extraordinary data appear rarely enough, then for the causal bases of the correctness of the calculation against the alternative that implies that an anomaly has occurred, we will use the Tukey approach [30]. From the previously observed data from D, we construct the set of order statistics and the sequence of disjoint regions of type (2) in descending order. This construction corresponds mainly to ordinary data, since extraordinary data is rare by assumption. If the IT result for the new data x generates extraordinary data from B^+, then we find the position x in the constructed order. If this place x corresponds to frequently occurring data, then the extraordinary result is possibly the result of an anomaly. If data x corresponds to the area of rarely appearing data in the set of order statistics, then there is no contradiction between x and extraordinary output data. This can be an argument that rare inputs produce rare, extraordinary IT results.

5 Conclusion

This paper is devoted to the use of evaluation of causal bases for increasing trust in the results of computing. Trust in computing results is not an accurately computable characteristic. Therefore, the results of the paper are not focused on accurate estimates of causal bases.

It should be emphasized that approximate studies of causal bases do not set a goal to accurately describe the causes of the studied characteristic. Approximate estimates aim to independently interpret and increase trust in computer results. Accompanying complex calculations by these estimates of causal bases allows for interpretation of computer calculations results in terms of inconsistency of input and output data.

References

1. DARPA Sets Up Fast Track for Third Wave AI. - 26 Jul 2018. https://defence.pk/pdf/threads/darpa-sets-up-fast-track-for-third-wave-ai.569563/. Accessed 4 May 2021
2. IJCAI 2017 Workshop on Explainable Artificial Intelligence (XAI). http://home.earthlink.net/~dwaha/research/meetings/ijcai17-xai/. Accessed 4 May 2021
3. Gunning, D., Aha, D.: DARPA's explainable artificial intelligence (XAI) program. J. AI Mag. **40**(2), 44–58 (2019)
4. NSCAI Final Report (January 2021, Draft). https://www.nscai.gov/wp-content/uploads/2021/01/NSCAI-Draft-Final-Report-1.19.21.pdf
5. Cohen, P.: DARPA's big mechanism program. J. Phys. Biol. **12**(4), 1–9 (2015)
6. Explanation. New philosophical encyclopedia. Electronic Library of the Institute of Philosophy of the Russian Academy of Sciences. https://iphlib.ru/greenstone3/library/collection/newphilenc/document/HASH0147b3e4f487b73bea51af47. Accessed 4 May 2021
7. Hempel, C.G.: Studies in the Logic of Confirmation. J. Mind. **54**, Part I: 1–26; Part II: 97–121 (1945)
8. Hempel, C.G.: Philosophy of Natural Sciences. Prentice-Hall, Hoboken (1966)
9. Peirce, C.S.: Collected papers. In: Hartshorne, C., Weiss, P., Burks, W. (eds.) vol. 1–8, The Belknap Press of Harvard University Press, Cambridge (1965–1967)
10. Peirce, C.S.: Reasoning and logic of things. In: Lectures for the 1898 Cambridge Conferences. RSUH, Moscow (2005). (in Russian)
11. Tarski, A.: The semantic conception of truth and the foundations of semantics. J. Philos. Phenomenological Res. **4**(3), 341–375 (1944)
12. Yessenin-Volpin, A.S.: On the anti-traditional (ultra-intuitionistic) program of the foundations of mathematics and natural science thinking. J. Questions Philos. **8**, 100–136 (1996). (in Russian)
13. Holyoak, K.J., Morrison, R.J. (eds.): The Oxford Handbook of Thinking and Reasoning. Oxford University Press, Oxford (2012)
14. Domingos, P., Guestrin, C., Mooney, R., Dietterich, T., Kautz, H., Tenenbaum, J.: Final report: a Unified Approach to Abductive Inference. Washington University, WA (2014)
15. Finn, V.K.: Mill's inductive methods in artificial intelligence systems. J. Sci. Tech. Inf. Process. Part I: **38**(6), 385–302 (2011); Part II: **39**(5), 241–260 (2012). (in Russian)
16. Finn, V.K.: Artificial intelligence: Methodology, applications, and philosophy. Krasand, Moscow (2011).(in Russian)
17. Grusho, A., Zabezhailo, M., Timonina, E.: On causal representativeness of training samples of precedents in diagnostic type tasks. J. Inform. Primen. **14**(1), 80–86 (2020). (in Russian)
18. Jain, S., Osherson, D., Royer, J.S., Sharma, A.: Systems That Learn. An Introduction to Learning Theory. 2nd edn. The MIT Press, Cambridge (1999)
19. Kubat, M.: An Introduction to Machine Learning, 2nd edn. Springer, Cham (2017). https://doi.org/10.1007/978-3-319-63913-0
20. Zhuravlev, Y.I.: Local algorithms for computing information I, II. J. Cybernetics. I: (1), 2–19 (1965); II: (2), 1–11 (1966). (in Russian)
21. Rudakov, K.V.: On algebraic theory of universal and local constraints for classification problems. In: Zhuravlev, Y.I. (ed.) Pattern Recogn. Classif. Forecast. Science, Moscow (1989) (in Russian)

22. Zhuravlev, Y.I.: Correct algebras over sets of incorrect (heuristic) algorithms. I-III. J. Cybern. I: (4), 5–17 (1977); II: (6), 21–27 (1977); III: (2), 35–43 (1978). (in Russian)
23. Rudakov, K.V.: Completeness and universal limitations in the problem of correction of heuristic classification algorithms. J. Cybern. **3**, 106–108 (1987). (in Russian)
24. Finn, V.K.: On intelligent data analysis. J. Artif. Intell. News **3**, 3–19 (2004)
25. Zabezhailo, M.I.: To the question of the adequacy of the grounds for accepting the results of data mining using the JSM-method. J. Sci. Tech. Inf. Process. **1**, 1–9 (2015)
26. Finn, V.K.: Distributive lattices of inductive JSM procedures. J. Autom. Doc. Math. Linguist. **48**, 265–295 (2014)
27. Grusho, A.A, Zabezhailo M.I., Smirnov D.V., Timonina E.E., Shorgin S.Y.: Mathematical statistics in the task of identifying hostile insiders. J. Inform. Primen. **14**(3), 71–75 (2020). (in Russian)
28. Grusho, A.A., Grusho, N.A., Zabezhailo, M.I., Smirnov, D.V., Timonina, E.E.: Parametrization in applied problems of search of the empirical reasons. J. Inform. Primen. **12**(3), 62–66 (2018). (in Russian)
29. Grusho, A.A.: Statistical significance criteria for cluster structures based on pairwise proximity measures. J. OPPM Surv. Appl. Ind. Math. **3**(1), 43–46 (1996). (in Russian)
30. Tukey, J.W.: Exploratory Data Analysis. Addison Wesley, Boston (1977)

Approaches for Creating a Digital Ecosystem of an Industrial Holding

A. E. Tyulin, A. A. Chursin, A. V. Yudin, and P. Yu. Grosheva(✉)

RUDN University, 6 Miklukho-Maklaya Street, Moscow 117198, Russian Federation
{tyulin-ane,chursin-aa,yudin-av,grosheva-pyu}@rudn.ru

Abstract. The article is devoted to the creation of a digital ecosystem of an industrial holding. It contains the set of tools for solving problems associated with building digital ecosystems of an industrial holding, its scientific and technological capacity and methods of its transformation into a new product, taking into account the sufficiency of resource provision. These tools and methods can be used comprehensively in the process of creating and manufacturing heterogeneous products, developing new management processes based on big data analytics.

Keywords: Unique products · Intelligent systems · Digital economy · Systemic approach · Digital ecosystem

1 Introduction

Currently, the economies of most countries are undergoing digital transformation. This process is related to the emergence of new economic categories and paradigms, as well as the formation of new approaches to the management of economic processes at equal levels with the establishment of new subjects of management. In this regard, in Russia it was proposed to establish a new subject of regulation at the state level—a digital platform or ecosystem and to establish qualifying features for them at the legislative level, as well as to define a regulatory body.

This understanding of the digital ecosystem corresponds to the general concept of the Innovation Triple Helix, which represents the relations of such key participants and stakeholders of the innovation ecosystem as the State, business and scientific organizations [1]. This concept is currently being actively developed [2,3].

In the modern world, trends in world economies not only influence each other, but also enter into insoluble contradictions [4,5]. We can observe their obvious transformation, the redistribution of the roles of the leading countries and macroregions in the global economic and geopolitical arena, as well as the redistribution of spheres of influence in crucial market segments driving the

This paper has been supported by the RUDN University Strategic Academic Leadership Program.

V. M. Vishnevskiy et al. (Eds.): DCCN 2021, CCIS 1552, pp. 433–444, 2022.
https://doi.org/10.1007/978-3-030-97110-6_34

economy. At the same time, technology holdings and corporations, which are becoming increasingly important, are beginning to dictate the rules of the game in the markets. In this regard, national businesses find themselves under the strong influence of large technology giants, and on the other hand, they are forced to survive in the current economic situation. All this leads to the formation of large business ecosystems.

Sberbank has been developing the largest ecosystem in Russia since 2017. To do this, it buys other players of the financial and non-financial market. For three years, Sberbank has spent $ 1 billion on its development, or 3% of net profit. Now the Sber ecosystem includes about 30 companies and services in various areas: e-commerce, ready-made food delivery, taxi, car sharing, media and entertainment, a mobile operator, healthcare, real estate transactions. Ecosystems among Russian big business are developed by Yandex, Mail.ru Group, Sberbank and Tinkoff-Bank. The Yandex ecosystem includes 87 services.

The development trends of digital platforms are already widely developed abroad. For example, in China, digital platforms in the manufacturing sector are already considered critical for modernizing industries, increasing productivity, optimizing resource allocation and increasing employment.

China's digital industrial platforms are beginning to compete globally. One of the most important platforms is INDICS, created by China Aerospace Science & Industry Corporation Limited (CASIC), a state-owned missile manufacturer under the direct control of the central government. Two other corporate giants have created even more influential platforms: Haier, a manufacturer of household appliances and electronics, and Alibaba, an e-commerce giant.

The promotion of the Chinese economy of digital platforms is located in the ecosystem of major policy initiatives, "Internet+", "Made in China 2025" and "China Standards 2035", seeking to standardize advanced technologies, such as artificial intelligence, cloud computing, the Internet of Things and big data .

Other countries, especially Germany with its strong industrial base and rich experience in the field of "Industry 4.0" can benefit. German companies such as Siemens, SAP and Bosch are already involved in the Chinese economy of new digital industrial platforms.

2 Theoretical Basis for the Formation of a Digital Ecosystem of an Industrial Holding

The formation of a digital ecosystem focused on the manufacture of diversified promising products using the same production capacities, basic technologies and competencies (with the addition of new ones if necessary for the manufacture of certain products) for different market segments and different consumer groups provides the manufacturer with economic stability through the successful sale of products that are competitive in quality and price on the market. In this case, when building digital ecosystems, it is necessary to determine the optimal structure of the organization of production, in which the introduction of digital production will be most effective, taking into account its characteristics [6,7]. To this end, it is necessary to determine the optimal structure of its own production, the list of parts and components that will use the created flexible automated production at full capacity, and the work that will be transferred to related organizations that will produce certain parts, components, units of high quality in a shorter time based on their competencies and technological capabilities, while ensuring the lowest cost of production, which is carried out for a given competitiveness.

The development of digital ecosystems of the industrial holding, as one of the key factors and sources of economic development, is carried out at several levels, corresponding to the stages of the life cycle of radically new products. The development of the ecosystem is not a one-time, but a continuous process, the course of which is determined by the flexibility of the company and the influence of external factors-trends in global technological development.

The initial stages of the development of the digital ecosystem of an industrial holding and the creation of competitive advantages that ensure its stable development are shown in Fig. 1.

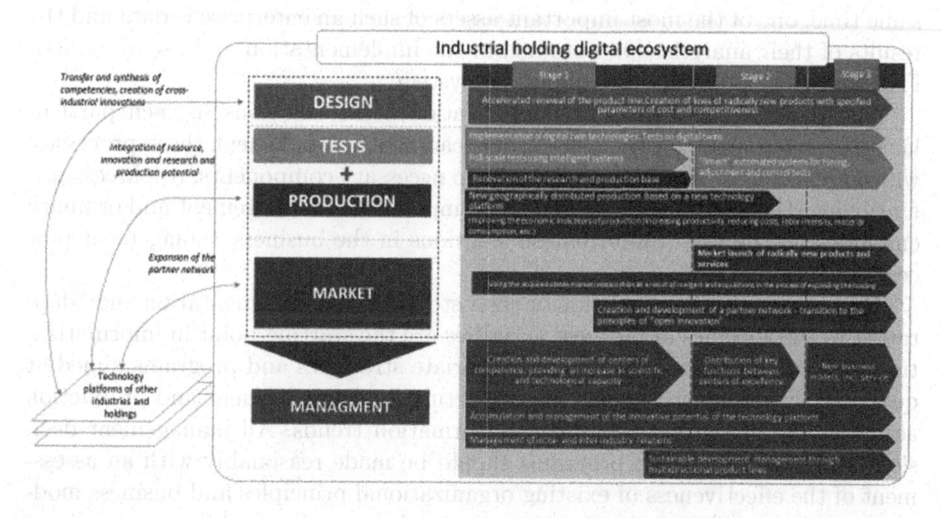

Fig. 1. The initial stages of the development of the digital ecosystem of an industrial holding.

The main factor in creating new products aimed at long-term satisfaction of needs is the continuous growth of the innovative capacity of the participants of the digital ecosystem, related to a large volume of basic research, extracting valuable information from large volumes, mastering and developing a set of competencies, technologies and equipment based on this valuable information and this research, improving management methods [8].

The development of such processes in economic systems is related to the effect of the economic law of advanced development, which states that the developed products must have consumer utility (value) that increases the needs of society, leading to the emergence of new markets and creating a stable economic development of the producer. In the context of globalization, outdated economic mechanisms and technological solutions are stagnating, and only the most innovatively developed organizational and economic structures find new economic and technological niches due to the previously created competitive advantages and are able to create conditions for the establishment and development of markets for innovative goods.

Production management processes are rapidly changing under the influence of advanced digital technologies integrated into the entire business process of the organization. Digital technologies stimulate new business models aimed at creating and producing goods of the future that can put the company on a path of advanced development. The emergence of new players on the market with fundamentally new products entails the need to organize such production management, which will create highly competitive products and services with the best technical characteristics and quality on the market at the lowest cost.

The solution to this problem is based on the transformation of the organizational and economic system of the enterprise into a digital ecosystem. At the same time, one of the most important assets of such an enterprise is data and the results of their analytics [9], obtained in the implementation and use of modern IT solutions when building a digital ecosystem.

The pace of distribution and application of IT solutions in each particular ecosystem depends primarily on its readiness to implement these processes, which are not feasible without at least two necessary components: resources (primarily logistical and financial) and the competence of management and ordinary employees in the use of information solutions in the business management process.

Basic elements of automation or their step-by-step implementation and adaptation to the specifics of business activities are the starting point in informatization and the basis for developing appropriate strategies and programs aimed at qualitatively changing the existing "structures" of management and production activities associated with digital transformation trends. All management decisions within development programs should be made reasonably with an assessment of the effectiveness of existing organizational principles and business models, as well as predicting the return on the implementation of the most promising approaches and methods aimed at improving the quality of business processes due to their digitalization, increasing control over their implementation and optimization through the use of information solutions. When choosing information solutions, the main task is to create the solution that will be maximally adapted and interconnected with the competence component and the business competency management system.

As a result, informatization penetrates into every organization striving to comply with market trends, as well as to work in the advanced development mode, demonstrating its leading positions in the world market, covering all spheres of society and becoming a global phenomenon, and at the same time a tool for stimulating economic development.

In addressing the development of an organizational and economic system in modern conditions, the most important information source for managing them should be considered the data of the global information space, advanced knowledge and prospects for the development of fundamental science, as well as the results of intellectual processing of this information by modern mathematical methods (neural networks, machine learning, discrete mathematics). As part of the digitalization of business, a new organizational and economic environment is being formed, integrated into the information space, in which many aspects of activities related to the organization of the release of new products and services are developing.

Currently, most researchers of the role of informatization in the development of the economy consider the global information space as a set of information resources and infrastructures that make up State and inter-State computer networks, telecommunication systems and public networks, as well as other cross-border information transmission channels [10].

The active process of innovation and technological convergence that began in the XXI century in the world economy, due to the rapid evolution of the global information space stimulated the development of mechanisms for interpenetration and convergence of competencies, when the boundaries between individual competencies and technologies based on them erase, and the final results of the application of competencies are manifested in interdisciplinary research and development at the junction of various fields of science, resulting in radical competencies which cover the intersectoral level. The synergetic effect caused by the development of convergent information solutions and the expansion of the global information space contributes to obtaining a significant economic and production effect in the economy as a whole by generating synergetic innovations.

Convergent information solutions contribute to building the bank's key radical competencies, which make it possible to create a product that can potentially meet the needs of society and take a dominant position in the market if there is a sufficient amount of other resources. Such a product will be created according to the principle of "semi-finished product", when its final image will be formed taking into account the requirements of a specific consumer. Such a mechanism can be implemented as part of the new M2C business model (manufacturer to customer) and the reverse C2M model, in which personalized production is implemented, which involves the manufacture of a product that has the necessary (or desirable) original properties for a given consumer. The closer the producer is to the consumer, the more opportunities they will have to generate income. In the context of global informatization, the "germination" of competencies into various branches of science and their integration, as well as the accumulation of intellectual capital by both manufacturing companies and society, the distance between the manufacturer and the consumer of products is reduced at all stages of the product life cycle, and especially at the stage of their operation, as a result of the increasing spread of service models of the economy.

The considered effects of the evolution of the global information space and the processes of convergence of competencies and technologies lead to the following postulate describing the creation of unique products in the context of the evolution of the global information space:

The global information space is developing due to its replenishment with new knowledge underlying building key competencies and radical innovations. The global information space stimulates innovation and technological convergence, in which the boundaries of various competencies and technologies disappear, and innovation is the result of interdisciplinary research, a "semi-finished product" that a manufacturer, given the availability of resource opportunities, can bring to the level of a radically new unique product.

The formulated postulate is proposed to be considered as one of the bases for further theoretical studies of the advanced satisfaction of prospective needs occurring under the influence of the information environment and digital technologies to ensure the economic growth of business.

The active development of the information environment and infrastructure that provides the basis for the digital ecosystem for the implementation of the production activities of a company leads to a significant variety of products and services manufacured through the organization of personalized production, the development of new forms of marketing (for example, targeted marketing) and increased labor productivity through automation and robotization.

The economic growth achieved by these and other factors is justified by modern models of economic growth, which explain it by the increasing variety of products and services and the active development of the service sector.

In terms of economic growth and development of holdings, one global information space is not able to become a source of economic growth without calculating the effectiveness of its use and without taking into account the level of key competencies.

Therefore, an important issue in the context of managing the economic development of a company and creating unique products that meet current and future needs is the implementation of targeted entry into the global information space for a company interested in using it to build and develop necessary key competencies. In this case, the global information space is considered as a set of services (information is provided as a service for a fee), where the consumer purchases them specifically, given their interests in solving specific production and economic problems.

Herewith, there is such a phenomenon as recombination of innovations.

Recombination of innovations involves the intelligent integration of various technologies in one solution, which allows the creation of new digital business assets. These new digital assets are key opportunities that support the real digital transformation of digital enterprise business models, creating proactive products.

Recombinant innovation allows using the existing technological and competence capacity of the organization, represented by a set of key technologies and competencies that can be reconfigured using an integrated digital platform in new ways to create new ideas.

The holding's digital ecosystems also have an important impact on its competitiveness and ability to transition to advanced development through the production of goods capable of creating new markets or significantly expanding the boundaries of existing ones based on the competitiveness of products in terms of quality and price.

Due to the redistribution of resources and the use of the capacity of the digital ecosystem, it becomes possible to neutralize the negative impact of uncertainty factors on the sustainable development of the holding. Based on the redistribution of resources, the holding can form the direction of the strategic choice of its development:

- creation of competitive advantages that will strengthen positions in a certain market segment or enter new ones;
- organization of new types of activities able to generate significant income;
- creation of new sales markets based on the development of radically new products previously not available on the market;
- creating conditions for the holding's transition to advanced development based on the formation of a set of indicators of economic stability, competitiveness and resource provision.

In order to avoid the loss of economic stability, as well as to ensure the breakthrough creation of innovative capacity and adapt to work in a new technological platform in modern conditions, the holding and its companies need to implement a comprehensive approach to the development of radical competencies, provided by the transformation of the principles of work from all spheres of activity. In this regard, the following scheme is proposed for creating conditions for ensuring the accelerated growth of innovative capacity and the development of the holding's digital ecosystem in the digital economy (Fig. 2).

The implementation of the steps shown in Fig. 2 by holdings, in practice, will provide opportunities for the accelerated growth of their innovative capacity, allowing them to develop radical competencies and products, to maintain long-term competitive leadership in the market, and accelerate the industry's advanced transition to the creation of new technological platforms [11].

Here is a formal description of the mechanisms underlying the procedures for the coordinated development of the digital ecosystem of a holding and the companies operating within its perimeter.

3 Economic and Mathematical Model for the Development of a Digital Ecosystem of a Holding

To describe the development model of the digital ecosystem of the holding, we will consider the problem of optimal management, taking into account the active evolution of its companies, striving to achieve their own goals and operating in the scientific, technological and innovative framework of the main platform. We will use and develop the models suggested in [12].

Let P_0 (the digital ecosystem of the holding) strives to achieve the highest value of efficiency $f_0(xi, u)$, which can be understood, for example, as the profit

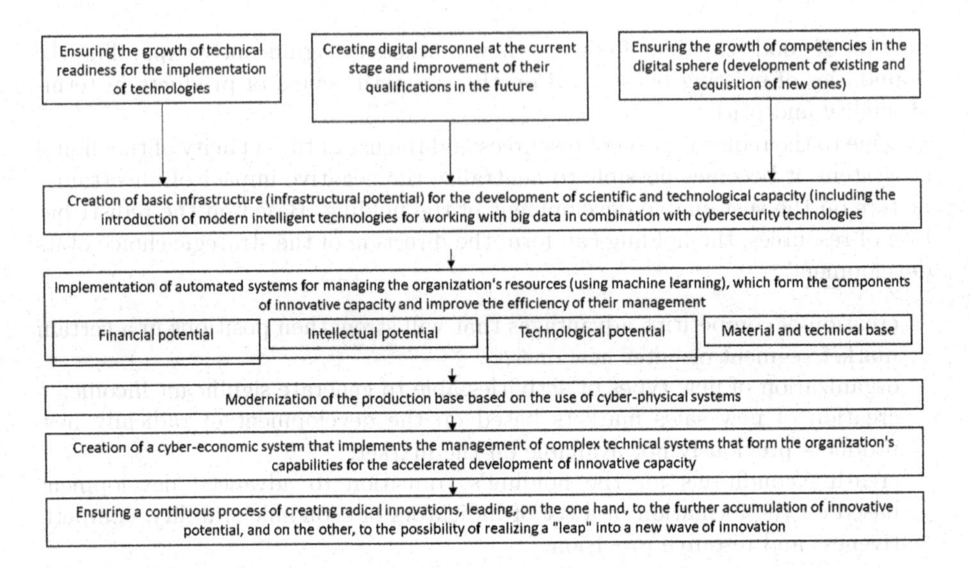

Fig. 2. Scheme of creating conditions for ensuring the development of the holding's digital ecosystem in the digital economy.

from the sale of high-tech products of companies, where u is the innovative capacity of the digital ecosystem, and x_i is the competitiveness of companies, $x_i \in X$, $x = (x_1, ..., x_n)$. The companies of the holding, in turn, strive to increase their own efficiency $f_i(x_i, u_i)$, $i = 1, 2, ..., n$, which can be understood as its profitability. Let's consider several possible mechanisms for the coordinated development of the digital ecosystem and its companies.

Mechanisms of the 1st type (direct). The digital ecosystem does not control the competitiveness of companies, but freely provides them with the competencies and knowledge of its own core. The best values of the control variables are determined from the solution of the problem:

$$G_1 = \sup_{u \in U} \min_{x_i \in B_i^1} f_0(x, u),$$

where B_1 is the set of optimal controls for the competitiveness of companies:

$$B_i^1 = \{x_i \in X_i | f_i(x_i, u_i) = \max_{y_i \in X_i} f_i(y_i, u_i)\}.$$

Mechanisms of the 2nd type (closed-loop). The digital ecosystem relies on the use of the innovative capacity of companies in the formation of its own optimal management strategy, and formulates it as functions $u_i = u_i(x_i)$. Then

$$B_i^2 = \{x_i \in X_i | f_i(x_i, u_i) = \sup_{y_i \in X_i} f_i(y_i, u_i') - \delta_i(u_i')\}, \delta_i(u_i') \geq 0,$$

and the greatest guaranteed result (efficiency of functioning) of the digital ecosystem is

$$G_2 = \sup_{u' \in U} \inf_{x_i \in B_i^2(u_i')} f_0(x, u').$$

In this case, the effectiveness of the holding companies is determined from the condition

$$sup_{(x_i,u_i) \in D_i}, D_i = \{x_i \in X_i, u_i \in U_i | f_i(x_i, u_i) > \max_{x_i \in X_i} \min_{u_i \in U_i} f_i(x_i, u_i)\},$$

where the efficiency $L = \max\min f_i(x_i, u_i)$ of company i is guaranteed.

As a result of such mechanisms, the increase in the efficiency of the holding companies determines the increase in the efficiency of the entire ecosystem as a whole.

To implement one or more areas of strategic choice that ensure advanced development, it is necessary to assess the capacity for the development of the holding's digital ecosystem and implement the first preparatory stage of its transition to the widespread use of digital technologies, based on two main directions:

- identification and systematization of advanced digital methods and technologies already used by the holding, assessment of their comparability or the possibility of using them as a reference point for creating a digital transformation platform;
- creating conditions for providing a sufficient number of resources, competencies and ensuring stable links between departments in the development and manufac-ture of competitive products.

In a general form, the algorithm for the creation of the digital ecosystem of an industrial holding, due to the requirements of digital transformation and the creation of a single technological platform is presented as a block diagram in Fig. 3.

For advanced development and global leadership in the market, high-tech holdings need today to focus on the development of their own digital ecosystem, corresponding to the advanced wave of innovation.

The achievement of global leadership by the holding involves the creation of a set of conditions:

- management of the holding as a special self-organizing system interacting with other systems based on corporate (industry) technology platforms and through macroeconomic mechanisms;
- ensuring long-term financial and economic stability of the business;
- development and manufacture of fundamentally new products focused on meeting long-term needs, with new properties, functions and consumer values;
- organization of a full life cycle (from shaping the techno-economic image and the organization of mass production to operation and disposal) of radically new products created on the basis of the latest advances in technology;

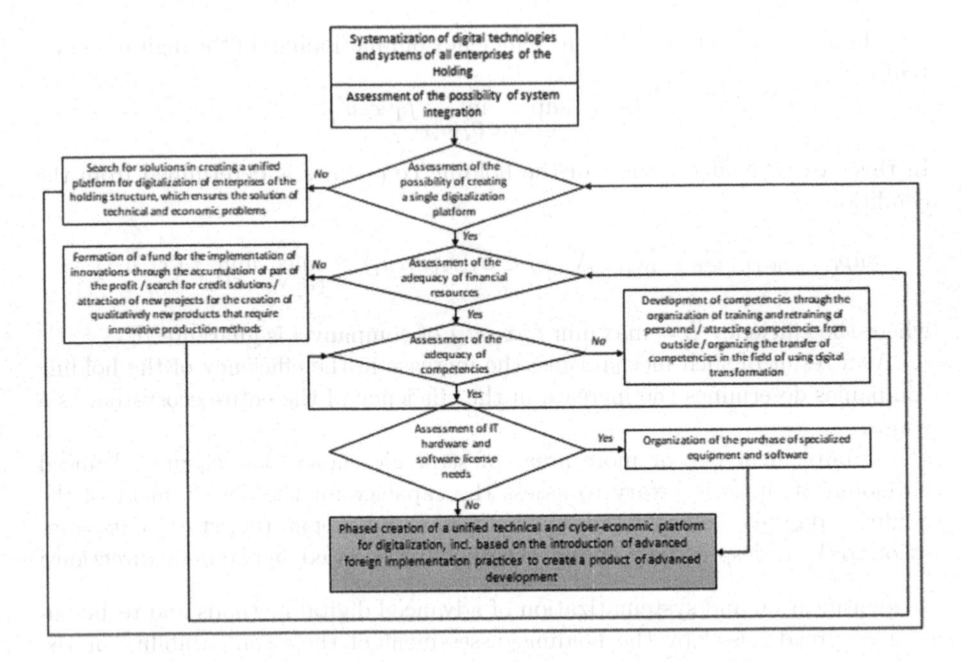

Fig. 3. Scheme of creating conditions for ensuring the development of the holding's digital ecosystem in the digital economy.

- continuous and systematic development of scientific and technological capacity and competencies based on the accumulation of knowledge and unique experience in the creation and manufacture of high-tech piece and small-scale products with unique technical characteristics and functions, as well as fundamental and applied research and development;
- building a high level of competitiveness (in the classical sense);
- active role in the development of industry technology platforms and centers of competence for building cross-industrial innovations and technological solutions at the global level;
- active development of mechanisms for intercorporate and intersectoral transfer of breakthrough innovative technologies to significantly reduce the time required to bring radically new products to the market.

Achieving the global competitiveness of the holding is closely related to the effective management of research, development, innovation, production, organizational and other processes, which determines building of scientific and technological capacity sufficient to create a completely new (radically new) product with high consumer properties, allowing to win a significant share of the existing market or create a new market for high-tech products.

In its very terms, with such a complex of interrelated activities, the holding should have a wide range of different competencies that are transformed into the creation of a new product through the functioning of the digital ecosystem.

As a rule practical development for the creation of complex high-tech products is carried out by several companies and their various design and technological divisions, which can be combined within the holding through a corporate (industry) digital ecosystem, which ensures the interaction between enterprises or organizations. The overall result of the implementation of the project to create a radically new product, i.e. compliance with the necessary techno-economic characteristics, allowing to maximally satisfy the needs of the customer and the market, or to ensure the creation of a new sales market for the developed products will depend on how these related organizations engaged in joint product development will fulfill their private technical tasks and create the final result with the necessary technical characteristics, materials, and components.

The developed products can be presented on the market only after their manufacture, therefore, there is the question about the need to create fundamentally new production facilities or modernize existing ones. The direction of development of the digital ecosystem should be determined based on the availability of the competencies of the holding structure, financial resources, timing of market entry, taking into account possible risks, including the risk that competitors may be ahead of schedule and release similar products with higher characteristics. It follows that when creating and developing corporate (sectoral) digital ecosystems, it is necessary to consider mutual production and technical ties, the possibility of technology transfer and the exchange of unique competencies, and to take measures to increase the scientific and technological capacity of the holding.

Effective use and development of unique competencies can provide global leadership and enable the creation of a market of fundamentally new products and technologies, and create new market segments based on them.

4 Conclusion

The solution to the problem of forming a digital ecosystem of an industrial holding is associated with building its scientific and technological capacity and methods of its transformation into a new product, taking into account the sufficiency of resource provision.

To solve this problem in creating a digital ecosystem, it is proposed to use the following tools and methods:

- a method for assessing the readiness of business units to increase scientific and technological capacity;
- tools for capacity transformation (its "transfer" from the scientific and theoretical plane to the applied one) related to the normative and methodological regulation of this process at various stages of the product life cycle;
- a model for assessing the impact of control actions on the components of scientific and technological capacity, as well as from the impact on the growth of the capacity and its transformation into radical properties of new products.

These tools and methods can be used comprehensively in the process of creating and manufacturing heterogeneous products. In real economic conditions, new products are created in companies that already are participants of the commodity markets and produce goods created within the digital ecosystem. The creation of radically new products is linked to the continuous development of the ecosystem itself due to the constant increase in scientific and technical potential and its transformation into highly competitive products that ensure the stability of the industrial holding in the market. It allows to build the trajectory of economic development of a high-tech industrial holding and determine the conditions under which the holding is in a state of advanced development and global technological superiority.

References

1. Etzkowitz, H., Leydesdorff, L.: The triple helix: university - industry - government relations a laboratory for knowledge based economic development. EASST Rev. **14**(1), 14–19 (1995)
2. Ranga, M., Etzkowitz, H.: Triple helix systems: analytical framework for innovation policy and practice in the Knowledge Society. Ind. High. Educ. **27**(4), 237–262 (2013)
3. Carayannis, E., Campbell, D.: "Mode 3" and "Quadruple Helix": toward a 21st century fractal innovation ecosystem. Int. J. Technol. Manag. **46**(3–4), 201–234 (2009)
4. Chursin, A., Tyulin, A.: Competence Management and Competitive Product Development: Concept and Implications for Practice. Springer, Cham (2018). https://doi.org/10.1007/978-3-319-75085-9
5. Tyulin, A., Chursin, A.: The New Economy of the Product Life Cycle: Innovation and Design in the Digital Era. Springer, Cham (2020). https://doi.org/10.1007/978-3-030-37814-1
6. Gavrilyuk, E.A., Mantserov, S.A., Ilichev, K.V., Turikov, M.I.: Information decision-support system on the basis of the method of diagnostics and control of the technical state of industrial equipment (conference paper). In: 12th International Scientific and Technical Conference "Dynamics of Systems, Mechanisms and Machines", Dynamics, vol. 20183, pp. 8601472. (2019)
7. Pimenov, V.I., Pimenov, I.V. Interpretation of a trained neural network based on genetic algorithms. Informatsionno-Upravliaiushchie Sistemy **6**, 12–20 (2020)
8. Boginskiy, A.I., Chursin, A.A., Nesterov, E.A., Tyulin, A.E.: Assessing the competitiveness of production of a high-tech corporation to ensure its advanced development. J. Adv. Res. Dyn. Control Syst. **11**(11 Special Issue), 73–81 (2019)
9. Boginsky, A.I., Chursin, A.A.: Digital models for optimization of production and technological processes. Bull. Mech. Eng. **2**, 63–67 (2020)
10. Kozoriz, N.L.: Problems of security and access to world information resources. Law State **6**, 103–107 (2013)
11. Chursin, A.A., Grosheva, P.Yu., Yudin, A.V.: Fundamentals of the economic growth of engineering enterprises in the face of challenges of the XXI century. IOP Conf. Ser. Mater. Sci. Eng. **862**(4), 042049 (2020)
12. Chursin, A.A., Dubina I.N., Carayannis E.G, Tyulin A.E, Yudin A.V.: Technological platforms as a tool for creating radical innovations. J. Knowl. Econ. (2021)

Intelligent Systems for Optimal Production Control of Unique Products

A. E. Tyulin[1], A. A. Chursin[1], I. N. Dubina[2], A. V. Yudin[1], and P. Yu. Grosheva[1(✉)]

[1] RUDN University, 6 Miklukho-Maklaya Street, Moscow 117198, Russian Federation
{tyulin-ane,chursin-aa,yudin-av,grosheva-pyu}@rudn.ru
[2] Altai State University, 61 Lenin Avenue, Barnaul 656049, Russian Federation

Abstract. The article proposes and discusses new approaches to the development and effective use of intelligent systems for optimal control of the production processes of unique technological products, that is, such products that are able to satisfy unmet demand and have no analogues in the world. The main results of this research are a conceptual model of an intelligent control system formation for the development and production of unique products.

Keywords: Unique products · Intelligent systems · Digital economy · Systematic approach · Economic and mathematical models · Optimal control

1 Introduction

The definition of a unique product as a product that has no world analogues and is able to satisfy unmet demand is relatively new [19,20], although the very problem of managing the manufacture of radically new products (radical innovations) and the establishment of an appropriate control system, including on the basis of intelligent systems and with the use of optimization models arises earlier (see, for example, [1,8,10,22]).

Control processes for the development and manufacture of unique products are rapidly changing with the development of the so-called new economy (creative, innovative, knowledge, digital) [9] and are influenced by the widespread use of advanced digital technologies integrated into all processes of product creation [5,7]. As a result, changes in these control and manufacture processes stimulate new approaches and principles aimed at creating and producing new products that can ensure the company's economic stability and competitiveness in the market [6,13,15].

The purpose of this research is to develop a system concept of an intelligent control system for the production of unique products and an economic and mathematical model of its functioning.

This paper has been supported by the RUDN University Strategic Academic Leadership Program.

V. M. Vishnevskiy et al. (Eds.): DCCN 2021, CCIS 1552, pp. 445–461, 2022.
https://doi.org/10.1007/978-3-030-97110-6_35

In accordance with this purpose, the article has the following structure.

The next section presents the authors' vision of the functioning of an intelligent production control system for unique products in the context of the information infrastructure of a digital enterprise.

Section 3 presents a systematic approach to the development of an intelligent computing platform for solving the problem of automated control of the development and production of unique goods.

Section 4 is divided into 3 subsections, which sequentially present: 1) a general model of the decision-making system and optimal production control at all stages of the life cycle of unique products; 2) economic and mathematical features of the formulation and solution of the optimization problem for managing the manufacture of unique products; 3) information basis of an intelligent control system for the development and manufacture of unique products.

In the final part, conclusions are drawn on the results of the paper as a whole.

2 The Main Components of an Intelligent Control System for the Development and Manufacture of Unique Products

A modern digital enterprise integrates many automated control systems on its platform, each of which has some intellectual properties: ERP, CRM, PLM, EMS, MES systems. (see Fig. 1).

An intelligent control system for the development and manufacture of unique products is part of the information infrastructure of a digital enterprise and is closely integrated with other information systems and databases of the enterprise. The close integration of various information systems on the digital enterprise platform is due to the use in these systems of a single basic toolkit, which include: 1) tools for working with data that ensure effective management and use of big data about the internal and external environment of the enterprise, on the basis of which the process of supporting effective management decisions can be built; 2) tools for ensuring the production process, allowing industrial enterprises to build the production process in the most efficient way and ensure a balance of "high quality - cost" in order to create a unique competitive product and 3) tools for interaction with the external environment of the enterprise.

The tools for working with data are represented by such technologies as artificial intelligence, fog computing, end-to-end, quantum and supercomputer technologies, identification technologies, blockchain, neural networks, etc. The combination of these technologies provides effective control and use of big data about the internal and external environment of the enterprise, on the basis of which the support for making effective management decisions can be built [23]. This toolkit makes it possible to use past experience to shape future decisions using progressive self-learning algorithms. In addition, decision-making models

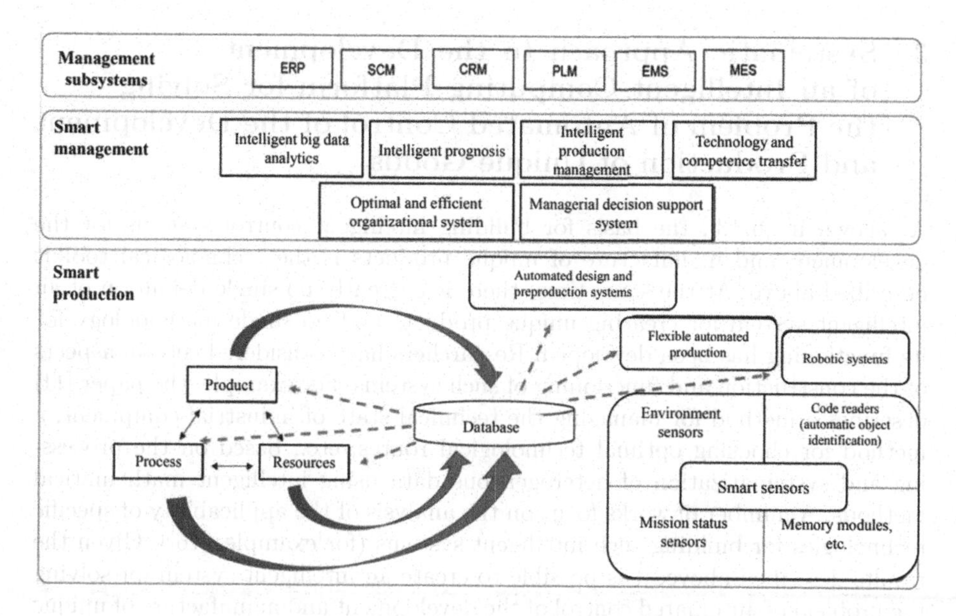

Fig. 1. Information systems of a digital enterprise.

independently adapt to new conditions when new data is obtained. An important advantage of using these technologies in data management is the achievement of high accuracy of their processing and minimization of errors that appear due to the human factor.

The tools for ensuring the production process include cyber-physical systems, 3D technologies (printing), robotics, additive technologies, open production technologies, etc. The use of these tools in practice allows industrial enterprises to build the production process in the most efficient way and ensure a balance of "high quality - cost" in order to create a unique and competitive product. In this case, the use of the listed technologies is aimed at minimizing the duration of the production cycle and reducing costs at all its stages while maintaining high technical parameters and creating competitive functionality.

The tools of interaction with the external environment of the enterprise are represented by such technologies as unmanned, paperless, mobile, biometric technologies, brain-computer technologies. This block of technological tools allows you to organize operational monitoring of changes in the external environment of the enterprise and analyze consumer preferences in order to form innovative ideas for creating new unique products, as well as to increase the duration of their life cycle in the market by organizing effective service that improves the image of the manufacturer in the consumer market.

3 Systematic Approach to the Development of an Intelligent Computing Platform for Solving the Problem of Automated Control of the Development and Production of Unique Goods

As shown in [6,13], the basis for building intelligent control systems for the development and manufacture of unique products is the technological toolkit described above. At the same time, there is currently no single definition of an intelligent system for creating unique products and no single methodology for its functioning has been developed. Researchers have considered certain aspects of the construction and functioning of such systems. For example, the paper [11] describes a method for managing the technical state of industrial equipment, a method for choosing optimal technological routes, etc., based on the processing and systematization of heterogeneous data using intelligent mathematical methods. A number of works focus on the analysis of the applicability of specific technologies for building such intelligent systems (for example, [16]). Given the results described above, it is possible to create an intelligent system for solving the problem of automated control of the development and manufacture of unique products, taking into account the need to maintain its competitive advantages for a long time (see Fig. 1).

The use of information methods and technologies for processing big data makes it possible to analyze and build forecasts, on the basis of which decisions are made to create the image of a unique product according to consumer preferences in order to achieve a high level of competitiveness and outperform competitors - market leaders. Currently, in order to dominate the market, enterprises create their own information computing platforms using advanced information methods and technologies that allow integrating two processes - analysis of the consumer market and creation of the image of promising products. The basis of such platforms is the creation and integration of information support systems for the development of techno-economic image and product design processes based on the use of information resources, technologies and competencies of the organization. The proposed intelligent control system for the development and production of unique products, built into the information infrastructure of a digital enterprise, provides a mechanism for flexible adaptation of the techno-economic image of unique products and the production system of an enterprise to changes in a competitive market environment, as well as to dynamically changing consumer expectations, which allows to maintain high competitive advantages of the manufactured unique products, and thereby contribute to the preservation of the manufacturer's dominant position in the market. The criterion for the effectiveness of the proposed intellectual system is the level of competitiveness of the manufactured product in real time.

In accordance with the block diagram of the proposed intelligent control system for the development and production of unique goods created using the advantages of modern information technologies (see Fig. 2), we describe its main subsystems (see Table 1).

Table 1. Subsystems as part of an intelligent control system for the development and production of unique goods.

№	Name of the subsystem	Subsystem task	Intellectual component	Influence on the creation of competitive advantages
1	Subsystem for shaping the image of a unique product	Shaping the techno-economic image based on monitoring prospective needs	Using methods and technologies for processing and intelligent analysis of big data obtained from the global information space to create competitive advantages of unique products	Shaping the image of a product, which in the future will have competitive advantages in the market, considering the forecast of the development of the needs of companies and society
2	Subsystem of automated control of key competencies	Management of scientific and technological capacity and key competencies	Intellectual monitoring of the global information space, aimed at identifying new knowledge and competencies among competing companies	Shaping radical innovation
3	Resource control subsystem	Effective resource control at all stages of creating unique products	Using intelligent expert systems to select the most optimal composition and quantity of all types of resources required to ensure the product life cycle	Creating competitive advantages by optimizing warehouse inventory, minimizing taxes, eliminating ineffective diversion of working capital, providing a stock of materials, raw materials and components necessary for production
4	Subsystem for managing the development of products for a given cost and competitiveness	Design, pre-production, providing high technical characteristics when designing for a given, competitive price in the market	Intelligent selection of the optimal ratio of techno-economic characteristics of the designed product sample in accordance with its techno-economic image. Using intellectual analysis and predicting the results of product testing at the design stage, identifying design errors. Intelligent analysis and selection of the most advanced production technologies that can provide high competitive advantages of products and minimize the time of their manufacture. Selecting the necessary competencies to design a product from the generated intelligent database. Testing products and receiving recommendations from the intelligent subsystem in terms of improving the prototype to create a better product	Ensuring competitive advantages through the selection of promising technologies and competencies of the enterprise for the manufacture of the product

(*continued*)

Table 1. (*continued*)

№	Name of the subsystem	Subsystem task	Intellectual component	Influence on the creation of competitive advantages
5	Subsystems for pre-production and production control	Pre-production and production	Building the production process based on the optimal selection of operations by an expert system that analyzes previous experience in the manufacture of promising products	Creating competitive advantages by ensuring a high level of production process, increasing the profitability of production
6	Sales control subsystem	Releasing the finished product on the market	Using intelligent methods to create advertising for a unique product and interaction with consumers. Application of automated order control systems. Building optimal logistics schemes. Evaluation of product malfunction using intelligent testing systems, which will then offer a multi-choice or the most optimal repair and maintenance option	Creating competitive advantages at the stage of receiving and processing orders, completing and preparing products for shipment to customers, shipping products to the vehicle and transportation to the destination, organizing payments. Creating competitive advantages at the stage of product sales and working with the consumers
7	Product update control subsystem given the evolution of needs	Effective control of product renewal processes at all stages of its life cycle	Assessment of the capacity of the enterprise, intellectual analysis and forecasting of future needs that will arise in the market in a few years. Continuous analysis of changing needs over time. Receiving, processing and analyzing data from the global information space. Developing recommendations for improving the product at all stages of its life cycle to maintain and create competitive advantages in the future	Maintaining competitive advantages by managing assortment policy, demand generation, sales promotion, and market research

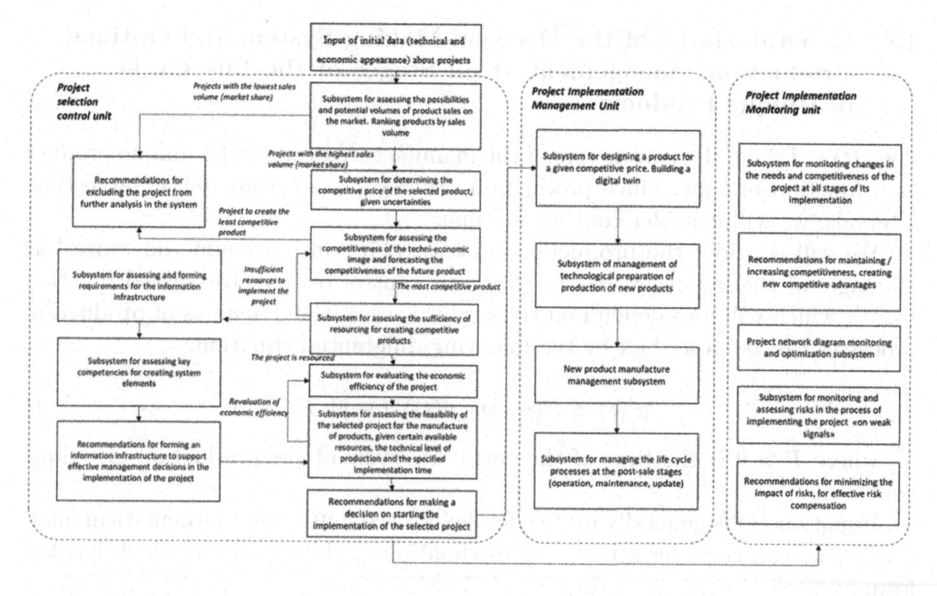

Fig. 2. Block diagram of an intelligent control system for the development and production of unique goods.

The criterion for the effectiveness of the proposed intellectual system is the level of competitiveness of the manufactured product in real time.

Due to the dynamic nature of the market situation, as well as a sharp and often unexpected change in the parameters of the competitiveness of products, there is a need for operational adjustments to the company. The proposed intelligent management system for the development and manufacture of unique products is well suited for managing these processes, since it has the properties of automatic adaptation to various changing parameters of the external and internal environment of the enterprise, providing the most effective mode of operation, that is, the system is adaptive, flexible and easily adaptable to changes in the external environment.

4 Systematic Approach to the Development of an Intelligent Computing Platform for Solving the Problem of Automated Control of the Development and Production of Unique Goods

Now we turn to a more detailed formal description of the intellectual control system for the development and production of unique goods – we will describe the economic and mathematical model of its functioning.

4.1 General Model of the Decision-Making System and Optimal Production Management at all Stages of the Life Cycle of Unique Products

Let $x(t) \in R^N$ be the parameters of the manufacture process of a unique product as a function of time. Since production processes are considered at large time intervals, we will consider continuous time.

We will consider the production process taking into account the control at all stages of the life cycle. We describe this control by a vector-valued function $\in R^M$, which will also depend on time. In this case, the dynamics of production processes can be described by the following differential equation:

$$x'(t) = F(t, x(t), u(t)), t \in [0, T] \tag{1}$$

where $T > 0$ is the time of the entire life cycle of the production of unique goods.

Equation (1) is generally quite complex for economic and mathematical analysis, so we can consider a linear approximation of this equation, which has the form

$$x'(t) = A(t) \cdot x(t) + B(t) \cdot u(t), \tag{2}$$

where $A(t)$ is an $N \times N$ matrix; $B(t)$ is an $N \times M$ matrix.

The goal of management when creating unique products is to minimize the target functionality

$$J(u) = |x_u(t) - x^*(t)| + \alpha \cdot \int a(t) \cdot |u(t)| dt, \tag{3}$$

where $x_u(t)$ is the solution of the Cauchy problem (with some initial data) for Eq. (1) or (2) in the case when control $u(t)$ is selected; $x^*(t)$ is the planned parameters of the production process; $a(t) \geq 0$ is the cost function of management decisions; $\alpha > 0$ is the dimension coefficient.

We will minimize the functional J on the set of admissible controls, which is defined as follows:

$$U = v(t) \in U(t) : t \in [0, T]. \tag{4}$$

Thus, we will assume that $u \in U$. Without loss of generality, we can assume that the sets $U(t) \subset R^M$ are bounded and closed, that is, compact.

The economic meaning of the functional J is that the first term shows the degree of divergence of the results of our control and the planned values, and the second term shows the "price" for making management decisions.

Thus, the optimization problem for managing production processes at all stages of the product life cycle consists in finding such a control $u^* \in U$ such that

$$J(u^*) = min J(u) : u \in U. \tag{5}$$

Problem (4) refers to the classical optimal control problem, and to solve it, one can apply the well-known methods of calculus of variations and optimal principles.

4.2 Economic and Mathematical Features of the Formulation and Solution of the Optimization Problem for the Control of the Production Processes of Unique Goods

Firstly, the sets $U(t)$ in problem (4) may be non-convex if economic solutions are considered. A common case is that the set $U(t)$ can consist of two points, which corresponds to the situation when making decisions "yes" or "no". Mathematically this circumstance can complicate the adoption of optimal decisions.

Secondly, when making control decisions in the production of unique goods, it is necessary to take into account the possible instability of decision behavior.

The economic basis for instability in making management decisions in the production is that, due to the uniqueness of products, non-optimal control decisions can radically change production processes.

As we have already noted, the specifics of the control of the production of unique products is that control decisions are unstable. Moreover, when formulating a dynamic problem, one should take into account not only the instability of control decisions, but also the uncertainty of the factors that exercise control. The differential Eq. (1) does not allow us to take into account these uncertainty factors, so we replace the considered dynamic model, described by a system of differential equations, with a dynamic model, described by differential inclusions [4,6,12,14,17,18,21].

To formulate the problem in the differential inclusions, we will consider multi-valued mappings.

Let us denote by Φ the set of subsets $F_s \subset R^N$, where each set F_s is a convex and compact set in R^N. In this case, $F == \cup F_s$. Now, using Φ, we will denote the function (mapping)

$$F : [0,T] \times R^N \times U \to \Phi. \tag{6}$$

Next, instead of the differential Eq. (1), we will consider the differential inclusion

$$x'(t) \in F(t, x(t), u(t)). \tag{7}$$

In the differential inclusion (7) instead of the derivative $x'(t)$, we can consider the contingent or paracontingence. A contingency is the set of all limit sequences

$$\frac{x(t_h) - x(t_0)}{t_h - t_0}, t_h \to t_0, t_h \neq t_0, h \to 0, \tag{8}$$

and paracontingency is the set of all limit points

$$\frac{x(t_h) - x(t_g)}{t_h - t_g}, t_h \to t_0, t_g \to t_0, t_h \neq t_g, h \to 0, g \to 0. \tag{9}$$

The use of differential inclusions instead of differential equations provides the following economic advantages of the economic-mathematical model:

- the ability to take into account the multivariance of the response to control decisions;
- the ability to take into account the instability of control when creating unique products;
- the ability to take into account uncertainty factors in the production of unique goods. Optimization of differential inclusions has certain features both theoretically and in terms of obtaining numerical solutions. Using differential inclusions, we can describe different scenarios for the production of unique goods at all stages of their life cycle. In this case, it is necessary to use the operational information that comes as a result of the implementation of production processes. Modern conditions for the manufacture of high-tech products are associated with the digitalization of production and the economic environment, so to make decisions based on operational information, it is necessary to use intelligent decision-making methods that fully work with the available information flows. To this end, we will consider the information bases of an intelligent control system for the development and production of unique goods.

Optimization of differential inclusions has certain features both theoretically and in terms of obtaining numerical solutions. Using differential inclusions, we can describe different scenarios for the production of unique goods at all stages of their life cycle. In this case, it is necessary to use the operational information that comes as a result of the implementation of production processes. Modern conditions for the manufacture of high-tech products are associated with the digitalization of production and the economic environment, so to make decisions based on operational information, it is necessary to use intelligent decision-making methods that fully work with the available information flows. To this end, we will consider the information bases of an intelligent control system for the development and production of unique goods.

4.3 Information Bases of an Intelligent Management System for the Development and Manufacture of Unique Products

The proposed intelligent management system for the development and manufacture of unique products is based on the use of the information principle when making economic decisions in conditions of uncertainty that arise as a result of the specifics of the manufacture of unique products. This information principle is based on a game-theoretic approach, similar to the game-theoretic approach in mathematical statistics. Let's formulate its mathematical foundations. We will assume that it is necessary to make one of the decisions, the set of which we denote by W, but according to the dynamic model of optimal production control of unique goods for each decision $w \in W$, a whole set of reactions is possible, since in the production of unique goods we are forced to make decisions under conditions of uncertainty Let $R(w)$ denote the set of reactions to the decision w. Each reaction will give a different economic result, which we will describe using dimensionless quantities, as is customary in game theory. To do this, we will consider the win function, which depends on the choice of $w \in W$ and each reaction $r \in R(w)$. Let $H(w, r)$ be an estimate of the result of the reaction of r to the decision made by w. Since $H(w, r)$ is a numerical function, the informational decision-making principle can be formulated as a maximin solution. By a maximin solution we mean $w^* \in W$ such that

$$min\{H(w^*, r) : r \in R(w^*)\} = max\{min\{H(w^*, r) : r \in R(w^*)\} : w \in W\}. \tag{10}$$

The maximin solution shows the assessment of the guaranteed result when making decisions under the conditions of uncertainty arising in the process of developing unique products. However, the use of maximin decisions is a deliberately pessimistic scenario that can be justified in simple antagonistic games, but when making economic decisions in production management it is required to use additional information about the state of production, which will make it possible to make an optimal decision. Thus, the informational principle of making economic decisions under conditions of uncertainty can be formulated as follows: when making economic decisions in conditions of uncertainty generated by the uniqueness of the product produced, it is necessary to clarify the non-maximin decision based on the available additional information about the possible reaction to the management decision. This principle allows building an intelligent management system for the development and manufacture of unique products in the context of digitalization of science-based industries.

According to the information principle of decision-making under uncertainty, the key issue is the probabilistic assessment of the onset of a particular reaction to a decision. Consider an intelligent method for obtaining such a probabilistic estimate.

Let the intelligent control system for the development and production of unique goods recommend a certain solution $w \in W$ at a certain stage of the life cycle of a unique product. Assume that the set of possible reactions to the w-set $R(w)$ is finite, i.e. $R(w) = \{r_1, r_2, \ldots, r_N\}$.

By $p_n, n = 1, 2, \ldots, N$, we denote the probabilities of the reaction r_n. Of course, the following condition must be met:

$$p_n \geq 0, p_1 + p_2 + \ldots + p_N = 1. \tag{11}$$

In the absence of any additional information, we will assume that the a priori probabilities of the reaction to the decision are given. In the case where we have certain information that can clarify the a priori probability of a reaction, we will denote this information by D. In this case, the probabilities of reactions can be expressed using the conditional probability

$$p_n = P(r = r_n | D), n = 1, 2, \ldots, N. \tag{12}$$

Here, $P(r = r_n|D)$ denotes the probability of the onset of the reaction r_n, provided that there is data (information) D.

To actually calculate the conditional probability $P(r = r_n|D)$, the Bayesian approach can be used, which is based on the formula of the same name. Using the Bayes formula, the conditional probability can be calculated as follows:

$$P(r = r_n|D) = (P(D|r = r_n) \cdot P(r = rn))/P(D). \tag{13}$$

Calculating the conditional probability $P(D|r = r_n)$ is a simpler task because, assuming a given system response, we can estimate the probability of data D. The process of constructing these conditional probabilities is also called a Bayesian machine learning process.

Normally the existing information is a collection of some facts. Therefore, we can write

$$D = (D_1, D_2, \ldots, D_M), \tag{14}$$

where D_m is an "elementary" fact from the available information at all stages of the life cycle of a unique product. Of course, different facts tend to be interdependent, so calculating the probability $P(D)$ may require knowledge of the correlation between different facts, which is a serious technical problem. To solve this problem, it is recommended to use a naive Bayesian approach [2], when an assumption is made about the independence of various facts. In this case, the probability $P(D)$ can be easily calculated using the following formula:

$$P(D) = P(D_1) \cdot P(D_2) \cdot \ldots \cdot P(D_M). \tag{15}$$

This assumption is not always the case, but the experience of using the Bayesian learning process shows that such an assumption leads mainly to correct reliable results, which, as our research shows, provides sufficient efficiency of solutions developed automatically by an control system. This efficiency ensures the implementation and operation of the proposed intelligent control system for the development and production of unique products at established and already functioning digital enterprises.

Let us demonstrate the application of the described approach based on a naive Bayesian classifier by the example of solving the problem of making a decision by an intelligent system to set a market price for a complex technical system (engine), which is one of the main units of an aircraft. The aircraft engine considered in the example has various technical characteristics determined by its image [25,26].

Based on this list of characteristics and similar characteristics of competing engines on the market, the relative competitiveness ratio Q_{eng} of this engine can be calculated using standard methods for determining competitiveness [3]. An important parameter on the basis of which the estimated value of the market price of the engine will be formed is the specific weight of the cost of the unit in question in the total cost of the aircraft. This assumption is due to the competition of aircraft from different manufacturers within the same segment in conditions of comparable technical and cost characteristics. A larger value of the competitiveness coefficient for a particular aircraft corresponds to higher values of the technical characteristics determined by its constituent units, or lower values of cost parameters.

The initial data for an intelligent system to make a decision to set a market price for an aircraft engine are:

- a set of aircraft that belong to the segment under consideration (list i);
- the competitive price of the aircraft (S_0, is determined by the market method based on a comparison of the characteristics of the aircraft on the market).
- coefficient of competitiveness of the aircraft (Q, known methods for assessing competitiveness);
- the share of the cost of the considered unit in the competitive price of the aircraft (a);
- the relative coefficient of the competitiveness of the engine, for the market leader the actual value is set equal to 1.

Based on these data, for the market leader, we calculate the coefficient E of the market-based price using the formula:

$$E = \frac{\sqrt{Q \cdot Q_{eng}}}{S + a \cdot S}, \qquad (16)$$

where S is the price of the aircraft minus the cost of the engine.

By varying the values of Q_{eng} and a in the proposed formula in the range of 10% of the actual values, we obtain an interval for the value E, and if the price of the considered unit falls within this interval, we will consider it reasonable. If it is impossible to identify a clear market leader, then to calculate the range of the value of E, the averaged initial values of the coefficients and quantities Q, S, Q_{eng} and a for several leading aircraft can be used.

Let's consider an example of calculating the coefficient of a market-based engine price. The initial data is presented in the Table 2.

Table 2. Calculation of the range for the coefficient of the market-based engine price.

Q	Q_{eng}	A	S	E
0,878	0,9	0,04	20	0,042737
0,878	0,9	0,05	20	0,04233
0,878	0,9	0,06	20	0,041931
0,878	1	0,04	20	0,045049
0,878	1	0,05	20	0,04462
0,878	1	0,06	20	0,044199
0,878	1,1	0,04	20	0,047248
0,878	1,1	0,05	20	0,046798
0,878	1,1	0,06	20	0,046356

Based on the calculation performed:

$$E \in (0,0423; 0,04373). \tag{17}$$

To assess the adequacy of the cost of the considered unit, we will use the machine learning method based on the application of a naive Bayesian classifier.

To apply Bayesian learning, it is necessary to determine the coefficients E for each aircraft from the segment based on the initial data (with the exception of the aircraft that includes the engine, the market price of which must be determined).

An example of calculating the E coefficient for each aircraft in the segment is presented in Table 3.

Table 3. Calculation of E.

Aircraft	Q	Q_{eng}	a	S	E	Note
Aircraft 1	0,878	1	0,04	20	0,04504	$E \in (0,0423; 0,04373)$
Aircraft 2	0,734	0,93	0,05	21	0,03747	$E \notin (0,0423; 0,04373)$
Aircraft 3	0,799	0,96	0,04	19,5	0,04318	$E \in (0,0423; 0,04373)$

Further, to build a naive Bayesian classifier based on the training sample (information about the aircraft of the segment), it is necessary to create a frequency table, which reflects the number of specific characteristics falling into certain ranges for the case when the value of E falls into the calculated interval and otherwise. The border of the two ranges can be defined as the median value of the corresponding characteristics of aircraft from the training sample (Table 4).

Table 4. A naive Bayesian classifier based on a training sample.

	$E_1(E \in (0,0423; 0,04373))$	$E_2(E \notin (0,0423; 0,04373))$
$a \leq 0,045$	2	0
$a > 0,045$	0	1
$Q \geq 0,0799$	1	0
$Q < 0,0799$	1	1
$Q_{unit} \geq 0,96$	2	0
$Q_{unit} < 0,96$	0	1

To determine the boundaries of the adequate cost of the unit in question, based on the constructed Bayesian classifier, it is necessary to calculate the probabilities of the aircraft on which the engine in question is placed in the selected classes (in the example, these are classes E_1 and E_2), depending on the share of the engine price in the market price of the aircraft, taking into account the known values of cost, competitiveness and technical characteristics of the engine in question.

The calculation using the formula (5) provides the following results:

$$P(E_1) = 0,14 \tag{18}$$

Accordingly, $P(E_2) = 0,86$.

Thus, the probability of falling into a class in which the price of the engine in question is 3 times higher than the probability of falling into a class corresponding to an adequate price for the engine.

The search for the market price of the engine is carried out by an intelligent system on the basis of a simulation model, in which the share of the engine price in the total price of the aircraft decreases. In this case, the competitiveness of the aircraft is recalculated as a result of a decrease in its price caused by a decrease in the cost of the engine under consideration.

Let the aircraft, for which the estimate of the market price of the engine is determined, has the following parameters entering the input of the naive Bayesian classifier:

- $a = 0.043$;
- $Q = 0.88$;
- $Q_{eng} = 0.9$.

For this source data::

$$P(E_1) = 0,81; P(E_2) = 1 - 0,81 = 0,19. \tag{19}$$

In this situation, when the price of the engine is 4.3% of the cost of the rest of the units, the probability that the price of the engine is adequate is 81%.

The obtained result can be interpreted using the Harrington universal verbal-numerical scale [24]. The numerical values of the Harrington scale gradation are obtained from the analysis of a large array of statistical data. Due to this, the Harrington scale is universal and is traditionally used to interpret various probability parameters (Table 5).

Table 5. Harrington scale for describing probability gradations.

Numerical value of probability	Description of the probability gradations
0,8–1,0	Very high
0,64–0,8	High
0,37–0,64	Average
0,2–0,37	Low
0,0–0,2	Very low

Thus, for the first example, in which the probability of an adequate estimate of the aggregate price is 0.14, it can be interpreted as very low, and for the second example, in which the corresponding probability is 0.81, it can be interpreted as very high.

Based on the analysis carried out, the intelligent system determines the management decision on setting the market price for the aircraft engine at the level of 4,3% of the price of the aircraft as a whole.

5 Conclusion

This paper presents economic and mathematical approaches to solving a pressing problem related to building an intelligent management system for the development and manufacture of unique products with high consumer properties. Based on the results of the research, taking into account the modern principles of digital modeling in shaping the techno-economic image of unique products, the development of methods for analyzing its competitiveness, as well as algorithms for evaluating the key competencies and scientific and technological capacity of the organization that develops innovative products, a scheme for the establishment of an intelligent management system for the development and manufacture of unique products is proposed.

Efficiently functioning intelligent systems allow decision-makers to use data, knowledge, objective and subjective analytical and simulation models for subsequent analysis and solution of semi-structured and unstructured problems arising in the process of creating unique products.

References

1. Carayannis, E.G., Chanaron, J.-J.: Managing Creative and Innovative People: The Art, Science and Craft of Fostering Creativity, Triggering Invention and Catalyzing Innovation. Praeger Publishers, London (2007)
2. Nikolenko S., Kadurin A., Arkhangelskaya E.: Deep learning. Peter, SPb (2018)
3. Chursin, A., Vlasov, Y.: Innovation as a Basis for Competitiveness: Theory and Practice. Springer, Heidelberg (2016). https://doi.org/10.1007/978-3-319-40600-8
4. Afanasova, M., Liou, Y.-C., Obukhovskii, V., Petrosyan, G.: On controllability for a system governed by a fractional-order semilinear functional differential inclusion in a Banach space. J. Nonlinear Convex Anal. **20**(9), 1919–1935 (2019)
5. Akberdina, V.V., Tyulin, A.E., Chursin, A.A., Yudin, A.V.: Influence of cross-industry information innovations of the space industry on the economic growth of the Russian regions. Econ. Region **16**(1), 228–241 (2020)
6. Boginskiy, A.I., Chursin, A.A., Nesterov, E.A., Tyulin, A.E.: Assessing the competitiveness of production of a high-tech corporation to ensure its advanced development. J. Adv. Res. Dyn. Control Syst. **11**(11 Special Issue), 73–81 (2019)
7. Chursin, A.A., Grosheva, P.Y., Yudin, A.V.: Fundamentals of the economic growth of engineering enterprises in the face of challenges of the XXI century. IOP Conf. Ser. Mater. Sci. Eng. **862**(4), 042049 (2020)
8. Dubina, I.N.: Optimising creativity management: problems and principles. Int. J. Manag. Decis. Mak. **7**(6), 677–691 (2006)

9. Dubina, I.N., Carayannis, E.G., Campbell, D.: Creativity economy and a crisis of the economy? coevolution of knowledge, innovation, and creativity, and of the knowledge economy and knowledge society. J. Knowl. Econ. **3**(1), 1–24 (2012)
10. Dubina, I.N., Oskorbin, N.M.: Game-theoretic models of incentive and control strategies in social and economic systems. Cybern. Syst. Int. J. **46**(5), 303–319 (2015)
11. Gavrilyuk, E.A., Mantserov, S.A., Ilichev, K.V., Turikov, M.I.: Information decision-support system on the basis of the method of diagnostics and control of the technical state of industrial equipment (conference paper). In: 12th International Scientific and Technical Conference "Dynamics of Systems, Mechanisms and Machines", Dynamics 2018, 3 January 2019, Paper Number 8601472, 12th International Scientific and Technical Conference "Dynamics of Systems, Mechanisms and Machines", Dynamics 2018, Omsk, Russian Federation, 13–15 November 2018. CFP18RAB-ART/144276 (2018)
12. Gliklikh, Y., Kornev, S., Obukhovskii, V.: Guiding potentials and periodic solutions of differential equations on manifolds. Glob. Stochastic Anal. **6**(1), 1–7 (2019)
13. Kabaldin, Y.G., Kolchin, P.V., Shatagin, D.A., Anosov, M.S., Chursin, A.A.: Digital twin for 3D printing on CNC machines. Russ. Eng. Res. **39**(10), 848–851 (2019). https://doi.org/10.3103/S1068798X19100101
14. Kamenskii, M., Obukhovskii, V., Petrosyan, G., Yao, J.-C.: Boundary value problems for semilinear differential inclusions of fractional order in a Banach space. Appl. Anal. **97**(4), 571–591 (2018)
15. Koptev, Y.N., Chursin, A.A.: Methods for calculating competitive indicators of high-tech products, considering the factors of product sales in the markets in aerospace industry. 2019 IOP Conf. Ser. Mater. Sci. Eng. **476**(1), 012015 (2018)
16. Pimenov, V.I., Pimenov, I.V.: Interpretation of a trained neural network based on genetic algorithms. Informatsionno-Upravliaiushchie Sistemy **6**, 12–20 (2020)
17. Shamin, R.V.: Approximation of evolution differential equations in scales of Hilbert spaces. Math. Notes **85**(1–2), 293–295 (2009)
18. Shamin, R.V.: On differential equations with nonlocal switch functionals. J. Math. Sci. (United States) **190**(1), 170–180 (2013)
19. Tyulin, A.E., Chursin, A.A.: Creating a smart management system for the manufacture of unique products. Russ. Eng. Res. **40**(11), 922–925 (2020). https://doi.org/10.3103/S1068798X20110192
20. Tyulin, A.E., Chursin, A.A., Drogovoz, P.A., Yudin, A.V.: Researching the processes determining the dominance of unique products in sales markets. In: AIP Conference Proceedings, vol. 2318, p. 070013 (2021)
21. Van Loi, N., Vu, M.Q., Hoai, N.T., Obukhovskii, V.A.: nonlocal problem for projected differential equations and inclusions with applications. Fixed Point Theory **20**(1), 233–244 (2019)
22. Yudin, A.V.: Development of a management system for the creation of radically new product. IOP Conf. Ser. Mater. Sci. Eng. **919**, 042024 (2020)
23. Boginsky, A.I., Grosheva, P.Y., Uchenov, A.A., Yudin, A.V.: Decision support methods for creating new products based on the analysis of consumer expectations. Innov. Invest. No. **8**, 62–69 (2019)
24. Harrington, E.C.: The desirable function. Ind. Qual. Control **21**(10), 494–498 (1965)
25. Pratt & Whitey. https://www.pwc.ca/en/products-and-services/products/helicopter-engines/pt6c, Accessed 11 Oct 2021
26. UEC Klimov. https://klimov.ru/production/helicopter/VK-2500/, Accessed 11 Oct 2021

Author Index

Abrosimov, Vyacheslav 293
Adou, Yves 36
Afanaseva, Olga 346
Akishin, Vladimir 148
Al Maqbali, K. A. K. 186
Alzaghir, Abbas 3
Andrabi, Umer 18
Astafiev, S. N. 396

Bai, Bo 220
Bérczes, Tamás 174
Bessonov, M. 123
Birukou, Aliaksandr 359
Blaginin, Alexey 333
Blanzieri, E. 214
Bondarchuk, A. S. 408
Buranova, Marina 62
Buzhin, I. 123

Cavecchia, V. 214
Chen, Li 220
Chursin, A. A. 36, 433, 445

Dagaev, Alexander 346
Danilyuk, Elena 233
Demidova, Anastasia V. 371
Dubina, I. N. 445

Ermolayeva, Anna 359

Farkhadov, M. P. 123
Fishchenko, E. A. 272

Glushkova, Anastasia G. 371
Gorbunova, A. V. 318
Grosheva, P. Yu. 433, 445
Grusho, Alexander 420
Grusho, Nikolai 420

Joshua, V. C. 186, 201

Kartashevskiy, Vyacheslav 62
Kirichek, Ruslan 346
Kislitsyn, A. A. 307
Kislyakov, Sergey 148
Kochetkov, Dmitry 359
Kopats, Dmitry 259
Kosarava, Katsiaryna 259
Koucheryavy, Andrey 3
Krishnamoorthy, Achyutha 186, 201
Kuki, Attila 174

Lapatin, Ivan 220, 333
Lebedev, A. V. 318
Lizyura, Olga 220

Markova, Ekaterina 36
Markovich, Natalia M. 284
Marochkina, Anastasia 111
Mathew, Nisha 201
Melikov, Agassi 163
Mironov, Y. 123
Moiseev, Alexander 220
Moiseeva, Svetlana 233
Mondal, Mainak 387

Nazarov, Anatoly 220, 233
Ndayikunda, Juvent 18
Nekrasova, Ruslana 247

Okunsky, Mikhail 387
Orlov, Yu. N. 307

Panteley, Ekaterina 293
Paramonov, Alexander 111
Pastorello, D. 214
Paul, Svetlana 220
Peng, Xi 220
Petrov, Dmitriy 18
Petrov, Iliyan 77, 92
Pham, Van Dai 346
Pristupa, Pavel 220

Rumyantsev, A. S. 214, 396
Ryzhov, Maksim S. 284

Sabbagh, Amani A. 48
Semenova, Daria 136

Sevastianov, Leonid A. 371
Shahmaliyev, Mamed 163
Shashev, D. V. 387, 408
Shcherbakov, Maxim V. 48
Shchetinin, Eugene Yu. 371
Shidlovskiy, S. V. 387, 408
Soldatenko, Aleksandr 136
Sotnikov, Alexander 148
Stepanov, Mikhail S. 18
Stepanov, Sergey N. 18
Sztrik, János 163, 174

Tatarnikova, Tatiana M. 111
Timonina, Elena 420
Tóth, Ádám 174
Tyulin, A. E. 433, 445

Yakovleva, Ekaterina 346
Yudin, A. V. 433, 445

Zabezhailo, Michael 420
Zatuliveter, Yu. S. 272

Printed in the United States
by Baker & Taylor Publisher Services

Printed in the United States
by Baker & Taylor Publisher Services